Differential Forms in Electromagnetics

IEEE PRESS SERIES ON ELECTROMAGNETIC WAVE THEORY

The IEEE Press Series on Electromagnetic Wave Theory consists of new titles as well as reprints and revisions of recognized classics in electromagnetic waves and applications which maintain long-term archival significance.

BOOKS IN THE IEEE PRESS SERIES ON ELECTROMAGNETIC WAVE THEORY

Christopoulos, C., *The Transmission-Line Modeling Methods. TLM*
Clemmow, P. C., *The Plane Wave Spectrum Representation of Electromagnetic Fields*
Collin, R. B., *Field Theory of Guided Waves,* Second Edition
Dudley, D. G., *Mathematical Foundations for Microwave Engineering,* Second Edition
Elliot, R. S., *Antenna Theory and Design,* Revised Edition
Elliot, R. S., *Electromagnetics: History, Theory, and Applications*
Felsen, L. B., and Marcuvitz, N., *Radiation and Scattering of Waves*
Harrington, R. F., *Field Computation of Moment Methods*
Harrington, R. F., *Time Harmonic Electromagnetic Fields*
Hansen et al., *Plane-Wave Theory of Time-Domain Fields. Near-Field Scanning Applications*
Ishimaru, A., *Wave Propagation and Scattering in Random Media*
Jones, D. S., *Methods in Electromagnetic Wave Propagation,* Second Edition
Lindell, I. V., *Methods for Electromagnetic Field Analysis*
Lindell, I. V., *Differential Forms in Electromagnetics*
Peterson et al., *Computational Methods for Electromagnetics*
Tai, C. T., *Generalized Vector and Dyadic Analysis: Applied Mathematics in Field Theory*
Tai, C. T., *Dyadic Green Functions in Electromagnetic Theory,* Second Edition
Van Bladel, J., *Singular Electromagnetic Fields and Sources*
Volakis et al., *Finite Element Method for Electromagnetics: Antennas, Microwave Circuits, and Scattering Applications*
Wait, J., *Electromagnetic Waves in Stratified Media*

Differential Forms in Electromagnetics

Ismo V. Lindell
Helsinki University of Technology, Finland

IEEE Antennas & Propagation Society, *Sponsor*

IEEE
IEEE PRESS

WILEY-INTERSCIENCE

A JOHN WILEY & SONS, INC., PUBLICATION

Published simultaneously in Canada.

Library of Congress Cataloging-in-Publication Data is available.

ISBN 0-471-64801-9

Printed in the United States of America.

10 9 8 7 6 5 4 3 2 1

Differential forms can be fun. Snapshot at the time of the 1978 URSI General Assembly in Helsinki Finland, showing Professor Georges A. Deschamps and the author disguised in fashionable sideburns.

This treatise is dedicated to the memory of Professor Georges A. Deschamps (1911–1998), the great proponent of differential forms to electromagnetics. He introduced this author to differential forms at the University of Illinois, Champaign-Urbana, where the latter was staying on a postdoctoral fellowship in 1972–1973. Actually, many of the dyadic operational rules presented here for the first time were born during that period. A later article by Deschamps [18] has guided this author in choosing the present notation.

IEEE Press
445 Hoes Lane
Piscataway, NJ 08855

Contents

Preface

The present text attempts to serve as an introduction to the differential form formalism applicable to electromagnetic field theory. A glance at Figure 1.2 on page 18, presenting the Maxwell equations and the medium equation in terms of differential forms, gives the impression that there cannot exist a simpler way to express these equations, and so differential forms should serve as a natural language for electromagnetism. However, looking at the literature shows that books and articles are almost exclusively written in Gibbsian vectors. Differential forms have been adopted to some extent by the physicists, an outstanding example of which is the classical book on gravitation by Misner, Thorne and Wheeler [58].

The reason why differential forms have not been used very much may be that, to be powerful, they require a toolbox of operational rules which so far does not appear to be well equipped. To understand the power of operational rules, one can try to imagine working with Gibbsian vectors without the bac cab rule $\mathbf{a} \times (\mathbf{b} \times \mathbf{c}) = \mathbf{b}(\mathbf{a} \cdot \mathbf{c}) - \mathbf{c}(\mathbf{a} \cdot \mathbf{b})$ which circumvents the need of expanding all vectors in terms of basis vectors. Differential-form formalism is based on an algebra of two vector spaces with a number of multivector spaces built upon each of them. This may be confusing at first until one realizes that different electromagnetic quantities are represented by different (dual) multivectors and the properties of the former follow from those of the latter. However, multivectors require operational rules to make their analysis effective. Also, there arises a problem of notation because there are not enough fonts for each multivector species. This has been solved here by introducing marking symbols (multihooks and multiloops), easy to use in handwriting like the overbar or arrow for marking Gibbsian vectors. It was not typographically possible to add these symbols to equations in the book. Instead, examples of their use have been given in figures showing some typical equations. The coordinate-free algebra of dyadics, which has been used in conjunction with Gibbsian vectors (actually, dyadics were introduced by J.W. Gibbs himself in the 1880s, [26–28]), has so far been missing from the differential-form formalism. In this book one of the main features is the introduction of an operational dyadic toolbox. The need is seen when considering problems involving general linear media which are defined by a set of medium dyadics. Also, some quantities which are represented by Gibbsian vectors become dyadics in differential-form representation. A collection of rules for multivectors and dyadics is given as an appendix at the end of the book. An advantage of differential forms when compared to Gibbsian vectors often brought forward lies in the geometrical content of different (dual) multivectors, best illustrated in the aforementioned book on gravitation. However, in the present book, the analytical aspect is emphasized because geometrical interpretations do not help very much in problem solving. Also, dyadics cannot be represented geometrically at all. For complex vectors associated with time-harmonic fields the geometry becomes complex.

It is assumed that the reader has a working knowledge on Gibbsian vectors and, perhaps, basic Gibbsian dyadics as given in [40]. Special attention has been made to introduce the differential-form formalism with a notation differing from that of Gibbsian notation as little as possible to make a step to differential forms manageable. This means balancing between notations used by mathematicians and electrical engineers in favor of the latter. Repetition of basics has not been avoided. In particular, dyadics will be introduced twice, in Chapters 1 and 2. The level of applications to electromagnetics has been left somewhat abstract because otherwise it would need a book of double or triple this size to cover all the aspects usually presented in books with Gibbsian vectors and dyadics. It is hoped such a book will be written by someone. Many details have been left as problems, with hints and solutions to some of them given as an appendix.

The text is an outgrowth of lecture material presented in two postgraduate courses at the Helsinki University of Technology. This author is indebted to two collaborators of the courses, Dr. Pertti Lounesto (a world-renown expert in Clifford algebras who sadly died during the preparation of this book) from Helsinki Institute of Technology, and Professor Bernard Jancewicz, from University of Wroclaw. Also thanks are due to the active students of the courses, especially Henrik Wallén. An early version of the present text has been read by professors Frank Olyslager (University of Ghent) and Kurt Suchy (University of Düsseldorf) and their comments have helped this author move forward.

ISMO V. LINDELL

Koivuniemi, Finland
January 2004

Differential Forms in Electromagnetics

1

Multivectors

1.1 THE GRASSMANN ALGEBRA

The exterior algebra associated with differential forms is also known as the Grassmann algebra. Its originator was Hermann Grassmann (1809–1877), a German mathematician and philologist who mainly acted as a high-school teacher in Stettin (presently Szczecin in Poland) without ever obtaining a university position.[1] His father, Justus Grassmann, also a high-school teacher, authored two textbooks on elementary mathematics, *Raumlehre* (*Theory of the Space*, 1824) and *Trigonometrie* (1835). They contained footnotes where Justus Grassmann anticipated an algebra associated with geometry. In his view, a parallelogram was a geometric product of its sides whereas a parallelepiped was a product of its height and base parallelogram. This must have had an effect on Hermann Grassmann's way of thinking and eventually developed into the algebra carrying his name.

In the beginning of the 19th century, the classical analysis based on Cartesian coordinates appeared cumbersome for many simple geometric problems. Because problems in planar geometry could also be solved in a simple and elegant way in terms of complex variables, this inspired a search for a three-dimensional complex analysis. The generalization seemed, however, to be impossible.

To show his competence for a high-school position, Grassmann wrote an extensive treatise(over 200 pages), *Theorie der Ebbe und Flut* (*Theory of Tidal Movement*, 1840). There he introduced a geometrical analysis involving addition and differentiation of oriented line segments (Strecken), or vectors in modern language. By

[1]This historical review is based mainly on reference 15. See also references 22, 37 and 39.

generalizing the idea given by his father, he defined the geometrical product of two vectors as the area of a parallelogram and that of three vectors as the volume of a parallelepiped. In addition to the geometrical product, Grassmann defined also a linear product of vectors (the dot product). This was well before the famous day, Monday October 16, 1843, when William Rowan Hamilton (1805-1865) discovered the four-dimensional complex numbers, the quaternions.

During 1842–43 Grassmann wrote the book *Lineale Ausdehnungslehre* (*Linear Extension Theory*, 1844), in which he generalized the previous concepts. The book was a great disappointment: it hardly sold at all, and finally in 1864 the publisher destroyed the remaining stock of 600 copies. *Ausdehnungslehre* contained philosophical arguments and thus was extremely hard to read. This was seen from the fact that no one would write a review of the book. Grassmann considered algebraic quantities which could be numbers, line segments, oriented areas, and so on, and defined 16 relations between them. He generalized everything to a space of n dimensions, which created more difficulties for the reader.

The geometrical product of the previous treatise was renamed as outer product. For example, in the outer product ab of two vectors (line segments) a and b the vector a was moved parallel to itself to a distance defined by the vector b, whence the product ab defined a parallelogram with an orientation. The orientation was reversed when the order was reversed: $ab = -ba$. If the parallelogram ab was moved by the vector c, the product abc gave a parallelepiped with an orientation. The outer product was more general than the geometric product, because it could be extended to a space of n dimensions. Thus it could be applied to solving a set of linear equations without a geometric interpretation.

During two decades the scientific world took the *Ausdehnungslehre* with total silence, although Grassmann had sent copies of his book to many well-known mathematicians asking for their comments. Finally, in 1845, he had to write a summary of his book by himself.

Only very few scientists showed any interest during the 1840s and 1850s. One of them was Adhemar-Jean-Claude de Saint-Venant, who himself had developed a corresponding algebra. In his article "Sommes et différences géométriques pour simplifier la mecanique" (Geometrical sums and differences for the simplification of mechanics, 1845), he very briefly introduced addition, subtraction, and differentiation of vectors and a similar outer product. Also, Augustin Cauchy had in 1853 developed a method to solve linear algebraic equations in terms of anticommutative elements ($ij = -ji$), which he called "clefs algébraiques" (algebraic keys). In 1852 Hamilton obtained a copy of Grassmann's book and expressed first his admiration which later turned to irony ("the greater the extension, the smaller the intention"). The afterworld has, however, considered the *Ausdehnungslehre* as a first classic of linear algebra, followed by Hamilton's book *Lectures on Quaternions* (1853).

During 1844–1862 Grassmann authored books and scientific articles on physics, philology (he is still a well-known authority in Sanscrit) and folklore (he published a collection of folk songs). However, his attempts to get a university position were not succesful, although in 1852 he was granted the title of Professor. Eventually, Grassmann published a completely rewritten version of his book, *Vollständige Aus-*

$a = \frac{dH}{dy} - \frac{dG}{dz}$ $b = \frac{dF}{dz} - \frac{dH}{dx}$ $c = \frac{dG}{dx} - \frac{dF}{dy}$	(A)	$\mathbf{B} = \nabla \times \mathbf{A}$
$P = c\frac{dy}{dt} - b\frac{dz}{dt} - \frac{dF}{dt} - \frac{d\psi}{dx}$ $Q = a\frac{dz}{dt} - c\frac{dx}{dt} - \frac{dG}{dt} - \frac{d\psi}{dy}$ $R = b\frac{dx}{dt} - a\frac{dy}{dt} - \frac{dH}{dt} - \frac{d\psi}{dz}$	(B)	$\mathbf{E} = \mathbf{v} \times \mathbf{B} - \frac{\partial \mathbf{A}}{\partial t} - \nabla \phi$
$X = vc - wb$ $Y = wa - uc$ $Z = ub - va$	(C)	$\mathbf{F} = \mathbf{J} \times \mathbf{B}$
$a = \alpha + 4\pi A$ $b = \beta + 4\pi B$ $c = \gamma + 4\pi C$	(D)	$\mathbf{B} = \mu_o \mathbf{H} + \mathbf{M}$
$4\pi u = \frac{d\gamma}{dy} - \frac{d\beta}{dz}$ $4\pi v = \frac{d\alpha}{dz} - \frac{d\gamma}{dx}$ $4\pi w = \frac{d\beta}{dx} - \frac{d\alpha}{dy}$	(E)	$\mathbf{J} = \nabla \times \mathbf{H}$
$\mathfrak{D} = \frac{1}{4\pi} K \mathfrak{E}$	(F)	$\mathbf{D} = \epsilon \mathbf{E}$
$\mathfrak{K} = C \mathfrak{E}$	(G)	$\mathbf{J}_c = \sigma \mathbf{E}$
$\mathfrak{C} = \mathfrak{K} + \dot{\mathfrak{D}}$	(H)	$\mathbf{J} = \mathbf{J}_c + \frac{\partial \mathbf{D}}{\partial t}$
$u = p + \frac{df}{dt}$ $v = q + \frac{dg}{dt}$ $w = r + \frac{dh}{dt}$	(H*)	$\mathbf{J} = \mathbf{J}_c + \frac{\partial \mathbf{D}}{\partial t}$
$\mathfrak{C} = (C + \frac{1}{4\pi} K \frac{d}{dt}) \mathfrak{E}$	(I)	$\mathbf{J} = \sigma \mathbf{E} + \epsilon \frac{\partial \mathbf{E}}{\partial t}$
$u = CP + \frac{1}{4\pi} K \frac{dP}{dt}$ $v = CQ + \frac{1}{4\pi} K \frac{dQ}{dt}$ $w = CR + \frac{1}{4\pi} K \frac{dR}{dt}$	(I*)	$\mathbf{J} = \sigma \mathbf{E} + \epsilon \frac{\partial \mathbf{E}}{\partial t}$
$\rho = \frac{df}{dx} + \frac{dg}{dy} + \frac{dh}{dz}$	(J)	$\varrho = \nabla \cdot \mathbf{D}$
$\sigma = lf + mg + nh + l'f' + m'g' + n'h'$	(K)	$\varrho_s = \mathbf{n} \cdot (\mathbf{D}_1 - \mathbf{D}_2)$
$\mathfrak{B} = \mu \mathfrak{H}$	(L)	$\mathbf{B} = \mu \mathbf{H}$

Fig. 1.1 *The original set of equations (A)–(L) as labeled by Maxwell in his Treatise (1873), with their interpretation in modern Gibbsian vector notation. The simplest equations were also written in vector form.*

dehnungslehre (*Complete Extension Theory*), on which he had started to work in 1854. The foreword bears the date 29 August 1861. Grassmann had it printed on his own expense in 300 copies by the printer Enslin in Berlin in 1862 [29]. In its preface he complained the poor reception of the first version and promised to give his arguments in Euclidean rigor in the present version.[2] Indeed, instead of relying on philosophical and physical arguments, the book was based on mathematical theorems. However, the reception of the second version was similar to that of the first one. Only in 1867 Hermann Hankel wrote a comparative article on the Grassmann algebra and quaternions, which started an interest in Grassmann's work. Finally there was also growing interest in the first edition of the *Ausdehnungslehre*, which made the publisher release a new printing in 1879, after Grassmann's death. Toward the end of his life, Grassmann had, however, turned his interest from mathematics to philology, which brought him an honorary doctorate among other signs of appreciation.

Although Grassmann's algebra could have become an important new mathematical branch during his lifetime, it did not. One of the reasons for this was the difficulty in reading his books. The first one was not a normal mathematical monograph with definitions and proofs. Grassmann gave his views on the new concepts in a very abstract way. It is true that extended quantities (Ausdehnungsgrösse) like multivectors in a space of n dimensions were very abstract concepts, and they were not easily digestible. Another reason for the poor reception for the Grassmann algebra is that Grassmann worked in a high school instead of a university where he could have had a group of scientists around him. As a third reason, we might recall that there was no great need for a vector algebra before the the arrival of Maxwell's electromagnetic theory in the 1870s, which involved interactions of many vector quantites. Their representation in terms of scalar quantites, as was done by Maxwell himself, created a messy set of equations which were understood only by a few scientist of his time (Figure 1.1).

After a short success period of Hamilton's quaternions in 1860-1890, the vector notation created by J. Willard Gibbs (1839–1903) and Oliver Heaviside (1850–1925) for the three-dimensional space overtook the analysis in physics and electromagnetics during the 1890s. Einstein's theory of relativity and Minkowski's space of four dimensions brought along the tensor calculus in the early 1900s. William Kingdon Clifford (1845–1879) was one of the first mathematicians to know both Hamilton's quaternions and Grassmann's analysis. A combination of these presently known as the Clifford algebra has been applied in physics to some extent since the 1930's [33,54]. Élie Cartan (1869–1951) finally developed the theory of differential forms based on the outer product of the Grassmann algebra in the early 1900s. It was adopted by others in the 1930s. Even if differential forms are generally applied in physics, in electromagnetics the Gibbsian vector algebra is still the most common method of notation. However, representation of the Maxwell equations in terms of differential forms has remarkable simple form in four-dimensional space-time (Figure 1.2).

[2] This book was only very recently translated into English [29] based on an edited version which appeared in the collected works of Grassmann.

$$\begin{array}{rcl} \mathbf{d} \wedge \mathbf{\Psi} & = & \gamma \\ \mathbf{d} \wedge \mathbf{\Phi} & = & 0 \\ \mathbf{\Psi} & = & \overline{\overline{\mathsf{M}}}|\mathbf{\Phi} \end{array}$$

Fig. 1.2 *The two Maxwell equations and the medium equation in differential-form formalism. Symbols will be explained in Chapter 4.*

Grassmann had hoped that the second edition of *Ausdehnungslehre* would raise interest in his contemporaries. Fearing that this, too, would be of no avail, his final sentences in the foreword were addressed to future generations [15, 75]:

> ... But I know and feel obliged to state (though I run the risk of seeming arrogant) that even if this work should again remain unused for another seventeen years or even longer, without entering into actual development of science, still that time will come when it will be brought forth from the dust of oblivion, and when ideas now dormant will bring forth fruit. I know that if I also fail to gather around me in a position (which I have up to now desired in vain) a circle of scholars, whom I could fructify with these ideas, and whom I could stimulate to develop and enrich further these ideas, nevertheless there will come a time when these ideas, perhaps in a new form, will rise anew and will enter into living communication with contemporary developments. For truth is eternal and divine, and no phase in the development of the truth divine, and no phase in the development of truth, however small may be region encompassed, can pass on without leaving a trace; truth remains, even though the garments in which poor mortals clothe it may fall to dust.
>
> Stettin, 29 August 1861

1.2 VECTORS AND DUAL VECTORS

1.2.1 Basic definitions

Vectors are elements of an n-dimensional vector space denoted by $\mathbb{E}_1(n)$, and they are in general denoted by boldface lowercase Latin letters $\mathbf{a}$, $\mathbf{b}$,... Most of the analysis is applicable to any dimension n but special attention is given to three-dimensional Euclidean (Eu3) and four-dimensional Minkowskian (Mi4) spaces (these concepts will be explained in terms of metric dyadics in Section 2.5). A set of linearly independent vectors $\{\mathbf{e}_i\} = \mathbf{e}_1, \mathbf{e}_2, ..., \mathbf{e}_n$ forms a basis if any vector $\mathbf{a}$ can be uniquely expressed in terms of the basis vectors as

$$\mathbf{a} = \sum_{i=1}^{n} a_i \mathbf{e}_i, \tag{1.1}$$

where the a_i are scalar coefficients (real or complex numbers).

Dual vectors are elements of another n-dimensional vector space denoted by $\mathbb{F}_1(n)$, and they are in general denoted by boldface Greek letters $\boldsymbol{\alpha}$, $\boldsymbol{\beta}$,... A dual vector basis is denoted by $\{\boldsymbol{\varepsilon}_i\} = \boldsymbol{\varepsilon}_1,...,\boldsymbol{\varepsilon}_n$. Any dual vector $\boldsymbol{\alpha}$ can be uniquely expressed in terms of the dual basis vectors as

$$\boldsymbol{\alpha} = \sum_{i=1}^{n} \alpha_i \boldsymbol{\varepsilon}_i, \tag{1.2}$$

with scalar coefficients α_i. Many properties valid for vectors are equally valid for dual vectors and conversely. To save space, in obvious cases, this fact is not explicitly stated.

Working with two different types of vectors is one factor that distinguishes the present analysis from the classical Gibbsian vector analysis [28]. Vector-like quantities in physics can be identified by their nature to be either vectors or dual vectors, or, rather, multivectors or dual multivectors to be discussed below. The disadvantage of this division is, of course, that there are more quantities to memorize. The advantage is, however, that some operation rules become more compact and valid for all space dimensions. Also, being a multivector or a dual multivector is a property similar to the dimension of a physical quantity which can be used in checking equations with complicated expressions. One could include additional properties to multivectors, not discussed here, which make one step forward in this direction. In fact, multivectors could be distinguished as being either true or pseudo multivectors, and dual multivectors could be distinguished as true or pseudo dual multivectors [36]. This would double the number of species in the zoo of multivectors.

Vectors and dual vectors can be given geometrical interpretations in terms of arrows and sets of parallel planes, and this can be extended to multivectors and dual multivectors. Actually, this has given the possibility to geometrize all of physics [58]. However, our goal here is not visualization but developing analytic tools applicable to electromagnetic problems. This is why the geometric content is passed by very quickly.

1.2.2 Duality product

The vector space[3] $\mathbb{E}_1$ and the dual vector space $\mathbb{F}_1$ can be associated so that every element of the dual vector space $\mathbb{F}_1$ defines a linear mapping of the elements of the vector space $\mathbb{E}_1$ to real or complex numbers. Similarly, every element of the vector space $\mathbb{E}_1$ defines a linear mapping of the elements of the dual vector space $\mathbb{F}_1$. This mutual linear mapping can be expressed in terms of a symmetric product called the duality product (inner product or contraction) which, following Deschamps [18], is denoted by the sign $|$

$$\boldsymbol{\alpha},\ \mathbf{a} \rightarrow \boldsymbol{\alpha}|\mathbf{a} = \mathbf{a}|\boldsymbol{\alpha}. \tag{1.3}$$

A vector $\mathbf{a}$ and a dual vector $\boldsymbol{\alpha}$ can be called orthogonal (or, rather, annihilating) if they satisfy $\mathbf{a}|\boldsymbol{\alpha} = 0$. The vector and dual vector bases $\{\mathbf{e}_i\}$, $\{\boldsymbol{\varepsilon}_i\}$ are called

[3] When the dimension n is general or has an agreed value, iwe write $\mathbb{E}_1$ instead of $\mathbb{E}_1(n)$ for simplicity.

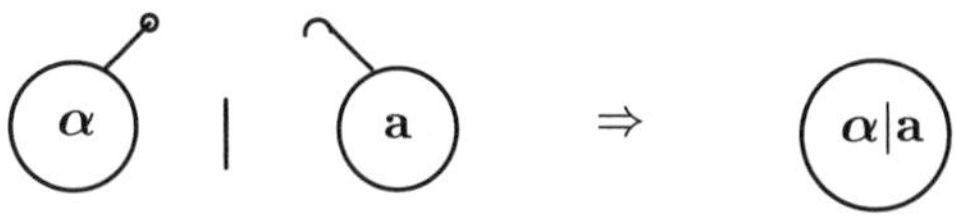

Fig. 1.3 *Hook and eye serve as visual aid to distinguish between vectors and dual vectors. The hook and the eye cancel each other in the duality product.*

reciprocal [21,28] (dual in [18]) to one another if they satisfy

$$\boldsymbol{\varepsilon}_i|\mathbf{e}_j = \mathbf{e}_j|\boldsymbol{\varepsilon}_i = \delta_{ij}. \tag{1.4}$$

Here δ_{ij} is the Kronecker symbol, $\delta_{ij} = 0$ when $i \neq j$ and $= 1$ when $i = j$. Given a basis of vectors or dual vectors the reciprocal basis can be constructed as will be seen in Section 2.4. In terms of the expansions (1.1), (1.2) in the reciprocal bases, the duality product of a vector $\mathbf{a}$ and a dual vector $\boldsymbol{\alpha}$ can be expressed as

$$\boldsymbol{\alpha}|\mathbf{a} = \sum_{i,j}(\alpha_i\boldsymbol{\varepsilon}_i)|(a_j\mathbf{e}_j) = \sum_i \alpha_i a_i. \tag{1.5}$$

The duality product must not be mistaken for the scalar product (dot product) of the vector space, denoted by $\mathbf{a}\cdot\mathbf{b}$, to be introduced in Section 2.5. The elements of the duality product are from two different spaces while those of the dot product are from the same space.

To distinguish between different quantities it is helpful to have certain suggestive mental aids, for example, hooks for vectors and eyes for dual vectors as in Figure 1.3. In the duality product the hook of a vector is fastened to the eye of the dual vector and the result is a scalar with neither a hook nor an eye left free. This has an obvious analogy in atoms forming molecules.

1.2.3 Dyadics

Linear mappings from a vector to a vector can be conveniently expressed in the coordinate-free dyadic notation. Here we consider only the basic notation and leave more detailed properties to Chapter 2. Dyadic product of a vector $\mathbf{c}$ and a dual vector $\boldsymbol{\gamma}$ is denoted by $\mathbf{c}\boldsymbol{\gamma}$. The "no-sign" dyadic multiplication originally introduced by Gibbs [28,40] is adopted here instead of the sign $\otimes$ preferred by the mathematicians. Also, other signs for the dyadic product have been in use since Gibbs,— for example, the colon [53].

The dyadic product can be defined by considering the expression

$$\mathbf{b} = \mathbf{c}(\boldsymbol{\gamma}|\mathbf{a}) = (\mathbf{c}\boldsymbol{\gamma})|\mathbf{a}, \tag{1.6}$$

which extends the associative law (order of the two multiplications as shown by the brackets). The dyad $\mathbf{c}\boldsymbol{\gamma}$ acts as a linear mapping from a vector $\mathbf{a}$ to another vector $\mathbf{b}$.

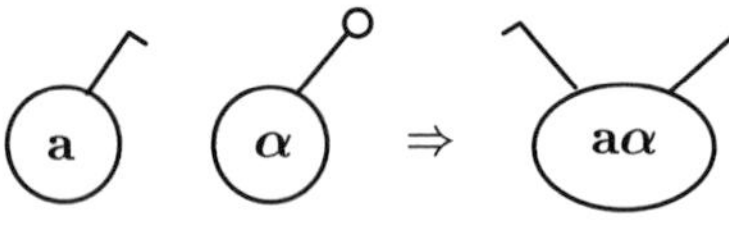

Fig. 1.4 *Dyadic product (no sign) of a vector and a dual vector in this order produces an object which can be visualized as having a hook on the left and an eye on the right.*

Similarly, the dyadic product $\boldsymbol{\gamma}\mathbf{c}$ acts as a linear mapping from a dual vector $\boldsymbol{\alpha}$ to $\boldsymbol{\beta}$ as

$$\boldsymbol{\beta} = \boldsymbol{\gamma}(\mathbf{c}|\boldsymbol{\alpha}) = (\boldsymbol{\gamma}\mathbf{c})|\boldsymbol{\alpha}. \tag{1.7}$$

The dyadic product $\mathbf{a}\boldsymbol{\alpha}$ can be pictured as an ordered pair of quantities glued back-to-back so that the hook of the vector $\mathbf{a}$ points to the left and the eye of the dual vector $\boldsymbol{\alpha}$ points to the right (Figure 1.4).

Any linear mapping within each vector space $\mathbb{E}_1$ and $\mathbb{F}_1$ can be performed through dyadic polynomials, or dyadics in short. Whenever possible, dyadics are denoted by capital sans-serif characters with two overbars, otherwise by standard symbols with two overbars:

$$\mathbf{b} = \overline{\overline{\mathsf{A}}}|\mathbf{a}, \quad \overline{\overline{\mathsf{A}}} = \sum \mathbf{c}_i\boldsymbol{\gamma}_i = \mathbf{c}_1\boldsymbol{\gamma}_1 + \cdots + \mathbf{c}_n\boldsymbol{\gamma}_n, \tag{1.8}$$

$$\boldsymbol{\beta} = \overline{\overline{\mathsf{A}}}^T|\boldsymbol{\alpha}, \quad \overline{\overline{\mathsf{A}}}^T = \sum \boldsymbol{\gamma}_i\mathbf{c}_i = \boldsymbol{\gamma}_1\mathbf{c}_1 + \cdots + \boldsymbol{\gamma}_n\mathbf{c}_n. \tag{1.9}$$

Here, ^{T} denotes the transpose operation: $(\mathbf{c}\boldsymbol{\gamma})^T = \boldsymbol{\gamma}\mathbf{c}$. Mapping of a vector by a dyadic can be pictured as shown in Figure 1.5.

Let us denote the space of dyadics of the type $\overline{\overline{\mathsf{A}}}$ above by $\mathbb{E}_1\mathbb{F}_1$ (short for $\mathbb{E}_1 \times \mathbb{F}_1$) and, that of the type $\overline{\overline{\mathsf{A}}}^T$ by $\mathbb{F}_1\mathbb{E}_1$ ($\mathbb{F}_1 \times \mathbb{E}_1$). An element of the space $\mathbb{E}_1\mathbb{F}_1$ maps the vector space $\mathbb{E}_1$ onto itself (from the right, from the left it maps the space $\mathbb{F}_1$ onto itself). If a given dyadic $\overline{\overline{\mathsf{A}}}$ maps the space $\mathbb{E}_1$ onto itself, i.e., any vector basis $\{\mathbf{e}_i\}$ to another vector basis $\{\mathbf{e}'_i\}$, the dyadic is called *complete* and there exists a unique inverse dyadic $\overline{\overline{\mathsf{A}}}^{-1}$. The dyadic is incomplete if it maps $\mathbb{E}_1$ only to a subspace of $\mathbb{E}_1$. Such a dyadic does not have a unique inverse. The dimensions of the dyadic spaces $\mathbb{F}_1\mathbb{E}_1$ and $\mathbb{E}_1\mathbb{F}_1$ are n^2.

The dyadic product does not commute. Actually, as was seen above, the transpose operation ^{T} maps dyadics $\mathbb{E}_1\mathbb{F}_1$ to another space $\mathbb{F}_1\mathbb{E}_1$. There are no concepts like symmetry and antisymmetry applicable to dyadics in these spaces. Later we will encounter other dyadic spaces $\mathbb{E}_1\mathbb{E}_1$, $\mathbb{F}_1\mathbb{F}_1$ containing symmetric and antisymmetric dyadics.

The unit dyadic $\overline{\overline{\mathsf{I}}}$ maps any vector to itself: $\overline{\overline{\mathsf{I}}}|\mathbf{a} = \mathbf{a}$. Thus, it also maps any $\mathbb{E}_1\mathbb{F}_1$ dyadic to itself: $\overline{\overline{\mathsf{I}}}|\overline{\overline{\mathsf{A}}} = \overline{\overline{\mathsf{A}}}$. Because any vector $\mathbf{a}$ can be expressed in terms of a basis $\{\mathbf{e}_i\}$ and its reciprocal dual basis $\{\boldsymbol{\varepsilon}_i\}$ as

$$\mathbf{a} = \sum \mathbf{e}_i(\boldsymbol{\varepsilon}_i|\mathbf{a}) = \sum(\mathbf{e}_i\boldsymbol{\varepsilon}_i)|\mathbf{a}, \tag{1.10}$$

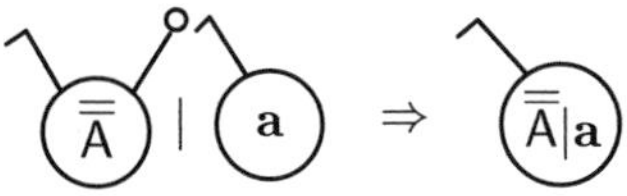

Fig. 1.5 *Dyadic $\overline{\overline{\mathsf{A}}}$ maps a vector* $\mathbf{a}$ *to the vector* $\overline{\overline{\mathsf{A}}}|\mathbf{a}$.

the unit dyadic can be expanded as

$$\overline{\overline{\mathsf{I}}} = \sum \mathbf{e}_i \boldsymbol{\varepsilon}_i = \mathbf{e}_1 \boldsymbol{\varepsilon}_1 + \mathbf{e}_2 \boldsymbol{\varepsilon}_2 + \cdots + \mathbf{e}_n \boldsymbol{\varepsilon}_n. \tag{1.11}$$

The form is not unique because we can choose one of the reciprocal bases $\{\mathbf{e}_i\}$, $\{\boldsymbol{\varepsilon}_j\}$ arbitrarily. The transposed unit dyadic

$$\overline{\overline{\mathsf{I}}}^T = \sum \boldsymbol{\varepsilon}_i \mathbf{e}_i = \boldsymbol{\varepsilon}_1 \mathbf{e}_1 + \boldsymbol{\varepsilon}_2 \mathbf{e}_2 + \cdots + \boldsymbol{\varepsilon}_n \mathbf{e}_n \tag{1.12}$$

serves as the unit dyadic for the dual vectors satisfying $\overline{\overline{\mathsf{I}}}^T|\boldsymbol{\alpha} = \boldsymbol{\alpha}$ for any dual vector $\boldsymbol{\alpha}$. We can also write $\boldsymbol{\alpha}|\overline{\overline{\mathsf{I}}} = \boldsymbol{\alpha}$ and $\mathbf{a}|\overline{\overline{\mathsf{I}}}^T = \mathbf{a}$.

Problems

1.2.1 Given a basis of vectors $\{\mathbf{a}_i\}$ and a basis of dual vectors $\{\boldsymbol{\beta}_j\}$, find the basis of dual vectors $\{\boldsymbol{\alpha}_j\}$ dual to $\{\mathbf{a}_i\}$ in terms of the basis $\{\boldsymbol{\beta}_j\}$.

1.2.2 Show that, in a space of n dimensions, any dyadic $\overline{\overline{\mathsf{A}}}$ can be expressed as a sum of n dyads $\mathbf{a}_i \boldsymbol{\alpha}_i$.

1.3 BIVECTORS

1.3.1 Wedge product

The wedge product (outer product) between any two elements $\mathbf{a}$ and $\mathbf{b}$ of the vector space $\mathbb{E}_1$ and elements $\boldsymbol{\alpha}, \boldsymbol{\beta}$ of the dual vector space $\mathbb{F}_1$ is defined to satisfy the anticommutative law:

$$\mathbf{a} \wedge \mathbf{b} = -\mathbf{b} \wedge \mathbf{a}, \quad \boldsymbol{\alpha} \wedge \boldsymbol{\beta} = -\boldsymbol{\beta} \wedge \boldsymbol{\alpha}. \tag{1.13}$$

Anticommutativity implies that the wedge product of any element with itself vanishes:

$$\mathbf{a} \wedge \mathbf{a} = 0, \quad \boldsymbol{\alpha} \wedge \boldsymbol{\alpha} = 0. \tag{1.14}$$

Actually, (1.14) implies (1.13), because we can expand

$$(\mathbf{a} + \mathbf{b}) \wedge (\mathbf{a} + \mathbf{b}) = \mathbf{a} \wedge \mathbf{a} + \mathbf{a} \wedge \mathbf{b} + \mathbf{b} \wedge \mathbf{a} + \mathbf{b} \wedge \mathbf{b}$$

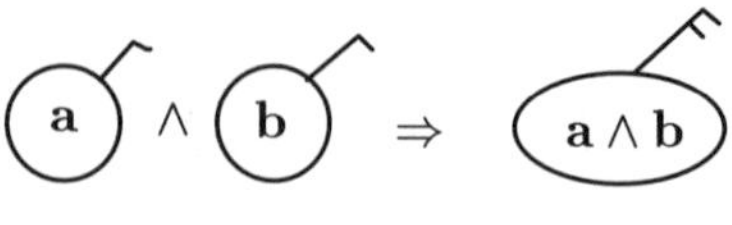

Fig. 1.6 *Visual aid for forming the wedge product of two vectors. The bivector has a double hook and, the dual bivector, a double eye.*

$$= \mathbf{a} \wedge \mathbf{b} + \mathbf{b} \wedge \mathbf{a} = 0. \tag{1.15}$$

A scalar factor can be moved outside the wedge product:

$$\mathbf{a} \wedge (\lambda \mathbf{b}) = \lambda(\mathbf{a} \wedge \mathbf{b}). \tag{1.16}$$

Wedge product between a vector and a dual vector is not defined.

1.3.2 Basis bivectors

The wedge product of two vectors is neither a vector nor a dyadic but a bivector,[4] or 2-vector, which is an element of another space $\mathbb{E}_2$. Correspondingly, the wedge product of two dual vectors is a dual bivector, an element of the space $\mathbb{F}_2$. A bivector can be visualized by a double hook as in Figure 1.6 and, a dual bivector, by a double eye. Whenever possible, bivectors are denoted by boldface Roman capital letters like $\mathbf{A}$, and dual bivectors are denoted by boldface Greek capital letters like $\mathbf{\Phi}$. However, in many cases we have to follow the classical notation of the electromagnetic literature.

A bivector of the form $\mathbf{a} \wedge \mathbf{b}$ is called a simple bivector [33]. General elements of the bivector space $\mathbb{E}_2$ are linear combinations of simple bivectors,

$$\mathbf{A} = \mathbf{a}_1 \wedge \mathbf{b}_1 + \mathbf{a}_2 \wedge \mathbf{b}_2 + \cdots = \sum \mathbf{a}_i \wedge \mathbf{b}_i. \tag{1.17}$$

The basis elements in the spaces $\mathbb{E}_2$ and $\mathbb{F}_2$ can be expanded in terms of the respective basis elements of $\mathbb{E}_1$ and $\mathbb{F}_1$. The basis bivectors and dual bivectors are denoted by lowercase letters with double indices as

$$\mathbf{e}_{ij} = \mathbf{e}_i \wedge \mathbf{e}_j = -\mathbf{e}_{ji}, \tag{1.18}$$

$$\boldsymbol{\varepsilon}_{ij} = \boldsymbol{\varepsilon}_i \wedge \boldsymbol{\varepsilon}_j = -\boldsymbol{\varepsilon}_{ji}. \tag{1.19}$$

Due to antisymmetry of the wedge product, the bi-index ij has some redundancy since the basis elements with indices of the form ii are zero and the elements corresponding to the bi-index ij equal the negative of those with the bi-index ji. Thus, instead of n^2, the dimension of the spaces $\mathbb{E}_2(n)$ and $\mathbb{F}_2(n)$ is only $n(n-1)/2$. For the two-, three-, and four-dimensional vector spaces, the respective dimensions of the bivector spaces are one, three, and six.

[4]Note that, originally, J.W. Gibbs called complex vectors of the form $\mathbf{a} + j\mathbf{b}$ bivectors. This meaning is still occasionally encountered in the literature [9].

The wedge product of two vector expansions

$$\mathbf{a} = \sum a_i \mathbf{e}_i, \quad \mathbf{b} = \sum b_j \mathbf{e}_j \tag{1.20}$$

gives the bivector expansion

$$\begin{aligned} \mathbf{a} \wedge \mathbf{b} &= \sum a_i \mathbf{e}_i \wedge \sum b_j \mathbf{e}_j = \sum_{i,j} a_i b_j \mathbf{e}_{ij} & (1.21) \\ &= a_1 b_2 \mathbf{e}_{12} + a_2 b_1 \mathbf{e}_{21} + a_1 b_3 \mathbf{e}_{13} + a_3 b_1 \mathbf{e}_{31} + \cdots . & (1.22) \end{aligned}$$

Because of the redundancy, we can reduce the number of bi-indices ij by ordering, i.e., restricting to indices satisfying $i < j$:

$$\begin{aligned} \mathbf{a} \wedge \mathbf{b} &= \sum_{i<j} (a_i b_j - a_j b_i) \mathbf{e}_{ij} \\ &= (a_1 b_2 - a_2 b_1)\mathbf{e}_{12} + (a_1 b_3 - a_3 b_1)\mathbf{e}_{13} + \cdots + (a_1 b_n - a_n b_1)\mathbf{e}_{1n} \\ &+ (a_2 b_3 - a_3 b_2)\mathbf{e}_{23} + (a_2 b_4 - a_4 b_2)\mathbf{e}_{24} + \cdots + (a_2 b_n - a_n b_2)\mathbf{e}_{2n} \\ &+ \cdots + (a_{n-1} b_n - a_n b_{n-1})\mathbf{e}_{(n-1)n}. \end{aligned} \tag{1.23}$$

Euclidean and Minkowskian bivectors For a more symmetric representation, cyclic ordering of the bi-indices is often preferred in the three-dimensional Euclidean Eu3 space:

$$\mathbf{a} \wedge \mathbf{b} = (a_1 b_2 - a_2 b_1)\mathbf{e}_{12} + (a_2 b_3 - a_3 b_2)\mathbf{e}_{23} + (a_3 b_1 - a_1 b_3)\mathbf{e}_{31}. \tag{1.24}$$

The four-dimensional Minkowskian space Mi4 can be understood as Eu3 with an added dimension corresponding to the index 4. In this case, the ordering is usually taken cyclic in the indices 1,2,3 and the index 4 is written last as

$$\begin{aligned} \mathbf{a} \wedge \mathbf{b} &= (a_1 b_2 - a_2 b_1)\mathbf{e}_{12} + (a_2 b_3 - a_3 b_2)\mathbf{e}_{23} + (a_3 b_1 - a_1 b_3)\mathbf{e}_{31} \\ &+ (a_1 b_4 - a_4 b_1)\mathbf{e}_{14} + (a_2 b_4 - a_4 b_2)\mathbf{e}_{24} + (a_3 b_4 - a_4 b_3)\mathbf{e}_{34}. \end{aligned} \tag{1.25}$$

More generally, expressing Minkowskian vectors $\mathbf{a}_\mathrm{M}$ and dual vectors $\boldsymbol{\alpha}_\mathrm{M}$ as

$$\mathbf{a}_\mathrm{M} = \mathbf{a} + \mathbf{e}_4 a_4, \qquad \boldsymbol{\alpha}_\mathrm{M} = \boldsymbol{\alpha} + \boldsymbol{\varepsilon}_4 \alpha_4, \tag{1.26}$$

where $\mathbf{a}$ and $\boldsymbol{\alpha}$ are vector and dual vector components in the Euclidean Eu3 space, the wedge product of two Minkowskian vectors can be expanded as

$$\mathbf{a}_\mathrm{M} \wedge \mathbf{b}_\mathrm{M} = (\mathbf{a} + \mathbf{e}_4 a_4) \wedge (\mathbf{b} + \mathbf{e}_4 b_4) = \mathbf{a} \wedge \mathbf{b} + (\mathbf{a} b_4 - \mathbf{b} a_4) \wedge \mathbf{e}_4. \tag{1.27}$$

Thus, any bivector or dual bivector in the Mi4 space can be naturally expanded in the form

$$\mathbf{A}_\mathrm{M} = \mathbf{A} + \mathbf{a} \wedge \mathbf{e}_4, \qquad \mathbf{\Phi}_\mathrm{M} = \mathbf{\Phi} + \boldsymbol{\alpha} \wedge \boldsymbol{\varepsilon}_4, \tag{1.28}$$

where $\mathbf{A}$, $\mathbf{a}$, $\mathbf{\Phi}$, and $\boldsymbol{\alpha}$ denote the respective Euclidean bivector, vector, dual bivector, and dual vector components.

For two-dimensional vectors the dimension of the bivectors is 1 and all bivectors can be expressed as multiples of a single basis element $\mathbf{e}_{12}$. Because for the three-dimensional vector space the bivector space has the dimension 3, bivectors have a close relation to vectors. In the Gibbsian vector algebra, where the wedge product is replaced by the cross product, bivectors are identified with vectors. In the four-dimensional vector space, bivectors form a six-dimensional space, and they can be represented in terms of a combination of a three-dimensional vector and bivector, each of dimension 3.

In terms of basis bivectors, respective expansions for the general bivector $\mathbf{A} = \sum_{i,j} A_{ij}\mathbf{e}_{ij}$ in spaces of dimension $n = 2, 3,$ and 4 can be, respectively, written as

$$\mathbf{A} = A_{12}\mathbf{e}_{12}, \tag{1.29}$$

$$\mathbf{A} = A_{12}\mathbf{e}_{12} + A_{23}\mathbf{e}_{23} + A_{31}\mathbf{e}_{31}, \tag{1.30}$$

$$\mathbf{A} = A_{12}\mathbf{e}_{12} + A_{23}\mathbf{e}_{23} + A_{31}\mathbf{e}_{31} + A_{14}\mathbf{e}_{14} + A_{24}\mathbf{e}_{24} + A_{34}\mathbf{e}_{34}. \tag{1.31}$$

Similar expansions apply for the dual bivectors $\mathbf{\Phi} = \sum \Phi_{ij}\boldsymbol{\varepsilon}_{ij}$. It can be shown that any bivector $\mathbf{A}$ in the case $n = 3$ can be expressed in the form of a simple bivector $\mathbf{A} = \mathbf{a} \wedge \mathbf{b}$ in terms of two vectors $\mathbf{a}, \mathbf{b}$. The proof is left as an exercise. This decomposition is not unique since, for example, we can write $(\mathbf{a} + \lambda\mathbf{b}) \wedge \mathbf{b}$ instead of $\mathbf{a} \wedge \mathbf{b}$ with any scalar λ without changing the bivector. On the other hand, for $n = 4$, any bivector can be expressed as a sum of two simple bivectors, in the form $\mathbf{A} = \mathbf{a} \wedge \mathbf{b} + \mathbf{c} \wedge \mathbf{d}$ in terms of four vectors $\mathbf{a}, \mathbf{b}, \mathbf{c}, \mathbf{d}$. Again, this representation is not unique. The proof can be based on separating the fourth dimension as was done in (1.28).

1.3.3 Duality product

The duality product of a vector and a dual vector is straightforwardly generalized to that of a bivector and a dual bivector by defining the product for the reciprocal basis bivectors and dual bivectors as

$$\boldsymbol{\varepsilon}_{12}|\mathbf{e}_{12} = 1, \quad \boldsymbol{\varepsilon}_{12}|\mathbf{e}_{13} = 0, \quad \boldsymbol{\varepsilon}_{13}|\mathbf{e}_{13} = 1, ... \tag{1.32}$$

and more generally

$$\boldsymbol{\varepsilon}_J|\mathbf{e}_K = \delta_{JK}, \quad J = \{ij\}, \; K = \{k\ell\}. \tag{1.33}$$

Here, J and K are ordered bi-indices ($i < j$, $k < \ell$) and the symbol δ_{JK} has the value 1 only when both $ij = k\ell$, otherwise it is zero. Thus, we can write

$$\delta_{JK} = \delta_{\{ij\}\{k\ell\}} = \delta_{ik}\delta_{j\ell}. \tag{1.34}$$

The corresponding definition for nonordered indices J, K has to take also into account that $\delta_{JK} = -1$ when $ij = \ell k$, in which case (1.34) is generalized to

$$\delta_{JK} = \delta_{ik}\delta_{j\ell} - \delta_{i\ell}\delta_{jk}. \tag{1.35}$$

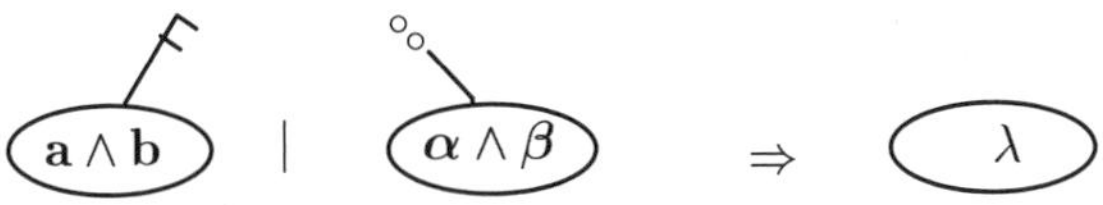

Fig. 1.7 *Duality product of a bivector* $\mathbf{a}\wedge\mathbf{b}$ *and a dual bivector* $\boldsymbol{\alpha}\wedge\boldsymbol{\beta}$ *gives the scalar* $\lambda = (\mathbf{a}\wedge\mathbf{b})|(\boldsymbol{\alpha}\wedge\boldsymbol{\beta})$.

Bivectors and dual bivectors can be pictured as objects with a respective double hook and double eye. In the duality product the double hook is fastened to the double eye to make an object with no free hooks or eyes, a scalar, Figure 1.7. The duality product of a bivector and a dual bivector can be expanded in terms of duality products of vectors and dual vectors. The expansion is based on the bivector identity

$$\begin{aligned}(\boldsymbol{\alpha}\wedge\boldsymbol{\beta})|(\mathbf{a}\wedge\mathbf{b}) &= (\boldsymbol{\alpha}|\mathbf{a})(\boldsymbol{\beta}|\mathbf{b}) - (\boldsymbol{\alpha}|\mathbf{b})(\boldsymbol{\beta}|\mathbf{a}) \\ &= \det\begin{pmatrix} (\boldsymbol{\alpha}|\mathbf{a}) & (\boldsymbol{\alpha}|\mathbf{b}) \\ (\boldsymbol{\beta}|\mathbf{a}) & (\boldsymbol{\beta}|\mathbf{b}) \end{pmatrix}\end{aligned} \quad (1.36)$$

which can be easily derived by first expanding the vectors and dual vectors in terms of the basis vectors $\mathbf{e}_i$ and the reciprocal dual basis vectors $\boldsymbol{\varepsilon}_j$. From the form of (1.36) it can be seen that all expressions change sign when $\mathbf{a}$ and $\mathbf{b}$ or $\boldsymbol{\alpha}$ and $\boldsymbol{\beta}$ are interchanged. By arranging terms, the identity (1.36) can be rewritten in two other forms

$$(\boldsymbol{\alpha}\wedge\boldsymbol{\beta})|(\mathbf{a}\wedge\mathbf{b}) = \boldsymbol{\alpha}|(\mathbf{ab}-\mathbf{ba})|\boldsymbol{\beta} = \mathbf{a}|(\boldsymbol{\alpha\beta}-\boldsymbol{\beta\alpha})|\mathbf{b}. \quad (1.37)$$

Here we have introduced the dyadic product of two vectors, $\mathbf{ab}$, and two dual vectors, $\boldsymbol{\alpha\beta}$, which are elements of the respective spaces $\mathbb{E}_1\mathbb{E}_1 (= \mathbb{E}_1 \times \mathbb{E}_1)$ and $\mathbb{F}_1\mathbb{F}_1 (= \mathbb{F}_1 \times \mathbb{F}_1)$. Any sum of dyadic products $\sum \mathbf{a}_i\mathbf{b}_i$ serves as a mapping of a dual vector to a vector $\mathbb{F}_1 \to \mathbb{E}_1$ as $\sum \mathbf{a}_i\mathbf{b}_i|\boldsymbol{\gamma} = \mathbf{c}$. In analogy to the double-dot product in the Gibbsian dyadic algebra [28,40], we can define the double-duality product $||$ between two dyadics, elements of the spaces $\mathbb{E}_1\mathbb{E}_1$ and $\mathbb{F}_1\mathbb{F}_1$ or $\mathbb{E}_1\mathbb{F}_1$ and $\mathbb{F}_1\mathbb{E}_1$:

$$\begin{aligned}(\boldsymbol{\alpha}|\mathbf{a})(\boldsymbol{\beta}|\mathbf{b}) &= (\boldsymbol{\alpha\beta})||(\mathbf{ab}) = (\mathbf{ab})||(\boldsymbol{\alpha\beta}) \\ &= (\mathbf{a}\boldsymbol{\beta})||(\boldsymbol{\alpha}\mathbf{b}) = (\boldsymbol{\alpha}\mathbf{b})||(\mathbf{a}\boldsymbol{\beta}).\end{aligned} \quad (1.38)$$

The result is a scalar. For two dyadics $\overline{\overline{\mathsf{A}}}, \overline{\overline{\mathsf{B}}} \in \mathbb{E}_1\mathbb{F}_1$ or $\in \mathbb{F}_1\mathbb{E}_1$ the double-duality product satisfies

$$\overline{\overline{\mathsf{A}}}||\overline{\overline{\mathsf{B}}}^T = \overline{\overline{\mathsf{B}}}^T||\overline{\overline{\mathsf{A}}} = \overline{\overline{\mathsf{A}}}^T||\overline{\overline{\mathsf{B}}} = \overline{\overline{\mathsf{B}}}||\overline{\overline{\mathsf{A}}}^T. \quad (1.39)$$

The identity (1.36) can be rewritten in the following forms:

$$\begin{aligned}(\boldsymbol{\alpha}\wedge\boldsymbol{\beta})|(\mathbf{a}\wedge\mathbf{b}) &= (\boldsymbol{\alpha\beta})||(\mathbf{ab}-\mathbf{ba}) = (\boldsymbol{\alpha\beta}-\boldsymbol{\beta\alpha})||(\mathbf{ab}) \\ &= \frac{1}{2}(\boldsymbol{\alpha\beta}-\boldsymbol{\beta\alpha})||(\mathbf{ab}-\mathbf{ba}).\end{aligned} \quad (1.40)$$

For two antisymmetric dyadics

$$\overline{\overline{\mathsf{A}}} = \sum(\mathbf{a}_i\mathbf{b}_i - \mathbf{b}_i\mathbf{a}_i), \quad \overline{\overline{\mathsf{B}}} = \sum(\boldsymbol{\alpha}_j\boldsymbol{\beta}_j - \boldsymbol{\beta}_j\boldsymbol{\alpha}_j) \tag{1.41}$$

this can be generalized as

$$\overline{\overline{\mathsf{A}}}||\overline{\overline{\mathsf{B}}} = 2\sum_{i,j}(\mathbf{a}_i \wedge \mathbf{b}_i)|(\boldsymbol{\alpha}_j \wedge \boldsymbol{\beta}_j). \tag{1.42}$$

The double-duality product of a symmetric and an antisymmetric dyadic always vanishes, as is seen from (1.39).

1.3.4 Incomplete duality product

Because the scalar $(\boldsymbol{\alpha} \wedge \boldsymbol{\beta})|(\mathbf{a} \wedge \mathbf{b})$ is a linear function of the dual vector $\boldsymbol{\alpha}$, there is must exist a vector $\mathbf{c}$ such that we can write

$$(\boldsymbol{\alpha} \wedge \boldsymbol{\beta})|(\mathbf{a} \wedge \mathbf{b}) = \boldsymbol{\alpha}|\mathbf{c}. \tag{1.43}$$

Since $\mathbf{c}$ is a linear function of the dual vector $\boldsymbol{\beta}$ and the bivector $\mathbf{a} \wedge \mathbf{b}$, we can express it in the form of their product as

$$\mathbf{c} = \boldsymbol{\beta}\rfloor(\mathbf{a} \wedge \mathbf{b}), \tag{1.44}$$

where $\rfloor$ is the sign of the incomplete duality product. Another way to express the same scalar quantity is

$$\mathbf{c}|\boldsymbol{\alpha} = -(\mathbf{a} \wedge \mathbf{b})|(\boldsymbol{\beta} \wedge \boldsymbol{\alpha}) = -((\mathbf{a} \wedge \mathbf{b})\lfloor\boldsymbol{\beta})|\boldsymbol{\alpha}, \tag{1.45}$$

which defines another product sign $\lfloor$. The relation between the two incomplete products is, thus,

$$\boldsymbol{\beta}\rfloor(\mathbf{a} \wedge \mathbf{b}) = -(\mathbf{a} \wedge \mathbf{b})\lfloor\boldsymbol{\beta}. \tag{1.46}$$

More generally, we can write

$$\boldsymbol{\alpha}\rfloor\mathbf{A} = -\mathbf{A}\lfloor\boldsymbol{\alpha}, \quad \mathbf{a}\rfloor\boldsymbol{\Phi} = -\boldsymbol{\Phi}\lfloor\mathbf{a}, \tag{1.47}$$

when $\mathbf{a}$ is a vector, $\boldsymbol{\alpha}$ is a dual vector, $\mathbf{A}$ is a bivector and $\boldsymbol{\Phi}$ is a dual bivector. The horizontal line points to the quantity which has lower grade. In these cases it is either a vector or a dual vector. In an incomplete duality product of a dual vector and a bivector one hook and one eye are eliminated as is shown in Figure 1.8.

The defining formulas can be easily memorized from the rules

$$\mathbf{a}|(\mathbf{b}\rfloor\boldsymbol{\Phi}) = (\mathbf{a} \wedge \mathbf{b})|\boldsymbol{\Phi}, \quad (\boldsymbol{\Phi}\lfloor\mathbf{b})|\mathbf{a} = \boldsymbol{\Phi}(\mathbf{b} \wedge \mathbf{a}), \tag{1.48}$$

valid for any two vectors $\mathbf{a}, \mathbf{b}$ and a dual bivector $\boldsymbol{\Phi}$. The dual form is

$$\boldsymbol{\alpha}|(\boldsymbol{\beta}\rfloor\mathbf{A}) = (\boldsymbol{\alpha} \wedge \boldsymbol{\beta})|\mathbf{A}, \quad (\mathbf{A}\lfloor\boldsymbol{\alpha})|\boldsymbol{\beta} = \mathbf{A}|(\boldsymbol{\alpha} \wedge \boldsymbol{\beta}). \tag{1.49}$$

Fig. 1.8 *The incomplete duality product between a dual vector $\boldsymbol{\beta}$ and a bivector $\mathbf{A}$ gives a vector $\boldsymbol{\beta}\rfloor\mathbf{A}$.*

Bac cab rule Applying (1.36), we obtain an important expansion rule for two vectors $\mathbf{b}, \mathbf{c}$ and a dual vector $\boldsymbol{\alpha}$, which, followed by its dual, can be written as

$$\boldsymbol{\alpha}\rfloor(\mathbf{b}\wedge\mathbf{c}) = \mathbf{b}(\boldsymbol{\alpha}|\mathbf{c}) - \mathbf{c}(\boldsymbol{\alpha}|\mathbf{b}) = -(\mathbf{b}\wedge\mathbf{c})\lfloor\boldsymbol{\alpha}, \tag{1.50}$$

$$\mathbf{a}\rfloor(\boldsymbol{\beta}\wedge\boldsymbol{\gamma}) = \boldsymbol{\beta}(\mathbf{a}|\boldsymbol{\gamma}) - \boldsymbol{\gamma}(\mathbf{a}|\boldsymbol{\beta}) = -(\boldsymbol{\beta}\wedge\boldsymbol{\gamma})\lfloor\mathbf{a}. \tag{1.51}$$

These identities can be easily memorized from their similarity to the "bac-cab" rule familiar for Gibbsian vectors:

$$\mathbf{a}\times(\mathbf{b}\times\mathbf{c}) = \mathbf{b}(\mathbf{a}\cdot\mathbf{c}) - \mathbf{c}(\mathbf{a}\cdot\mathbf{b}) = -(\mathbf{b}\times\mathbf{c})\times\mathbf{a}. \tag{1.52}$$

The incomplete duality product often corresponds to the cross product in Gibbsian vector algebra.

Decomposition of vector The bac-cab rule for Gibbsian vectors, (1.52), can be used to decompose any vector to components parallel and orthogonal to a given vector. Equation (1.50) defines a similar decomposition for a vector $\mathbf{b}$ in a component parallel to a given vector $\mathbf{a}$ and orthogonal to a given dual vector $\boldsymbol{\alpha}$ provided these satisfy the condition $\mathbf{a}|\boldsymbol{\alpha} \neq 0$,

$$\mathbf{b} = \mathbf{b}_{\|} + \mathbf{b}_{\perp}, \quad \mathbf{a}\wedge\mathbf{b}_{\|} = 0, \quad \boldsymbol{\alpha}|\mathbf{b}_{\perp} = 0. \tag{1.53}$$

In fact, writing (1.50) with $\mathbf{c} = \mathbf{a}$ as

$$\mathbf{b} = \frac{\mathbf{b}|\boldsymbol{\alpha}}{\mathbf{a}|\boldsymbol{\alpha}}\mathbf{a} + \frac{\boldsymbol{\alpha}\rfloor(\mathbf{b}\wedge\mathbf{a})}{\mathbf{a}|\boldsymbol{\alpha}}, \tag{1.54}$$

we can identify the components as

$$\mathbf{b}_{\|} = \frac{\mathbf{b}|\boldsymbol{\alpha}}{\mathbf{a}|\boldsymbol{\alpha}}\mathbf{a}, \quad \mathbf{b}_{\perp} = \frac{\boldsymbol{\alpha}\rfloor(\mathbf{b}\wedge\mathbf{a})}{\mathbf{a}|\boldsymbol{\alpha}}. \tag{1.55}$$

1.3.5 Bivector dyadics

Dyadic product $\mathbf{C}\boldsymbol{\Gamma}$ of a bivector $\mathbf{C}$ and a dual bivector $\boldsymbol{\Gamma}$ is defined as a linear mapping from a bivector $\mathbf{A}$ to a multiple of the bivector $\mathbf{C}$ ($\mathbb{E}_2 \rightarrow \mathbb{E}_2$):

$$(\mathbf{C}\boldsymbol{\Gamma})|\mathbf{A} = \mathbf{C}(\boldsymbol{\Gamma}|\mathbf{A}). \tag{1.56}$$

More generally, bivector dyadics of the form $\overline{\overline{\mathsf{A}}} = \sum \mathbf{C}_i \boldsymbol{\Gamma}_i$ are elements of the space $\mathbb{E}_2\mathbb{F}_2$. The dual counterpart is the dyadic $\sum \boldsymbol{\Gamma}_i \mathbf{C}_i$ of the space $\mathbb{F}_2\mathbb{E}_2$. In terms of reciprocal bivector and dual bivector bases $\{\mathbf{e}_{ik}\}$, $\{\boldsymbol{\varepsilon}_{j\ell}\}$, any bivector dyadic $\overline{\overline{\mathsf{A}}}$ can be expressed as

$$\overline{\overline{\mathsf{A}}} = \sum A_{IJ} \mathbf{e}_I \boldsymbol{\varepsilon}_J, \tag{1.57}$$

where I and J are two (ordered) bi-indices. Expanding any bivector $\mathbf{C}$ as

$$\mathbf{C} = \sum_{i<j} \mathbf{e}_{ij}(\boldsymbol{\varepsilon}_{ij}|\mathbf{C}) = \sum_{i<j} (\mathbf{e}_{ij}\boldsymbol{\varepsilon}_{ij})|\mathbf{C}, \tag{1.58}$$

we can identify the unit bivector dyadic denoted by $\overline{\overline{\mathsf{I}}}^{(2)}$, which maps a bivector to itself as

$$\overline{\overline{\mathsf{I}}}^{(2)} = \sum_{i<j} \mathbf{e}_{ij}\boldsymbol{\varepsilon}_{ij} = \mathbf{e}_{12}\boldsymbol{\varepsilon}_{12} + \mathbf{e}_{13}\boldsymbol{\varepsilon}_{13} + \mathbf{e}_{23}\boldsymbol{\varepsilon}_{23} + \cdots \tag{1.59}$$

The indices here are ordered, but we can also write it in terms of nonordered indices as

$$\overline{\overline{\mathsf{I}}}^{(2)} = \frac{1}{2}\sum_{i,j} \mathbf{e}_{ij}\boldsymbol{\varepsilon}_{ij} = \frac{1}{2}[\mathbf{e}_{12}\boldsymbol{\varepsilon}_{12} + \mathbf{e}_{21}\boldsymbol{\varepsilon}_{21} + \mathbf{e}_{13}\boldsymbol{\varepsilon}_{13} + \mathbf{e}_{31}\boldsymbol{\varepsilon}_{31} + \cdots]. \tag{1.60}$$

The bivector unit dyadic $\overline{\overline{\mathsf{I}}}^{(a)}$ maps any bivector to itself, and its dual equals its transpose,

$$\overline{\overline{\mathsf{I}}}^{(2)}|(\mathbf{a}\wedge\mathbf{b}) = \mathbf{a}\wedge\mathbf{b}, \quad \overline{\overline{\mathsf{I}}}^{(2)T}|(\boldsymbol{\alpha}\wedge\boldsymbol{\beta}) = \boldsymbol{\alpha}\wedge\boldsymbol{\beta}. \tag{1.61}$$

Double-wedge product We can introduce the double-wedge product $^\wedge_\wedge$ between two dyadic products of vectors and dual vectors,

$$(\mathbf{a}\boldsymbol{\alpha})^\wedge_\wedge(\mathbf{b}\boldsymbol{\beta}) = (\mathbf{a}\wedge\mathbf{b})(\boldsymbol{\alpha}\wedge\boldsymbol{\beta}), \tag{1.62}$$

and extend it to two dyadics as

$$\overline{\overline{\mathsf{A}}}^\wedge_\wedge\overline{\overline{\mathsf{B}}} = \sum(\mathbf{a}_i\boldsymbol{\alpha}_i)^\wedge_\wedge\sum(\mathbf{b}_j\boldsymbol{\beta}_j) = \sum_{i,j}(\mathbf{a}_i\wedge\mathbf{b}_j)(\boldsymbol{\alpha}_i\wedge\boldsymbol{\beta}_j). \tag{1.63}$$

Double-wedge product of two dyadics of the space $\mathbb{E}_1\mathbb{F}_1$ is in the space $\mathbb{E}_2\mathbb{F}_2$. From the anticommutation property of the wedge product, $\mathbf{a}\wedge\mathbf{b} = -\mathbf{b}\wedge\mathbf{a}$, it follows that the double-wedge product of two dyadics commutes:

$$\begin{aligned}\overline{\overline{\mathsf{A}}}^\wedge_\wedge\overline{\overline{\mathsf{B}}} &= \sum(\mathbf{a}_i\boldsymbol{\alpha}_i)^\wedge_\wedge\sum(\mathbf{b}_j\boldsymbol{\beta}_j) = \sum(\mathbf{a}_i\wedge\mathbf{b}_j)(\boldsymbol{\alpha}_i\wedge\boldsymbol{\beta}_j) \\ &= \sum(\mathbf{b}_i\wedge\mathbf{a}_j)(\boldsymbol{\beta}_i\wedge\boldsymbol{\alpha}_j) = \overline{\overline{\mathsf{B}}}^\wedge_\wedge\overline{\overline{\mathsf{A}}}.\end{aligned} \tag{1.64}$$

$^\wedge_\wedge$ resembles its Gibbsian counterpart $^\times_\times$, the double-cross product [28, 40], which is also commutative. Expanding the bivector unit dyadic as

$$\overline{\overline{\mathsf{I}}}^{(2)} = \frac{1}{2}\sum_{i,j}(\mathbf{e}_i\wedge\mathbf{e}_j)(\boldsymbol{\varepsilon}_i\wedge\boldsymbol{\varepsilon}_j) = \frac{1}{2}\sum(\mathbf{e}_i\boldsymbol{\varepsilon}_i)^\wedge_\wedge\sum(\mathbf{e}_j\boldsymbol{\varepsilon}_j), \tag{1.65}$$

we obtain an important relation between the unit dyadics in vector and bivector spaces:

$$\bar{\bar{I}}^{(2)} = \frac{1}{2}\bar{\bar{I}}{}^{\wedge}_{\wedge}\bar{\bar{I}}. \tag{1.66}$$

The transpose of this identity gives the corresponding relation in the space of dual vectors and bivectors.

Problems

1.3.1 From linear independence of the basis bivectors show that the condition $\mathbf{a}\wedge\mathbf{b} = 0$ with $\mathbf{a} \neq 0$ implies the existence of a scalar λ such that $\mathbf{b} = \lambda\mathbf{a}$.

1.3.2 Prove (1.36) (a) through the basis bivectors and dual bivectors and (b) by considering the possible resulting scalar function made up with the four different duality products $\boldsymbol{\alpha}|\mathbf{a}$, $\boldsymbol{\beta}|\mathbf{a}$, $\boldsymbol{\alpha}|\mathbf{b}$ and $\boldsymbol{\beta}|\mathbf{b}$.

1.3.3 Show that a bivector of the form $\mathbf{A} = A_{12}\mathbf{a}_1 \wedge \mathbf{a}_2 + A_{23}\mathbf{a}_2 \wedge \mathbf{a}_3 + A_{31}\mathbf{a}_3 \wedge \mathbf{a}_1$ can be written as a simple bivector, in the form $\mathbf{A} = \mathbf{b} \wedge \mathbf{c}$. This means that for $n = 3$ any bivector can be expressed as a simple bivector. *Hint*: The representation is not unique. Try $\mathbf{A} = (\alpha\mathbf{a}_1 + \beta\mathbf{a}_2) \wedge (\gamma\mathbf{a}_2 + \delta\mathbf{a}_3)$ and find the scalar coefficients $\alpha \cdots \delta$.

1.3.4 Show that for $n = 4$ any bivector $\mathbf{A}$ can be written as a sum of two simple bivectors, in the form $\mathbf{A} = \mathbf{a} \wedge \mathbf{b} + \mathbf{c} \wedge \mathbf{d}$.

1.3.5 Prove the rule (1.90).

1.3.6 Prove that if a dual vector $\boldsymbol{\alpha}$ satisfies $\mathbf{a}|\boldsymbol{\alpha} = 0$ for some vector $\mathbf{a}$, there exists a dual bivector $\boldsymbol{\Phi}$ such that $\boldsymbol{\alpha}$ can be expressed in the form $\boldsymbol{\alpha} = \mathbf{a}\rfloor\boldsymbol{\Phi}$.

1.4 MULTIVECTORS

Further consecutive wedge multiplications of vectors $\mathbf{a} \wedge \mathbf{b} \wedge \mathbf{c} \cdots$ give trivectors, quadrivectors, and multivectors of higher grade. This brings about a problem of notation because there does not exist enough different character types for each multivector and dual multivector species. To emphasize the grade of the multivector, we occasionally denote p-vectors and dual p-vectors by superscripts as $\mathbf{a}^p$ and $\boldsymbol{\gamma}^p$. However, in general, trivectors and quadrivectors will be denoted similarly as either vectors or bivectors, by boldface Latin letters like $\mathbf{a}$, $\mathbf{A}$. Dual trivectors and quadrivectors will be denoted by the corresponding Greek letters like $\boldsymbol{\alpha}$, $\boldsymbol{\Phi}$.

1.4.1 Trivectors

Because the wedge product is associative, it is possible to leave out the brackets in defining the trivector,

$$\mathbf{a} \wedge (\mathbf{b} \wedge \mathbf{c}) = (\mathbf{a} \wedge \mathbf{b}) \wedge \mathbf{c} = \mathbf{a} \wedge \mathbf{b} \wedge \mathbf{c}. \tag{1.67}$$

From the antisymmetry of the wedge product, we immediately obtain the permutation rule

$$\begin{aligned}\mathbf{a}\wedge\mathbf{b}\wedge\mathbf{c} &= \mathbf{b}\wedge\mathbf{c}\wedge\mathbf{a}=\mathbf{c}\wedge\mathbf{a}\wedge\mathbf{b}=-\mathbf{a}\wedge\mathbf{c}\wedge\mathbf{b}\\ &= -\mathbf{c}\wedge\mathbf{b}\wedge\mathbf{a}=-\mathbf{b}\wedge\mathbf{a}\wedge\mathbf{c}.\end{aligned} \tag{1.68}$$

This means that the trivector is invariant to cyclic permutations and otherwise changes sign. The space of trivectors is $\mathbb{E}_3$, and that of dual trivectors, $\mathbb{F}_3$. Because of antisymmetry, the product vanishes when two vectors in the trivector product are the same. More generally, if one of the vectors is a linear combination of the other two, the trivector product vanishes:

$$\mathbf{a}\wedge\mathbf{b}\wedge(\alpha\mathbf{a}+\beta\mathbf{b})=-\alpha\mathbf{a}\wedge\mathbf{a}\wedge\mathbf{b}+\beta\mathbf{a}\wedge\mathbf{b}\wedge\mathbf{b}=0. \tag{1.69}$$

The general trivector can be expressed as a sum of trivector products,

$$\mathbf{Q}=\sum\mathbf{a}_i\wedge\mathbf{b}_i\wedge\mathbf{c}_i=\mathbf{a}_1\wedge\mathbf{b}_1\wedge\mathbf{c}_1+\mathbf{a}_2\wedge\mathbf{b}_2\wedge\mathbf{c}_2+\cdots \tag{1.70}$$

The maximum number of linearly independent terms in this expansion equals the dimension of the trivector space which is $n(n-1)(n-2)/3!$. When $n=2$, the dimension of the space of trivectors is zero. In fact, there exist no trivectors based on the two-dimensional vector space $\mathbb{E}_1(2)$ because three vectors are always linearly dependent. In the three-dimensional space the trivectors form a one-dimensional space because they all are multiples of a single trivector $\mathbf{e}_1\wedge\mathbf{e}_2\wedge\mathbf{e}_3=\mathbf{e}_{123}$. In the four-dimensional vector space trivectors form another space of four dimensions. In particular, in the Minkowskian space Mi4 of vectors, trivectors $\mathbf{k}_{\mathrm{M}}$ can be expressed in terms of three-dimensional Euclidean trivectors $\mathbf{k}$ and bivectors $\mathbf{K}$ as

$$\mathbf{k}_{\mathrm{M}}=\mathbf{k}+\mathbf{e}_4\wedge\mathbf{K}. \tag{1.71}$$

As a memory aid, trivectors can be pictured as objects with triple hooks and dual trivectors as ones with triple eyes. In the duality product the triple hook is fastened to the triple eye and the result is a scalar with no free hooks or eyes. Geometrical interpretation will be discussed in Section 1.5.

1.4.2 Basis trivectors

A set of basis trivectors $\{\mathbf{e}_{ijk}\}$ can be constructed from basis vectors and bivectors as

$$\begin{aligned}\mathbf{e}_{ijk} &= \mathbf{e}_{ij}\wedge\mathbf{e}_k=\mathbf{e}_i\wedge\mathbf{e}_{jk}=\mathbf{e}_i\wedge\mathbf{e}_j\wedge\mathbf{e}_k\\ &= \mathbf{e}_{jki}=\mathbf{e}_{kij}=-\mathbf{e}_{ikj}=-\mathbf{e}_{kji}=-\mathbf{e}_{jik}.\end{aligned} \tag{1.72}$$

Similar expressions can be written for the dual case. The trivector and dual trivector bases are reciprocal if the underlying vector and dual vector bases are reciprocal:

$$\begin{aligned}\varepsilon_{ijk}|\mathbf{e}_{rst} &= \mathbf{e}_{rst}|\varepsilon_{ijk}=\delta_{\{rst\}\{ijk\}}=\delta_{ri}\delta_{sj}\delta_{tk}\\ &= (\mathbf{e}_r|\varepsilon_i)(\mathbf{e}_s|\varepsilon_j)(\mathbf{e}_t|\varepsilon_k).\end{aligned} \tag{1.73}$$

Here ijk and rst are ordered tri-indices.

Basis expansion gives us a possibility to prove the following simple theorem: If $\mathbf{a} \neq 0$ is a vector and $\mathbf{A}$ a bivector which satisfies $\mathbf{a} \wedge \mathbf{A} = 0$, there exists another vector $\mathbf{b}$ such that we can write $\mathbf{A} = \mathbf{a} \wedge \mathbf{b}$. In fact, choosing a basis with $\mathbf{e}_1 = \mathbf{a}$ and expressing $\mathbf{A} = \sum A_{ij}\mathbf{e}_{ij}$ with $i < j$ gives us

$$\mathbf{a} \wedge \mathbf{A} = \sum_{1<i<j} A_{ij}\mathbf{e}_1 \wedge \mathbf{e}_i \wedge \mathbf{e}_j = 0. \tag{1.74}$$

Because $\mathbf{e}_1 \wedge \mathbf{e}_i \wedge \mathbf{e}_j$ are linearly independent trivectors for $1 < i < j$, this implies $A_{ij} = 0$ when $i \neq 1$. Thus, the bivector $\mathbf{A}$ can be written in the form

$$\mathbf{A} = A_{12}\mathbf{e}_1 \wedge \mathbf{e}_2 + A_{13}\mathbf{e}_1 \wedge \mathbf{e}_3 + \cdots = \mathbf{a} \wedge \mathbf{b}, \quad \mathbf{b} = \sum A_{1j}\mathbf{e}_j. \tag{1.75}$$

1.4.3 Trivector identities

From the permutational property (1.72) we can derive the following basic identity for the duality product of a trivector and a dual trivector:

$$(\boldsymbol{\alpha} \wedge \boldsymbol{\beta} \wedge \boldsymbol{\gamma})|(\mathbf{a} \wedge \mathbf{b} \wedge \mathbf{c}) = \det\begin{pmatrix} \boldsymbol{\alpha}|\mathbf{a} & \boldsymbol{\alpha}|\mathbf{b} & \boldsymbol{\alpha}|\mathbf{c} \\ \boldsymbol{\beta}|\mathbf{a} & \boldsymbol{\beta}|\mathbf{b} & \boldsymbol{\beta}|\mathbf{c} \\ \boldsymbol{\gamma}|\mathbf{a} & \boldsymbol{\gamma}|\mathbf{b} & \boldsymbol{\gamma}|\mathbf{c} \end{pmatrix}. \tag{1.76}$$

As a simple check, it can be seen that the right-hand side vanishes whenever two of the vectors or dual vectors are scalar multiples of each other. Introducing the triple duality product between triadic products of vectors and dual vectors as

$$(\mathbf{abc})|||(\boldsymbol{\alpha\beta\gamma}) = (\mathbf{a}|\boldsymbol{\alpha})(\mathbf{b}|\boldsymbol{\beta})(\mathbf{c}|\boldsymbol{\gamma}) = (\boldsymbol{\alpha\beta\gamma})|||(\mathbf{abc}), \tag{1.77}$$

(1.76) can expressed as

$$\begin{aligned} &(\boldsymbol{\alpha} \wedge \boldsymbol{\beta} \wedge \boldsymbol{\gamma})|(\mathbf{a} \wedge \mathbf{b} \wedge \mathbf{c}) \\ &\quad = (\boldsymbol{\alpha\beta\gamma})|||(\mathbf{abc} + \mathbf{bca} + \mathbf{cab} - \mathbf{acb} - \mathbf{cba} - \mathbf{bac}) \\ &\quad = (\boldsymbol{\alpha\beta\gamma} + \boldsymbol{\beta\gamma\alpha} + \boldsymbol{\gamma\alpha\beta} - \boldsymbol{\alpha\gamma\beta} - \boldsymbol{\beta\alpha\gamma} - \boldsymbol{\gamma\beta\alpha})|||(\mathbf{abc}), \end{aligned} \tag{1.78}$$

which can be easily memorized. Triadics or polyadics of higher order will not otherwise be applied in the present analysis. Other forms are obtained by applying (1.40):

$$(\boldsymbol{\alpha} \wedge \boldsymbol{\beta} \wedge \boldsymbol{\gamma})|(\mathbf{a} \wedge \mathbf{b} \wedge \mathbf{c}) \tag{1.79}$$

$$\begin{aligned} &= (\boldsymbol{\alpha} \wedge \boldsymbol{\beta})|\Big((\mathbf{a} \wedge \mathbf{b})\mathbf{c} + (\mathbf{b} \wedge \mathbf{c})\mathbf{a} + (\mathbf{c} \wedge \mathbf{a})\mathbf{b}\Big)|\boldsymbol{\gamma} \\ &= \boldsymbol{\alpha}|\Big(\mathbf{a}(\mathbf{b} \wedge \mathbf{c}) + \mathbf{b}(\mathbf{c} \wedge \mathbf{a}) + \mathbf{c}(\mathbf{a} \wedge \mathbf{b})\Big)|(\boldsymbol{\beta} \wedge \boldsymbol{\gamma}) \\ &= (\mathbf{a} \wedge \mathbf{b})|\Big((\boldsymbol{\alpha} \wedge \boldsymbol{\beta})\boldsymbol{\gamma} + (\boldsymbol{\beta} \wedge \boldsymbol{\gamma})\boldsymbol{\alpha} + (\boldsymbol{\gamma} \wedge \boldsymbol{\alpha})\boldsymbol{\beta}\Big)|\mathbf{c}, \\ &= \mathbf{a}|\Big(\boldsymbol{\alpha}(\boldsymbol{\beta} \wedge \boldsymbol{\gamma}) + \boldsymbol{\beta}(\boldsymbol{\gamma} \wedge \boldsymbol{\alpha}) + \boldsymbol{\gamma}(\boldsymbol{\alpha} \wedge \boldsymbol{\beta})\Big)|(\mathbf{b} \wedge \mathbf{c}). \end{aligned} \tag{1.80}$$

The sum expressions involve dyadic products of vectors and bivectors or dual vectors and dual bivectors.

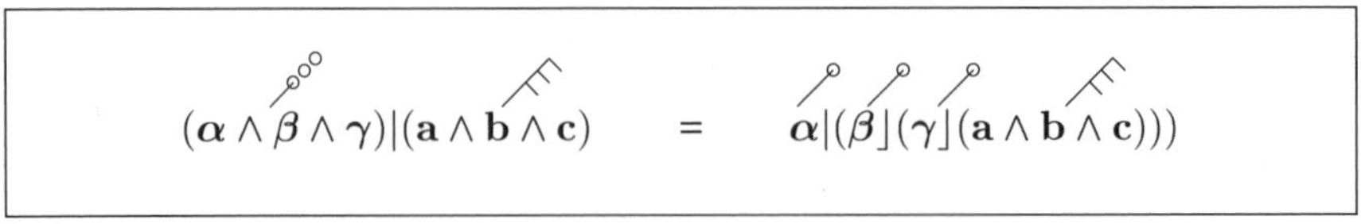

Fig. 1.9 *Visualization of a trivector product rule.*

Incomplete duality product In terms of (1.78), (1.80) we can expand the incomplete duality products between a dual bivector and a trivector defined by

$$(\alpha\wedge\beta\wedge\gamma)|(\mathbf{a}\wedge\mathbf{b}\wedge\mathbf{c}) = \alpha|((\beta\wedge\gamma)\rfloor(\mathbf{a}\wedge\mathbf{b}\wedge\mathbf{c}), \tag{1.81}$$

as

$$(\beta\wedge\gamma)\rfloor(\mathbf{a}\wedge\mathbf{b}\wedge\mathbf{c}) = \Big(\mathbf{a}(\mathbf{b}\wedge\mathbf{c}) + \mathbf{b}(\mathbf{c}\wedge\mathbf{a}) + \mathbf{c}(\mathbf{a}\wedge\mathbf{b})\Big)|(\beta\wedge\gamma). \tag{1.82}$$

A visualization of incomplete products arising from a trivector product is seen in Figure 1.9. Equation (1.82) shows us that $\mathbf{a}\wedge\mathbf{b}\wedge\mathbf{c} = 0$ implies linear dependence of the vector triple $\mathbf{a}, \mathbf{b}, \mathbf{c}$, because we then have

$$\mathbf{a}((\mathbf{b}\wedge\mathbf{c})|(\beta\wedge\gamma)) + \mathbf{b}((\mathbf{c}\wedge\mathbf{a})|(\beta\wedge\gamma)) + \mathbf{c}((\mathbf{a}\wedge\mathbf{b})|(\beta\wedge\gamma)) = 0, \tag{1.83}$$

valid for any dual vectors β, γ. Thus, in a space of three dimensions, three vectors $\mathbf{a}, \mathbf{b}, \mathbf{c}$ can make a basis only if $\mathbf{a}\wedge\mathbf{b}\wedge\mathbf{c} \neq 0$. Similarly, from

$$(\alpha\wedge\beta\wedge\gamma)|(\mathbf{a}\wedge\mathbf{b}\wedge\mathbf{c}) = (\alpha\wedge\beta)|\Big(\gamma\rfloor(\mathbf{a}\wedge\mathbf{b}\wedge\mathbf{c})\Big), \tag{1.84}$$

we can expand the incomplete duality product between a dual vector and a trivector as

$$\gamma\rfloor(\mathbf{a}\wedge\mathbf{b}\wedge\mathbf{c}) = \Big((\mathbf{a}\wedge\mathbf{b})\mathbf{c} + (\mathbf{b}\wedge\mathbf{c})\mathbf{a} + (\mathbf{c}\wedge\mathbf{a})\mathbf{b}\Big)|\gamma. \tag{1.85}$$

Different forms for this rule can be found through duality and multiplying from the right. Equation (1.85) shows us that $\mathbf{a}\wedge\mathbf{b}\wedge\mathbf{c} = 0$ also implies linear dependence of the three bivectors $(\mathbf{a}\wedge\mathbf{b}),\ (\mathbf{b}\wedge\mathbf{c}),\ (\mathbf{c}\wedge\mathbf{a})$.

The trivector unit dyadic $\overline{\overline{\mathsf{I}}}^{(3)}$ maps any trivector to itself, and its dual equals its transpose,

$$\overline{\overline{\mathsf{I}}}^{(3)}|(\mathbf{a}\wedge\mathbf{b}\wedge\mathbf{c}) = \mathbf{a}\wedge\mathbf{b}\wedge\mathbf{c}, \tag{1.86}$$

$$\overline{\overline{\mathsf{I}}}^{(3)T}|(\alpha\wedge\beta\wedge\gamma) = \alpha\wedge\beta\wedge\gamma. \tag{1.87}$$

The basis expansion is

$$\overline{\overline{\mathsf{I}}}^{(3)} = \sum_{i<j<k} \mathbf{e}_{ijk}\varepsilon_{ijk} = \mathbf{e}_{123}\varepsilon_{123} + \mathbf{e}_{124}\varepsilon_{124} + \mathbf{e}_{134}\varepsilon_{134} + \cdots \tag{1.88}$$

The trivector unit dyadic can be shown to satisfy the relations

$$\overline{\overline{\mathsf{I}}}^{(3)} = \frac{1}{3}\overline{\overline{\mathsf{I}}}^{(2)}{}^{\wedge}_{\wedge}\overline{\overline{\mathsf{I}}} = \frac{1}{3}\overline{\overline{\mathsf{I}}}{}^{\wedge}_{\wedge}\overline{\overline{\mathsf{I}}}^{(2)} = \frac{1}{3!}\overline{\overline{\mathsf{I}}}{}^{\wedge}_{\wedge}\overline{\overline{\mathsf{I}}}{}^{\wedge}_{\wedge}\overline{\overline{\mathsf{I}}}. \tag{1.89}$$

Bac cab rule A useful bac cab formula is obtained from (1.50) when replacing the vector $\mathbf{c}$ by a bivector $\mathbf{C}$. The rule (1.50) is changed to

$$\boldsymbol{\alpha}\rfloor(\mathbf{b}\wedge\mathbf{C}) = \mathbf{b}\wedge(\boldsymbol{\alpha}\rfloor\mathbf{C}) + \mathbf{C}(\boldsymbol{\alpha}|\mathbf{b}) = (\mathbf{b}\wedge\mathbf{C})\lfloor\boldsymbol{\alpha}, \tag{1.90}$$

while its dual (1.51) is changed to

$$\mathbf{a}\rfloor(\boldsymbol{\beta}\wedge\boldsymbol{\Gamma}) = \boldsymbol{\beta}\wedge(\mathbf{a}\rfloor\boldsymbol{\Gamma}) + \boldsymbol{\Gamma}(\mathbf{a}|\boldsymbol{\beta}) = (\boldsymbol{\beta}\wedge\boldsymbol{\Gamma})\lfloor\mathbf{a}. \tag{1.91}$$

Note the difference in sign in (1.50) and (1.90) and their duals (this is why we prefer calling them "bac cab rules" instead of "bac-cab rules" as is done for Gibbsian vectors [14]). The proof of (1.90) and some of its generalizations are left as an exercise. A futher generalization to multivectors will be given in Section 1.4.8. The rule (1.90) gives rise to a decomposition rule similar to that in (1.53). In fact, given a bivector $\mathbf{B}$ we can expand it in components parallel to another bivector $\mathbf{A}$ and orthogonal to a dual vector $\boldsymbol{\alpha}$ as

$$\mathbf{B} = \mathbf{B}_{\|} + \mathbf{B}_{\perp}, \quad \mathbf{a}\wedge\mathbf{B}_{\|} = 0, \quad \boldsymbol{\alpha}\rfloor\mathbf{B}_{\perp} = 0. \tag{1.92}$$

The components can be expressed as

$$\mathbf{B}_{\|} = -\frac{\mathbf{a}\wedge(\boldsymbol{\alpha}\rfloor\mathbf{B})}{\boldsymbol{\alpha}|\mathbf{a}}, \quad \mathbf{B}_{\perp} = \frac{\boldsymbol{\alpha}\rfloor(\mathbf{a}\wedge\mathbf{B})}{\boldsymbol{\alpha}|\mathbf{a}}. \tag{1.93}$$

1.4.4 p-vectors

Proceeding in the same manner as before, multiplying p vectors or dual vectors, quantities called p-vectors or dual p-vectors, elements of the space $\mathbb{E}_p$ or $\mathbb{F}_p$, are obtained. The dimension of these two spaces equals the binomial coefficient

$$C_p^n = \binom{n}{p} = \frac{n(n-1)..(n-p+1)}{p!} = \frac{n!}{p!(n-p)!}, \quad n \geq p. \tag{1.94}$$

The following table gives the dimension C_p^n corresponding to the grade p of the multivector and dimension n of the original vector space:

p	$n=2$	$n=3$	$n=4$
5	0	0	0
4	0	0	1
3	0	1	4
2	1	3	6
1	2	3	4
0	1	1	1

One may note that the dimension of $\mathbb{E}_p$ is largest for $p = n/2$ when n is even and for $p = (n \pm 1)/2$ when n is odd. The dimension C_p^n is the same when p is replaced by $n - p$. This is why the spaces $\mathbb{E}_p$ and $\mathbb{E}_{n-p}$ can be mapped onto one another. A mapping of this kind is denoted by $*$ (Hodge's star operator) in mathematical literature. Here we express it through Hodge's dyadics $\overline{\overline{\mathsf{H}}}_p$. If $\mathbf{x}^p$ is an element of $\mathbb{E}_p$, $*\mathbf{x}^p$ or $\overline{\overline{\mathsf{H}}}_p|\mathbf{x}^p$ is an element of $\mathbb{E}_{n-p}$. More on Hodge's dyadics will be found in Chapter 2.

The duality product can be generalized for any p-vectors and dual p-vectors through the reciprocal p-vector and dual p-vector bases satisfying the orthogonality

$$\boldsymbol{\varepsilon}_J|\mathbf{e}_K = \delta_{JK}. \tag{1.95}$$

Here $J = \{j_1 j_2 \cdots j_p\}$ and $K = \{k_1 k_2 \cdots k_p\}$ are two ordered p-indices with $j_1 < j_2 ... < j_p$, $k_1 < k_2 ... < k_p$ and

$$\delta_{JK} = \delta_{j_1 k_1}\delta_{j_2 k_2}\cdots\delta_{j_p k_p}. \tag{1.96}$$

The p-vector unit dyadic is defined through p unit dyadics $\overline{\overline{\mathsf{I}}}$ as

$$\overline{\overline{\mathsf{I}}}^{(p)} = \frac{1}{p!}\overline{\overline{\mathsf{I}}}{}^\wedge_\wedge\overline{\overline{\mathsf{I}}}{}^\wedge_\wedge\cdots{}^\wedge_\wedge\overline{\overline{\mathsf{I}}}. \tag{1.97}$$

If $\mathbf{x}^p$ is a p-vector and $\mathbf{y}^q$ is a q-vector, from the anticommutativity rule of the wedge product we can obtain the general commutation rule

$$\mathbf{x}^p \wedge \mathbf{y}^q = (-1)^{pq}\mathbf{y}^q \wedge \mathbf{x}^p. \tag{1.98}$$

This product commutes unless both p and q are odd numbers, in which case it anti-commutes. For example, if $\mathbf{y}^q$ is a bivector ($q = 2$), it commutes with any p-vector $\mathbf{x}^p$ in the wedge product.

1.4.5 Incomplete duality product

The incomplete duality product can be defined to a p-vector and a dual q-vector when $p \neq q$. The result is a $p - q$ vector if $p > q$ and a dual $q - p$ vector if $p < q$. In mathematics such a product is known as contraction, because it reduces the grade of the multivector or dual multivector. In the multiplication sign $\lfloor$ or $\rfloor$ the short line points to the multivector or dual multivector with smaller p or q. In reference 18 the duality product sign $|$ was defined to include also the incomplete duality product. Actually, since the different multiplication signs $\wedge$, $|$, $\lfloor$, $\rfloor$ apply in different environments, we could replace all of them by a single sign and interprete the operation with respect to the grades of the multiplicants. However, since this would make the analysis more vulnerable to errors, it does not appear wise to simplify the notation too much.

Considering a dual q-vector $\boldsymbol{\alpha}^q$ and a q-vector of the form $\mathbf{a}^p \wedge \mathbf{b}^{q-p}$ with $q > p$, their duality product can be expressed as

$$\boldsymbol{\alpha}^q|(\mathbf{a}^p \wedge \mathbf{b}^{q-p}) = (\boldsymbol{\alpha}^q\lfloor\mathbf{a}^p)|\mathbf{b}^{q-p}, \tag{1.99}$$

which defines the incomplete duality product of a p-vector and a dual q-vector. The dual case is

$$\mathbf{a}^q|(\boldsymbol{\alpha}^p \wedge \boldsymbol{\beta}^{q-p}) = (\mathbf{a}^q\lfloor\boldsymbol{\alpha}^p)|\boldsymbol{\beta}^{q-p}. \tag{1.100}$$

Again, we can use the memory aid by inserting p hooks in the p-vector $\mathbf{a}^p$ and q eyes in the dual q-vector $\boldsymbol{\alpha}^q$, as was done in Figure 1.9. If $p < q$, in the incomplete duality product p hooks are caught by p eyes and the object is left with $q-p$ free eyes, which makes it a dual $(q-p)$-vector $\boldsymbol{\alpha}^q\lfloor\mathbf{a}^p$.

From the symmetry of the duality product we can write

$$\begin{aligned}\boldsymbol{\alpha}_q|(\mathbf{a}_p\wedge\mathbf{b}_{q-p}) &= (\mathbf{a}_p\wedge\mathbf{b}_{q-p})|\boldsymbol{\alpha}_q\\ &= (-1)^{p(q-p)}(\mathbf{b}_{p-q}\wedge\mathbf{a}_p)|\boldsymbol{\alpha}_q\\ &= \mathbf{b}_{q-p}|((-1)^{p(q-p)}\mathbf{a}_p\rfloor\boldsymbol{\alpha}_q).\end{aligned}\tag{1.101}$$

Thus, we obtain the important commutation relations

$$\boldsymbol{\alpha}_q\lfloor\mathbf{a}_p = (-1)^{p(q-p)}\mathbf{a}_p\rfloor\boldsymbol{\alpha}_q,\qquad \mathbf{a}_q\lfloor\boldsymbol{\alpha}_p = (-1)^{p(q-p)}\boldsymbol{\alpha}_p\rfloor\mathbf{a}_q,\tag{1.102}$$

which can be remembered so that the power of -1 is the smaller of p and q multiplied by their difference. It is seen that the incomplete duality product is antisymmetric only when the smaller index p is odd and the larger index q is even. In all other cases it is symmetric.

1.4.6 Basis multivectors

In forming incomplete duality products between different elements of multivector spaces, it is often helpful to work with expansions, in which cases incomplete duality products of different basis elements are needed. Most conveniently, they can be expressed as a set of rules which can be derived following the example of

$$i\neq j,\quad (\varepsilon_{ij}\lfloor\mathbf{e}_i)|\mathbf{e}_k = \varepsilon_{ij}|\mathbf{e}_{ik} = \delta_{jk} = \varepsilon_j|\mathbf{e}_k \;\Rightarrow\; \varepsilon_{ij}\lfloor\mathbf{e}_i = \varepsilon_j.\tag{1.103}$$

Here we can use the antisymmetric property of the wedge product as $\varepsilon_{ij} = -\varepsilon_{ji}$ and the commutation rule (1.102) to obtain other variants:

$$i\neq j,\quad \varepsilon_{ij}\lfloor\mathbf{e}_i = \varepsilon_j,\quad \varepsilon_{ij}\lfloor\mathbf{e}_j = -\varepsilon_i,\tag{1.104}$$

$$i\neq j,\quad \varepsilon_i\rfloor\mathbf{e}_{ij} = -\mathbf{e}_j,\quad \varepsilon_j\rfloor\mathbf{e}_{ij} = \mathbf{e}_i.\tag{1.105}$$

The dual cases can be simply written by excanging vectors and dual vectors:

$$i\neq j,\quad \mathbf{e}_{ij}\lfloor\varepsilon_i = \mathbf{e}_j,\quad \mathbf{e}_{ij}\lfloor\varepsilon_j = -\mathbf{e}_i,\tag{1.106}$$

$$i\neq j,\quad \mathbf{e}_i\rfloor\varepsilon_{ij} = -\varepsilon_j,\quad \mathbf{e}_j\rfloor\varepsilon_{ij} = \varepsilon_i.\tag{1.107}$$

Generic formulas for trivectors and quadrivectors can be written as follows when $\mathbf{e}_{ijk}\neq 0$ and $\mathbf{e}_{ijk\ell}\neq 0$:

$$\varepsilon_{ijk}\lfloor\mathbf{e}_{ij} = \mathbf{e}_{ij}\rfloor\varepsilon_{ijk} = \varepsilon_k,\tag{1.108}$$

$$\varepsilon_{ijk}\lfloor\mathbf{e}_i = \mathbf{e}_i\rfloor\varepsilon_{ijk} = \varepsilon_{jk},\tag{1.109}$$

$$\varepsilon_{ijk\ell}\lfloor\mathbf{e}_i = -\mathbf{e}_i\rfloor\varepsilon_{ijk\ell} = \varepsilon_{jk\ell},\tag{1.110}$$

$$\varepsilon_{ijk\ell}\lfloor\mathbf{e}_{ij} = \mathbf{e}_{ij}\rfloor\varepsilon_{ijk\ell} = \varepsilon_{k\ell},\tag{1.111}$$

$$\varepsilon_{ijk\ell}\lfloor\mathbf{e}_{ijk} = -\mathbf{e}_{ijk}\rfloor\varepsilon_{ijk\ell} = \varepsilon_\ell,\tag{1.112}$$

which can be transformed to other forms using antisymmetry and duality. A good memory rule is that we can eliminate the first indices in the expressions of the form ($\varepsilon_{ijk\ell} \neq 0, \mathbf{e}_{ijk\ell} \neq 0$)

$$\varepsilon_{ijk\ell} \lfloor \mathbf{e}_{ijk} = \varepsilon_{jk\ell} \lfloor \mathbf{e}_{jk} = \varepsilon_{k\ell} \lfloor \mathbf{e}_k = \varepsilon_\ell, \tag{1.113}$$

and the last indices in the form

$$\varepsilon_{ijk} \rfloor \mathbf{e}_{ijk\ell} = -\varepsilon_{ijk} \rfloor \mathbf{e}_{\ell ijk} = -\varepsilon_{ij} \rfloor \mathbf{e}_{\ell ij} = -\varepsilon_i \rfloor \mathbf{e}_{\ell i} = -\mathbf{e}_\ell. \tag{1.114}$$

If $J = j_1 j_2 \cdots j_p$ is a p-index and $\mathbf{e}_J^p, \mathbf{e}_{K(J)}^{n-p}$ are a basis p-vector and its complementary basis $n-p$ vector,

$$\mathbf{e}_J^p = \mathbf{e}_{j_1} \wedge \mathbf{e}_{j_2} \wedge \cdots \wedge \mathbf{e}_{j_p}, \tag{1.115}$$

$$\mathbf{e}_{K(J)}^{n-p} = \mathbf{e}_1 \wedge \cdots \wedge \mathbf{e}_{j_1-1} \wedge \mathbf{e}_{j_1+1} \wedge \cdots \wedge \mathbf{e}_{j_p-1} \wedge \mathbf{e}_{j_p+1} \wedge \cdots \wedge \mathbf{e}_n, \tag{1.116}$$

we can derive the relations

$$\mathbf{e}_J^p \wedge \mathbf{e}_{K(J)}^{n-p} = (-1)^{\sigma(J)} \mathbf{e}_N, \quad \mathbf{e}_{K(J)}^{n-p} \wedge \mathbf{e}_J^p = (-1)^{p(n-p)} (-1)^{\sigma(J)} \mathbf{e}_N, \tag{1.117}$$

where we denote

$$\sigma(J) = \sum_{i=1}^{p} (j_i - i) = (j_1 - 1) + (j_2 - 2) + \cdots + (j_p - p). \tag{1.118}$$

Details are left as an exercise. Equation (1.117) implies the rules

$$\varepsilon_N \lfloor \mathbf{e}_J^p = (-1)^{\sigma(J)} \varepsilon_{K(J)}, \tag{1.119}$$

$$\varepsilon_N \lfloor \mathbf{e}_{K(J)}^{n-p} = (-1)^{p(n-p)} (-1)^{\sigma(J)} \varepsilon_J^p, \tag{1.120}$$

$$\mathbf{e}_N \lfloor (\varepsilon_N \lfloor \mathbf{e}_J^p) = (-1)^{\sigma(P)} \mathbf{e}_N \lfloor \varepsilon_{K(J)}^{n-p} = (-1)^{p(n-p)} \mathbf{e}_J^p, \tag{1.121}$$

$$\mathbf{e}_N \lfloor (\varepsilon_N \lfloor \mathbf{e}_{K(J)}^{n-p}) = (-1)^{p(n-p)} (-1)^{\sigma(P)} \mathbf{e}_N \lfloor \varepsilon_J^p = (-1)^{p(n-p)} \mathbf{e}_{K(J)}^{n-p}, \tag{1.122}$$

which can be easily written in their dual form.

Example As an example of applying the previous formulas let us expand the dual quadrivector $\varepsilon_{k\ell} \wedge (\varepsilon_{ijrs} \lfloor \mathbf{e}_{ij})$ in the case $n = 4$. From (1.111) we have for $\varepsilon_{ijrs} \neq 0$

$$\varepsilon_{k\ell} \wedge (\varepsilon_{ijrs} \lfloor \mathbf{e}_{ij}) = \varepsilon_{k\ell} \wedge \varepsilon_{rs} = \varepsilon_{k\ell rs}, \tag{1.123}$$

which is a multiple of ε_{1234}. Let us assume ordered pairs $k < \ell$ and $i < j$ and $r \neq s$. Now, obviously, ε_{ijrs} vanishes unless i and j are different from r and s. Thus, the result vanishes unless $k = i$ and $\ell = j$ and for $\varepsilon_{ijrs} \neq 0$ the result can be written as

$$\varepsilon_{k\ell} \wedge (\varepsilon_{ijrs} \lfloor \mathbf{e}_{ij}) = \delta_{\{k\ell\}\{ij\}} \varepsilon_{k\ell rs} = (\varepsilon_{k\ell} | \mathbf{e}_{ij}) \varepsilon_{ijrs}. \tag{1.124}$$

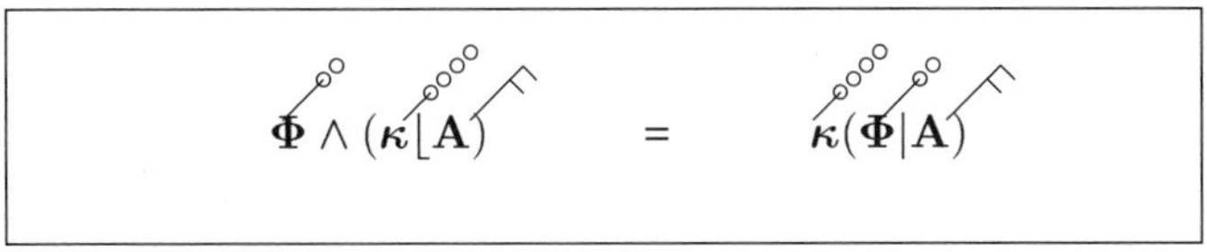

Fig. 1.10 *Visualization of the rule (1.125) valid for $n = 4$. One can check that the quantities on each side is a dual quadrivector.*

Let us generalize this result. Since in a four-dimensional space any dual quadrivector is a multiple of ε_{1234}, we can replace ε_{ijrs} by an arbitrary quadrivector $\boldsymbol{\kappa} = \kappa\varepsilon_{1234}$. Further, we can multiply the equation by a scalar A_{ij} and sum over i, j to obtain an arbitrary bivector $\mathbf{A} = \sum A_{ij}\mathbf{e}_{ij}$. Similarly, we can multiply both sides by the scalar $\Phi_{k\ell}$ and sum, whence an arbitrary dual bivector $\boldsymbol{\Phi} = \sum \Phi_{k\ell}\varepsilon_{k\ell}$ will arise. Thus, we arrive at the dual quadrivector identity

$$\boldsymbol{\Phi} \wedge (\boldsymbol{\kappa}\lfloor\mathbf{A}) = \boldsymbol{\kappa}(\boldsymbol{\Phi}|\mathbf{A}) \tag{1.125}$$

valid for any dual bivector $\boldsymbol{\Phi}$, dual quadrivector $\boldsymbol{\kappa}$ and bivector $\mathbf{A}$ in the four-dimensional space. A visualization of this is seen in Figure 1.10.

1.4.7 Generalized bac cab rule

The bac cab rules (1.50) and (1.90) can be generalized to the following identity, which, however, does not have the mnemonic "bac cab" form:

$$(\mathbf{a}^q \wedge \mathbf{a}^p)\lfloor\boldsymbol{\alpha} = (\mathbf{a}^q\lfloor\boldsymbol{\alpha}) \wedge \mathbf{a}^p + (-1)^q\mathbf{a}^q \wedge (\mathbf{a}^p\lfloor\boldsymbol{\alpha}). \tag{1.126}$$

Here, $\mathbf{a}^q$ is a q-vector, $\mathbf{a}^p$ is a p-vector, and $\boldsymbol{\alpha}$ is a dual vector. For the special case $q = 1$ we have the rule

$$(\mathbf{a} \wedge \mathbf{a}^p)\lfloor\boldsymbol{\alpha} = (\mathbf{a}|\boldsymbol{\alpha})\mathbf{a}^p - \mathbf{a} \wedge (\mathbf{a}^p\lfloor\boldsymbol{\alpha}). \tag{1.127}$$

When $p + q > n$, the left-hand side vanishes and we have

$$(\mathbf{a}^q\lfloor\boldsymbol{\alpha}) \wedge \mathbf{a}^p = (-1)^{pq}(\mathbf{a}^p\lfloor\boldsymbol{\alpha}) \wedge \mathbf{a}^q, \quad p + q > n. \tag{1.128}$$

The proof of (1.126) is left as a problem. Let us make a few checks of the identity. The inner consistency can be checked by changing the order of all wedge multiplications with associated sign changes. Equation (1.126), then, becomes

$$\begin{aligned}(-1)^{pq}&(\mathbf{a}^p \wedge \mathbf{a}^q)\lfloor\boldsymbol{\alpha} \\ &= (-1)^{p(q-1)}\mathbf{a}^p \wedge (\mathbf{a}^q\lfloor\boldsymbol{\alpha}) + (-1)^q(-1)^{q(p-1)}(\mathbf{a}^p\lfloor\boldsymbol{\alpha}) \wedge \mathbf{a}^q,\end{aligned} \tag{1.129}$$

which can be seen to reduce to (1.126) with q and p interchanged. As a second check we choose $\mathbf{a}^q = \mathbf{c}$ and $\mathbf{a}^p = \mathbf{b}$ as vectors with $p = q = 1$. Equation (1.126) then gives us the bac cab rule (1.50), because with proper change of product signs we arrive at

$$\begin{aligned}(\mathbf{c}\wedge\mathbf{b})\lfloor\boldsymbol{\alpha} &= \boldsymbol{\alpha}\rfloor(\mathbf{b}\wedge\mathbf{c}) = (\mathbf{c}|\boldsymbol{\alpha})\mathbf{b} - \mathbf{c}(\mathbf{b}|\boldsymbol{\alpha}) \\ &= \mathbf{b}(\boldsymbol{\alpha}|\mathbf{c}) - \mathbf{c}(\boldsymbol{\alpha}|\mathbf{b}). \end{aligned} \tag{1.130}$$

Finally, choosing $\mathbf{a}^q = \mathbf{b}$, a vector and $\mathbf{a}^p = \mathbf{C}$, a bivector, with $q = 1, p = 2$, the rule (1.90) is seen to follow from (1.126):

$$\begin{aligned}(\mathbf{b}\wedge\mathbf{C})\lfloor\boldsymbol{\alpha} &= \boldsymbol{\alpha}\rfloor(\mathbf{b}\wedge\mathbf{C}) = (\mathbf{b}|\boldsymbol{\alpha})\mathbf{C} - \mathbf{b}\wedge(\mathbf{C}\lfloor\boldsymbol{\alpha}) \\ &= \mathbf{b}\wedge(\boldsymbol{\alpha}\rfloor\mathbf{C}) + \mathbf{C}(\boldsymbol{\alpha}|\mathbf{b}). \end{aligned} \tag{1.131}$$

Thus, (1.126) could be called the mother of bac cab rules.

Equation (1.126) involves a single dual vector $\boldsymbol{\alpha}$ and it can be used to produce other useful rules by adding new dual vectors. In fact, multiplying (1.126) by a second dual vector as $\lfloor\boldsymbol{\beta}$ we obtain the following identity valid for $p, q \geq 2$:

$$\begin{aligned}&(\mathbf{a}^q\wedge\mathbf{a}^p)\lfloor(\boldsymbol{\alpha}\wedge\boldsymbol{\beta}) = \\ &\quad (\mathbf{a}^q\lfloor(\boldsymbol{\alpha}\wedge\boldsymbol{\beta}))\wedge\mathbf{a}^p - (-1)^q(\mathbf{a}^q\lfloor\boldsymbol{\alpha})\wedge(\mathbf{a}^p\lfloor\boldsymbol{\beta}) \\ &\quad +(-1)^q(\mathbf{a}^q\lfloor\boldsymbol{\beta})\wedge(\mathbf{a}^p\lfloor\boldsymbol{\alpha}) + \mathbf{a}^q\wedge(\mathbf{a}^p\lfloor(\boldsymbol{\alpha}\wedge\boldsymbol{\beta})). \end{aligned} \tag{1.132}$$

A further multiplication of (1.132) by another dual vector as $\lfloor\boldsymbol{\gamma}$ gives

$$\begin{aligned}&(\mathbf{a}^q\wedge\mathbf{a}^p)\lfloor(\boldsymbol{\alpha}\wedge\boldsymbol{\beta}\wedge\boldsymbol{\gamma}) \\ &= (\mathbf{a}^q\lfloor(\boldsymbol{\alpha}\wedge\boldsymbol{\beta}\wedge\boldsymbol{\gamma}))\wedge\mathbf{a}^p + (\mathbf{a}^q\lfloor\boldsymbol{\alpha})\wedge(\mathbf{a}^p\lfloor(\boldsymbol{\beta}\wedge\boldsymbol{\gamma})) \\ &+(\mathbf{a}^q\lfloor\boldsymbol{\beta})\wedge(\mathbf{a}^p\lfloor(\boldsymbol{\gamma}\wedge\boldsymbol{\alpha})) + (\mathbf{a}^q\lfloor\boldsymbol{\gamma})\wedge(\mathbf{a}^p\lfloor(\boldsymbol{\alpha}\wedge\boldsymbol{\beta})) \\ &+(-1)^q(\mathbf{a}^q\lfloor(\boldsymbol{\alpha}\wedge\boldsymbol{\beta}))\wedge(\mathbf{a}^p\lfloor\boldsymbol{\gamma}) + (-1)^q(\mathbf{a}^q\lfloor(\boldsymbol{\beta}\wedge\boldsymbol{\gamma}))\wedge(\mathbf{a}^p\lfloor\boldsymbol{\alpha}) \\ &+(-1)^p(\mathbf{a}^q\lfloor(\boldsymbol{\gamma}\wedge\boldsymbol{\alpha}))\wedge(\mathbf{a}^p\lfloor\boldsymbol{\beta}) + (-1)^q\mathbf{a}^q\wedge(\mathbf{a}^p\lfloor(\boldsymbol{\alpha}\wedge\boldsymbol{\beta}\wedge\boldsymbol{\gamma})), \end{aligned} \tag{1.133}$$

which now requires $p, q \geq 3$.

Decomposition rule As an example of using the generalized bac cab rule (1.126) we consider the possibility of decomposing a p-vector $\mathbf{a}^p$ in two components as

$$\mathbf{a}^p = \mathbf{a}^p_{\|} + \mathbf{a}^p_{\perp}, \tag{1.134}$$

with respect to a given vector $\mathbf{a}$ and a given dual vector $\boldsymbol{\alpha}$ assumed to satisfy $\mathbf{a}|\boldsymbol{\alpha} \neq 0$. Using terminology similar to that in decomposing vectors and bivectors as in (1.53) and (1.92), the component $\mathbf{a}^p_{\|}$ is called parallel to the vector $\mathbf{a}$, and the component $\mathbf{a}^p_{\perp}$ orthogonal to $\boldsymbol{\alpha}$. These concepts are defined by the respective conditions

$$\mathbf{a}\wedge\mathbf{a}^p_{\|} = 0, \quad \boldsymbol{\alpha}\rfloor\mathbf{a}^p_{\perp} = 0. \tag{1.135}$$

Applying the bac cab rule (1.127), the decomposition can be readily written as

$$\mathbf{a}^p = \frac{\mathbf{a}\wedge(\mathbf{a}^p\lfloor\boldsymbol{\alpha})}{\mathbf{a}|\boldsymbol{\alpha}} - \frac{(\mathbf{a}\wedge\mathbf{a}^p)\lfloor\boldsymbol{\alpha}}{\mathbf{a}|\boldsymbol{\alpha}}, \tag{1.136}$$

from which the two p-vector components can be identified as

$$\mathbf{a}^p_{\|} = \frac{\mathbf{a} \wedge (\mathbf{a}^p \lfloor \boldsymbol{\alpha})}{\mathbf{a}|\boldsymbol{\alpha}}, \qquad \mathbf{a}^p_{\perp} = -\frac{(\mathbf{a} \wedge \mathbf{a}^p) \lfloor \boldsymbol{\alpha}}{\mathbf{a}|\boldsymbol{\alpha}}. \tag{1.137}$$

Problems

1.4.1 Show that we can express

$$(\mathbf{a} \wedge \mathbf{b} \wedge \mathbf{c})|(\boldsymbol{\alpha} \wedge \boldsymbol{\beta} \wedge \boldsymbol{\gamma}) = (\mathbf{a}|\boldsymbol{\alpha})(\mathbf{b} \wedge \mathbf{c})|(\boldsymbol{\beta} \wedge \boldsymbol{\gamma})$$
$$+(\mathbf{a}|\boldsymbol{\beta})(\mathbf{b} \wedge \mathbf{c})|(\boldsymbol{\gamma} \wedge \boldsymbol{\alpha}) + (\mathbf{a}|\boldsymbol{\gamma})(\mathbf{b} \wedge \mathbf{c})|(\boldsymbol{\alpha} \wedge \boldsymbol{\beta}),$$

and

$$(\mathbf{a} \wedge \mathbf{b} \wedge \mathbf{c})|(\boldsymbol{\alpha} \wedge \boldsymbol{\beta} \wedge \boldsymbol{\gamma})$$
$$= (\mathbf{a}|\boldsymbol{\alpha})(\mathbf{b} \wedge \mathbf{c})|(\boldsymbol{\beta} \wedge \boldsymbol{\gamma}) - (\mathbf{a}\rfloor(\boldsymbol{\beta} \wedge \boldsymbol{\gamma}))|(\boldsymbol{\alpha}\rfloor(\mathbf{b} \wedge \mathbf{c})).$$

1.4.2 The bac cab rule can be written as

$$\boldsymbol{\alpha}\rfloor(\mathbf{a} \wedge \mathbf{b}) = (\mathbf{ab} - \mathbf{ba})|\boldsymbol{\alpha}.$$

Derive its generalizations

$$\boldsymbol{\alpha}\rfloor(\mathbf{a} \wedge \mathbf{b} \wedge \mathbf{c}) = ((\mathbf{a} \wedge \mathbf{b})\mathbf{c} + (\mathbf{b} \wedge \mathbf{c})\mathbf{a} + (\mathbf{c} \wedge \mathbf{a})\mathbf{b})|\boldsymbol{\alpha}$$

and

$$\boldsymbol{\alpha}\rfloor(\mathbf{a} \wedge \mathbf{b} \wedge \mathbf{c} \wedge \mathbf{d})$$
$$= ((\mathbf{a} \wedge \mathbf{b} \wedge \mathbf{c})\mathbf{d} - (\mathbf{b} \wedge \mathbf{c} \wedge \mathbf{d})\mathbf{a} + (\mathbf{c} \wedge \mathbf{d} \wedge \mathbf{a})\mathbf{b} - (\mathbf{d} \wedge \mathbf{a} \wedge \mathbf{b})\mathbf{c})|\boldsymbol{\alpha}$$

1.4.3 Show that if a bivector $\mathbf{A}$ in a four-dimensional space $n = 4$ satisfies $\mathbf{A} \wedge \mathbf{A} = 0$, it can be represented as a simple bivector, in the form $\mathbf{A} = \mathbf{a} \wedge \mathbf{b}$.

1.4.4 Derive (1.76)

1.4.5 Prove that the space of trivectors has the dimension $n(n-1)(n-2)/3!$.

1.4.6 Derive the general commutation rule (1.98).

1.4.7 Given a basis of vectors $\{\mathbf{e}_i\}$, $i = 1 \cdots n$ and defining the complementary $(n-1)$-vectors $\mathbf{e}_{K(i)} = \mathbf{e}_1 \wedge \cdots \mathbf{e}_{i-1} \wedge \mathbf{e}_{i+1} \cdots \wedge \mathbf{e}_n$ and the n-vector $\mathbf{e}_N = \mathbf{e}_1 \wedge \cdots \wedge \mathbf{e}_n$, prove the identities

$$\mathbf{e}_i \wedge \mathbf{e}_{K(i)} = (-1)^{i-1}\mathbf{e}_N, \qquad \mathbf{e}_{K(i)} \wedge \mathbf{e}_i = (-1)^{n-i}\mathbf{e}_N.$$

1.4.8 Defining the dual n-vector $\boldsymbol{\varepsilon}_N = \boldsymbol{\varepsilon}_1 \wedge \cdots \wedge \boldsymbol{\varepsilon}_n$ corresponding to the basis reciprocal to $\{\mathbf{e}_i\}$, prove the identities

$$\boldsymbol{\varepsilon}_N \lfloor \mathbf{e}_i = (-1)^{i-1}\boldsymbol{\varepsilon}_{K(i)}, \qquad \boldsymbol{\varepsilon}_N \lfloor \mathbf{e}_{K(i)} = (-1)^{n-i}\boldsymbol{\varepsilon}_i.$$

1.4.9 Prove the identity

$$(\boldsymbol{\alpha}\rfloor\mathbf{a}_N)\wedge\mathbf{a}=(\boldsymbol{\alpha}|\mathbf{a})\mathbf{a}_N,$$

where $\mathbf{a}$ is a vector, $\boldsymbol{\alpha}$ is a dual vector, and $\mathbf{a}_N$ is an n-vector.

1.4.10 If $\mathbf{A}$ is a bivector and $\mathbf{a}\neq 0$ is a vector, show that vanishing of the trivector $\mathbf{a}\wedge\mathbf{A}=0$ implies the existence of a vector $\mathbf{b}$ such that we can write $\mathbf{A}=\mathbf{a}\wedge\mathbf{b}$.

1.4.11 Show that $\mathbf{a}\wedge\mathbf{b}\wedge\mathbf{c}\wedge\mathbf{d}=0$ implies a linear relation between the four vectors $\mathbf{a}\cdots\mathbf{d}$.

1.4.12 Derive (1.117) and (1.118).

1.4.13 Starting from the last result of problem 1.4.2 and applying (1.85), show that the bac cab rule (1.90) can be written as

$$\boldsymbol{\alpha}\rfloor(\mathbf{b}\wedge\mathbf{C})=\mathbf{b}\wedge(\boldsymbol{\alpha}\rfloor\mathbf{C})-\mathbf{C}(\boldsymbol{\alpha}|\mathbf{b}),$$

when $\mathbf{C}$ is a trivector, $\mathbf{b}$ is a vector, and $\boldsymbol{\alpha}$ is a dual vector. Write the special case for $n=3$.

1.4.14 Show that the identity in the previous problem can be written as

$$(\mathbf{b}\wedge\mathbf{C})\lfloor\boldsymbol{\alpha}=\mathbf{C}(\boldsymbol{\alpha}|\mathbf{b})-(\boldsymbol{\alpha}\rfloor\mathbf{C})\wedge\mathbf{b}$$

and check that the same form applies for $\mathbf{C}$ being a bivector as well as a trivector. Actually it works for a quadrivector as well, so it may be valid for any p-vector, although a proof was not yet found.

1.4.15 Starting from (1.82), derive the following rule between vectors $\mathbf{a},\mathbf{b},\mathbf{c}$ and dual vectors $\boldsymbol{\beta},\boldsymbol{\gamma}$:

$$(\boldsymbol{\beta}\wedge\boldsymbol{\gamma})\rfloor(\mathbf{a}\wedge\mathbf{b}\wedge\mathbf{c})=\mathbf{a}(\mathbf{b}\wedge\mathbf{c})|(\boldsymbol{\beta}\wedge\boldsymbol{\gamma})-(\mathbf{b}\wedge\mathbf{c})\lfloor\Big(\mathbf{a}\rfloor(\boldsymbol{\beta}\wedge\boldsymbol{\gamma})\Big),$$

whose dual can be expressed in terms of a bivector $\mathbf{A}$ and a dual bivector $\boldsymbol{\Gamma}$ as

$$\mathbf{A}\rfloor(\boldsymbol{\beta}\wedge\boldsymbol{\Gamma})=\boldsymbol{\beta}(\mathbf{A}|\boldsymbol{\Gamma})+\boldsymbol{\Gamma}\lfloor(\mathbf{A}\lfloor\boldsymbol{\beta}).$$

This is another bac cab rule.

1.4.16 Prove the identity

$$\mathbf{A}\wedge(\mathbf{B}\lfloor\boldsymbol{\alpha})+\mathbf{B}\wedge(\mathbf{A}\lfloor\boldsymbol{\alpha})=(\mathbf{A}\wedge\mathbf{B})\lfloor\boldsymbol{\alpha}$$

where $\mathbf{A}$ and $\mathbf{B}$ are two bivectors and $\boldsymbol{\alpha}$ a dual vector.

1.4.17 Prove the identity

$$\mathbf{a}\wedge(\mathbf{e}_N\lfloor\boldsymbol{\Gamma})=-\mathbf{e}_N\lfloor(\boldsymbol{\Gamma}\lfloor\mathbf{a}),$$

where $\mathbf{a}$ is a vector, $\boldsymbol{\Gamma}$ is a dual bivector, and $\mathbf{e}_N$ is an n-vector.

1.4.18 Prove the identity

$$(\mathbf{a}\rfloor\mathbf{\Gamma})\rfloor\mathbf{A} = \mathbf{A}\lfloor(\mathbf{\Gamma}\lfloor\mathbf{a}) = (\mathbf{a}\wedge\mathbf{A})\lfloor\mathbf{\Gamma} - \mathbf{a}(\mathbf{A}|\mathbf{\Gamma})$$

where $\mathbf{A}$ is a bivector, $\mathbf{\Gamma}$ is a dual bivector, and $\mathbf{a}$ is a vector.

1.4.19 Starting from the expansion

$$(\mathbf{a}_1\wedge\mathbf{a}_2\wedge\cdots\wedge\mathbf{a}_p)\lfloor\boldsymbol{\alpha} = -\boldsymbol{\alpha}|\sum_{i=1}^{p}(-1)^i\mathbf{a}_i\mathbf{a}_{K(i)},$$

with $\mathbf{a}_{K(i)} = \mathbf{a}_1\wedge\cdots\wedge\mathbf{a}_{i-1}\wedge\mathbf{a}_{i+1}\wedge\cdots\wedge\mathbf{a}_p$, $1\le p\le n$, prove the identity

$$(\mathbf{a}\wedge\mathbf{a}^q)\lfloor\boldsymbol{\alpha} = (\mathbf{a}|\boldsymbol{\alpha})\mathbf{a}^q - \mathbf{a}\wedge(\mathbf{a}^q\lfloor\boldsymbol{\alpha})$$

where $\mathbf{a}^q$ is a q-vector, $\mathbf{a}$ is a vector, and $\boldsymbol{\alpha}$ is a dual vector, $1\le q\le n$.

1.4.20 Show that by inserting $\mathbf{a}^q = \mathbf{b}\wedge\mathbf{a}^p$ in the previous identity and denoting the bivector $\mathbf{a}\wedge\mathbf{b} = \mathbf{A}$ we can derive the identity

$$(\mathbf{A}\wedge\mathbf{a}^p)\lfloor\boldsymbol{\alpha} = (\mathbf{A}\lfloor\boldsymbol{\alpha})\wedge\mathbf{a}^p + \mathbf{A}\wedge(\mathbf{a}^p\lfloor\boldsymbol{\alpha})$$

1.4.21 As a generalization of the two previous identities, we anticipate the identity

$$(\mathbf{a}^q\wedge\mathbf{a}^p)\lfloor\boldsymbol{\alpha} = (\mathbf{a}^q\lfloor\boldsymbol{\alpha})\wedge\mathbf{a}^p + (-1)^q\mathbf{a}^q\wedge(\mathbf{a}^p\lfloor\boldsymbol{\alpha})$$

to be valid. Assuming this and writing $\mathbf{a}^p = \mathbf{b}\wedge\mathbf{a}^{p-1}$, $\mathbf{a}^q\wedge\mathbf{b} = \mathbf{a}^{q+1}$, show that the identity takes the form

$$(\mathbf{a}^{q+1}\wedge\mathbf{a}^{p-1})\lfloor\boldsymbol{\alpha} = (\mathbf{a}^{q+1}\lfloor\boldsymbol{\alpha})\wedge\mathbf{a}^{p-1} + (-1)^{q+1}\mathbf{a}^{q+1}\wedge(\mathbf{a}^{p-1}\lfloor\boldsymbol{\alpha}).$$

From this we can conclude that, if the anticipated identity is valid for $q = 1$ and $q = 2$, it will be valid for any q.

1.4.22 Derive (1.132) and (1.133) in detail.

1.4.23 Prove the generalized form of (1.125) for general n,

$$\mathbf{\Phi}\wedge(\boldsymbol{\kappa}_N\lfloor\mathbf{A}) = \boldsymbol{\kappa}_N(\mathbf{\Phi}|\mathbf{A}),$$

when $\mathbf{A}$ is a bivector, $\mathbf{\Phi}$ is a dual bivector and $\boldsymbol{\kappa}_N$ is a dual n-vector.

1.4.24 Prove that if a dual bivector $\mathbf{\Phi}$ satisfies $\mathbf{a}\rfloor\mathbf{\Phi} = 0$, for some vector $\mathbf{a}$, there exists a dual trivector $\boldsymbol{\gamma}\in\mathbb{F}_3$ such that $\mathbf{\Phi}$ can be expressed in the form $\mathbf{\Phi} = \mathbf{a}\rfloor\boldsymbol{\gamma}$.

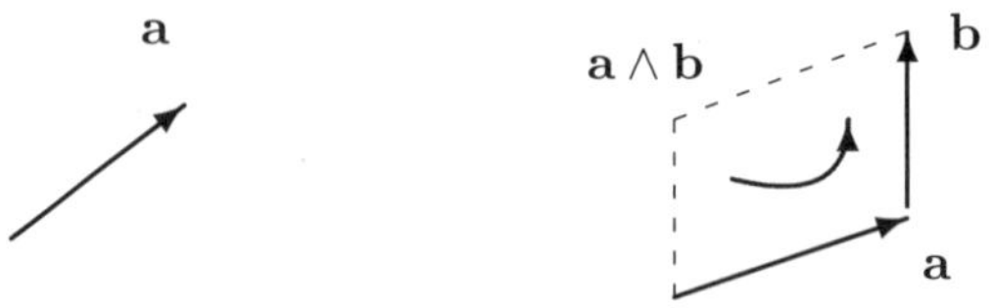

Fig. 1.11 *Geometric interpretation of a three-dimensional vector as a directed line segment and a bivector as an oriented area. The orientation is defined as a sense of circulation on the area.*

1.5 GEOMETRIC INTERPRETATION

Multivector algebra is based on the geometric foundations introduced by Grassmann, and its elements can be given geometric interpretations. While appealing to the eye, they do not very much help in problem solving. The almost naive algebraic memory aid with hooks and eyes appears to be sufficient in most cases for checking algebraic expressions against obvious errors. It is interesting to note that the article by Deschamps [18] introducing differential forms to electromagnetics did not contain a single geometric figure. On the other hand, there are splendid examples of geometric interpretation of multivectors and dual multivectors [58]. In the following we give a simplified overview on the geometric aspects.

1.5.1 Vectors and bivectors

A three-dimensional real vector is generally interpreted as an oriented line segment (arrow) in space. The corresponding interpretation for a bivector is an oriented area. For example, the bivector $\mathbf{a} \wedge \mathbf{b}$ defines a parallelogram defined by the vectors $\mathbf{a}$ and $\mathbf{b}$ (Figure 1.11). The order of the two vectors defines an orientation of the bivector as the sense of circulation around the parallelogram and the area of the parallelogram is proportional to the magnitude of the bivector. Change of order of $\mathbf{a}$ and $\mathbf{b}$ does not change the area but reverses the orientation. Parallel vectors defined by a linear relation of the form $\mathbf{a} = a\mathbf{b}$, $a \neq 0$ give zero area, $\mathbf{a} \wedge \mathbf{b} = 0$. Orthogonality at this stage is only defined between a vector and a dual vector as $\mathbf{a}|\boldsymbol{\alpha} = 0$. Orthogonality of two vectors or two dual vectors is a concept which depends on the definition of a metric dyadic (see Section 2.5).

For example, taking two basis vectors $\mathbf{u}_x$ and $\mathbf{u}_y$, the wedge product of $\mathbf{a} = a\mathbf{u}_x$ and $\mathbf{b} = b\mathbf{u}_y$ gives the bivector

$$\mathbf{a} \wedge \mathbf{b} = ab\mathbf{u}_x \wedge \mathbf{u}_y, \tag{1.138}$$

which is a multiple of the basis bivector $\mathbf{u}_x \wedge \mathbf{u}_y$. Rotating the vectors by an angle θ as $\mathbf{a} = a(\mathbf{u}_x \cos\theta + \mathbf{u}_y \sin\theta)$ and $\mathbf{b} = b(\mathbf{u}_x \sin\theta - \mathbf{u}_y \cos\theta)$, the bivector becomes

$$\mathbf{a} \wedge \mathbf{b} = ab(\cos^2\theta + \sin^2\theta)\mathbf{u}_x \wedge \mathbf{u}_y = ab\mathbf{u}_x \wedge \mathbf{u}_y, \tag{1.139}$$

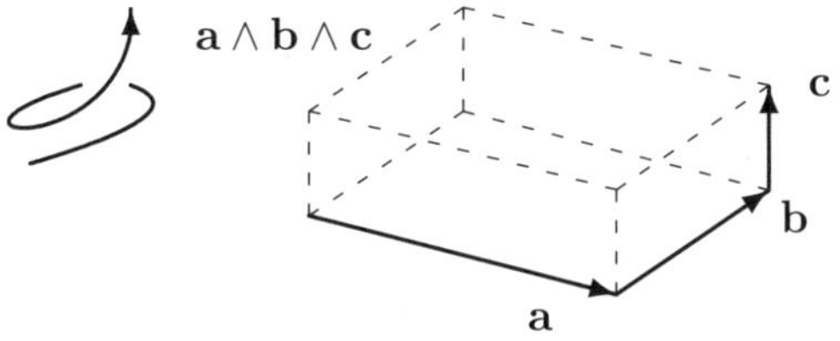

Fig. 1.12 *Geometric interpretation of a trivector as an oriented volume. Orientation is defined as handedness, which in this case is that of a right-handed screw.*

which coincides with the previous one. Thus, the bivector is invariant to the rotation. Also, it is invariant if we multiply $\mathbf{a}$ and divide $\mathbf{b}$ by the same real number λ.

1.5.2 Trivectors

Three real vectors $\mathbf{a}$, $\mathbf{b}$, and $\mathbf{c}$ define a parallelepiped. The trivector $\mathbf{a}\wedge\mathbf{b}\wedge\mathbf{c}$ represents its volume V with orientation. In component form we can write

$$\begin{aligned}&\mathbf{a}\wedge\mathbf{b}\wedge\mathbf{c}\\ &= [a_x(b_yc_z - b_zc_y) + a_y(b_zc_x - b_xc_z) + a_z(b_xc_y - b_yc_x)]\mathbf{u}_x\wedge\mathbf{u}_y\wedge\mathbf{u}_z\\ &= \det\begin{pmatrix} a_x & a_y & a_z\\ b_x & b_y & b_z\\ c_x & c_y & c_z\end{pmatrix}\mathbf{u}_x\wedge\mathbf{u}_y\wedge\mathbf{u}_z. \qquad (1.140)\end{aligned}$$

The determinant of the 3×3 matrix defined by the components of the vector equals the Gibbsian scalar triple product $\mathbf{a}\times\mathbf{b}\cdot\mathbf{c}$ and it vanishes when the three vectors are linearly dependent. In this case the vectors lie in the same plane and the volume of the parallepiped is zero. The orientation of a given trivector $\mathbf{k}$ is given in terms of its handedness. Taken after one another, three nonplanar vectors $\mathbf{a}, \mathbf{b}, \mathbf{c}$ define a screw in space which can be right or left handed, Figure 1.12. Changing the order of any two vectors changes the handedness of the trivector. Handedness of a trivector can also be determined with respect to a given reference dual trivector ε_{123}. If $\mathbf{k}|\varepsilon_{123}$ is positive, $\mathbf{k}$ has the same handedness as ε_{123}, otherwise it has the opposite handedness. If the coordinate system x, y, z is labeled as right handed, the trivector $\mathbf{u}_x\wedge\mathbf{u}_y\wedge\mathbf{u}_z$ corresponds to a unit cube with right-handed orientation. If the expression in the square brackets in (1.140) is positive, it represents the volume of a right-handed parallelepiped. With negative sign, its magnitude gives the volume of a left-handed parallelepiped. For complex vectors the simple geometric interpretation breaks. Complex vectors can be interpreted in terms of oriented ellipses [28,40], but bivectors and trivectors cannot be easily given a mental picture.

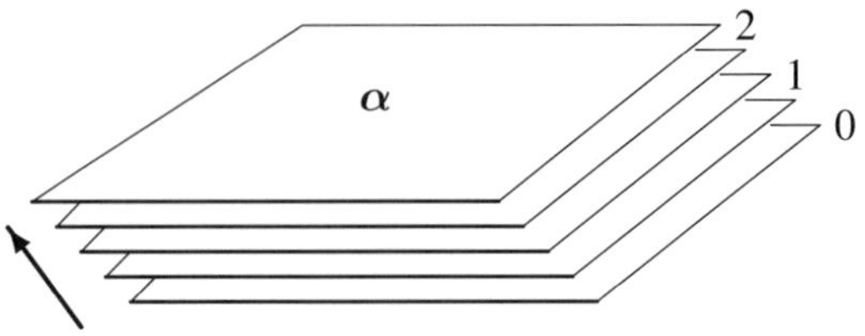

Fig. 1.13 *Geometric interpretation of a dual vector as a set of parallel planes with orientation and density.*

1.5.3 Dual vectors

Dual vectors are objects which map vectors onto scalars through the duality product. Elements of dual vectors can be represented in the three-dimensional Euclidean vector space by considering the set of vectors which are mapped onto the same scalar value. For example, the dual vector $\boldsymbol{\alpha} = \alpha\varepsilon_1$ maps all vectors $\mathbf{r} = \sum x_i \mathbf{e}_i$ onto the scalar value a when the tip of the vector $\mathbf{r}$ lies on the plane $x_1 = a$. Thus, the dual vector $\boldsymbol{\alpha}$ is like a directed measuring tape with parallel planes labeled by the distance in the perpendicular direction of the tape. The magnitude ("length´´) of the dual vector is the density of planes (labeling by the inch or centimeter on the tape) and the direction of increasing scalar values represents the orientation of the dual vector. Of course, in the space of dual vectors, a dual vector appears as a directed line segment and a vector, which is the dual of a dual vector, as a set of parallel planes.

A field of vectors can be represented by field lines, which give the orientation of the vector at each point in the physical space. The magnitude of the vector at each position can be given by a set of surfaces corresponding to a constant value. The field lines need not be orthogonal to these surfaces. The dual vectors can then be represented by surfaces tangent to the dual-vector planes at each position, FigureForm153. The density of the planes gives the magnitude of the dual vector.

The sum of two vectors is represented through the well-known parallelogram construction. The sum of two dual vectors $\boldsymbol{\alpha} + \boldsymbol{\beta} = \boldsymbol{\gamma}$ can be constructed as follows. Since $\boldsymbol{\alpha}$ and $\boldsymbol{\beta}$ are represented by two families of parallel planes, let us consider two planes of each. Unless the planes are parallel, in which case the addition is simple, they cut in four common parallel lines which define two diagonal planes not parallel to either of the original ones. One of these new planes defines a set of parallel planes corresponding to the sum, and the other one the difference, of $\boldsymbol{\alpha}$ and $\boldsymbol{\beta}$. The sum corresponds to the pair of planes whose positive orientation has positive components on both original sets of planes.

The duality product of a vector and a dual vector $\boldsymbol{\alpha}|\mathbf{a}$ equals the number of planes pierced by the vector $\mathbf{a}$. A good mental picture is obtained through the method of measuring a javelin throw in prewar Olympic games where the lines of constant

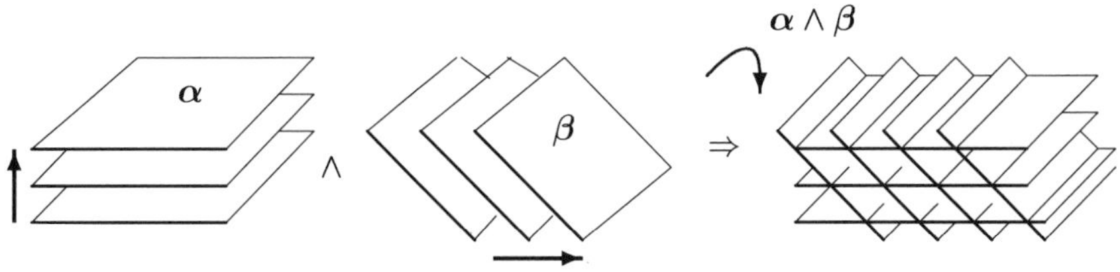

Fig. 1.14 *Geometric interpretation of a dual bivector as a set of parallel tubes with orientation (sense of rotation) and density.*

distance used to be parallel[5] to the line behind which the throw was done. If the vector is parallel to the planes, there is no piercing and the inner product is zero. In this case, the vector $\mathbf{a}$ and the dual vector $\boldsymbol{\alpha}$ can be called orthogonal. The number of planes pierced is largest when the vector is perpendicular to them.

A three-dimensional vector basis $\{\mathbf{e}_i\}$ can be formed by any three non-coplanar vectors. They need not have the same length or be perpendicular to one another. The associated dual basis $\{\varepsilon_j\}$ consists of three families of planes which cut the space in closed cells, that is, they do not form open tubes. The densities of the dual basis vectors depend on the vectors $\mathbf{e}_i$. Because, for example, we have $\varepsilon_2|\mathbf{e}_1 = 0$ and $\varepsilon_3|\mathbf{e}_1 = 0$, the vector $\mathbf{e}_1$ must be parallel to the common lines of the plane families ε_2 and ε_3. $\mathbf{e}_1$ is not orthogonal to the family of planes ε_1. However, the density of ε_1 is such that $\mathbf{e}_1$ pierces exactly one interval of planes.

1.5.4 Dual bivectors and trivectors

The bivectors in the dual space, $\boldsymbol{\alpha} \wedge \boldsymbol{\beta}$, can be pictured as a family of tubes defined by two families of planes and the magnitude equals the density of tubes, Figure 1.14. The inner product $(\boldsymbol{\alpha} \wedge \boldsymbol{\beta})|(\mathbf{a} \wedge \mathbf{b})$ represents the number of tubes enclosed by the parallelogram defined by the bivector $\mathbf{a} \wedge \mathbf{b}$.

The dual trivector $\boldsymbol{\alpha}\wedge\boldsymbol{\beta}\wedge\boldsymbol{\gamma}$ represents the density of cells defined by three families of planes. The inner product $(\boldsymbol{\alpha}\wedge\boldsymbol{\beta}\wedge\boldsymbol{\gamma})|(\mathbf{a}\wedge\mathbf{b}\wedge\mathbf{c})$ gives the number of cells of the dual trivector $\boldsymbol{\alpha}\wedge\boldsymbol{\beta}\wedge\boldsymbol{\gamma}$ enclosed by the parallelepiped of the trivector $\mathbf{a}\wedge\mathbf{b}\wedge\mathbf{c}$. The result can also be a negative number if the handedness of the cells is opposite to that of the parallelepiped. Similar ideas can be applied to p-vectors and dual p-vectors in spaces of higher dimension, but their visualization must rely on analogies.

Problems

1.5.1 Find a geometrical meaning for summing two bivectors, $\mathbf{A} + \mathbf{B}$, in the three-dimensional space. *Hint*: Bivectors are oriented areas on a certain plane. The form of their contour line can be changed. Assume the planes of $\mathbf{A}$ and $\mathbf{B}$ cut

[5] In the 1930s this method was changed to one with circular lines.

along a line containing the vector $\mathbf{c}$. Then there exist vectors $\mathbf{a}$ and $\mathbf{b}$ such that $\mathbf{A} = \mathbf{a} \wedge \mathbf{c}$ and $\mathbf{B} = \mathbf{b} \wedge \mathbf{c}$.

1.5.2 Verify the geometric construction of the sum $\boldsymbol{\alpha} + \boldsymbol{\beta}$ and difference $\boldsymbol{\alpha} - \boldsymbol{\beta}$ of two dual vectors $\boldsymbol{\alpha}$ and $\boldsymbol{\beta}$.

1.5.3 Interprete geometrically the bivector relation

$$\mathbf{a}_1 \wedge \mathbf{a}_2 = \mathbf{a}_1 \wedge (\mathbf{a}_1 + \mathbf{a}_2)$$

by considering the parallelograms defined by the bivectors on each side.

1.5.4 Interprete geometrically the planar bivector relation

$$\mathbf{a}_1 \wedge \mathbf{a}_2 + (\mathbf{a}_1 + \mathbf{a}_2) \wedge \mathbf{a}_3 = \mathbf{a}_1 \wedge \mathbf{a}_2 + \mathbf{a}_1 \wedge \mathbf{a}_3 + \mathbf{a}_2 \wedge \mathbf{a}_3$$

by considering the areas of the different triangles.

2

Dyadics

The dyadic notation and some basic concepts of dyadics were already considered in the previous chapter. Here we will concentrate on more fundamental properties of the dyadic algebra. In analogy to the Gibbsian dyadics, multivector dyadics make a powerful tool for electromagnetic analysis in matching the coordinate-free dyadic notation of linear mappings to coordinate-free multivectors and the differential-form formalism in general. However, before being able to work with dyadics in their full power, one has to learn some basic identities and operations to avoid the need to go back to the coordinate-dependent notation during the analysis.

2.1 PRODUCTS OF DYADICS

2.1.1 Basic notation

A dyadic can be formally defined as a sum of dyadic products of multivectors and/or dual multivectors. Dyadics represent linear mappings between spaces of multivectors and/or dual multivectors. They arise naturally from multivector expressions through associativity, that is, by changing the order of multiplications, an example of which was seen in (1.6). As another example, consider the expression

$$\mathbf{b} = \mathbf{a}(\mathbf{A}|\mathbf{\Phi}) = (\mathbf{a}\mathbf{A})|\mathbf{\Phi}, \tag{2.1}$$

where $\mathbf{a}, \mathbf{b}$ are vectors, $\mathbf{A}$ is a bivector, and $\mathbf{\Phi}$ is a dual bivector. Obviously, here we have a mapping from a dual bivector $\mathbf{\Phi}$ to a vector $\mathbf{b}$ through the dyad $\mathbf{aA}$ which is an element of a space denoted by $\mathbb{E}_1\,\mathbb{E}_2$. General elements of that space are polynomials

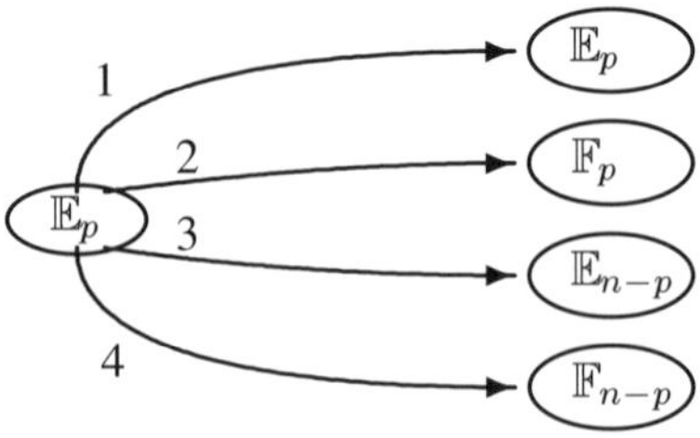

Fig. 2.1 *Four possible mappings from space of p-vectors $\mathbb{E}_p$ to spaces of multivectors or dual multivectors of the same dimension can be expressed through dyadics belonging to four different spaces. Mappings 2 correspond to metric dyadics and mappings 3 to Hodge dyadics. For the dual case, $\mathbb{E}$ is changed to $\mathbb{F}$ and conversely.*

of similar dyads and can be expressed as the sum

$$\sum \mathbf{a}_i \mathbf{A}_i = \mathbf{a}_1 \mathbf{A}_1 + \mathbf{a}_2 \mathbf{A}_2 + \mathbf{a}_3 \mathbf{A}_3 + \cdots. \tag{2.2}$$

The space of dyadics mapping q-vectors to p-vectors consists of dyadics of the form $\sum \mathbf{a}_i^p \boldsymbol{\alpha}_i^q$ where $\mathbf{a}_i^p \in \mathbb{E}_p$ are p-vectors and $\boldsymbol{\alpha}_i^q \in \mathbb{F}_q$ are dual q-vectors. The space of such dyadics is denoted by $\mathbb{E}_p\mathbb{F}_q = \mathbb{E}_p \times \mathbb{F}_q$. Other basic classes of dyadics are denoted by $\mathbb{E}_p\mathbb{E}_q$, $\mathbb{F}_p\mathbb{F}_q$ and $\mathbb{F}_p\mathbb{E}_q$, where the first index tells us the space of the precedants (first factor in the dyadic product) while the latter tells us that of the antecedants (last factor of the dyadic product). For example, in the case $\mathbb{F}_p\mathbb{E}_q$ the dyadics consisting of a sum of dyadic products of dual p-vectors and q-vectors represent linear mappings from q-vectors $\boldsymbol{\alpha}^q \in \mathbb{F}_q$ to p-vectors $\mathbf{a}^p \in \mathbb{F}_p$. Basic symbol for dyadics is a capital sans-serif letter with a double overline, $\overline{\overline{\mathsf{A}}}$. However, we soon have to make exceptions for common quantities like electromagnetic medium dyadics.

Obviously, main attention must be concentrated on dyadics which represent mappings between spaces of the same dimension. This requires either $p = q$ or $p = n-q$. In matrix algebra the corresponding matrices are square. Attention will be first given to dyadics belonging to the spaces $\mathbb{E}_p\mathbb{F}_p$ and $\mathbb{F}_p\mathbb{E}_p$ which correspond to mapping 1 in Figure 2.1. Dyadics 2 belonging to the spaces $\mathbb{F}_p\mathbb{F}_p$ and $\mathbb{E}_p\mathbb{E}_p$ are called metric dyadics while those (3) belonging to the spaces $\mathbb{E}_p\mathbb{F}_{n-p}$ and $\mathbb{F}_p\mathbb{E}_{n-p}$ are called Hodge dyadics. The fourth type $\mathbb{E}_p\mathbb{E}_{n-p}$ and $\mathbb{F}_p\mathbb{F}_{n-p}$ of such dyadic spaces has no special name.

One should note that many concepts familiar from Gibbsian dyadic algebra are valid only for certain classes of dyadics. Because the transpose operation ^{T} maps a dyadic from the space $\mathbb{E}_p\mathbb{F}_q$ to the space $\mathbb{F}_q\mathbb{E}_p$ as

$$\overline{\overline{\mathsf{A}}}^T = \sum (\mathbf{a}_i^p \boldsymbol{\alpha}_i^q)^T = \sum \boldsymbol{\alpha}_i^q \mathbf{a}_i^p, \tag{2.3}$$

properties of symmetry and antisymmetry make sense only in the metric dyadic spaces of the form $\mathbb{F}_p\mathbb{F}_p$ and $\mathbb{E}_p\mathbb{E}_p$. On the other hand, eigenvalue problems and dyadic powers can be only defined in dyadic spaces of the form $\mathbb{F}_p\mathbb{E}_p$ and $\mathbb{F}_p\mathbb{E}_p$ which map elements to the same space. Inverse of a dyadic in the classical sense is also limited to these two spaces, but it can be generalized so that the result lies in a different space. So, for example, the inverse of a dyadic in the space $\mathbb{E}_p\mathbb{E}_{n-p}$ lies in the space $\mathbb{F}_{n-p}\mathbb{F}_p$.

2.1.2 Duality product

The duality product of two $\mathbb{E}_p\mathbb{F}_p$ dyadics is understood as

$$\overline{\overline{\mathsf{A}}}|\overline{\overline{\mathsf{B}}} = \sum(\mathbf{a}_i^p\boldsymbol{\alpha}_i^p)|\sum(\mathbf{b}_j^p\boldsymbol{\beta}_j^p) = \sum(\boldsymbol{\alpha}_i^p|\mathbf{b}_j^p)\mathbf{a}_i^p\boldsymbol{\beta}_j^p. \tag{2.4}$$

The result is another linear mapping $\mathbb{E}_p \to \mathbb{E}_p$, an element of the same dyadic space $\mathbb{E}_p\mathbb{F}_p$. The duality product defines an associative and noncommutative algebra similar to that of the matrix product,

$$(\overline{\overline{\mathsf{A}}}|\overline{\overline{\mathsf{B}}})|\overline{\overline{\mathsf{C}}} = \overline{\overline{\mathsf{A}}}|(\overline{\overline{\mathsf{B}}}|\overline{\overline{\mathsf{C}}}), \tag{2.5}$$

$$\overline{\overline{\mathsf{A}}}|\overline{\overline{\mathsf{B}}} \neq \overline{\overline{\mathsf{B}}}|\overline{\overline{\mathsf{A}}} \quad \text{(in general)}. \tag{2.6}$$

The square and cube of a dyadic are defined by

$$\overline{\overline{\mathsf{A}}}^2 = \overline{\overline{\mathsf{A}}}|\overline{\overline{\mathsf{A}}}, \quad \overline{\overline{\mathsf{A}}}^3 = \overline{\overline{\mathsf{A}}}|\overline{\overline{\mathsf{A}}}|\overline{\overline{\mathsf{A}}}, \tag{2.7}$$

and similarly for higher powers $\overline{\overline{\mathsf{A}}}^m$. The zeroth power of any dyadic can be understood as the unit (identity) dyadic for p-vectors, $\overline{\overline{\mathsf{A}}}^0 = \overline{\overline{\mathsf{I}}}^{(p)}$.

As was seen in Chapter 1, the unit dyadic in the space of $\mathbb{E}_1\mathbb{F}_1$ can be expanded in any vector basis and its complementary dual basis as

$$\overline{\overline{\mathsf{I}}} = \sum \mathbf{e}_i\boldsymbol{\varepsilon}_i = \mathbf{e}_1\boldsymbol{\varepsilon}_1 + \mathbf{e}_2\boldsymbol{\varepsilon}_2 + \cdots + \mathbf{e}_n\boldsymbol{\varepsilon}_n. \tag{2.8}$$

Negative powers of a $\mathbb{E}_1\mathbb{F}_1$ dyadic can also be defined if there exists the inverse dyadic $\overline{\overline{\mathsf{A}}}^{-1}$ satisfying

$$\overline{\overline{\mathsf{A}}}^{-1}|\overline{\overline{\mathsf{A}}} = \overline{\overline{\mathsf{A}}}|\overline{\overline{\mathsf{A}}}^{-1} = \overline{\overline{\mathsf{I}}}. \tag{2.9}$$

Construction of the inverse dyadic is considered in Section 2.4.

The duality product corresponds, in most respects, to the product of $n \times n$ matrices. The correspondence can be seen through basis expansions of the form $\overline{\overline{\mathsf{A}}} = \sum A_{ij}\mathbf{e}_i\boldsymbol{\varepsilon}_j$ by considering the rules for the coefficient matrices $[A_{ij}]$.

2.1.3 Double-duality product

The double-duality product $||$ between a dyadic and a transposed dyadic in the space $\mathbb{E}_1\mathbb{F}_1$ is a scalar defined by

$$\begin{aligned}\sum(\mathbf{a}_i\boldsymbol{\alpha}_i)||\sum(\mathbf{b}_j\boldsymbol{\beta}_j)^T &= \sum(\mathbf{a}_i|\boldsymbol{\beta}_j)(\boldsymbol{\alpha}_i|\mathbf{b}_j)\\ &= \sum(\mathbf{a}_i\boldsymbol{\alpha}_i)^T||\sum(\mathbf{b}_j\boldsymbol{\beta}_j),\end{aligned} \tag{2.10}$$

and it is straightforwardly extended to multidyadic spaces $\mathbb{E}_p\mathbb{F}_p$, $\mathbb{F}_p$, $\mathbb{E}_p$. The trace of a dyadic is defined as the double-duality product of the dyadic and the transposed unit dyadic:

$$\mathrm{tr}\overline{\overline{\mathsf{A}}} = \mathrm{tr}\sum \mathbf{a}_i\boldsymbol{\alpha}_i = \sum \mathbf{a}_i|\boldsymbol{\alpha}_i = \overline{\overline{\mathsf{A}}}||\overline{\overline{\mathsf{I}}}^T = \overline{\overline{\mathsf{A}}}^T||\overline{\overline{\mathsf{I}}}. \tag{2.11}$$

To see this result we expand

$$\boldsymbol{\alpha}|\mathbf{a} = \boldsymbol{\alpha}|\overline{\overline{\mathsf{I}}}|\mathbf{a} = \sum(\boldsymbol{\alpha}|\mathbf{e}_i)(\boldsymbol{\varepsilon}_i|\mathbf{a}) = \boldsymbol{\alpha}\mathbf{a}||\sum \mathbf{e}_i\boldsymbol{\varepsilon}_i = \boldsymbol{\alpha}\mathbf{a}||\overline{\overline{\mathsf{I}}}. \tag{2.12}$$

Thus, the trace operation simply inserts the duality product sign | between the factors of the dyadic product. It is easy to verify that $\mathrm{tr}\overline{\overline{\mathsf{I}}} = n$, the dimension of the vector space $\mathbb{E}_1$.

Any $\mathbb{E}_1\mathbb{F}_1$ dyadic can be uniquely decomposed in a multiple of the unit dyadic and its trace-free part $\overline{\overline{\mathsf{A}}}_\circ$ as

$$\overline{\overline{\mathsf{A}}} = \frac{1}{n}(\overline{\overline{\mathsf{A}}}||\overline{\overline{\mathsf{I}}}^T)\overline{\overline{\mathsf{I}}} + \overline{\overline{\mathsf{A}}}_\circ, \tag{2.13}$$

where $\mathrm{tr}\overline{\overline{\mathsf{A}}}_\circ = 0$.

2.1.4 Double-wedge product

Any $\mathbb{E}_1\mathbb{F}_1$ dyadic $\overline{\overline{\mathsf{A}}}$ induces a linear mapping in the space of bivectors $\mathbb{E}_2$. This is based on the following identity

$$(\overline{\overline{\mathsf{A}}}|\mathbf{a}) \wedge (\overline{\overline{\mathsf{A}}}|\mathbf{b}) = \frac{1}{2}(\overline{\overline{\mathsf{A}}}{}^{\wedge}_{\wedge}\overline{\overline{\mathsf{A}}})|(\mathbf{a}\wedge\mathbf{b}), \tag{2.14}$$

where $\mathbf{a}$ and $\mathbf{b}$ are vectors. Equation (2.14) is proved in a more general form below. The double-wedge product between two $\mathbb{E}_1\mathbb{F}_1$ dyadics defined by

$$\sum(\mathbf{a}_i\boldsymbol{\alpha}_i){}^{\wedge}_{\wedge}\sum(\mathbf{b}_j\boldsymbol{\beta}_j) = \sum(\mathbf{a}_i\wedge\mathbf{b}_j)(\boldsymbol{\alpha}_i\wedge\boldsymbol{\beta}_j) \tag{2.15}$$

is an element of the space $\mathbb{E}_2\mathbb{F}_2$, a bivector dyadic which maps bivectors to bivectors.

From the associativity and anticommutativity of the wedge product for vectors or dual vectors one can easily show that the double-wedge product is associative[1] and commutative:

$$(\overline{\overline{\mathsf{A}}}{}^{\wedge}_{\wedge}\overline{\overline{\mathsf{B}}}){}^{\wedge}_{\wedge}\overline{\overline{\mathsf{C}}} = \overline{\overline{\mathsf{A}}}{}^{\wedge}_{\wedge}(\overline{\overline{\mathsf{B}}}{}^{\wedge}_{\wedge}\overline{\overline{\mathsf{C}}}), \tag{2.16}$$

$$\overline{\overline{\mathsf{A}}}{}^{\wedge}_{\wedge}\overline{\overline{\mathsf{B}}} = \overline{\overline{\mathsf{B}}}{}^{\wedge}_{\wedge}\overline{\overline{\mathsf{A}}}. \tag{2.17}$$

Because of associativity, there is no need to write brackets in a row of dyadics with only double-wedge multiplications. Also, because of commutativity, the dyadics in such a product can be taken in any order, whence the result is invariant to any permutation of dyadics. Thus, we can write, for example, between three dyadics of the same dyadic space the rule

$$\overline{\overline{\mathsf{A}}}{}^{\wedge}_{\wedge}\overline{\overline{\mathsf{B}}}{}^{\wedge}_{\wedge}\overline{\overline{\mathsf{C}}} = \overline{\overline{\mathsf{B}}}{}^{\wedge}_{\wedge}\overline{\overline{\mathsf{C}}}{}^{\wedge}_{\wedge}\overline{\overline{\mathsf{A}}} = \overline{\overline{\mathsf{C}}}{}^{\wedge}_{\wedge}\overline{\overline{\mathsf{A}}}{}^{\wedge}_{\wedge}\overline{\overline{\mathsf{B}}} = \overline{\overline{\mathsf{A}}}{}^{\wedge}_{\wedge}\overline{\overline{\mathsf{C}}}{}^{\wedge}_{\wedge}\overline{\overline{\mathsf{B}}} = \overline{\overline{\mathsf{C}}}{}^{\wedge}_{\wedge}\overline{\overline{\mathsf{B}}}{}^{\wedge}_{\wedge}\overline{\overline{\mathsf{A}}} = \overline{\overline{\mathsf{B}}}{}^{\wedge}_{\wedge}\overline{\overline{\mathsf{A}}}{}^{\wedge}_{\wedge}\overline{\overline{\mathsf{C}}}. \tag{2.18}$$

[1]Note that the corresponding double-cross product in the Gibbsian dyadic algebra is not associative [40].

Identities Identity (2.14) is a special case of a more general one which can be derived by applying (1.36) to the following expansion:

$$
\begin{aligned}
&(\sum_i \mathbf{a}_i\boldsymbol{\alpha}_i {}^\wedge_\wedge \sum_j \mathbf{b}_j\boldsymbol{\beta}_j)|(\mathbf{c}\wedge\mathbf{d}) \\
&= \sum_{i,j}(\mathbf{a}_i\wedge\mathbf{b}_j)\{(\boldsymbol{\alpha}_i\wedge\boldsymbol{\beta}_j)|(\mathbf{c}\wedge\mathbf{d})\} \\
&= \sum_{i,j}(\mathbf{a}_i\wedge\mathbf{b}_j)[(\boldsymbol{\alpha}_i|\mathbf{c})(\boldsymbol{\beta}_j|\mathbf{d}) - (\boldsymbol{\beta}_j|\mathbf{c})(\boldsymbol{\alpha}_i|\mathbf{d})] \\
&= (\sum_i \mathbf{a}_i\boldsymbol{\alpha}_i|\mathbf{c})\wedge(\sum_j \mathbf{b}_j\boldsymbol{\beta}_j|\mathbf{d}) + (\sum_j \mathbf{b}_j\boldsymbol{\beta}_j|\mathbf{c})\wedge(\sum_i \mathbf{a}_i\boldsymbol{\alpha}_i|\mathbf{d}). \qquad (2.19)
\end{aligned}
$$

Denoting the dyadic expansions by $\overline{\overline{\mathsf{A}}}$ and $\overline{\overline{\mathsf{B}}}$ leads to the following identity:

$$
(\overline{\overline{\mathsf{A}}} {}^\wedge_\wedge \overline{\overline{\mathsf{B}}})|(\mathbf{a}\wedge\mathbf{b}) = (\overline{\overline{\mathsf{A}}}|\mathbf{a})\wedge(\overline{\overline{\mathsf{B}}}|\mathbf{b}) + (\overline{\overline{\mathsf{B}}}|\mathbf{a})\wedge(\overline{\overline{\mathsf{A}}}|\mathbf{b}), \qquad (2.20)
$$

which is satisfied by any two dyadics $\overline{\overline{\mathsf{A}}}, \overline{\overline{\mathsf{B}}} \in \mathbb{E}_1\mathbb{F}_1$ and vectors $\mathbf{a},\mathbf{b} \in \mathbb{E}_1$. The corresponding dual form is obtained when replacing the dyadics by their transposes and vectors by dual vectors:

$$
(\overline{\overline{\mathsf{A}}}^T {}^\wedge_\wedge \overline{\overline{\mathsf{B}}}^T)|(\boldsymbol{\alpha}\wedge\boldsymbol{\beta}) = (\overline{\overline{\mathsf{A}}}^T|\boldsymbol{\alpha})\wedge(\overline{\overline{\mathsf{B}}}^T|\boldsymbol{\beta}) + (\overline{\overline{\mathsf{B}}}^T|\boldsymbol{\alpha})\wedge(\overline{\overline{\mathsf{A}}}^T|\boldsymbol{\beta}). \qquad (2.21)
$$

Obviously, (2.14) is a special case of (2.20) when $\overline{\overline{\mathsf{B}}} = \overline{\overline{\mathsf{A}}}$.

From (2.20) we can form another identity by multiplying it dyadically from the right by the bivector $(\boldsymbol{\alpha}\wedge\boldsymbol{\beta})$ and using the definition of a double-wedge product:

$$
\begin{aligned}
(\overline{\overline{\mathsf{A}}} {}^\wedge_\wedge \overline{\overline{\mathsf{B}}})|(\mathbf{a}\wedge\mathbf{b})(\boldsymbol{\alpha}\wedge\boldsymbol{\beta}) &= (\overline{\overline{\mathsf{A}}} {}^\wedge_\wedge \overline{\overline{\mathsf{B}}})|(\mathbf{a}\boldsymbol{\alpha} {}^\wedge_\wedge \mathbf{b}\boldsymbol{\beta}) \\
&= (\overline{\overline{\mathsf{A}}}|\mathbf{a}\boldsymbol{\alpha}) {}^\wedge_\wedge (\overline{\overline{\mathsf{B}}}|\mathbf{b}\boldsymbol{\beta}) + (\overline{\overline{\mathsf{B}}}|\mathbf{a}\boldsymbol{\alpha}) {}^\wedge_\wedge (\overline{\overline{\mathsf{A}}}|\mathbf{b}\boldsymbol{\beta}). \qquad (2.22)
\end{aligned}
$$

Since this is a linear relation valid for any dyads $\mathbf{a}\boldsymbol{\alpha}$ and $\mathbf{b}\boldsymbol{\beta}$, it is valid for any dyadic polynomials $\overline{\overline{\mathsf{C}}} = \sum \mathbf{a}_i\boldsymbol{\alpha}_i$ and $\overline{\overline{\mathsf{D}}} = \sum \mathbf{b}_j\boldsymbol{\beta}_j$, whence we can write the following identity valid for any four $\mathbb{E}_1\mathbb{F}_1$ dyadics

$$
(\overline{\overline{\mathsf{A}}} {}^\wedge_\wedge \overline{\overline{\mathsf{B}}})|(\overline{\overline{\mathsf{C}}} {}^\wedge_\wedge \overline{\overline{\mathsf{D}}}) = (\overline{\overline{\mathsf{A}}}|\overline{\overline{\mathsf{C}}}) {}^\wedge_\wedge (\overline{\overline{\mathsf{B}}}|\overline{\overline{\mathsf{D}}}) + (\overline{\overline{\mathsf{B}}}|\overline{\overline{\mathsf{C}}}) {}^\wedge_\wedge (\overline{\overline{\mathsf{A}}}|\overline{\overline{\mathsf{D}}}). \qquad (2.23)
$$

From duality, it is also valid for any four $\mathbb{F}_1\mathbb{E}_1$ dyadics. As a simple check, it is seen that the right-hand side is invariant in the changes $\overline{\overline{\mathsf{A}}} \leftrightarrow \overline{\overline{\mathsf{B}}}$ and/or $\overline{\overline{\mathsf{C}}} \leftrightarrow \overline{\overline{\mathsf{D}}}$, which follows also from the commutation property of the two double-wedge products on the left-hand side.

2.1.5 Double-wedge square

The double-wedge square of a $\mathbb{E}_1\mathbb{F}_1$ dyadic is defined and denoted similarly as the Gibbsian double-cross square [26, 40, 65],

$$
\overline{\overline{\mathsf{A}}}^{(2)} = \frac{1}{2}\overline{\overline{\mathsf{A}}} {}^\wedge_\wedge \overline{\overline{\mathsf{A}}}. \qquad (2.24)
$$

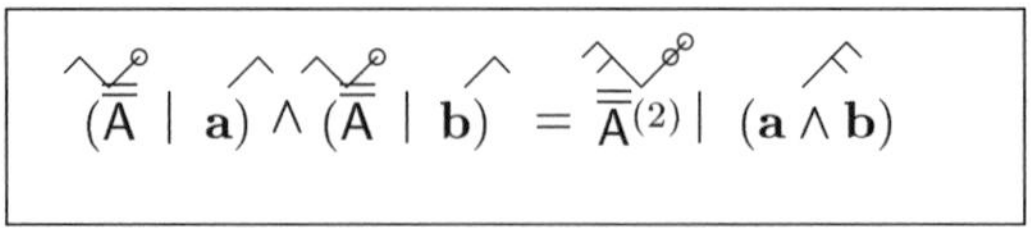

Fig. 2.2 *Visualizing the mapping property (2.24) of the double-wedge square of a dyadic.*

From (2.14) it follows that if two vectors $\mathbf{a}, \mathbf{b}$ are mapped through the dyadic $\overline{\overline{\mathsf{A}}}$, their wedge product $\mathbf{a}\wedge\mathbf{b}$ is mapped through the double-wedge square $\overline{\overline{\mathsf{A}}}^{(2)}$:

$$(\overline{\overline{\mathsf{A}}}|\mathbf{a})\wedge(\overline{\overline{\mathsf{A}}}|\mathbf{b}) = \overline{\overline{\mathsf{A}}}^{(2)}|(\mathbf{a}\wedge\mathbf{b}). \tag{2.25}$$

The double-wedge square of the unit dyadic $\overline{\overline{\mathsf{I}}}$ is

$$\begin{aligned} \overline{\overline{\mathsf{I}}}^{(2)} &= \frac{1}{2}\sum_{i=1}^{n}\sum_{j=1}^{n}(\mathbf{e}_i\wedge\mathbf{e}_j)(\boldsymbol{\varepsilon}_i\wedge\boldsymbol{\varepsilon}_j) \\ &= \sum \mathbf{e}_J\boldsymbol{\varepsilon}_J, \qquad J=\{ij\},\ i<j. \end{aligned} \tag{2.26}$$

As was seen in the previous chapter, $\overline{\overline{\mathsf{I}}}^{(2)}$ serves as the unit dyadic for the bivector space $\mathbb{E}_2$:

$$\overline{\overline{\mathsf{I}}}^{(2)}|(\mathbf{a}\wedge\mathbf{b}) = \mathbf{a}\wedge\mathbf{b}. \tag{2.27}$$

Correspondingly, its transpose is the unit dyadic for dual bivectors:

$$\overline{\overline{\mathsf{I}}}^{(2)T}|(\boldsymbol{\alpha}\wedge\boldsymbol{\beta}) = \boldsymbol{\alpha}\wedge\boldsymbol{\beta}. \tag{2.28}$$

The trace operation can be defined for dyadics of the space $\overline{\overline{\mathsf{A}}}\in\mathbb{E}_2\mathbb{F}_2$ as

$$\mathrm{tr}\,\overline{\overline{\mathsf{A}}} = \overline{\overline{\mathsf{A}}}||\overline{\overline{\mathsf{I}}}^{(2)T}, \tag{2.29}$$

and as $\overline{\overline{\mathsf{A}}}||\overline{\overline{\mathsf{I}}}^{(2)}$ for the transposed dyadic space, $\overline{\overline{\mathsf{A}}}\in\mathbb{F}_2\mathbb{E}_2$.

Properties of the double-wedge square If the double-wedge square of a dyadic vanishes, $\overline{\overline{\mathsf{A}}}^{(2)}=0$, from (2.25) we see that any two vectors $\mathbf{a}, \mathbf{b}$ are mapped by the dyadic $\overline{\overline{\mathsf{A}}}$ so that their wedge product vanishes: $(\overline{\overline{\mathsf{A}}}|\mathbf{a})\wedge(\overline{\overline{\mathsf{A}}}|\mathbf{b})=0$. This means that all vectors mapped by the dyadic $\overline{\overline{\mathsf{A}}}$ become parallel — that is, scalar multiples of the same vector, say $\mathbf{c}$. Thus, for any vector $\mathbf{a}$ we can write

$$\overline{\overline{\mathsf{A}}}|\mathbf{a} = \lambda(\mathbf{a})\mathbf{c}, \tag{2.30}$$

where $\lambda(\mathbf{a})$ is a linear scalar function of $\mathbf{a}$. Since such a function is of the form $\boldsymbol{\gamma}|\mathbf{a}$ where $\boldsymbol{\gamma}$ is some dual vector, the dyadic with vanishing double-wedge square can be expressed as a single dyad, in the form $\overline{\overline{\mathsf{A}}}=\mathbf{c}\boldsymbol{\gamma}$. Following Gibbs [28], such dyadics will be called linear[2] because, geometrically, a linear dyadic $\mathbf{c}\boldsymbol{\gamma}$ maps all vectors parallel to a line defined by the vector $\mathbf{c}$. The condition $\overline{\overline{\mathsf{A}}}^{(2)} = 0$ can be taken as the definition of the linear dyadic $\overline{\overline{\mathsf{A}}}$.

The following identity is seen to follow as a special case from (2.23):

$$(\overline{\overline{\mathsf{A}}}|\overline{\overline{\mathsf{B}}})^{(2)} = \overline{\overline{\mathsf{A}}}^{(2)}|\overline{\overline{\mathsf{B}}}^{(2)}, \quad \overline{\overline{\mathsf{A}}}, \overline{\overline{\mathsf{B}}} \in \mathbb{E}_1\mathbb{F}_1 \text{ or } \in \mathbb{F}_1\mathbb{E}_1. \tag{2.31}$$

If one of $\overline{\overline{\mathsf{A}}}$, $\overline{\overline{\mathsf{B}}}$ is a linear dyadic, we have $(\overline{\overline{\mathsf{A}}}|\overline{\overline{\mathsf{B}}})^{(2)} = 0$ and, thus, the product is also a linear dyadic. This is also evident from the form $\mathbf{c}\boldsymbol{\gamma}$ of the linear dyadic.

From (1.36) written in the form

$$\begin{aligned}
&\sum_{i,j}(\mathbf{a}_i \wedge \mathbf{b}_j)|(\boldsymbol{\alpha}_i \wedge \boldsymbol{\beta}_j) \\
&= \mathrm{tr}(\sum \mathbf{a}_i\boldsymbol{\alpha}_i {}^\wedge_\wedge \sum \mathbf{b}_j\boldsymbol{\beta}_j) \\
&= (\sum \mathbf{a}_i\boldsymbol{\alpha}_i||\overline{\overline{\mathsf{I}}}^T)(\sum \mathbf{b}_j\boldsymbol{\beta}_j||\overline{\overline{\mathsf{I}}}^T) - (\sum \mathbf{a}_i\boldsymbol{\alpha}_i)||(\sum \boldsymbol{\beta}_j\mathbf{b}_j).
\end{aligned} \tag{2.32}$$

we can write by denoting the series expressions as $\overline{\overline{\mathsf{A}}}$, $\overline{\overline{\mathsf{B}}}$ the identity

$$\mathrm{tr}(\overline{\overline{\mathsf{A}}} {}^\wedge_\wedge \overline{\overline{\mathsf{B}}}) = (\mathrm{tr}\,\overline{\overline{\mathsf{A}}})(\mathrm{tr}\,\overline{\overline{\mathsf{B}}}) - \mathrm{tr}(\overline{\overline{\mathsf{A}}}|\overline{\overline{\mathsf{B}}}). \tag{2.33}$$

This has the special case

$$\mathrm{tr}\,\overline{\overline{\mathsf{A}}}^{(2)} = \frac{1}{2}[(\mathrm{tr}\,\overline{\overline{\mathsf{A}}})^2 - \mathrm{tr}(\overline{\overline{\mathsf{A}}}^2)]. \tag{2.34}$$

Inserting $\overline{\overline{\mathsf{A}}} = \overline{\overline{\mathsf{I}}}$ and $\mathrm{tr}\,\overline{\overline{\mathsf{I}}} = n$, we obtain

$$\mathrm{tr}\,\overline{\overline{\mathsf{I}}}^{(2)} = \frac{1}{2}(n^2 - n) = \frac{n(n-1)}{2} = C_2^n, \tag{2.35}$$

which equals the dimension of the space of bivectors.

2.1.6 Double-wedge cube

If vectors are mapped through the dyadic $\overline{\overline{\mathsf{A}}}$, trivectors are mapped as

$$\begin{aligned}
(\overline{\overline{\mathsf{A}}}|\mathbf{a}) \wedge (\overline{\overline{\mathsf{A}}}|\mathbf{b}) \wedge (\overline{\overline{\mathsf{A}}}|\mathbf{c}) &= [\overline{\overline{\mathsf{A}}}^{(2)}|(\mathbf{a} \wedge \mathbf{b})] \wedge (\overline{\overline{\mathsf{A}}}|\mathbf{c}) \\
&= \overline{\overline{\mathsf{A}}}^{(3)}|(\mathbf{a} \wedge \mathbf{b} \wedge \mathbf{c}).
\end{aligned} \tag{2.36}$$

The double-wedge cube of a dyadic can be defined as

$$\overline{\overline{\mathsf{A}}}^{(3)} = \frac{1}{3!}\overline{\overline{\mathsf{A}}} {}^\wedge_\wedge \overline{\overline{\mathsf{A}}} {}^\wedge_\wedge \overline{\overline{\mathsf{A}}} = \frac{1}{6}\overline{\overline{\mathsf{A}}} {}^\wedge_\wedge \overline{\overline{\mathsf{A}}} {}^\wedge_\wedge \overline{\overline{\mathsf{A}}}. \tag{2.37}$$

[2]Note that the concept 'linear dyadic' introduced here is totally different from that of 'linear mapping'.

As an example, in the three-dimensional space ($n = 3$) any dyadics can be expressed as a sum of three dyads and the double-wedge cube becomes a single trivector dyad:

$$\overline{\overline{\mathsf{A}}} = \mathbf{a}_1\boldsymbol{\alpha}_1 + \mathbf{a}_2\boldsymbol{\alpha}_2 + \mathbf{a}_3\boldsymbol{\alpha}_3, \quad \Rightarrow \quad \overline{\overline{\mathsf{A}}}^{(3)} = (\mathbf{a}_1 \wedge \mathbf{a}_2 \wedge \mathbf{a}_3)(\boldsymbol{\alpha}_1 \wedge \boldsymbol{\alpha}_2 \wedge \boldsymbol{\alpha}_3). \tag{2.38}$$

Identity (2.36) is a special case of

$$\begin{aligned}
&(\overline{\overline{\mathsf{A}}}{}^{\wedge}_{\wedge}\overline{\overline{\mathsf{B}}}{}^{\wedge}_{\wedge}\overline{\overline{\mathsf{C}}})|(\mathbf{a} \wedge \mathbf{b} \wedge \mathbf{c}) = \\
&(\overline{\overline{\mathsf{A}}}|\mathbf{a}) \wedge (\overline{\overline{\mathsf{B}}}|\mathbf{b}) \wedge (\overline{\overline{\mathsf{C}}}|\mathbf{c}) + (\overline{\overline{\mathsf{A}}}|\mathbf{b}) \wedge (\overline{\overline{\mathsf{B}}}|\mathbf{c}) \wedge (\overline{\overline{\mathsf{C}}}|\mathbf{a}) \\
&+(\overline{\overline{\mathsf{A}}}|\mathbf{c}) \wedge (\overline{\overline{\mathsf{B}}}|\mathbf{a}) \wedge (\overline{\overline{\mathsf{C}}}|\mathbf{b}) - (\overline{\overline{\mathsf{A}}}|\mathbf{a}) \wedge (\overline{\overline{\mathsf{B}}}|\mathbf{c}) \wedge (\overline{\overline{\mathsf{C}}}|\mathbf{b}) \\
&-(\overline{\overline{\mathsf{A}}}|\mathbf{c}) \wedge (\overline{\overline{\mathsf{B}}}|\mathbf{b}) \wedge (\overline{\overline{\mathsf{C}}}|\mathbf{a}) - (\overline{\overline{\mathsf{A}}}|\mathbf{b}) \wedge (\overline{\overline{\mathsf{B}}}|\mathbf{a}) \wedge (\overline{\overline{\mathsf{C}}}|\mathbf{c}),
\end{aligned} \tag{2.39}$$

whose derivation is left as an exercise. For a simple check, it can be seen that the right-hand side is invariant in all permutations of the three dyadics, which is also confirmed from the commutation invariance of the double-wedge product on the left-hand side. Setting $\overline{\overline{\mathsf{A}}} = \overline{\overline{\mathsf{I}}}$ in (2.36) yields the special case

$$\overline{\overline{\mathsf{I}}}^{(3)}|(\mathbf{a} \wedge \mathbf{b} \wedge \mathbf{c}) = \mathbf{a} \wedge \mathbf{b} \wedge \mathbf{c}, \tag{2.40}$$

whence $\overline{\overline{\mathsf{I}}}^{(3)}$ serves as the unit dyadic for trivectors.

Identity (2.39) can be derived similarly as (2.20), by using (1.76). Multiplying (2.39) by the dual trivector $(\boldsymbol{\alpha} \wedge \boldsymbol{\beta} \wedge \boldsymbol{\gamma})$ from the right and using the definition of the double-wedge product we can write the left-hand side as $(\overline{\overline{\mathsf{A}}}{}^{\wedge}_{\wedge}\overline{\overline{\mathsf{B}}}{}^{\wedge}_{\wedge}\overline{\overline{\mathsf{C}}})|(\mathbf{a}\boldsymbol{\alpha}{}^{\wedge}_{\wedge}\mathbf{b}\boldsymbol{\beta}{}^{\wedge}_{\wedge}\mathbf{c}\boldsymbol{\gamma})$ and the six terms on the right-hand side accordingly. Since the identity is linear in the dyad $\mathbf{a}\boldsymbol{\alpha}$, it is valid when $\mathbf{a}\boldsymbol{\alpha}$ is replaced by an arbitrary dyadic $\overline{\overline{\mathsf{D}}}$. Similarly, we can replace $\mathbf{b}\boldsymbol{\beta}$ by $\overline{\overline{\mathsf{E}}}$ and $\mathbf{c}\boldsymbol{\gamma}$ by $\overline{\overline{\mathsf{F}}}$ and the identity takes on the following very general form:

$$\begin{aligned}
&(\overline{\overline{\mathsf{A}}}{}^{\wedge}_{\wedge}\overline{\overline{\mathsf{B}}}{}^{\wedge}_{\wedge}\overline{\overline{\mathsf{C}}})|(\overline{\overline{\mathsf{D}}}{}^{\wedge}_{\wedge}\overline{\overline{\mathsf{E}}}{}^{\wedge}_{\wedge}\overline{\overline{\mathsf{F}}}) = \\
&(\overline{\overline{\mathsf{A}}}|\overline{\overline{\mathsf{D}}}){}^{\wedge}_{\wedge}(\overline{\overline{\mathsf{B}}}|\overline{\overline{\mathsf{E}}}){}^{\wedge}_{\wedge}(\overline{\overline{\mathsf{C}}}|\overline{\overline{\mathsf{F}}}) + (\overline{\overline{\mathsf{A}}}|\overline{\overline{\mathsf{E}}}){}^{\wedge}_{\wedge}(\overline{\overline{\mathsf{B}}}|\overline{\overline{\mathsf{F}}}){}^{\wedge}_{\wedge}(\overline{\overline{\mathsf{C}}}|\overline{\overline{\mathsf{D}}}) \\
&+(\overline{\overline{\mathsf{A}}}|\overline{\overline{\mathsf{F}}}){}^{\wedge}_{\wedge}(\overline{\overline{\mathsf{B}}}|\overline{\overline{\mathsf{D}}}){}^{\wedge}_{\wedge}(\overline{\overline{\mathsf{C}}}|\overline{\overline{\mathsf{E}}}) + (\overline{\overline{\mathsf{A}}}|\overline{\overline{\mathsf{D}}}){}^{\wedge}_{\wedge}(\overline{\overline{\mathsf{B}}}|\overline{\overline{\mathsf{F}}}){}^{\wedge}_{\wedge}(\overline{\overline{\mathsf{C}}}|\overline{\overline{\mathsf{E}}}) \\
&+(\overline{\overline{\mathsf{A}}}|\overline{\overline{\mathsf{F}}}){}^{\wedge}_{\wedge}(\overline{\overline{\mathsf{B}}}|\overline{\overline{\mathsf{E}}}){}^{\wedge}_{\wedge}(\overline{\overline{\mathsf{C}}}|\overline{\overline{\mathsf{D}}}) + (\overline{\overline{\mathsf{A}}}|\overline{\overline{\mathsf{E}}}){}^{\wedge}_{\wedge}(\overline{\overline{\mathsf{B}}}|\overline{\overline{\mathsf{D}}}){}^{\wedge}_{\wedge}(\overline{\overline{\mathsf{C}}}|\overline{\overline{\mathsf{F}}}).
\end{aligned} \tag{2.41}$$

Setting $\overline{\overline{\mathsf{A}}} = \overline{\overline{\mathsf{B}}} = \overline{\overline{\mathsf{C}}}$, (2.41) yields the special case

$$\overline{\overline{\mathsf{A}}}^{(3)}|(\overline{\overline{\mathsf{D}}}{}^{\wedge}_{\wedge}\overline{\overline{\mathsf{E}}}{}^{\wedge}_{\wedge}\overline{\overline{\mathsf{F}}}) = (\overline{\overline{\mathsf{A}}}|\overline{\overline{\mathsf{D}}}){}^{\wedge}_{\wedge}(\overline{\overline{\mathsf{A}}}|\overline{\overline{\mathsf{E}}}){}^{\wedge}_{\wedge}(\overline{\overline{\mathsf{A}}}|\overline{\overline{\mathsf{F}}}). \tag{2.42}$$

Setting further $\overline{\overline{\mathsf{D}}} = \overline{\overline{\mathsf{E}}} = \overline{\overline{\mathsf{F}}}$ and denoting this dyadic by $\overline{\overline{\mathsf{B}}}$, we arrive at another special case,

$$(\overline{\overline{\mathsf{A}}}|\overline{\overline{\mathsf{B}}})^{(3)} = \overline{\overline{\mathsf{A}}}^{(3)}|\overline{\overline{\mathsf{B}}}^{(3)}. \tag{2.43}$$

This can be actually continued to any number of dyadics as

$$(\overline{\overline{\mathsf{A}}}_1|\overline{\overline{\mathsf{A}}}_2|\cdots|\overline{\overline{\mathsf{A}}}_q)^{(3)} = \overline{\overline{\mathsf{A}}}_1{}^{(3)}|\overline{\overline{\mathsf{A}}}_2{}^{(3)}|\cdots|\overline{\overline{\mathsf{A}}}_q{}^{(3)}. \tag{2.44}$$

Properties of the double-wedge cube If the double-wedge cube of a $\mathbb{E}_1\mathbb{F}_1$ dyadic $\overline{\overline{\mathsf{A}}}$ vanishes, from (2.36) we have

$$\overline{\overline{\mathsf{A}}}^{(3)} = 0 \quad \Rightarrow \quad (\overline{\overline{\mathsf{A}}}|\mathbf{a}) \wedge (\overline{\overline{\mathsf{A}}}|\mathbf{b}) \wedge (\overline{\overline{\mathsf{A}}}|\mathbf{c}) = 0 \tag{2.45}$$

for any possible vectors $\mathbf{a}, \mathbf{b}, \mathbf{c}$. This means that the three vectors $\overline{\overline{\mathsf{A}}}|\mathbf{a}$, $\overline{\overline{\mathsf{A}}}|\mathbf{b}$ and $\overline{\overline{\mathsf{A}}}|\mathbf{c}$ must be linearly dependent. Obviously in this case the dyadic $\overline{\overline{\mathsf{A}}}$ maps all vectors to a two-dimensional subspace (plane), whence $\overline{\overline{\mathsf{A}}}$ can be called a planar dyadic. If the dimension of the vector space is $n = 2$, all dyadics are planar because they satisfy $\overline{\overline{\mathsf{A}}}^{(3)} = 0$. Because a plane is defined by two vectors, any planar dyadic can be reduced to a sum of two dyads, to the form $\overline{\overline{\mathsf{A}}} = \mathbf{c}\boldsymbol{\gamma} + \mathbf{d}\boldsymbol{\delta}$. Conversely, every dyadic of this form satisfies $\overline{\overline{\mathsf{A}}}^{(3)} = 0$. When $n > 2$, a planar dyadic maps all vectors to the same plane and, thus, does not have an inverse.

The trace of the double-wedge cube can be expanded through (1.76) as

$$\operatorname{tr}\overline{\overline{\mathsf{A}}}^{(3)} = \frac{1}{6}\sum_{i,j,k}(\mathbf{a}_i \wedge \mathbf{a}_j \wedge \mathbf{a}_k)|(\boldsymbol{\alpha}_i \wedge \boldsymbol{\alpha}_j \wedge \boldsymbol{\alpha}_k)$$

$$= \frac{1}{6}\Big[\sum_i(\mathbf{a}_i|\boldsymbol{\alpha}_i)\sum_j(\mathbf{a}_j|\boldsymbol{\alpha}_j)\sum_k(\mathbf{a}_k|\boldsymbol{\alpha}_k) + \sum_{i,j,k}(\mathbf{a}_i|\boldsymbol{\alpha}_j)(\mathbf{a}_j|\boldsymbol{\alpha}_k)(\mathbf{a}_k|\boldsymbol{\alpha}_i)$$

$$+ \sum_{i,j,k}(\mathbf{a}_i|\boldsymbol{\alpha}_k)(\mathbf{a}_j|\boldsymbol{\alpha}_i)(\mathbf{a}_k|\boldsymbol{\alpha}_j) - \sum_{i,j,k}(\mathbf{a}_i|\boldsymbol{\alpha}_i)(\mathbf{a}_j|\boldsymbol{\alpha}_k)(\mathbf{a}_k|\boldsymbol{\alpha}_j)$$

$$- \sum_{i,j,k}(\mathbf{a}_i|\boldsymbol{\alpha}_k)(\mathbf{a}_j|\boldsymbol{\alpha}_j)(\mathbf{a}_k|\boldsymbol{\alpha}_i) - \sum_{i,j,k}(\mathbf{a}_i|\boldsymbol{\alpha}_j)(\mathbf{a}_j|\boldsymbol{\alpha}_i)(\mathbf{a}_k|\boldsymbol{\alpha}_k)\Big]$$

$$= \frac{1}{6}[(\operatorname{tr}\overline{\overline{\mathsf{A}}})^3 + 2\operatorname{tr}(\overline{\overline{\mathsf{A}}}^3) - 3(\operatorname{tr}\overline{\overline{\mathsf{A}}})\operatorname{tr}(\overline{\overline{\mathsf{A}}}^2)]. \tag{2.46}$$

Inserting $\overline{\overline{\mathsf{A}}} = \overline{\overline{\mathsf{I}}}$ and $\operatorname{tr}\overline{\overline{\mathsf{I}}} = n$, we obtain

$$\operatorname{tr}\overline{\overline{\mathsf{I}}}^{(3)} = \frac{1}{6}(n^3 + 2n - 3n^2) = \frac{n(n-1)(n-2)}{3!} = C_3^n, \tag{2.47}$$

which equals the dimension of trivectors.

As an example, in the three-dimensional space ($n = 3$) the double-wedge cube of a dyadic expanded in terms of basis vectors as

$$\overline{\overline{\mathsf{A}}} = \sum_{i=1}^{3}\sum_{j=1}^{3} A_{ij}\mathbf{e}_i\boldsymbol{\varepsilon}_j \tag{2.48}$$

can be shown to take the form

$$\overline{\overline{\mathsf{A}}}^{(3)} = \det[A_{ij}]\mathbf{e}_{123}\boldsymbol{\varepsilon}_{123} = \det[A_{ij}]\overline{\overline{\mathsf{I}}}^{(3)}, \tag{2.49}$$

which is proportional to the determinant of the coordinate matrix of the dyadic. Let us call this quantity determinant of the dyadic $\overline{\overline{\mathsf{A}}}$ and denote it by $\det\overline{\overline{\mathsf{A}}} = \det[A_{ij}]$.

Taking the trace from both sides and noting that for $n = 3$, $\mathrm{tr}\,\overline{\overline{\mathsf{I}}}^{(3)} = 1$, we simply have

$$\det\overline{\overline{\mathsf{A}}} = \mathrm{tr}\,\overline{\overline{\mathsf{A}}}^{(3)}, \quad n = 3. \tag{2.50}$$

Because the right-hand side is independent of the chosen basis, so is the left-hand side and, in particular, $\det[A_{ij}]$ even though the coefficients A_{ij} depend on the basis.

2.1.7 Higher double-wedge powers

The previous concepts can be generalized to the pth double-wedge power of any $\mathbb{E}_1\,\mathbb{F}_1$ dyadic $\overline{\overline{\mathsf{A}}}$, defined as

$$\overline{\overline{\mathsf{A}}}^{(p)} = \left(\sum \mathbf{a}_i\boldsymbol{\alpha}_i\right)^{(p)} = \sum \mathbf{a}_J\boldsymbol{\alpha}_J = \frac{1}{p!}\overline{\overline{\mathsf{A}}}{}^{\wedge}_{\wedge}\overline{\overline{\mathsf{A}}}{}^{\wedge}_{\wedge}\ldots{}^{\wedge}_{\wedge}\overline{\overline{\mathsf{A}}}. \tag{2.51}$$

Here the multi-index J has p indices $J = \{i_1 i_2 ... i_p\}$ taken out of the total of n indices in numerical order. The divisor $p!$ equals the number of different permutations of p indices. The mapping formula for the double-wedge cube (2.36) has the obvious generalization

$$\overline{\overline{\mathsf{A}}}^{(p)}|(\mathbf{a}_1 \wedge \mathbf{a}_2 \wedge \cdots \wedge \mathbf{a}_p) = (\overline{\overline{\mathsf{A}}}|\mathbf{a}_1) \wedge (\overline{\overline{\mathsf{A}}}|\mathbf{a}_2) \wedge \cdots \wedge (\overline{\overline{\mathsf{A}}}|\mathbf{a}_p), \tag{2.52}$$

which again shows us that $\overline{\overline{\mathsf{I}}}^{(p)}$ serves as the unit dyadic for p-vectors.

When p equals the dimension of the space n, there is only one ordered multi-index $J = N = 12\cdots n$. In this case, $\overline{\overline{\mathsf{A}}}^{(n)}$ becomes a multiple of the dyadic product of a n-vector $\mathbf{k}_N$ and a dual n-vector $\boldsymbol{\kappa}_N$. Thus, all dyadics of the form $\overline{\overline{\mathsf{A}}}^{(n)}$ are multiples of each other and, in particular, of the unit dyadic which can be expressed as $\overline{\overline{\mathsf{I}}}^{(n)} = \mathbf{k}_N\boldsymbol{\kappa}_N/(\mathbf{k}_N|\boldsymbol{\kappa}_N)$ or $\overline{\overline{\mathsf{I}}}^{(n)} = \mathbf{e}_N\boldsymbol{\varepsilon}_N$. The determinant of a $\mathbb{E}_1\,\mathbb{F}_1$ dyadic $\overline{\overline{\mathsf{A}}}$ in the space of n dimensions can now be defined as the scalar $\det\overline{\overline{\mathsf{A}}}$ satisfying

$$\overline{\overline{\mathsf{A}}}^{(n)} = \det\overline{\overline{\mathsf{A}}}\,\overline{\overline{\mathsf{I}}}^{(n)}. \tag{2.53}$$

Generalizing (2.47) to

$$\mathrm{tr}\,\overline{\overline{\mathsf{I}}}^{(p)} = C_p^n = \frac{n!}{p!(n-p)!}, \quad p \le n, \tag{2.54}$$

$\mathrm{tr}\,\overline{\overline{\mathsf{I}}}^{(p)}$ gives the dimension of the space of p-vectors in an n-dimensional vector space. Inserting $\mathrm{tr}\,\overline{\overline{\mathsf{I}}}^{(n)} = 1$ in (2.53), the rule (2.50) corresponding to $n = 3$ can be generalized to

$$\det\overline{\overline{\mathsf{A}}} = \mathrm{tr}\,\overline{\overline{\mathsf{A}}}^{(n)}. \tag{2.55}$$

2.1.8 Double-incomplete duality product

The incomplete duality product $\rfloor$ between a vector and a bivector, defined through (1.48), generates the corresponding double-incomplete product $\rfloor\rfloor$ between dyadics by writing

$$(\mathbf{a}\rfloor\boldsymbol{\Phi})(\boldsymbol{\alpha}\rfloor\mathbf{A}) = (\mathbf{a}\boldsymbol{\alpha})\rfloor\rfloor(\boldsymbol{\Phi}\mathbf{A}), \tag{2.56}$$

where $\mathbf{a}$ and $\boldsymbol{\alpha}$ denote a vector and a dual vector while $\mathbf{A}$ and $\boldsymbol{\Phi}$ denote a bivector and a dual bivector. Another way to define the double-incomplete duality product is to consider four $\mathbb{E}_1\mathbb{F}_1$ dyadics $\overline{\overline{\mathsf{A}}}, \overline{\overline{\mathsf{B}}}, \overline{\overline{\mathsf{C}}}$ and $\overline{\overline{\mathsf{D}}}$ and the scalar $(\overline{\overline{\mathsf{A}}}{}^\wedge_\wedge\overline{\overline{\mathsf{B}}})||(\overline{\overline{\mathsf{C}}}{}^\wedge_\wedge\overline{\overline{\mathsf{D}}})^T$ as a linear function of the dyadic $\overline{\overline{\mathsf{A}}}$. Expressing it in the form

$$(\overline{\overline{\mathsf{A}}}{}^\wedge_\wedge\overline{\overline{\mathsf{B}}})||(\overline{\overline{\mathsf{C}}}{}^\wedge_\wedge\overline{\overline{\mathsf{D}}})^T = \overline{\overline{\mathsf{A}}}||(\overline{\overline{\mathsf{B}}}\rfloor\rfloor(\overline{\overline{\mathsf{C}}}{}^\wedge_\wedge\overline{\overline{\mathsf{D}}})^T), \tag{2.57}$$

it defines $\overline{\overline{\mathsf{B}}}\rfloor\rfloor(\overline{\overline{\mathsf{C}}}{}^\wedge_\wedge\overline{\overline{\mathsf{D}}})^T$ as a $\mathbb{F}_1\mathbb{E}_1$ dyadic. Similarly, the other product $\lfloor\lfloor$ is defined from

$$(\overline{\overline{\mathsf{C}}}{}^\wedge_\wedge\overline{\overline{\mathsf{D}}})^T||(\overline{\overline{\mathsf{A}}}{}^\wedge_\wedge\overline{\overline{\mathsf{B}}}) = ((\overline{\overline{\mathsf{C}}}{}^\wedge_\wedge\overline{\overline{\mathsf{D}}})^T\lfloor\lfloor\overline{\overline{\mathsf{B}}})||\overline{\overline{\mathsf{A}}}. \tag{2.58}$$

Because these expressions give the same scalar, we obtain the commutation relation

$$\overline{\overline{\mathsf{B}}}\rfloor\rfloor(\overline{\overline{\mathsf{C}}}{}^\wedge_\wedge\overline{\overline{\mathsf{D}}})^T = (\overline{\overline{\mathsf{C}}}{}^\wedge_\wedge\overline{\overline{\mathsf{D}}})^T\lfloor\lfloor\overline{\overline{\mathsf{B}}}. \tag{2.59}$$

Problems

2.1.1 Derive (2.20) using (2.14) by replacing $\overline{\overline{\mathsf{A}}}$ in (2.14) by $\overline{\overline{\mathsf{A}}} + \overline{\overline{\mathsf{B}}}$.

2.1.2 Using a similar technique as when deriving (2.33), derive the dyadic identity

$$\begin{aligned}\mathrm{tr}(\overline{\overline{\mathsf{A}}}{}^\wedge_\wedge\overline{\overline{\mathsf{B}}}{}^\wedge_\wedge\overline{\overline{\mathsf{C}}}) &= \mathrm{tr}\,\overline{\overline{\mathsf{A}}}\,\mathrm{tr}\,\overline{\overline{\mathsf{B}}}\,\mathrm{tr}\,\overline{\overline{\mathsf{C}}} + \mathrm{tr}(\overline{\overline{\mathsf{A}}}|\overline{\overline{\mathsf{B}}}|\overline{\overline{\mathsf{C}}}) + \mathrm{tr}(\overline{\overline{\mathsf{A}}}|\overline{\overline{\mathsf{C}}}|\overline{\overline{\mathsf{B}}}) \\ &\quad -\mathrm{tr}\,\overline{\overline{\mathsf{A}}}\,\mathrm{tr}(\overline{\overline{\mathsf{B}}}|\overline{\overline{\mathsf{C}}}) - \mathrm{tr}\,\overline{\overline{\mathsf{B}}}\,\mathrm{tr}(\overline{\overline{\mathsf{C}}}|\overline{\overline{\mathsf{A}}}) - \mathrm{tr}\,\overline{\overline{\mathsf{C}}}\,\mathrm{tr}(\overline{\overline{\mathsf{A}}}|\overline{\overline{\mathsf{B}}})\end{aligned}$$

where $\overline{\overline{\mathsf{A}}}, \overline{\overline{\mathsf{B}}}, \overline{\overline{\mathsf{C}}}$ are three dyadics in the space $\mathbb{E}_1\mathbb{F}_1$.

2.1.3 Using (2.23), derive the identity

$$(\overline{\overline{\mathsf{A}}}|\overline{\overline{\mathsf{B}}}){}^\wedge_\wedge(\overline{\overline{\mathsf{B}}}|\overline{\overline{\mathsf{A}}}) = (\overline{\overline{\mathsf{A}}}{}^\wedge_\wedge\overline{\overline{\mathsf{B}}})^2 - \overline{\overline{\mathsf{A}}}^2{}^\wedge_\wedge\overline{\overline{\mathsf{B}}}^2$$

2.1.4 Derive the identity (2.39).

2.1.5 Derive (2.41) in detail.

2.1.6 Show that the double-wedge square of a planar dyadic is a linear dyadic.

2.1.7 Consider the eigenproblem

$$\overline{\overline{\mathsf{A}}}|\mathbf{a} = \lambda\mathbf{a} \quad \rightarrow \quad (\overline{\overline{\mathsf{A}}} - \lambda\overline{\overline{\mathsf{I}}})|\mathbf{a} = 0,$$

when $\overline{\overline{\mathsf{A}}}$ is a $\mathbb{E}_1\mathbb{F}_1$-dyadic. Show that the equation for the eigenvalues λ can be expressed as

$$(\overline{\overline{\mathsf{A}}} - \lambda\overline{\overline{\mathsf{I}}})^{(n)} = \sum_{i=0}^{n}(-\lambda)^i\mathrm{tr}\,\overline{\overline{\mathsf{A}}}^{(n-i)} = 0,$$

when n is the dimension of the vector space.

2.1.8 Using the identity of problem 2.1.2 above, derive the identity

$$\mathrm{tr}(\overline{\overline{\mathsf{I}}}{}^{\wedge}_{\wedge}\overline{\overline{\mathsf{A}}}^{(2)}) = (n-2)\mathrm{tr}\,\overline{\overline{\mathsf{A}}}^{(2)},$$

where $\overline{\overline{\mathsf{A}}} \in \mathbb{E}_1\mathbb{F}_1$, and check the special case $\overline{\overline{\mathsf{A}}} = \overline{\overline{\mathsf{I}}}$.

2.1.9 Derive the Cayley–Hamilton equation in two dimensions:

$$\overline{\overline{\mathsf{A}}}^2 - (\mathrm{tr}\,\overline{\overline{\mathsf{A}}})\overline{\overline{\mathsf{A}}} + (\det\overline{\overline{\mathsf{A}}})\overline{\overline{\mathsf{I}}} = 0.$$

Hint: Multiplying by $\overline{\overline{\mathsf{A}}}^{-1}$, we have the equivalent equation

$$(\det\overline{\overline{\mathsf{A}}})\overline{\overline{\mathsf{A}}}^{-1} = (\mathrm{tr}\,\overline{\overline{\mathsf{A}}})\overline{\overline{\mathsf{I}}} - \overline{\overline{\mathsf{A}}},$$

with $\det\overline{\overline{\mathsf{A}}} = \mathrm{tr}[\overline{\overline{\mathsf{A}}}^{(2)}]$. Prove this by applying the expression of the inverse of a two-dimensional dyadic.

2.1.10 Using the identity of problem 2.1.2, derive the identity

$$\mathrm{tr}(\overline{\overline{\mathsf{I}}}{}^{\wedge}_{\wedge}\overline{\overline{\mathsf{A}}}^{(2)}) = (n-2)\mathrm{tr}\,\overline{\overline{\mathsf{A}}}^{(2)},$$

where $\overline{\overline{\mathsf{A}}} \in \mathbb{E}_1\mathbb{F}_1$, and make a check by setting $\overline{\overline{\mathsf{A}}} = \overline{\overline{\mathsf{I}}}$.

2.1.11 Prove the identity

$$\mathrm{tr}(\overline{\overline{\mathsf{I}}}^T\rfloor\rfloor\overline{\overline{\mathsf{A}}}^{(2)}) = 2\mathrm{tr}\,\overline{\overline{\mathsf{A}}}^{(2)}.$$

2.1.12 Prove the identity

$$\mathrm{tr}(\overline{\overline{\mathsf{A}}}^T\rfloor\rfloor(\overline{\overline{\mathsf{B}}}{}^{\wedge}_{\wedge}\overline{\overline{\mathsf{C}}})) = (\mathrm{tr}\,\overline{\overline{\mathsf{A}}})\mathrm{tr}(\overline{\overline{\mathsf{B}}}{}^{\wedge}_{\wedge}\overline{\overline{\mathsf{C}}}) - \mathrm{tr}(\overline{\overline{\mathsf{A}}}{}^{\wedge}_{\wedge}\overline{\overline{\mathsf{B}}}{}^{\wedge}_{\wedge}\overline{\overline{\mathsf{C}}}).$$

2.1.13 Prove the identity $\mathrm{tr}(\overline{\overline{\mathsf{A}}}{}^{\wedge}_{\wedge}\overline{\overline{\mathsf{I}}}) = (n-1)\mathrm{tr}\,\overline{\overline{\mathsf{A}}}$.

2.2 DYADIC IDENTITIES

Effective use of dyadics in Gibbsian vector algebra requires a number of basic identities, some of which have been presented above. Their use makes expansions in coordinates obsolete and, thus, tends to shorten the analysis. To understand this one should imagine using coordinate expansions instead of the convenient "bac cab formula" every time a Gibbsian vector expression of the type $\mathbf{a}\times(\mathbf{b}\times\mathbf{c})$ appears. The number of useful dyadic identites in multivector algebra is obviously larger than that in the Gibbsian vector algebra, but still manageable. In the following, derivation of some identities is demonstrated. More identities are given without proof in the Appendix A and can be derived by the reader as an exercise.

A dyadic identity can be written in the form $F(\overline{\overline{\mathsf{A}}}, \overline{\overline{\mathsf{B}}}, \cdots) = 0$, where F is a function of dyadic arguments. To be an identity, the equation must be valid for any argument dyadics. A dyadic identity is multilinear if the function F is linear in each of its dyadic arguments. An example of a bilinear dyadic identity is $\overline{\overline{\mathsf{A}}}{}^\wedge_\wedge\overline{\overline{\mathsf{B}}} - \overline{\overline{\mathsf{B}}}{}^\wedge_\wedge\overline{\overline{\mathsf{A}}} = 0$. As other examples, (2.23) is a four-linear and (2.41) a six-linear identity. In deriving new dyadic identities the following property appears useful:

> *If a multilinear identity is valid for arbitrary linear dyadics, it is also valid for arbitrary dyadics.*

Thus, if a multilinear identity is valid in the form[3] $F(..., \mathbf{a}\boldsymbol{\alpha}, ...) = 0$ so that its jth dyadic argument is a linear dyadic $\mathbf{a}\boldsymbol{\alpha}$, it is also valid in the form $F(..., \overline{\overline{\mathsf{A}}}, ...) = 0$ for any dyadic $\overline{\overline{\mathsf{A}}}$, element of the space $\mathbb{E}_1\,\mathbb{F}_1$, as its jth argument. To prove this, it is sufficient to write $\overline{\overline{\mathsf{A}}} = \sum \mathbf{a}_i\boldsymbol{\alpha}_i$ and note that from linearity we have

$$F(..., \overline{\overline{\mathsf{A}}}, ...) = F(..., \sum \mathbf{a}_i\boldsymbol{\alpha}_i, ...) = \sum_i F(..., \mathbf{a}_i\boldsymbol{\alpha}_i, ...) = 0. \tag{2.60}$$

Since the index j is arbitrary, the same is valid with respect to all arguments. Thus, a dyadic identity can be derived from a multivector identity by grouping multivectors to form linear dyadics. An example of this was given already in forming the identity (2.33).

We can also derive dyadic identities from vector identities of the form $\overline{\overline{\mathsf{F}}}|\mathbf{a} = 0$. Here $\overline{\overline{\mathsf{F}}}$ is a dyadic expression which maps a multivector $\mathbf{a}$ to a space of multivectors of the same dimension as $\mathbf{a}$. If $\overline{\overline{\mathsf{F}}}|\mathbf{a} = 0$ is valid for any $\mathbf{a}$ of the space in question, a more general identity arises in the form $\overline{\overline{\mathsf{F}}} = 0$. As a consequence of this, if for any vector $\mathbf{a} \in \mathbb{E}_1$ and any dual vector $\boldsymbol{\alpha} \in \mathbb{F}_1$ we have $\overline{\overline{\mathsf{F}}}||\boldsymbol{\alpha}\mathbf{a} = 0$, where $\overline{\overline{\mathsf{F}}} \in \mathbb{E}_1\,\mathbb{F}_1$ is a dyadic expression, we have $\overline{\overline{\mathsf{F}}} = 0$. As a further consequence, the same is true if $\overline{\overline{\mathsf{F}}}||\overline{\overline{\mathsf{A}}}^T = 0$ for any $\overline{\overline{\mathsf{A}}} \in \mathbb{E}_1\,\mathbb{F}_1$. This gives us a possibility to transform scalar identities with dyadic arguments to dyadic identites with dyadic arguments. It is, of course, of interest to find as general identities as possible.

Example As an example of the last procedure, let us generalize the scalar-valued identity (2.33). For that we use the definition of the double-incomplete duality product $\rfloor\rfloor$ in (2.57) by first writing

$$\mathrm{tr}(\overline{\overline{\mathsf{A}}}{}^\wedge_\wedge\overline{\overline{\mathsf{B}}}) = (\overline{\overline{\mathsf{A}}}{}^\wedge_\wedge\overline{\overline{\mathsf{B}}})||\overline{\overline{\mathsf{I}}}^{(2)T} = \overline{\overline{\mathsf{B}}}||(\overline{\overline{\mathsf{A}}}\rfloor\rfloor\overline{\overline{\mathsf{I}}}^{(2)T}). \tag{2.61}$$

Equating this with the right-hand side of (2.33) we obtain

$$\overline{\overline{\mathsf{B}}}||(\overline{\overline{\mathsf{A}}}\rfloor\rfloor\overline{\overline{\mathsf{I}}}^{(2)T}) = \overline{\overline{\mathsf{B}}}||\overline{\overline{\mathsf{I}}}^T(\mathrm{tr}\,\overline{\overline{\mathsf{A}}}) - \overline{\overline{\mathsf{B}}}||\overline{\overline{\mathsf{A}}}^T. \tag{2.62}$$

Since this is valid for any dyadic $\overline{\overline{\mathsf{B}}}$, the following identity is valid for any dyadic $\overline{\overline{\mathsf{A}}}$:

$$\overline{\overline{\mathsf{A}}}\rfloor\rfloor\overline{\overline{\mathsf{I}}}^{(2)T} = \overline{\overline{\mathsf{I}}}^T(\mathrm{tr}\,\overline{\overline{\mathsf{A}}}) - \overline{\overline{\mathsf{A}}}^T. \tag{2.63}$$

[3]Here the linear dyadic is assumed to be of the form $\mathbf{a}\boldsymbol{\alpha}$. However, the method works equally well for linear dyadics of the form $\boldsymbol{\alpha}\mathbf{a}$, $\mathbf{ab}$, or $\boldsymbol{\alpha\beta}$. Also, $\mathbf{a}$ may be any multivector and $\boldsymbol{\alpha}$ any dual multivector.

This corresponds to the Gibbsian dyadic identity [40]

$$\overline{\overline{\mathsf{A}}}{}^{\times}_{\times}\overline{\overline{\mathsf{I}}} = \overline{\overline{\mathsf{I}}}(\mathrm{tr}\,\overline{\overline{\mathsf{A}}}) - \overline{\overline{\mathsf{A}}}^T. \tag{2.64}$$

2.2.1 Gibbs' identity in three dimensions

In the classical three-dimensional vector algebra one encounters the four-linear vector identity

$$\mathbf{a}(\mathbf{b}\times\mathbf{c}\cdot\mathbf{d}) + \mathbf{b}(\mathbf{c}\times\mathbf{a}\cdot\mathbf{d}) + \mathbf{c}(\mathbf{a}\times\mathbf{b}\cdot\mathbf{d}) = \mathbf{d}(\mathbf{a}\times\mathbf{b}\cdot\mathbf{c}), \tag{2.65}$$

known as Gibbs' identity [28, 40]. This can be reformulated for three-dimensional multivectors in the following dyadic form (a proof in a more general n-dimensional form is given in the subsequent section):

$$\mathbf{a}(\mathbf{b}\wedge\mathbf{c}\wedge\mathbf{d}) + \mathbf{b}(\mathbf{c}\wedge\mathbf{a}\wedge\mathbf{d}) + \mathbf{c}(\mathbf{a}\wedge\mathbf{b}\wedge\mathbf{d}) = \mathbf{d}(\mathbf{a}\wedge\mathbf{b}\wedge\mathbf{c}). \tag{2.66}$$

Because each term of (2.66) is a $\mathbb{E}_1\,\mathbb{E}_3$ dyadic, the identity can be reduced to vector form by multiplying from the right by an arbitrary dual trivector $\boldsymbol{\kappa}_N \neq 0$ ($N = 123$):

$$\mathbf{a}(\mathbf{b}\wedge\mathbf{c}\wedge\mathbf{d})|\boldsymbol{\kappa}_N + \mathbf{b}(\mathbf{c}\wedge\mathbf{a}\wedge\mathbf{d})|\boldsymbol{\kappa}_N + \mathbf{c}(\mathbf{a}\wedge\mathbf{b}\wedge\mathbf{d})|\boldsymbol{\kappa}_N = \mathbf{d}(\mathbf{a}\wedge\mathbf{b}\wedge\mathbf{c})|\boldsymbol{\kappa}_N. \tag{2.67}$$

Linear dependence The last form of Gibbs' identity (2.67) is very useful in defining linear relations between vectors in the three-dimensional space. Let us look at different cases.

1. If $\mathbf{a}\wedge\mathbf{b} = 0$, (2.67) reduces to an equation of the form $\alpha\mathbf{a} + \beta\mathbf{b} = 0$ with scalar numbers α and β. Thus, in this case $\mathbf{a}$ and $\mathbf{b}$ are linearly related (parallel vectors).

2. If $\mathbf{a}\wedge\mathbf{b}\wedge\mathbf{c} = 0$, (2.67) reduces to an equation of the form $\alpha\mathbf{a} + \beta\mathbf{b} + \gamma\mathbf{c} = 0$ with scalar numbers α, β and γ. Thus, in this case $\mathbf{a}$, $\mathbf{b}$, and $\mathbf{c}$ are linearly related (coplanar) vectors.

3. If $\boldsymbol{\kappa}_N|(\mathbf{a}\wedge\mathbf{b}\wedge\mathbf{c}) \neq 0$, any given vector $\mathbf{d}$ can be expressed in the form $\alpha\mathbf{a} + \beta\mathbf{b} + \gamma\mathbf{c}$.

Actually, in the third case the three-dimensional vectors $\mathbf{a}, \mathbf{b}, \mathbf{c}$ form a basis and for any dual trivector $\boldsymbol{\kappa}_N$ we can express any given vector $\mathbf{d}$ in the form

$$\mathbf{d} = \mathbf{a}\frac{(\mathbf{b}\wedge\mathbf{c}\wedge\mathbf{d})|\boldsymbol{\kappa}_N}{(\mathbf{a}\wedge\mathbf{b}\wedge\mathbf{c})|\boldsymbol{\kappa}_N} + \mathbf{b}\frac{(\mathbf{c}\wedge\mathbf{a}\wedge\mathbf{d})|\boldsymbol{\kappa}_N}{(\mathbf{a}\wedge\mathbf{b}\wedge\mathbf{c})|\boldsymbol{\kappa}_N} + \mathbf{c}\frac{(\mathbf{a}\wedge\mathbf{b}\wedge\mathbf{d})|\boldsymbol{\kappa}_N}{(\mathbf{a}\wedge\mathbf{b}\wedge\mathbf{c})|\boldsymbol{\kappa}_N}. \tag{2.68}$$

Unit dyadic Another form of the identity (2.66) is

$$[\mathbf{a}(\mathbf{b}\wedge\mathbf{c}) + \mathbf{b}(\mathbf{c}\wedge\mathbf{a}) + \mathbf{c}(\mathbf{a}\wedge\mathbf{b})]\wedge\mathbf{d} = \mathbf{d}(\mathbf{a}\wedge\mathbf{b}\wedge\mathbf{c}). \tag{2.69}$$

The dyadic in square brackets belongs to the space $\mathbb{E}_1\,\mathbb{E}_2$, and it obviously vanishes if $\mathbf{a}\wedge\mathbf{b}\wedge\mathbf{c} = 0$, in which case $\mathbf{a}, \mathbf{b}, \mathbf{c}$ are linearly dependent. Because we can write

$(\mathbf{u}\wedge\mathbf{v}\wedge\mathbf{w})|\boldsymbol{\kappa}_N = ((\mathbf{u}\wedge\mathbf{v})\rfloor\boldsymbol{\kappa}_N)|\mathbf{w}$ for any dual trivector $\boldsymbol{\kappa}_N \neq 0$ and vectors $\mathbf{u},\mathbf{v},\mathbf{w}$, multiplying (2.69) by $|\boldsymbol{\kappa}_N$ it can be further written as

$$\{(\mathbf{a}(\mathbf{b}\wedge\mathbf{c}) + \mathbf{b}(\mathbf{c}\wedge\mathbf{a}) + \mathbf{c}(\mathbf{a}\wedge\mathbf{b}))\rfloor\boldsymbol{\kappa}_N\}|\mathbf{d} = ((\mathbf{a}\wedge\mathbf{b}\wedge\mathbf{c})|\boldsymbol{\kappa}_N)\bar{\bar{\mathsf{I}}}|\mathbf{d}. \tag{2.70}$$

Because this is valid for any $\mathbf{d}$, the dyadics in front must be the same. This leads to the identity

$$(\mathbf{a}(\mathbf{b}\wedge\mathbf{c}) + \mathbf{b}(\mathbf{c}\wedge\mathbf{a}) + \mathbf{c}(\mathbf{a}\wedge\mathbf{b}))\rfloor\boldsymbol{\kappa}_N = ((\mathbf{a}\wedge\mathbf{b}\wedge\mathbf{c})|\boldsymbol{\kappa}_N)\bar{\bar{\mathsf{I}}}. \tag{2.71}$$

Assuming the vectors $\mathbf{a},\mathbf{b},\mathbf{c}$ form a basis, we have $(\mathbf{a}\wedge\mathbf{b}\wedge\mathbf{c})|\boldsymbol{\kappa}_N \neq 0$, and an expression for the unit dyadic in the three-dimensional space can be written in terms of a given vector basis $(\mathbf{a},\mathbf{b},\mathbf{c})$ and a any dual trivector $\boldsymbol{\kappa}_N$ as

$$\bar{\bar{\mathsf{I}}} = \mathbf{a}\frac{(\mathbf{b}\wedge\mathbf{c})\rfloor\boldsymbol{\kappa}_N}{(\mathbf{a}\wedge\mathbf{b}\wedge\mathbf{c})|\boldsymbol{\kappa}_N} + \mathbf{b}\frac{(\mathbf{c}\wedge\mathbf{a})\rfloor\boldsymbol{\kappa}_N}{(\mathbf{a}\wedge\mathbf{b}\wedge\mathbf{c})|\boldsymbol{\kappa}_N} + \mathbf{c}\frac{(\mathbf{a}\wedge\mathbf{b})\rfloor\boldsymbol{\kappa}_N}{(\mathbf{a}\wedge\mathbf{b}\wedge\mathbf{c})|\boldsymbol{\kappa}_N}. \tag{2.72}$$

Because the unit dyadic $\bar{\bar{\mathsf{I}}}$ can be expressed in terms of a vector basis and its reciprocal dual-vector basis in the form (1.11), (2.72) defines the reciprocal of the three-dimensional basis of vectors $\mathbf{a},\mathbf{b},\mathbf{c}$ in terms of an arbitrary dual trivector $\boldsymbol{\kappa}_N$.

2.2.2 Gibbs' identity in n dimensions

Gibbs' identity (2.66) can be straighforwardly extended to the space of n dimensions. It involves a set of n linearly independent vectors $\mathbf{a}_1,\mathbf{a}_2\cdots\mathbf{a}_n$ satisfying $\mathbf{a}_N = \mathbf{a}_1\wedge\mathbf{a}_2\wedge\cdots\wedge\mathbf{a}_n \neq 0$. Let us form a set of $(n-1)$-vectors complementary to $\{\mathbf{a}_i\}$ as

$$\mathbf{a}_{K(i)} = \mathbf{a}_1\wedge\mathbf{a}_2\wedge\cdots\wedge\mathbf{a}_{i-1}\wedge\mathbf{a}_{i+1}\wedge\cdots\wedge\mathbf{a}_n, \quad 1\le i\le n. \tag{2.73}$$

They satisfy $\mathbf{a}_{K(i)} \neq 0$ because of $\mathbf{a}_N \neq 0$. Here, $K(i)$ is the $(n-1)$-index complementary to i,

$$K(i) = 12..(i-1)(i+1)..n. \tag{2.74}$$

The following two orthogonality conditions are obviously valid:

$$\mathbf{a}_j\wedge\mathbf{a}_{K(i)} = (-1)^{i-1}\mathbf{a}_N\delta_{ij}, \quad \mathbf{a}_{K(i)}\wedge\mathbf{a}_j = (-1)^{n-i}\mathbf{a}_N\delta_{ij}. \tag{2.75}$$

Now let us consider the linear combination

$$\mathbf{d} = \sum_j d_j\mathbf{a}_j, \tag{2.76}$$

where the d_j are scalar coefficients. Applying the above orthogonality, we can write

$$(-1)^{n-i}\mathbf{a}_{K(i)}\wedge\mathbf{d} = (-1)^{n-i}d_i\mathbf{a}_{K(i)}\wedge\mathbf{a}_i = d_i\mathbf{a}_N. \tag{2.77}$$

Multiplying both sides dyadically by $\mathbf{a}_i$ and summing $i = 1\cdots n$ gives us finally the dyadic form for Gibbs' identity in n dimensions as

$$\sum_i(-1)^{n-i}\mathbf{a}_i(\mathbf{a}_{K(i)}\wedge\mathbf{d}) = \sum_i d_i\mathbf{a}_i\mathbf{a}_N = \mathbf{d}\mathbf{a}_N. \tag{2.78}$$

Writing the sum expression out explicitly, Gibbs' identity can be expressed in the following easily memorizable form:

$$\begin{aligned}\mathbf{d}(\mathbf{a}_1 \wedge \mathbf{a}_2 \wedge \cdots \wedge \mathbf{a}_n) &= \mathbf{a}_1(\mathbf{d} \wedge \mathbf{a}_2 \wedge \mathbf{a}_3 \wedge \cdots \wedge \mathbf{a}_n) \\ &+\mathbf{a}_2(\mathbf{a}_1 \wedge \mathbf{d} \wedge \mathbf{a}_3 \wedge \cdots \wedge \mathbf{a}_n) \\ &+ \cdots + \mathbf{a}_n(\mathbf{a}_1 \wedge \mathbf{a}_2 \wedge \cdots \wedge \mathbf{a}_{n-1} \wedge \mathbf{d}). \quad (2.79)\end{aligned}$$

For $n = 3$, (2.79) gives (2.66) as a special case.

Again, multiplying from the right by an arbitrary dual n-vector $\boldsymbol{\kappa}_N \neq 0$, we obtain a vector identity which allows us to expand any given vector $\mathbf{d}$ in terms of n vectors $\{\mathbf{a}_i\}$ satisfying $\mathbf{a}_N \neq 0$:

$$\begin{aligned}\mathbf{d} &= \mathbf{a}_1 \frac{(\mathbf{d} \wedge \mathbf{a}_2 \wedge \mathbf{a}_3 \wedge \cdots \wedge \mathbf{a}_n)|\boldsymbol{\kappa}_N}{\mathbf{a}_N|\boldsymbol{\kappa}_N} + \mathbf{a}_2 \frac{(\mathbf{a}_1 \wedge \mathbf{d} \wedge \mathbf{a}_3 \wedge \cdots \wedge \mathbf{a}_n)|\boldsymbol{\kappa}_N}{\mathbf{a}_N|\boldsymbol{\kappa}_N} \\ &+ \cdots + \mathbf{a}_n \frac{(\mathbf{a}_1 \wedge \mathbf{a}_2 \wedge \cdots \wedge \mathbf{a}_{n-1} \wedge \mathbf{d})|\boldsymbol{\kappa}_N}{\mathbf{a}_N|\boldsymbol{\kappa}_N} \\ &= \sum_i (-1)^{n-i} \mathbf{a}_i \frac{(\mathbf{a}_{K(i)} \wedge \mathbf{d})|\boldsymbol{\kappa}_N}{\mathbf{a}_N|\boldsymbol{\kappa}_N}. \quad (2.80)\end{aligned}$$

This allows us to express the unit dyadic in n dimensions as

$$\overline{\overline{\mathsf{I}}} = \sum_i (-1)^{n-i} \mathbf{a}_i \frac{\boldsymbol{\kappa}_N \lfloor \mathbf{a}_{K(i)}}{\mathbf{a}_N|\boldsymbol{\kappa}_N}. \quad (2.81)$$

Gibbs' identity in the space of n dimensions also allows us to define the reciprocal basis for a given basis of n vectors $\{\mathbf{a}_i\}$. This will be postponed to Section 2.4.

2.2.3 Constructing identities

Let us finish this Section by considering four examples elucidating different methods of constructing identities between dyadics and multivectors.

Example 1 Writing the "bac cab rule" (1.50) in two forms as

$$\boldsymbol{\alpha}\rfloor(\mathbf{b} \wedge \mathbf{c}) = \boldsymbol{\alpha}|\overline{\overline{\mathsf{I}}}\rfloor(\mathbf{b} \wedge \mathbf{c}) = \boldsymbol{\alpha}|(\mathbf{cb} - \mathbf{bc}), \quad (2.82)$$

$$(\mathbf{bc} - \mathbf{cb})|\boldsymbol{\alpha} = -(\mathbf{b} \wedge \mathbf{c})\lfloor\boldsymbol{\alpha} = -(\mathbf{b} \wedge \mathbf{c})\lfloor\overline{\overline{\mathsf{I}}}^T|\boldsymbol{\alpha}, \quad (2.83)$$

which are valid for any dual vector $\boldsymbol{\alpha}$, we obtain the combined dyadic identity

$$\overline{\overline{\mathsf{I}}}\rfloor(\mathbf{b} \wedge \mathbf{c}) = (\mathbf{b} \wedge \mathbf{c})\lfloor\overline{\overline{\mathsf{I}}}^T = \mathbf{cb} - \mathbf{bc}. \quad (2.84)$$

This resembles the Gibbsian dyadic formula [40]

$$\overline{\overline{\mathsf{I}}} \times (\mathbf{b} \times \mathbf{c}) = (\mathbf{b} \times \mathbf{c}) \times \overline{\overline{\mathsf{I}}} = \mathbf{cb} - \mathbf{bc}. \quad (2.85)$$

In terms of (2.84) we can show that any antisymmetric dyadic $\overline{\overline{\mathsf{A}}}$ of the space $\mathbb{E}_1\mathbb{E}_1$ can be expressed as

$$\overline{\overline{\mathsf{A}}} = \overline{\overline{\mathsf{I}}}\rfloor\mathbf{A} = \mathbf{A}\lfloor\overline{\overline{\mathsf{I}}}^T, \tag{2.86}$$

in terms of some bivector $\mathbf{A}$. This, again, has the Gibbsian counterpart that any antisymmetric dyadic can be expressed in terms of a vector $\mathbf{a}$ as $\mathbf{a}\times\overline{\overline{\mathsf{I}}}$.

Expressing a given dyadic $\overline{\overline{\mathsf{A}}} = \sum \mathbf{b}_i\mathbf{c}_i$, which is known to be antisymmetric, in the form

$$\overline{\overline{\mathsf{A}}} = \frac{1}{2}(\overline{\overline{\mathsf{A}}} - \overline{\overline{\mathsf{A}}}^T) = \frac{1}{2}\sum(\mathbf{b}_i\mathbf{c}_i - \mathbf{c}_i\mathbf{b}_i) \tag{2.87}$$

and using (2.84), we can identify the bivector $\mathbf{A}$ from the expansion of the dyadic as

$$\mathbf{A} = \frac{1}{2}\sum(\mathbf{b}_i\wedge\mathbf{c}_i - \mathbf{c}_i\wedge\mathbf{b}_i) = \sum \mathbf{b}_i\wedge\mathbf{c}_i. \tag{2.88}$$

Thus, the bivector of a given antisymmetric dyadic $\in \mathbb{E}_1\mathbb{E}_1$ is found by inserting the wedge product sign between the vectors in the dyadic products. The expansion (2.86) for the general antisymmetric dyadic is valid in the space of n dimensions.

Example 2 Starting again from the 'bac cab rule' (1.50) and its dual we combine them to the form of a dyadic rule

$$\begin{aligned}&(\boldsymbol{\alpha}\rfloor(\mathbf{b}\wedge\mathbf{c}))(\mathbf{a}\rfloor(\boldsymbol{\beta}\wedge\boldsymbol{\gamma}))\\ &= \ (\mathbf{b}(\boldsymbol{\alpha}|\mathbf{c}) - \mathbf{c}(\boldsymbol{\alpha}|\mathbf{b}))\,(\boldsymbol{\beta}(\mathbf{a}|\boldsymbol{\gamma}) - \boldsymbol{\gamma}(\mathbf{a}|\boldsymbol{\beta})).\end{aligned} \tag{2.89}$$

Expressing both sides in terms of double products we have

$$\begin{aligned}&\boldsymbol{\alpha}\mathbf{a}\rfloor\rfloor(\mathbf{b}\boldsymbol{\beta}{}^\wedge_\wedge\mathbf{c}\boldsymbol{\gamma})\\ &= \ \mathbf{b}\boldsymbol{\beta}(\boldsymbol{\alpha}\mathbf{a}||\mathbf{c}\boldsymbol{\gamma}) + \mathbf{c}\boldsymbol{\gamma}(\boldsymbol{\alpha}\mathbf{a}||\mathbf{b}\boldsymbol{\beta}) - \mathbf{b}\boldsymbol{\beta}|\mathbf{a}\boldsymbol{\alpha}|\mathbf{c}\boldsymbol{\gamma} - \mathbf{c}\boldsymbol{\gamma}|\mathbf{a}\boldsymbol{\alpha}|\mathbf{b}\boldsymbol{\beta}.\end{aligned} \tag{2.90}$$

This is a trilinear identity in three linear dyadics, $\mathbf{a}\boldsymbol{\alpha}, \mathbf{b}\boldsymbol{\beta}$ and $\mathbf{c}\boldsymbol{\gamma}$, whence we can replace them by arbitrary dyadics $\overline{\overline{\mathsf{A}}}, \overline{\overline{\mathsf{B}}}$ and $\overline{\overline{\mathsf{C}}}$ of $\mathbb{E}_1\mathbb{F}_1$. Thus we have arrived at the identity

$$\overline{\overline{\mathsf{A}}}^T\rfloor\rfloor(\overline{\overline{\mathsf{B}}}{}^\wedge_\wedge\overline{\overline{\mathsf{C}}}) = \overline{\overline{\mathsf{B}}}(\overline{\overline{\mathsf{A}}}^T||\overline{\overline{\mathsf{C}}}) + \overline{\overline{\mathsf{C}}}(\overline{\overline{\mathsf{A}}}^T||\overline{\overline{\mathsf{B}}}) - \overline{\overline{\mathsf{B}}}|\overline{\overline{\mathsf{A}}}|\overline{\overline{\mathsf{C}}} - \overline{\overline{\mathsf{C}}}|\overline{\overline{\mathsf{A}}}|\overline{\overline{\mathsf{B}}}. \tag{2.91}$$

This corresponds to the Gibbsian dyadic identity

$$\overline{\overline{\mathsf{A}}}{}^\times_\times(\overline{\overline{\mathsf{B}}}{}^\times_\times\overline{\overline{\mathsf{C}}}) = \overline{\overline{\mathsf{B}}}(\overline{\overline{\mathsf{A}}}:\overline{\overline{\mathsf{C}}}) + \overline{\overline{\mathsf{C}}}(\overline{\overline{\mathsf{A}}}:\overline{\overline{\mathsf{B}}}) - \overline{\overline{\mathsf{B}}}\cdot\overline{\overline{\mathsf{A}}}^T\cdot\overline{\overline{\mathsf{C}}} - \overline{\overline{\mathsf{C}}}\cdot\overline{\overline{\mathsf{A}}}^T\cdot\overline{\overline{\mathsf{B}}}, \tag{2.92}$$

given in reference 40. A consequence of (2.91) is

$$\overline{\overline{\mathsf{A}}}^T\rfloor\rfloor\overline{\overline{\mathsf{B}}}^{(2)} = \overline{\overline{\mathsf{B}}}(\overline{\overline{\mathsf{A}}}^T||\overline{\overline{\mathsf{B}}}) - \overline{\overline{\mathsf{B}}}|\overline{\overline{\mathsf{A}}}|\overline{\overline{\mathsf{B}}}. \tag{2.93}$$

Identities (2.91) and (2.93) are also valid for metric dyadics: $\overline{\overline{\mathsf{A}}} \in \mathbb{E}_1\mathbb{E}_1$, $\overline{\overline{\mathsf{B}}}, \overline{\overline{\mathsf{C}}} \in \mathbb{F}_1\mathbb{F}_1$ or $\overline{\overline{\mathsf{A}}} \in \mathbb{F}_1\mathbb{F}_1$, $\overline{\overline{\mathsf{B}}}, \overline{\overline{\mathsf{C}}} \in \mathbb{E}_1\mathbb{E}_1$.

Example 3 From (2.75) we can write for the reciprocal basis vectors and dual basis vectors

$$1 = (-1)^{n-i}(\varepsilon_{K(i)} \wedge \varepsilon_i)|\mathbf{e}_N = (-1)^{n-i}\varepsilon_{K(i)}|(\varepsilon_i\rfloor\mathbf{e}_N), \tag{2.94}$$

which gives

$$\varepsilon_i\rfloor\mathbf{e}_N = (-1)^{n-i}\mathbf{e}_{K(i)}. \tag{2.95}$$

Thus, multiplying by the vector $\mathbf{a} = \sum a_j\mathbf{e}_j$, we can write

$$(\varepsilon_i\rfloor\mathbf{e}_N) \wedge \mathbf{a} = (-1)^{n-i}\sum_j \mathbf{e}_{K(i)} \wedge \mathbf{e}_j a_j = \mathbf{e}_N a_i = (\varepsilon_i|\mathbf{a})\mathbf{e}_N. \tag{2.96}$$

Because the basis is arbitrary, we can more generally write

$$(\boldsymbol{\alpha}\rfloor\mathbf{k}_N) \wedge \mathbf{a} = (\boldsymbol{\alpha}|\mathbf{a})\mathbf{k}_N, \tag{2.97}$$

where $\mathbf{a}$ and $\boldsymbol{\alpha}$ are an arbitrary vector and dual vector, and $\mathbf{k}_N$ is any n-vector. From $\boldsymbol{\alpha}|\overline{\overline{\mathsf{A}}} = 0$ for every $\boldsymbol{\alpha}$ we conclude $\overline{\overline{\mathsf{A}}} = 0$ and, thus, obtain the dyadic identity

$$(\overline{\overline{\mathsf{I}}}\rfloor\mathbf{k}_N) \wedge \mathbf{a} = \mathbf{a}\mathbf{k}_N. \tag{2.98}$$

Identity (2.98) can be transformed into various forms. For example, for the dimension $n = 4$, we can replace the arbitrary quadrivector $\mathbf{k}_N$ by $\mathbf{A} \wedge \mathbf{B}$, where $\mathbf{A}$ and $\mathbf{B}$ are two bivectors and the result is

$$(\overline{\overline{\mathsf{I}}}\rfloor\mathbf{A} \wedge \mathbf{B}) \wedge \mathbf{a} = \mathbf{a}(\mathbf{A} \wedge \mathbf{B}). \tag{2.99}$$

The dual counterparts of this and its transpose are

$$\{\overline{\overline{\mathsf{I}}}^T\rfloor(\boldsymbol{\Phi} \wedge \boldsymbol{\Psi})\} \wedge \boldsymbol{\alpha} = \boldsymbol{\alpha}(\boldsymbol{\Phi} \wedge \boldsymbol{\Psi}), \tag{2.100}$$

$$\boldsymbol{\alpha} \wedge \{(\boldsymbol{\Phi} \wedge \boldsymbol{\Psi})\lfloor\overline{\overline{\mathsf{I}}}\} = -(\boldsymbol{\Phi} \wedge \boldsymbol{\Psi})\boldsymbol{\alpha}, \tag{2.101}$$

where $\boldsymbol{\alpha}$ is a dual vector and $\boldsymbol{\Phi}$ and $\boldsymbol{\Psi}$ are two dual bivectors in the space of four dimensions. These identities have application in Section 5.5 where $\boldsymbol{\Psi}$ and $\boldsymbol{\Phi}$ are electromagnetic two-forms (dual bivector fields).

Example 4 As yet another identity useful to electromagnetics in the space of four dimensions we prove

$$\boldsymbol{\Psi} \wedge \overline{\overline{\mathsf{I}}}^T\rfloor\boldsymbol{\Phi} + \boldsymbol{\Phi} \wedge \overline{\overline{\mathsf{I}}}^T\rfloor\boldsymbol{\Psi} = (\boldsymbol{\Psi} \wedge \boldsymbol{\Phi})\lfloor\overline{\overline{\mathsf{I}}}, \tag{2.102}$$

where $\boldsymbol{\Psi}$ and $\boldsymbol{\Phi}$ are arbitrary four-dimensional dual bivectors and $\overline{\overline{\mathsf{I}}}$ is the four-dimensional unit dyadic. We start by expanding Gibbs' identity (2.79) for $n = 4$ as

$$\begin{aligned}
&\mathbf{b}(\mathbf{a}_1 \wedge \mathbf{a}_2 \wedge \mathbf{a}_3 \wedge \mathbf{a}_4) \\
&\quad= \mathbf{a}_1(\mathbf{b} \wedge \mathbf{a}_2 \wedge \mathbf{a}_3 \wedge \mathbf{a}_4) + \mathbf{a}_2(\mathbf{a}_1 \wedge \mathbf{b} \wedge \mathbf{a}_3 \wedge \mathbf{a}_4) \\
&\quad\quad +\mathbf{a}_3(\mathbf{a}_1 \wedge \mathbf{a}_2 \wedge \mathbf{b} \wedge \mathbf{a}_4) + \mathbf{a}_4(\mathbf{a}_1 \wedge \mathbf{a}_2 \wedge \mathbf{a}_3 \wedge \mathbf{b}) \\
&\quad= \big[(\mathbf{a}_2\mathbf{a}_1 - \mathbf{a}_1\mathbf{a}_2) \wedge (\mathbf{a}_3 \wedge \mathbf{a}_4) + (\mathbf{a}_4\mathbf{a}_3 - \mathbf{a}_3\mathbf{a}_4) \wedge (\mathbf{a}_1 \wedge \mathbf{a}_2)\big] \wedge \mathbf{b}.
\end{aligned}$$

Fig. 2.3 *Visualization of identity (2.106).*

Applying (2.84) this can be written as

$$\mathbf{b}(\mathbf{a}_1 \wedge \mathbf{a}_2 \wedge \mathbf{a}_3 \wedge \mathbf{a}_4)$$
$$= \quad \big[(\mathbf{a}_1 \wedge \mathbf{a}_2)\lfloor \bar{\bar{\mathsf{I}}}^T \wedge (\mathbf{a}_3 \wedge \mathbf{a}_4) + (\mathbf{a}_3 \wedge \mathbf{a}_4)\lfloor \bar{\bar{\mathsf{I}}}^T \wedge (\mathbf{a}_1 \wedge \mathbf{a}_2)\big] \wedge \mathbf{b}. \tag{2.103}$$

Since this identity is linear in the bivectors $\mathbf{a}_1 \wedge \mathbf{a}_2$ and $\mathbf{a}_3 \wedge \mathbf{a}_4$, these can be replaced by arbitrary bivectors $\mathbf{A}$ and $\mathbf{B}$:

$$\mathbf{b}(\mathbf{A} \wedge \mathbf{B}) = (\mathbf{A}\lfloor \bar{\bar{\mathsf{I}}}^T \wedge \mathbf{B} + \mathbf{B}\lfloor \bar{\bar{\mathsf{I}}}^T \wedge \mathbf{A}) \wedge \mathbf{b}. \tag{2.104}$$

Finally, we apply (2.98) where $\mathbf{k}_N$ is replaced by the quadrivector $\mathbf{A} \wedge \mathbf{B}$:

$$(\bar{\bar{\mathsf{I}}}\rfloor(\mathbf{A} \wedge \mathbf{B})) \wedge \mathbf{b} = (\mathbf{A}\lfloor \bar{\bar{\mathsf{I}}}^T \wedge \mathbf{B} + \mathbf{B}\lfloor \bar{\bar{\mathsf{I}}}^T \wedge \mathbf{A}) \wedge \mathbf{b}. \tag{2.105}$$

Since this is valid for any $\mathbf{b}$, the following dyadic identity finally results (Figure 2.3)

$$\bar{\bar{\mathsf{I}}}\rfloor(\mathbf{A} \wedge \mathbf{B}) = \mathbf{A}\lfloor \bar{\bar{\mathsf{I}}}^T \wedge \mathbf{B} + \mathbf{B}\lfloor \bar{\bar{\mathsf{I}}}^T \wedge \mathbf{A}. \tag{2.106}$$

The identity (2.102) to be proved is the transposed dual counterpart of (2.106).

Identity (2.102) can also be written for $n = 3$, in which case dual quadrivectors like $\boldsymbol{\Psi} \wedge \boldsymbol{\Phi}$ vanish. In this case the identity becomes

$$\boldsymbol{\Psi} \wedge \bar{\bar{\mathsf{I}}}^T\rfloor\boldsymbol{\Phi} = -\boldsymbol{\Phi} \wedge \bar{\bar{\mathsf{I}}}^T\rfloor\boldsymbol{\Psi}, \tag{2.107}$$

implying

$$\boldsymbol{\Psi} \wedge \bar{\bar{\mathsf{I}}}^T\rfloor\boldsymbol{\Psi} = 0, \quad \mathbf{A} \wedge \bar{\bar{\mathsf{I}}}\rfloor\mathbf{A} = 0 \tag{2.108}$$

for any bivector $\mathbf{A}$ and dual bivector $\boldsymbol{\Psi}$. Here, $\bar{\bar{\mathsf{I}}}$ denotes the three-dimensional unit dyadic. These identities are needed when considering the magnetic force on a current source.

Problems

2.2.1 Prove Gibbs' identity (2.66) through a basis expansion of different terms.

2.2.2 Prove (2.97) through basis expansion of different terms.

2.2.3 For a set of vectors $\{\mathbf{a}_i\}$ in an n-dimensional space, derive the identity

$$\bar{\bar{\mathsf{I}}}\rfloor\mathbf{a}_N = \sum_i (-1)^{n-i}\mathbf{a}_i\mathbf{a}_{K(i)}.$$

2.2.4 Derive the following identity between four $\mathbb{E}_1\mathbb{F}_1$ dyadics

$$\begin{aligned}(\overline{\overline{\mathsf{A}}}{}^{\wedge}_{\wedge}\overline{\overline{\mathsf{B}}})||(\overline{\overline{\mathsf{C}}}{}^{\wedge}_{\wedge}\overline{\overline{\mathsf{D}}})^T &= (\overline{\overline{\mathsf{A}}}||\overline{\overline{\mathsf{C}}}^T)(\overline{\overline{\mathsf{B}}}||\overline{\overline{\mathsf{D}}}^T) + (\overline{\overline{\mathsf{A}}}||\overline{\overline{\mathsf{D}}}^T)(\overline{\overline{\mathsf{B}}}||\overline{\overline{\mathsf{C}}}^T) \\ &\quad -(\overline{\overline{\mathsf{A}}}|\overline{\overline{\mathsf{D}}})||(\overline{\overline{\mathsf{B}}}|\overline{\overline{\mathsf{C}}})^T - (\overline{\overline{\mathsf{A}}}|\overline{\overline{\mathsf{C}}})||(\overline{\overline{\mathsf{B}}}|\overline{\overline{\mathsf{D}}})^T.\end{aligned}$$

2.2.5 Starting from the previous identity derive the following ones:

$$(\overline{\overline{\mathsf{A}}}{}^{\wedge}_{\wedge}\overline{\overline{\mathsf{B}}})\lfloor\lfloor\overline{\overline{\mathsf{C}}}^T = (\overline{\overline{\mathsf{A}}}||\overline{\overline{\mathsf{C}}}^T)\overline{\overline{\mathsf{B}}} + (\overline{\overline{\mathsf{B}}}||\overline{\overline{\mathsf{C}}}^T)\overline{\overline{\mathsf{A}}} - \overline{\overline{\mathsf{A}}}|\overline{\overline{\mathsf{C}}}|\overline{\overline{\mathsf{B}}} - \overline{\overline{\mathsf{B}}}|\overline{\overline{\mathsf{C}}}|\overline{\overline{\mathsf{A}}},$$

$$\overline{\overline{\mathsf{A}}}^{(2)}||\overline{\overline{\mathsf{B}}}^{(2)T} = (\overline{\overline{\mathsf{A}}}||\overline{\overline{\mathsf{B}}}^T)^2 - (\overline{\overline{\mathsf{A}}}|\overline{\overline{\mathsf{B}}}^T)||(\overline{\overline{\mathsf{B}}}|\overline{\overline{\mathsf{A}}}^T),$$

$$\overline{\overline{\mathsf{A}}}^{(2)}\lfloor\lfloor\overline{\overline{\mathsf{I}}}^T = (\mathrm{tr}\,\overline{\overline{\mathsf{A}}})\overline{\overline{\mathsf{A}}} - \overline{\overline{\mathsf{A}}}^2.$$

$$\overline{\overline{\mathsf{A}}}^T\rfloor\rfloor\overline{\overline{\mathsf{I}}}^{(2)} = \mathrm{tr}\,\overline{\overline{\mathsf{A}}}\,\overline{\overline{\mathsf{I}}} - \overline{\overline{\mathsf{A}}}$$

2.2.6 Check the dimensions of the dyadics on both sides of the expansion (2.86).

2.2.7 Derive the Cayley–Hamilton equation for $n = 2$:

$$\overline{\overline{\mathsf{A}}}^2 - (\mathrm{tr}\,\overline{\overline{\mathsf{A}}})\overline{\overline{\mathsf{A}}} + (\det\overline{\overline{\mathsf{A}}})\overline{\overline{\mathsf{I}}} = 0.$$

Hint: Multiplying by $\overline{\overline{\mathsf{A}}}^{-1}$, we have the equivalent equation

$$(\det\overline{\overline{\mathsf{A}}})\overline{\overline{\mathsf{A}}}^{-1} = (\mathrm{tr}\,\overline{\overline{\mathsf{A}}})\overline{\overline{\mathsf{I}}} - \overline{\overline{\mathsf{A}}},$$

with $\det\overline{\overline{\mathsf{A}}} = \mathrm{tr}[\overline{\overline{\mathsf{A}}}^{(2)}]$. Prove this by applying the expression of the inverse of a two-dimensional dyadic.

2.2.8 Prove the identity

$$\overline{\overline{\mathsf{A}}}^{(3)}\lfloor\lfloor\overline{\overline{\mathsf{I}}}^{(2)T} = \overline{\overline{\mathsf{A}}}^3 - \overline{\overline{\mathsf{A}}}^2\mathrm{tr}\,\overline{\overline{\mathsf{A}}} + \overline{\overline{\mathsf{A}}}\,\mathrm{tr}\,\overline{\overline{\mathsf{A}}}^{(2)}.$$

Show that for $n = 3$ this gives the Cayley-Hamilton equation

$$\overline{\overline{\mathsf{A}}}^3 - \overline{\overline{\mathsf{A}}}^2\,\mathrm{tr}\,\overline{\overline{\mathsf{A}}} + \overline{\overline{\mathsf{A}}}\,\mathrm{tr}\,\overline{\overline{\mathsf{A}}}^{(2)} - \overline{\overline{\mathsf{I}}}\det\overline{\overline{\mathsf{A}}} = 0$$

and the inverse dyadic can be written as

$$\overline{\overline{\mathsf{A}}}^{-1} = \frac{1}{\det\overline{\overline{\mathsf{A}}}}(\overline{\overline{\mathsf{A}}}^2 - \overline{\overline{\mathsf{A}}}\,\mathrm{tr}\,\overline{\overline{\mathsf{A}}} + \overline{\overline{\mathsf{I}}}\,\mathrm{tr}\,\overline{\overline{\mathsf{A}}}^{(2)}).$$

2.2.9 From the identity (problem 2.1.2)

$$\begin{aligned}\mathrm{tr}(\overline{\overline{\mathsf{A}}}{}^{\wedge}_{\wedge}\overline{\overline{\mathsf{B}}}{}^{\wedge}_{\wedge}\overline{\overline{\mathsf{C}}}) &= \mathrm{tr}\,\overline{\overline{\mathsf{A}}}\,\mathrm{tr}\,\overline{\overline{\mathsf{B}}}\,\mathrm{tr}\,\overline{\overline{\mathsf{C}}} + \mathrm{tr}(\overline{\overline{\mathsf{A}}}|\overline{\overline{\mathsf{B}}}|\overline{\overline{\mathsf{C}}}) + \mathrm{tr}(\overline{\overline{\mathsf{A}}}|\overline{\overline{\mathsf{C}}}|\overline{\overline{\mathsf{B}}}) \\ &\quad -\mathrm{tr}\,\overline{\overline{\mathsf{A}}}\,\mathrm{tr}(\overline{\overline{\mathsf{B}}}|\overline{\overline{\mathsf{C}}}) - \mathrm{tr}\,\overline{\overline{\mathsf{B}}}\,\mathrm{tr}(\overline{\overline{\mathsf{C}}}|\overline{\overline{\mathsf{A}}}) - \mathrm{tr}\,\overline{\overline{\mathsf{C}}}\,\mathrm{tr}(\overline{\overline{\mathsf{A}}}|\overline{\overline{\mathsf{B}}})\end{aligned}$$

derive

$$(\overline{\overline{\mathsf{B}}}{}^{\wedge}_{\wedge}\overline{\overline{\mathsf{C}}})^T\rfloor\rfloor\bar{\bar{\mathsf{I}}}^{(3)} = \bar{\bar{\mathsf{I}}}\,\mathrm{tr}\,\overline{\overline{\mathsf{B}}}\,\mathrm{tr}\,\overline{\overline{\mathsf{C}}} + \overline{\overline{\mathsf{B}}}|\overline{\overline{\mathsf{C}}} + \overline{\overline{\mathsf{C}}}|\overline{\overline{\mathsf{B}}} \\ -\bar{\bar{\mathsf{I}}}\,\mathrm{tr}(\overline{\overline{\mathsf{B}}}|\overline{\overline{\mathsf{C}}}) - \overline{\overline{\mathsf{C}}}\,\mathrm{tr}\,\overline{\overline{\mathsf{B}}} - \overline{\overline{\mathsf{B}}}\,\mathrm{tr}\,\overline{\overline{\mathsf{C}}},$$

and

$$\overline{\overline{\mathsf{A}}}{}^{(2)T}\rfloor\rfloor\bar{\bar{\mathsf{I}}}^{(3)} = \overline{\overline{\mathsf{A}}}{}^2 - \overline{\overline{\mathsf{A}}}\,\mathrm{tr}\,\overline{\overline{\mathsf{A}}} + \bar{\bar{\mathsf{I}}}\,\mathrm{tr}\,\overline{\overline{\mathsf{A}}}{}^{(2)},$$

which for $n = 3$ can be written as (see previous problem)

$$\overline{\overline{\mathsf{A}}}{}^{(2)T}\rfloor\rfloor\bar{\bar{\mathsf{I}}}^{(3)} = (\det\overline{\overline{\mathsf{A}}})\overline{\overline{\mathsf{A}}}{}^{-1}$$

2.3 EIGENPROBLEMS

2.3.1 Left and right eigenvectors

Any dyadic $\overline{\overline{\mathsf{A}}} = \sum \mathbf{a}_i\boldsymbol{\alpha}_i \in \mathbb{E}_1\mathbb{F}_1$ defines right and left eigenvectors through the respective equations

$$\overline{\overline{\mathsf{A}}}|\mathbf{x}_i = \lambda_i\mathbf{x}_i, \quad \boldsymbol{\xi}_j|\overline{\overline{\mathsf{A}}} = \mu_j\boldsymbol{\xi}_j, \quad i,j = 1\cdots n. \tag{2.109}$$

From

$$\boldsymbol{\xi}_j|\overline{\overline{\mathsf{A}}}|\mathbf{x}_i = \mu_j\boldsymbol{\xi}_j|\mathbf{x}_i = \lambda_i\boldsymbol{\xi}_j|\mathbf{x}_i \tag{2.110}$$

we have

$$(\mu_j - \lambda_i)(\boldsymbol{\xi}_j|\mathbf{x}_i) = 0, \tag{2.111}$$

whence either the right and left eigenvectors are orthogonal:

$$\boldsymbol{\xi}_j|\mathbf{x}_i = 0, \tag{2.112}$$

or the eigenvalues coincide:

$$\mu_j = \lambda_i, \quad \boldsymbol{\xi}_j|\mathbf{x}_i \neq 0. \tag{2.113}$$

Let us arrange the indices i, j so that we denote $\mu_j = \lambda_j$ and assume the eigenvectors normalized to satisfy

$$\boldsymbol{\xi}_j|\mathbf{x}_i = \delta_{ij}. \tag{2.114}$$

The eigenproblems of the double-wedge square dyadic $\overline{\overline{\mathsf{A}}}{}^{(2)}$ follow from those above. Applying (2.24) we have

$$\overline{\overline{\mathsf{A}}}{}^{(2)}|(\mathbf{x}_i \wedge \mathbf{x}_j) = (\overline{\overline{\mathsf{A}}}|\mathbf{x}_i) \wedge (\overline{\overline{\mathsf{A}}}|\mathbf{x}_j) = \lambda_i\lambda_j\mathbf{x}_i \wedge \mathbf{x}_j, \tag{2.115}$$

$$(\boldsymbol{\xi}_i \wedge \boldsymbol{\xi}_j)|\overline{\overline{\mathsf{A}}}{}^{(2)} = \lambda_i\lambda_j\boldsymbol{\xi}_i \wedge \boldsymbol{\xi}_j. \tag{2.116}$$

This can be generalized to any double-wedge power

$$\overline{\overline{\mathsf{A}}}{}^{(p)}|(\mathbf{x}_{i_1} \wedge \cdots \wedge \mathbf{x}_{i_p}) = \lambda_{i_1}\cdots\lambda_{i_p}(\mathbf{x}_{i_1} \wedge \cdots \wedge \mathbf{x}_{i_p}), \tag{2.117}$$

with eigenvalues λ_{i_j} in the product corresponding to the different eigenvectors $\mathbf{x}_{i_j}$ in the p-vector. Similar rules apply to the left-hand eigenvector cases. Finally, we have

$$\overline{\overline{\mathsf{A}}}^{(n)}|(\mathbf{x}_1 \wedge \cdots \wedge \mathbf{x}_n) = \lambda_1 \cdots \lambda_n(\mathbf{x}_1 \wedge \cdots \wedge \mathbf{x}_n). \tag{2.118}$$

Because of $\overline{\overline{\mathsf{A}}}^{(n)} = \det \overline{\overline{\mathsf{A}}}\ \overline{\overline{\mathsf{I}}}^{(n)}$ we can identify the determinant with the product of eigenvalues,

$$\det \overline{\overline{\mathsf{A}}} = \lambda_1 \cdots \lambda_n. \tag{2.119}$$

For $\det \overline{\overline{\mathsf{A}}} \neq 0$, none of the n eigenvalues vanishes. In this case the n normalized eigenvectors are linearly independent and form a basis. Thus, we can expand

$$\overline{\overline{\mathsf{I}}} = \sum \mathbf{x}_i \boldsymbol{\xi}_i. \tag{2.120}$$

Also, the dyadic $\overline{\overline{\mathsf{A}}}$ can be expressed in terms of its eigenvectors and eigenvalues as

$$\overline{\overline{\mathsf{A}}} = \sum_{i=1}^{n} \lambda_i \mathbf{x}_i \boldsymbol{\xi}_i, \tag{2.121}$$

because $\overline{\overline{\mathsf{A}}}$ and the sum expression map the eigenvector basis in the same manner. Corresponding expansions for the double-wedge powers of the dyadic are

$$\overline{\overline{\mathsf{A}}}^{(2)} = \sum_{i<j} \lambda_i \lambda_j (\mathbf{x}_i \wedge \mathbf{x}_j)(\boldsymbol{\xi}_i \wedge \boldsymbol{\xi}_j), \tag{2.122}$$

$$\overline{\overline{\mathsf{A}}}^{(3)} = \sum_{i<j<k} \lambda_i \lambda_j \lambda_k (\mathbf{x}_i \wedge \mathbf{x}_j \wedge \mathbf{x}_k)(\boldsymbol{\xi}_i \wedge \boldsymbol{\xi}_j \wedge \boldsymbol{\xi}_k), \tag{2.123}$$

$$\overline{\overline{\mathsf{A}}}^{(n)} = \lambda_1 \cdots \lambda_n (\mathbf{x}_1 \wedge \cdots \wedge \mathbf{x}_n)(\boldsymbol{\xi}_1 \wedge \cdots \wedge \boldsymbol{\xi}_n). \tag{2.124}$$

When all eigenvalues are nonzero, the inverse of the dyadic has the obvious form

$$\overline{\overline{\mathsf{A}}}^{-1} = \sum_{i=1}^{n} \frac{1}{\lambda_i} \mathbf{x}_i \boldsymbol{\xi}_i, \tag{2.125}$$

because from orthogonality (2.114) it satisfies

$$\overline{\overline{\mathsf{A}}}^{-1}|\overline{\overline{\mathsf{A}}} = \overline{\overline{\mathsf{A}}}|\overline{\overline{\mathsf{A}}}^{-1} = \sum \frac{\lambda_j}{\lambda_i} \mathbf{x}_i \boldsymbol{\xi}_i | \mathbf{x}_j \boldsymbol{\xi}_j = \sum \mathbf{x}_i \boldsymbol{\xi}_i = \overline{\overline{\mathsf{I}}}. \tag{2.126}$$

2.3.2 Eigenvalues

If any of the eigenvalues is zero, from (2.119) the determinant vanishes. This can be expressed as

$$\overline{\overline{\mathsf{A}}}|\mathbf{x} = 0, \quad \mathbf{x} \neq 0 \quad \Rightarrow \quad \overline{\overline{\mathsf{A}}}^{(n)} = 0, \tag{2.127}$$

$$\boldsymbol{\xi}|\overline{\overline{\mathsf{A}}} = 0, \quad \boldsymbol{\xi} \neq 0, \quad \Rightarrow \quad \overline{\overline{\mathsf{A}}}^{(n)} = 0. \tag{2.128}$$

As a consequence, any eigenvalue λ_i satisfies

$$(\overline{\overline{\mathsf{A}}} - \lambda_i\overline{\overline{\mathsf{I}}})|\mathbf{x}_i = 0, \quad \mathbf{x}_i \neq 0, \quad \Rightarrow \quad (\overline{\overline{\mathsf{A}}} - \lambda_i\overline{\overline{\mathsf{I}}})^{(n)} = 0, \tag{2.129}$$

Expanding the power expression leads to the Cayley–Hamilton equation satisfied by any eigenvalue $\lambda = \lambda_i$:

$$\overline{\overline{\mathsf{A}}}^{(n)} - \lambda\overline{\overline{\mathsf{A}}}^{(n-1)}{}^\wedge_\wedge\overline{\overline{\mathsf{I}}} + \lambda^2\overline{\overline{\mathsf{A}}}^{(n-2)}{}^\wedge_\wedge\overline{\overline{\mathsf{I}}}^{(2)} + \cdots + (-\lambda)^n\overline{\overline{\mathsf{I}}}^{(n)} = 0, \tag{2.130}$$

which is equivalent to a scalar equation since all terms are multiples of the dyadic $\overline{\overline{\mathsf{I}}}^{(n)}$. Defining the trace of a p-dyadic as

$$\overline{\overline{\mathsf{A}}}^{(p)}{}^\wedge_\wedge\overline{\overline{\mathsf{I}}}^{(n-p)} = \mathrm{tr}\,\overline{\overline{\mathsf{A}}}^{(p)}\,\overline{\overline{\mathsf{I}}}^{(n)}, \tag{2.131}$$

the Cayley–Hamilton equation (2.130) can be written in true scalar form as

$$\mathrm{tr}\,\overline{\overline{\mathsf{A}}}^{(n)} - \lambda\,\mathrm{tr}\,\overline{\overline{\mathsf{A}}}^{(n-1)} + \lambda^2\mathrm{tr}\,\overline{\overline{\mathsf{A}}}^{(n-2)} + \cdots + (-\lambda)^{n-1}\mathrm{tr}\,\overline{\overline{\mathsf{A}}} + (-\lambda)^n = 0. \tag{2.132}$$

Expressing this as

$$(\lambda_1 - \lambda)(\lambda_2 - \lambda)\cdots(\lambda_n - \lambda) = 0, \tag{2.133}$$

we can identify

$$\mathrm{tr}\,\overline{\overline{\mathsf{A}}} = \lambda_1 + \cdots + \lambda_n, \tag{2.134}$$

$$\mathrm{tr}\,\overline{\overline{\mathsf{A}}}^{(2)} = \lambda_1\lambda_2 + \lambda_1\lambda_3 + \cdots + \lambda_{n-1}\lambda_n, \tag{2.135}$$

$$\mathrm{tr}\,\overline{\overline{\mathsf{A}}}^{(n)} = \det\overline{\overline{\mathsf{A}}} = \lambda_1\lambda_2\cdots\lambda_n. \tag{2.136}$$

The eigenvalues $\lambda_1 \cdots \lambda_n$ can be found by solving (2.132), an algebraic equation of nth order.

2.3.3 Eigenvectors

After finding one of the eigenvalues, λ_i, of the dyadic $\overline{\overline{\mathsf{A}}}$, the next task is to find the corresponding eigenvectors $\mathbf{x}_i$ and $\boldsymbol{\xi}_i$. Let us assume that the eigenvalue λ_i does not correspond to a multiple root of (2.132). In this case the dyadic $(\overline{\overline{\mathsf{A}}} - \lambda_i\overline{\overline{\mathsf{I}}})^{(n-1)}$ does not vanish and it can be expressed as

$$\begin{aligned}(\overline{\overline{\mathsf{A}}} - \lambda_i\overline{\overline{\mathsf{I}}})^{(n-1)} &= \left(\prod_{j\neq i}^{n}(\lambda_j - \lambda_i)\right)\mathbf{x}_{K(i)}\boldsymbol{\xi}_{K(i)} \\ &= \mathrm{tr}\left((\overline{\overline{\mathsf{A}}} - \lambda_i\overline{\overline{\mathsf{I}}})^{(n-1)}\right)\mathbf{x}_{K(i)}\boldsymbol{\xi}_{K(i)}.\end{aligned} \tag{2.137}$$

Here again we denote

$$\mathbf{x}_{K(i)} = \mathbf{x}_1 \wedge \mathbf{x}_2 \wedge \cdots \wedge \mathbf{x}_{i-1} \wedge \mathbf{x}_{i+1} \wedge \cdots \mathbf{x}_n, \tag{2.138}$$

and similarly for $\boldsymbol{\xi}_{K(i)}$. Because of the properties satisfied by reciprocal basis vectors and dual basis vectors

$$\mathbf{e}_{K(i)}\rfloor\varepsilon_N = (-1)^{i-1}\varepsilon_i, \quad \varepsilon_{K(i)}\rfloor\mathbf{e}_N = (-1)^{i-1}\mathbf{e}_i, \tag{2.139}$$

we can write

$$\mathbf{x}_{K(i)}\boldsymbol{\xi}_{K(i)}\rfloor\rfloor\boldsymbol{\xi}_N\mathbf{x}_N = \mathbf{x}_{K(i)}\boldsymbol{\xi}_{K(i)}\rfloor\rfloor\bar{\bar{\mathsf{I}}}^{(n)T} = \boldsymbol{\xi}_i\mathbf{x}_i. \tag{2.140}$$

Thus, the left and right eigenvectors corresponding to the given eigenvalue λ_i can be obtained as

$$\boldsymbol{\xi}_i\mathbf{x}_i = \gamma(\bar{\bar{\mathsf{A}}} - \lambda_i\bar{\bar{\mathsf{I}}})^{(n-1)}\rfloor\rfloor\bar{\bar{\mathsf{I}}}^{(n)T}, \tag{2.141}$$

with some scalar γ whose value can be found through orthogonality (2.114). Since the expression in (2.141) only requires the knowledge of a nonmultiple eigenvalue λ_i, it gives a method to determine the right and left eigenvectors corresponding to that eigenvalue. The scalar factor γ is not essential, it can be chosen to set $\boldsymbol{\xi}_i|\mathbf{x}_i = 1$.

The preceding construction fails when $\lambda_i = \lambda_j$ is a double eigenvalue because in this case $(\bar{\bar{\mathsf{A}}} - \lambda_i\bar{\bar{\mathsf{I}}})^{(n-1)} = 0$. However, if it is not a triple eigenvalue, we have $(\bar{\bar{\mathsf{A}}} - \lambda_i\bar{\bar{\mathsf{I}}})^{(n-2)} \neq 0$ and

$$\gamma(\bar{\bar{\mathsf{A}}} - \lambda_i\bar{\bar{\mathsf{I}}})^{(n-2)}\rfloor\rfloor\bar{\bar{\mathsf{I}}}^{(n)T} = (\boldsymbol{\xi}_i \wedge \boldsymbol{\xi}_j)(\mathbf{x}_i \wedge \mathbf{x}_j), \tag{2.142}$$

gives a bivector dyadic relation for the eigenvectors corresponding to the same eigenvalue. The bivector $\mathbf{x}_i \wedge \mathbf{x}_j$ can be solved from this equation. It does not, however, uniquely define the eigenvectors $\mathbf{x}_i$ and $\mathbf{x}_j$ but, instead, a two-dimensional subspace containing all their linear combinations. In fact, any linear combination of the two possible eigenvectors can be an eigenvector. Obviously, this method can be continued to find subspaces of eigenvectors corresponding to multiple eigenvalues.

Problems

2.3.1 Find eigenvalues and eigenvectors of the dyadic

$$\bar{\bar{\mathsf{A}}} = a\bar{\bar{\mathsf{I}}} + \mathbf{a}\boldsymbol{\alpha},$$

where a is a scalar, $\mathbf{a}$ is a vector, and $\boldsymbol{\alpha}$ satisfying $\mathbf{a}|\boldsymbol{\alpha} \neq 0$, $\mathbf{a}|\boldsymbol{\alpha} \neq a$ is a dual vector, both in a space of n dimensions. Use the solution formula (2.141) for the eigenvectors. You may need the identities

$$\bar{\bar{\mathsf{A}}}\rfloor\bar{\bar{\mathsf{I}}}^{(2)T} = \mathrm{tr}\,\bar{\bar{\mathsf{A}}}\,\bar{\bar{\mathsf{I}}}^{T} - \bar{\bar{\mathsf{A}}}^{T}, \quad \bar{\bar{\mathsf{A}}}\rfloor\bar{\bar{\mathsf{I}}}^{(2)T} = \mathrm{tr}\,\bar{\bar{\mathsf{A}}}\,\bar{\bar{\mathsf{I}}}^{T} - \bar{\bar{\mathsf{A}}}^{T}$$

2.3.2 Prove through basis expansions the identity

$$\bar{\bar{\mathsf{I}}}^{(n-1)}{}^{\wedge}_{\wedge}\mathbf{a}\boldsymbol{\alpha} = (\mathbf{a}|\boldsymbol{\alpha})\bar{\bar{\mathsf{I}}}^{(n)}$$

valid for $\bar{\bar{\mathsf{A}}} \in \mathbb{E}_1\mathbb{F}_1$ and $\mathbf{a} \in \mathbb{E}_1$, $\boldsymbol{\alpha} \in \mathbb{F}_1$.

2.3.3 Find the eigenvalues and eigenvectors of the dyadic

$$\overline{\overline{\mathsf{A}}} = a\overline{\overline{\mathsf{I}}} + \mathbf{a}\boldsymbol{\alpha}, \qquad \mathbf{a}|\boldsymbol{\alpha} = 0,$$

using the identity of the previous problem.

2.4 INVERSE DYADIC

It is possible to construct the inverse of a dyadic without knowing its eigenvectors in explicit form after finding a rule to construct the reciprocal basis of a given basis of vectors or dual vectors.

2.4.1 Reciprocal basis

Gibbs' identity in three dimensions leads to the definition of the unit dyadic (2.72) in terms of any basis of vectors $\mathbf{a}, \mathbf{b}, \mathbf{c}$ and any dual trivector $\boldsymbol{\kappa}_N$, $N = 123$,

$$\overline{\overline{\mathsf{I}}} = \mathbf{a}\frac{\boldsymbol{\kappa}_N\lfloor(\mathbf{b}\wedge\mathbf{c})}{\boldsymbol{\kappa}_N|(\mathbf{a}\wedge\mathbf{b}\wedge\mathbf{c})} + \mathbf{b}\frac{\boldsymbol{\kappa}_N\lfloor(\mathbf{c}\wedge\mathbf{a})}{\boldsymbol{\kappa}_N|(\mathbf{a}\wedge\mathbf{b}\wedge\mathbf{c})} + \mathbf{c}\frac{\boldsymbol{\kappa}_N\lfloor(\mathbf{a}\wedge\mathbf{b})}{\boldsymbol{\kappa}_N|(\mathbf{a}\wedge\mathbf{b}\wedge\mathbf{c})}. \tag{2.143}$$

As a check, the trace of the unit dyadic can be expanded as

$$\mathrm{tr}\,\overline{\overline{\mathsf{I}}} = \frac{\mathbf{a}|(\boldsymbol{\kappa}_N\lfloor(\mathbf{b}\wedge\mathbf{c}))}{\boldsymbol{\kappa}_N|(\mathbf{a}\wedge\mathbf{b}\wedge\mathbf{c})} + \frac{\mathbf{b}|(\boldsymbol{\kappa}_N\lfloor(\mathbf{c}\wedge\mathbf{a}))}{\boldsymbol{\kappa}_N|(\mathbf{a}\wedge\mathbf{b}\wedge\mathbf{c})} + \frac{\mathbf{c}|(\boldsymbol{\kappa}_N\lfloor(\mathbf{a}\wedge\mathbf{b}))}{\boldsymbol{\kappa}_N|(\mathbf{a}\wedge\mathbf{b}\wedge\mathbf{c})} = 3, \tag{2.144}$$

when using the property of the incomplete duality product

$$\mathbf{a}|(\boldsymbol{\kappa}_N\lfloor(\mathbf{b}\wedge\mathbf{c})) = (\boldsymbol{\kappa}_N\lfloor(\mathbf{b}\wedge\mathbf{c}))|\mathbf{a} = \boldsymbol{\kappa}_N|(\mathbf{b}\wedge\mathbf{c}\wedge\mathbf{a}). \tag{2.145}$$

The same procedure can now be applied when constructing the unit dyadic in terms of a given basis $\{\mathbf{e}_i\}$ in the n-dimensional vector space. In fact, replacing $\mathbf{a}_i$ by $\mathbf{e}_i$ in (2.81) we obtain

$$\overline{\overline{\mathsf{I}}} = \sum_{i=1}^{n}(-1)^{n-i}\mathbf{e}_i\frac{\boldsymbol{\kappa}_N\lfloor\mathbf{e}_{K(i)}}{\boldsymbol{\kappa}_N|\mathbf{e}_N} \tag{2.146}$$

or, in a more explicit form,

$$\begin{aligned}\overline{\overline{\mathsf{I}}} &= (-1)^{n-1}\mathbf{e}_1\frac{\boldsymbol{\kappa}_N\lfloor(\mathbf{e}_2\wedge\mathbf{e}_3\wedge\cdots\wedge\mathbf{e}_n)}{\boldsymbol{\kappa}_N|\mathbf{e}_N}\\ &\quad+(-1)^{n-2}\mathbf{e}_2\frac{\boldsymbol{\kappa}_N\lfloor(\mathbf{e}_1\wedge\mathbf{e}_3\wedge\cdots\wedge\mathbf{e}_n)}{\boldsymbol{\kappa}_N|\mathbf{e}_N}\\ &\quad+\cdots+\mathbf{e}_n\frac{\boldsymbol{\kappa}_N\lfloor(\mathbf{e}_1\wedge\mathbf{e}_2\wedge\cdots\wedge\mathbf{e}_{n-1})}{\boldsymbol{\kappa}_N|\mathbf{e}_N}.\end{aligned} \tag{2.147}$$

Because the unit dyadic has the general form

$$\overline{\overline{\mathsf{I}}} = \sum_{i=1}^{n}\mathbf{e}_i\boldsymbol{\varepsilon}_i, \tag{2.148}$$

where ε_i are dual basis vectors complementary to the $\mathbf{e}_i$, from (2.146) we can identify the reciprocal dual basis vectors as

$$\varepsilon_i = (-1)^{n-i} \frac{\boldsymbol{\kappa}_N \lfloor \mathbf{e}_{K(i)}}{\boldsymbol{\kappa}_N | \mathbf{e}_N}. \tag{2.149}$$

Applying the known condition

$$(\boldsymbol{\kappa}_N \lfloor \mathbf{e}_{K(i)}) | \mathbf{e}_j = \boldsymbol{\kappa}_N | (\mathbf{e}_{K(i)} \wedge \mathbf{e}_j) = (-1)^{n-i} \delta_{ij} \boldsymbol{\kappa}_N | \mathbf{e}_N, \tag{2.150}$$

we can check the orthogonality

$$\varepsilon_i | \mathbf{e}_j = (-1)^{n-i} \frac{\boldsymbol{\kappa}_N | (\mathbf{e}_{K(i)} \wedge \mathbf{e}_j)}{\boldsymbol{\kappa}_N | \mathbf{e}_N} = \frac{\boldsymbol{\kappa}_N | \mathbf{e}_N}{\boldsymbol{\kappa}_N | \mathbf{e}_N} \delta_{ij} = \delta_{ij}. \tag{2.151}$$

Correspondingly, one can form the dual of the dual basis $\{\varepsilon_i\}$ as

$$\mathbf{e}_i = (-1)^{n-i} \frac{\mathbf{k}_N \lfloor \varepsilon_{K(i)}}{\mathbf{k}_N | \varepsilon_N}, \tag{2.152}$$

where $\mathbf{k}_N \neq 0$ is an arbitrary n-vector.

2.4.2 The inverse dyadic

The construction of the dual basis reciprocal to a given vector basis (2.149) makes it possible to find the inverse $\overline{\overline{\mathsf{B}}} = \overline{\overline{\mathsf{A}}}^{-1}$ of a given dyadic $\overline{\overline{\mathsf{A}}} \in \mathbb{E}_1 \mathbb{F}_1$. Let us assume its vector expansion as

$$\overline{\overline{\mathsf{A}}} = \sum_{i=1}^{n} \mathbf{a}_i \boldsymbol{\alpha}_i, \tag{2.153}$$

in which either $\{\mathbf{a}_i\}$ or $\{\boldsymbol{\alpha}_i\}$ can be any given basis in which case the other set of vectors or dual vectors defines the dyadic. In fact, we can write $\overline{\overline{\mathsf{A}}} = \overline{\overline{\mathsf{I}}} | \overline{\overline{\mathsf{A}}}$ or $\overline{\overline{\mathsf{A}}} = \overline{\overline{\mathsf{A}}} | \overline{\overline{\mathsf{I}}}$ and use an expansion for the unit dyadic corresponding to (2.148). For the existence of the inverse we must assume $\det \overline{\overline{\mathsf{A}}} \neq 0$ or the equivalent condition

$$\overline{\overline{\mathsf{A}}}^{(n)} = (\mathbf{a}_1 \wedge \mathbf{a}_2 \wedge \cdots \wedge \mathbf{a}_n)(\boldsymbol{\alpha}_1 \wedge \boldsymbol{\alpha}_2 \wedge \cdots \wedge \boldsymbol{\alpha}_n) \neq 0, \tag{2.154}$$

which means that $\{\mathbf{a}_i\}$ and $\{\boldsymbol{\alpha}_i\}$ are two bases, not reciprocal to one another in general.

Let us express the inverse dyadic $\overline{\overline{\mathsf{B}}}$ similarly in terms of two bases $\{\mathbf{b}_j\}$, $\{\boldsymbol{\beta}_j\}$ as

$$\overline{\overline{\mathsf{B}}} = \sum_{j=1}^{n} \mathbf{b}_j \boldsymbol{\beta}_j. \tag{2.155}$$

The condition for solving the inverse dyadic $\overline{\overline{\mathsf{B}}}$ is then

$$\overline{\overline{\mathsf{B}}} | \overline{\overline{\mathsf{A}}} = \sum_{i,j=1}^{n} \mathbf{b}_i (\boldsymbol{\beta}_i | \mathbf{a}_j) \boldsymbol{\alpha}_j = \overline{\overline{\mathsf{I}}}. \tag{2.156}$$

From the form of (2.156) one can immediately see that it is sufficient to require that $\{\boldsymbol{\beta}_j\}$ be reciprocal to the vector basis $\{\mathbf{a}_i\}$ and $\{\mathbf{b}_i\}$ be reciprocal to the dual vector basis $\{\boldsymbol{\alpha}_i\}$. In fact, requiring

$$\boldsymbol{\beta}_i|\mathbf{a}_j = \delta_{ij}, \qquad \sum_{i=1}^{n} \mathbf{b}_i\boldsymbol{\alpha}_i = \overline{\overline{\mathsf{I}}}, \tag{2.157}$$

(2.156) is obviously satisfied.

Because from the previous section we know how to form a basis dual to a given basis, we are now able to construct an explicit formula for the inverse $\overline{\overline{\mathsf{B}}}$ of the dyadic $\overline{\overline{\mathsf{A}}}$. Choosing $\mathbf{e}_i = \mathbf{a}_i$, $\boldsymbol{\kappa}_N = \boldsymbol{\alpha}_N$ in (2.149) and $\boldsymbol{\varepsilon}_i = \boldsymbol{\alpha}_i$, $\mathbf{k}_N = \mathbf{a}_N$ in (2.152), the inverse dyadic can be directly written as

$$\overline{\overline{\mathsf{B}}} = \sum_{i=1}^{n} \mathbf{b}_i\boldsymbol{\beta}_i = \sum_{i=1}^{n} \frac{(\mathbf{a}_N \lfloor \boldsymbol{\alpha}_{K(i)})(\boldsymbol{\alpha}_N \lfloor \mathbf{a}_{K(i)})}{(\mathbf{a}_N|\boldsymbol{\alpha}_N)^2}. \tag{2.158}$$

In terms of the incomplete double duality product $\lfloor\lfloor$ we can more compactly write

$$\overline{\overline{\mathsf{B}}} = \frac{\mathbf{a}_N\boldsymbol{\alpha}_N}{(\boldsymbol{\alpha}_N|\mathbf{a}_N)^2} \lfloor\lfloor \sum_{i=1}^{n} \boldsymbol{\alpha}_{K(i)}\mathbf{a}_{K(i)}. \tag{2.159}$$

Because the nth and $(n-1)$th double-wedge powers of $\overline{\overline{\mathsf{A}}}$ can be expressed as

$$\overline{\overline{\mathsf{A}}}^{(n)} = \mathbf{a}_N\boldsymbol{\alpha}_N = (\mathbf{a}_N|\boldsymbol{\alpha}_N)\mathbf{e}_N\boldsymbol{\varepsilon}_N = (\operatorname{tr}\overline{\overline{\mathsf{A}}}^{(n)})\overline{\overline{\mathsf{I}}}^{(n)}, \tag{2.160}$$

$$\overline{\overline{\mathsf{A}}}^{(n-1)} = \sum_{i=1}^{n} \mathbf{a}_{K(i)}\boldsymbol{\alpha}_{K(i)}, \tag{2.161}$$

the inverse dyadic takes on the simple final form

$$\overline{\overline{\mathsf{A}}}^{-1} = \frac{\overline{\overline{\mathsf{I}}}^{(n)} \lfloor\lfloor \overline{\overline{\mathsf{A}}}^{(n-1)T}}{\operatorname{tr}\overline{\overline{\mathsf{A}}}^{(n)}}. \tag{2.162}$$

Alternatively, we can denote

$$\overline{\overline{\mathsf{A}}}^{-1} = \frac{\operatorname{Adj}\overline{\overline{\mathsf{A}}}}{\det\overline{\overline{\mathsf{A}}}}, \tag{2.163}$$

when the adjoint dyadic and determinant are defined by

$$\operatorname{Adj}\overline{\overline{\mathsf{A}}} = \overline{\overline{\mathsf{I}}}^{(n)} \lfloor\lfloor \overline{\overline{\mathsf{A}}}^{(n-1)T}, \tag{2.164}$$

$$\det\overline{\overline{\mathsf{A}}} = \operatorname{tr}\overline{\overline{\mathsf{A}}}^{(n)} = \mathbf{a}_N|\boldsymbol{\alpha}_N = (\mathbf{a}_1 \wedge \cdots \wedge \mathbf{a}_n)|(\boldsymbol{\alpha}_i \wedge \cdots \wedge \boldsymbol{\alpha}_n). \tag{2.165}$$

The inverse of a dyadic $\overline{\overline{\mathsf{A}}} \in \mathbb{F}_1\mathbb{E}_1$ can be derived similarly by interchanging vectors and dual vectors, and the result becomes the same (2.162).

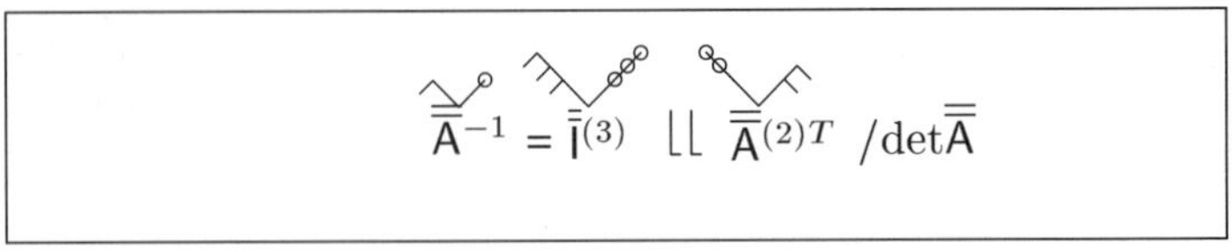

Fig. 2.4 *Checking the grades of the inverse formula (2.170) for $n = 3$.*

Inverse of a metric dyadic A dyadic $\overline{\overline{A}}$ of the space $\mathbb{E}_1\,\mathbb{E}_1$ will be called a metric dyadic, see Section 2.5. The inverse of such a dyadic must lie in the space $\mathbb{F}_1\,\mathbb{F}_1$, and the expression (2.162) for the inverse must be modified since unit dyadics do not exist in this case. Expressing $\overline{\overline{A}} \in \mathbb{E}_1\,\mathbb{E}_1$ in terms of two sets of basis vectors $\{\mathbf{a}_i\}, \{\mathbf{b}_i\}$ as

$$\overline{\overline{A}} = \sum_i^n \mathbf{a}_i \mathbf{b}_i, \tag{2.166}$$

its inverse can be written in terms of two sets of dual vectors $\{\boldsymbol{\alpha}_i\}, \{\boldsymbol{\beta}_i\}$:

$$\overline{\overline{A}}^{-1} = \sum_i^n \boldsymbol{\beta}_i \boldsymbol{\alpha}_i. \tag{2.167}$$

Requiring

$$\overline{\overline{A}}|\overline{\overline{A}}^{-1} = \sum_{i,j=1}^n \mathbf{a}_i \boldsymbol{\alpha}_j (\mathbf{b}_i|\boldsymbol{\beta}_j) = \overline{\overline{I}}, \quad \overline{\overline{A}}^{-1}|\overline{\overline{A}} = \overline{\overline{I}}^T, \tag{2.168}$$

this is obviously satisfied if $\{\boldsymbol{\alpha}_i\}$ is the reciprocal basis of $\{\mathbf{a}_i\}$ and $\{\boldsymbol{\beta}_i\}$ that of $\{\mathbf{b}_i\}$. From (2.149) we can now write

$$\overline{\overline{A}}^{-1} = \sum_{i=1}^n \frac{\boldsymbol{\kappa}_N \boldsymbol{\kappa}_N \lfloor\lfloor \mathbf{b}_{K(i)} \mathbf{a}_{K(i)}}{(\boldsymbol{\kappa}_N|\mathbf{a}_N)(\boldsymbol{\kappa}_N|\mathbf{b}_N)} = \frac{\boldsymbol{\kappa}_N \boldsymbol{\kappa}_N \lfloor\lfloor \overline{\overline{A}}^{(n-1)T}}{\boldsymbol{\kappa}_N \boldsymbol{\kappa}_N || \overline{\overline{A}}^{(n)}}, \tag{2.169}$$

where $\boldsymbol{\kappa}_N$ is any dual n-vector. Note that there is no unit dyadic in the space $\mathbb{E}_1\,\mathbb{E}_1$ and $\det\overline{\overline{A}}$ is not defined. For a dyadic $\overline{\overline{A}}$ of the space $\mathbb{F}_1\,\mathbb{F}_1$, in the inverse (2.169) $\boldsymbol{\kappa}_N$ must be replaced by replaced by a n-vector $\mathbf{k}_N$.

2.4.3 Inverse in three dimensions

Let us check the result by expanding the formula (2.162) in the three-dimensional space, in which case it has the form (Figure 2.4)

$$\overline{\overline{A}}^{-1} = \frac{\overline{\overline{I}}^{(3)} \lfloor\lfloor \overline{\overline{A}}^{(2)T}}{\det\overline{\overline{A}}}. \tag{2.170}$$

This bears some similarity to the corresponding Gibbsian dyadic formula $\overline{\overline{A}}^{-1} = \overline{\overline{A}}^{(2)T}/\det\overline{\overline{A}}$ [28,40]. To compare with the familiar form known for the corresponding matrix $[A_{ij}]$, let us express the dyadic in terms of reciprocal bases $\{\mathbf{e}_i\}$, $\{\boldsymbol{\varepsilon}_j\}$

$$\overline{\overline{\mathsf{A}}} = \sum_{i,j} A_{ij}\mathbf{e}_i\boldsymbol{\varepsilon}_j, \tag{2.171}$$

and substitute in (2.162) the expansions

$$\overline{\overline{\mathsf{A}}}^{(2)} = \frac{1}{2}\sum_{i,j}\sum_{k,\ell} A_{ij}A_{k\ell}(\mathbf{e}_i \wedge \mathbf{e}_k)(\boldsymbol{\varepsilon}_j \wedge \boldsymbol{\varepsilon}_\ell), \tag{2.172}$$

$$\overline{\overline{\mathsf{A}}}^{(3)} = \det[A_{ij}]\overline{\overline{\mathsf{I}}}^{(3)} = \det[A_{ij}]\mathbf{e}_{123}\boldsymbol{\varepsilon}_{123}, \tag{2.173}$$

$$\begin{aligned}\det[A_{ij}] &= A_{11}A_{22}A_{33} + A_{13}A_{32}A_{21} + A_{12}A_{23}A_{31} \\ &\quad -A_{11}A_{23}A_{32} - A_{22}A_{13}A_{31} - A_{33}A_{12}A_{21},\end{aligned} \tag{2.174}$$

$$\begin{aligned}\mathrm{Adj}\,\overline{\overline{\mathsf{A}}} &= \overline{\overline{\mathsf{I}}}^{(3)}\lfloor\lfloor\overline{\overline{\mathsf{A}}}^{(2)T} \\ &= \frac{1}{2}\sum_{i,j}\sum_{k,\ell} A_{ij}A_{k\ell}(\mathbf{e}_{123}\lfloor\boldsymbol{\varepsilon}_j \wedge \boldsymbol{\varepsilon}_\ell)(\boldsymbol{\varepsilon}_{123}\lfloor\mathbf{e}_i \wedge \mathbf{e}_k) \\ &= (A_{33}A_{22} - A_{32}A_{23})\mathbf{e}_1\boldsymbol{\varepsilon}_1 + (A_{32}A_{13} - A_{33}A_{12})\mathbf{e}_1\boldsymbol{\varepsilon}_2 \\ &\quad +(A_{12}A_{23} - A_{13}A_{22})\mathbf{e}_1\boldsymbol{\varepsilon}_3 + (A_{23}A_{31} - A_{21}A_{33})\mathbf{e}_2\boldsymbol{\varepsilon}_1 \\ &\quad +(A_{33}A_{11} - A_{31}A_{13})\mathbf{e}_2\boldsymbol{\varepsilon}_2 + (A_{13}A_{21} - A_{11}A_{23})\mathbf{e}_2\boldsymbol{\varepsilon}_3 \\ &\quad +(A_{21}A_{32} - A_{22}A_{31})\mathbf{e}_3\boldsymbol{\varepsilon}_1 + (A_{31}A_{12} - A_{32}A_{11})\mathbf{e}_3\boldsymbol{\varepsilon}_2 \\ &\quad +(A_{11}A_{22} - A_{12}A_{21})\mathbf{e}_3\boldsymbol{\varepsilon}_3.\end{aligned} \tag{2.175}$$

The inverse formula (2.170) can be verified directly by forming the duality product of $\mathrm{Adj}\,\overline{\overline{\mathsf{A}}}$ and $\overline{\overline{\mathsf{A}}}$.

As a simple check let us replace $\overline{\overline{\mathsf{A}}}$ by the simple diagonal dyadic

$$\overline{\overline{\mathsf{D}}} = D_1\mathbf{e}_1\boldsymbol{\varepsilon}_1 + D_2\mathbf{e}_2\boldsymbol{\varepsilon}_2 + D_3\mathbf{e}_3\boldsymbol{\varepsilon}_3. \tag{2.176}$$

The adjoint dyadic reduces now to the diagonal form

$$\mathrm{Adj}\,\overline{\overline{\mathsf{D}}} = D_2D_3\mathbf{e}_1\boldsymbol{\varepsilon}_1 + D_3D_1\mathbf{e}_2\boldsymbol{\varepsilon}_2 + D_1D_2\mathbf{e}_3\boldsymbol{\varepsilon}_3, \tag{2.177}$$

and the determinant reduces to

$$\det\overline{\overline{\mathsf{D}}} = D_1D_2D_3, \tag{2.178}$$

whence the inverse has the easily recognizable form

$$\overline{\overline{\mathsf{D}}}^{-1} = D_1^{-1}\mathbf{e}_1\boldsymbol{\varepsilon}_1 + D_2^{-1}\mathbf{e}_2\boldsymbol{\varepsilon}_2 + D_3^{-1}\mathbf{e}_3\boldsymbol{\varepsilon}_3. \tag{2.179}$$

Example Let us demonstrate the use of the formula (2.162) to find an analytical expression for the inverse dyadic in the space of n dimensions for a dyadic of the special form

$$\overline{\overline{\mathsf{A}}} = \overline{\overline{\mathsf{I}}} + \mathbf{a}\boldsymbol{\alpha}. \tag{2.180}$$

For this we need a few identities involving the unit dyadics and a dyadic $\overline{\overline{\mathsf{B}}} \in \mathbb{E}_1\mathbb{F}_1$:

$$\overline{\overline{\mathsf{I}}}^{(n-1)}{}^\wedge_\wedge\overline{\overline{\mathsf{B}}} = (\mathrm{tr}\overline{\overline{\mathsf{B}}})\overline{\overline{\mathsf{I}}}^{(n)}, \tag{2.181}$$

$$\overline{\overline{\mathsf{I}}}^{(2)}\lfloor\lfloor\overline{\overline{\mathsf{B}}}^T = (\mathrm{tr}\overline{\overline{\mathsf{B}}})\overline{\overline{\mathsf{I}}} - \overline{\overline{\mathsf{B}}}, \tag{2.182}$$

$$\overline{\overline{\mathsf{I}}}^{(n)}\lfloor\lfloor\overline{\overline{\mathsf{I}}}^{(n-p)T} = \overline{\overline{\mathsf{I}}}^{(p)}, \quad 1 \le p \le n-1, \tag{2.183}$$

whose proofs are left as exercises. Expanding

$$\begin{aligned}\overline{\overline{\mathsf{A}}}^{(n)} &= \frac{1}{n!}(\overline{\overline{\mathsf{I}}} + \mathbf{a}\boldsymbol{\alpha})^\wedge_\wedge(\overline{\overline{\mathsf{I}}} + \mathbf{a}\boldsymbol{\alpha})^\wedge_\wedge\cdots{}^\wedge_\wedge(\overline{\overline{\mathsf{I}}} + \mathbf{a}\boldsymbol{\alpha}) \\ &= \overline{\overline{\mathsf{I}}}^{(n)} + \overline{\overline{\mathsf{I}}}^{(n-1)}{}^\wedge_\wedge\mathbf{a}\boldsymbol{\alpha} \\ &= (1 + \mathbf{a}|\boldsymbol{\alpha})\overline{\overline{\mathsf{I}}}^{(n)} = \det\overline{\overline{\mathsf{A}}}\,\overline{\overline{\mathsf{I}}}^{(n)},\end{aligned} \tag{2.184}$$

we see that the condition for the inverse to exist is

$$\det\overline{\overline{\mathsf{A}}} = 1 + \mathbf{a}|\boldsymbol{\alpha} \neq 0, \quad \Rightarrow \quad \mathbf{a}|\boldsymbol{\alpha} \neq -1. \tag{2.185}$$

With the help of the identities above we can evaluate

$$\begin{aligned}\mathrm{Adj}\,\overline{\overline{\mathsf{A}}} &= \overline{\overline{\mathsf{I}}}^{(n)}\lfloor\lfloor\overline{\overline{\mathsf{A}}}^{(n-1)T} \\ &= \overline{\overline{\mathsf{I}}}^{(n)}\lfloor\lfloor\overline{\overline{\mathsf{I}}}^{(n-1)T} + \overline{\overline{\mathsf{I}}}^{(n)}\lfloor\lfloor(\overline{\overline{\mathsf{I}}}^{(n-2)T}{}^\wedge_\wedge\boldsymbol{\alpha}\mathbf{a}) \\ &= \overline{\overline{\mathsf{I}}} + (\overline{\overline{\mathsf{I}}}^{(n)}\lfloor\lfloor\overline{\overline{\mathsf{I}}}^{(n-2)T})\lfloor\lfloor\boldsymbol{\alpha}\mathbf{a} = \overline{\overline{\mathsf{I}}} + \overline{\overline{\mathsf{I}}}^{(2)}\lfloor\lfloor\boldsymbol{\alpha}\mathbf{a}.\end{aligned} \tag{2.186}$$

Finally, from (2.162) and (2.182), the inverse dyadic can be written as

$$(\overline{\overline{\mathsf{I}}} + \mathbf{a}\boldsymbol{\alpha})^{-1} = \frac{\mathrm{Adj}(\overline{\overline{\mathsf{I}}} + \mathbf{a}\boldsymbol{\alpha})}{\det(\overline{\overline{\mathsf{I}}} + \mathbf{a}\boldsymbol{\alpha})} = \frac{\overline{\overline{\mathsf{I}}} + \overline{\overline{\mathsf{I}}}^{(2)}\lfloor\lfloor\boldsymbol{\alpha}\mathbf{a}}{1 + \mathbf{a}|\boldsymbol{\alpha}} = \overline{\overline{\mathsf{I}}} - \frac{\mathbf{a}\boldsymbol{\alpha}}{1 + \mathbf{a}|\boldsymbol{\alpha}}. \tag{2.187}$$

The result can be easily checked as

$$(\mathrm{Adj}\,\overline{\overline{\mathsf{A}}})|\overline{\overline{\mathsf{A}}} = ((1 + \mathbf{a}|\boldsymbol{\alpha})\overline{\overline{\mathsf{I}}} - \mathbf{a}\boldsymbol{\alpha})|(\overline{\overline{\mathsf{I}}} + \mathbf{a}\boldsymbol{\alpha}) = (1 + \mathbf{a}|\boldsymbol{\alpha})\overline{\overline{\mathsf{I}}} = \det\overline{\overline{\mathsf{A}}}\,\overline{\overline{\mathsf{I}}}. \tag{2.188}$$

2.4.4 Isotropic and uniaxial dyadics

Dyadics $\overline{\overline{\mathsf{A}}} \in \mathbb{E}_1\mathbb{F}_1$ which do not change the polarization of a vector are called isotropic. Two non-null vectors $\mathbf{a}_1, \mathbf{a}_2$ have the same polarization if they are parallel, satisfying $\mathbf{a}_1 \wedge \mathbf{a}_2 = 0$. This is equivalent to the fact that there exist a scalar α such that $\mathbf{a}_2 = \alpha\mathbf{a}_1$. If the dyadic $\overline{\overline{\mathsf{A}}}$ maps $\mathbf{a}_1$ to $\mathbf{a}_2$, the condition of isotropy is thus

$$\overline{\overline{\mathsf{A}}}|\mathbf{a}_1 = \alpha\mathbf{a}_1, \tag{2.189}$$

which must be valid for any vector $\mathbf{a}_1$. The scalar α must be independent of $\mathbf{a}_1$ because the left-hand side is a linear function of $\mathbf{a}_1$. This implies $(\overline{\overline{\mathsf{A}}} - \alpha\overline{\overline{\mathsf{I}}})|\mathbf{a}_1 = 0$

for all $\mathbf{a}_1$, whence we must have $\overline{\overline{\mathsf{A}}} = \alpha\overline{\overline{\mathsf{I}}}$. The conclusion is that isotropic dyadics are multiples of the unit dyadic. The concept of isotropy can be extended to dyadics in the space $\mathbb{E}_p\mathbb{F}_p$ with the obvious result that the isotropic p-dyadics must be multiples of $\overline{\overline{\mathsf{I}}}^{(p)}$.

If there exists a special direction in the space of n dimensions, defined by a basis vector labeled as $\mathbf{e}_n$, vectors parallel to $\mathbf{e}_n$ correspond to that special direction. All other basis vectors $\mathbf{e}_1 \cdots \mathbf{e}_{n-1}$ make a $(n-1)$-dimensional subspace. Since the $(n-1)$-vector $\mathbf{e}_{K(n)}$ is independent of the particular choice of the vectors $\mathbf{e}_1 \cdots \mathbf{e}_{n-1}$ except for a scalar coefficient, we can define the reciprocal dual basis vector $\boldsymbol{\varepsilon}_n$ uniquely from (2.149), where $i = n$. Now we can call a dyadic $\overline{\overline{\mathsf{A}}} \in \mathbb{E}_1\mathbb{F}_1$ uniaxial if it maps the vector $\mathbf{e}_n$ to a vector parallel to itself and vectors $\mathbf{a}$ satisfying $\boldsymbol{\varepsilon}_n|\mathbf{a} = 0$ to vectors parallel to themselves. Such vectors $\mathbf{a}$ form an $(n-1)$-dimensional subspace in the n-dimensional vector space. Obviously, any uniaxial dyadic is a sum of a multiple of the unit dyadic corresponding to the vector subspace and a multiple of the dyad $\mathbf{e}_n\boldsymbol{\varepsilon}_n$:

$$\overline{\overline{\mathsf{A}}} = \alpha\sum_{i=1}^{n-1}\mathbf{e}_i\boldsymbol{\varepsilon}_i + \beta\mathbf{e}_n\boldsymbol{\varepsilon}_n. \tag{2.190}$$

In particular, in the Minkowskian four-dimensional space there is one special basis vector $\mathbf{e}_4$ which corresponds to the time and the orthogonal three-dimensional subspace corresponds to the Euclidean space. Thus, the uniaxial dyadics in the Minkowskian space can be expressed as

$$\overline{\overline{\mathsf{A}}} = \alpha\overline{\overline{\mathsf{I}}}_{\mathbf{E}} + \beta\mathbf{e}_4\boldsymbol{\varepsilon}_4, \tag{2.191}$$

where $\overline{\overline{\mathsf{I}}}_E$ is the three-dimensional Euclidean unit dyadic. Extending this to dyadics $\overline{\overline{\mathsf{A}}} \in \mathbb{E}_2\mathbb{F}_2$ gives the result that the corresponding uniaxial bivector dyadics must be of the form

$$\overline{\overline{\mathsf{A}}} = \alpha\overline{\overline{\mathsf{I}}}_{\mathbf{E}}^{(2)} + \beta\overline{\overline{\mathsf{I}}}_{\mathbf{E}}{}^{\wedge}_{\wedge}\mathbf{e}_4\boldsymbol{\varepsilon}_4. \tag{2.192}$$

Problems

2.4.1 Solve the inverse of the dyadic $\overline{\overline{\mathsf{A}}} = \overline{\overline{\mathsf{I}}} + \mathbf{a}\boldsymbol{\alpha}$ directly by solving for $\mathbf{x}$ in the equation $\overline{\overline{\mathsf{A}}}|\mathbf{x} = \mathbf{y}$, without using the formula (2.162).

2.4.2 Prove the dyadic identity

$$\overline{\overline{\mathsf{I}}}^{(n)}\lfloor\lfloor\overline{\overline{\mathsf{I}}}^{(n-1)T} = \overline{\overline{\mathsf{I}}}$$

2.4.3 Prove the generalization of the previous identity

$$\overline{\overline{\mathsf{I}}}^{(n)}\lfloor\lfloor\overline{\overline{\mathsf{I}}}^{(n-p)T} = \overline{\overline{\mathsf{I}}}^{(p)}, \quad 0 < p < n.$$

Hint: Because the resulting dyadic must be in the space $\mathbb{E}_p\mathbb{F}_p$ and there are no other dyadics involved, the result must be of the form $\alpha\overline{\overline{\mathsf{I}}}^{(p)}$. Find the scalar α by using the identity

$$\overline{\overline{\mathsf{I}}}^{(n-p)}{}^{\wedge}_{\wedge}\overline{\overline{\mathsf{I}}}^{(p)} = (\mathrm{tr}\,\overline{\overline{\mathsf{I}}}^{(p)})\overline{\overline{\mathsf{I}}}^{(n)},$$

which is easier to prove.

2.4.4 Prove the identity

$$\overline{\overline{\mathsf{I}}}^{(2)}\lfloor\lfloor\boldsymbol{\alpha}\mathbf{a} = (\boldsymbol{\alpha}|\mathbf{a})\overline{\overline{\mathsf{I}}} - \mathbf{a}\boldsymbol{\alpha},$$

and its generalization

$$\overline{\overline{\mathsf{I}}}^{(2)}\lfloor\lfloor\overline{\overline{\mathsf{A}}}^T = (\mathrm{tr}\,\overline{\overline{\mathsf{A}}})\overline{\overline{\mathsf{I}}} - \overline{\overline{\mathsf{A}}}.$$

2.4.5 Prove

$$\overline{\overline{\mathsf{I}}}^{(p)}{}^\wedge_\wedge\overline{\overline{\mathsf{I}}}^{(n-p)} = \mathrm{tr}\,\overline{\overline{\mathsf{I}}}^{(p)}\,\overline{\overline{\mathsf{I}}}^{(n)}.$$

2.4.6 Prove the identity

$$\overline{\overline{\mathsf{I}}}^{(2)}\lfloor\lfloor\boldsymbol{\alpha}\mathbf{a} = (\boldsymbol{\alpha}|\mathbf{a})\overline{\overline{\mathsf{I}}} - \mathbf{a}\boldsymbol{\alpha},$$

which leads to

$$\overline{\overline{\mathsf{I}}}^{(2)}\lfloor\lfloor\overline{\overline{\mathsf{A}}}^T = (\mathrm{tr}\,\overline{\overline{\mathsf{A}}})\overline{\overline{\mathsf{I}}} - \overline{\overline{\mathsf{A}}}.$$

2.4.7 Verify that the $\mathbb{E}_1\mathbb{F}_1$-dyadic $\overline{\overline{\mathsf{B}}} = \overline{\overline{\mathsf{A}}} + \mathbf{a}\boldsymbol{\alpha}$ has the inverse

$$\overline{\overline{\mathsf{B}}}^{-1} = \overline{\overline{\mathsf{A}}}^{-1} - \frac{\overline{\overline{\mathsf{A}}}^{-1}|\mathbf{a}\boldsymbol{\alpha}|\overline{\overline{\mathsf{A}}}^{-1}}{1 + \boldsymbol{\alpha}|\overline{\overline{\mathsf{A}}}^{-1}|\mathbf{a}},$$

which is known if $\overline{\overline{\mathsf{A}}}^{-1}$ is known.

2.4.8 Show that

$$\overline{\overline{\mathsf{A}}}^{(p)}{}^\wedge_\wedge\overline{\overline{\mathsf{I}}}^{(n-p)} = [\mathrm{tr}\,\overline{\overline{\mathsf{A}}}^{(p)}]\overline{\overline{\mathsf{I}}}^{(n)}.$$

2.4.9 Show that

$$(\overline{\overline{\mathsf{A}}}^{-1})^{(p)} = (\overline{\overline{\mathsf{A}}}^{(p)})^{-1}.$$

2.4.10 Expand $(\overline{\overline{\mathsf{A}}}{}^\wedge_\wedge\overline{\overline{\mathsf{I}}}^{(n-2)})\rfloor\rfloor\overline{\overline{\mathsf{I}}}^{(n)T}$, where $\overline{\overline{\mathsf{A}}} \in \mathbb{E}_1\mathbb{F}_1$.

2.4.11 Derive the following expression for the inverse of a Euclidean ($n = 3$) bivector dyadic $\overline{\overline{\mathsf{A}}} \in \mathbb{E}_2\mathbb{F}_2$:

$$\overline{\overline{\mathsf{A}}}^{-1} = \frac{1}{\mathrm{Det}\overline{\overline{\mathsf{A}}}}(\overline{\overline{\mathsf{A}}}^T\rfloor\rfloor\overline{\overline{\mathsf{I}}}^{(3)})^{(2)}, \quad \mathrm{Det}\,\overline{\overline{\mathsf{A}}} \neq 0,$$

where $\mathrm{Det}\overline{\overline{\mathsf{A}}}$ is the determinant of a bivector dyadic and defined by

$$\mathrm{Det}\,\overline{\overline{\mathsf{A}}} = \det(\overline{\overline{\mathsf{A}}}^T\rfloor\rfloor\overline{\overline{\mathsf{I}}}^{(3)}) = (\overline{\overline{\mathsf{A}}}^T\rfloor\rfloor\overline{\overline{\mathsf{I}}}^{(3)})^{(3)}||\overline{\overline{\mathsf{I}}}^{(3)T}.$$

Hint: Show that a basis of bivectors $\mathbf{A}_1, \mathbf{A}_2, \mathbf{A}_3$ defines the reciprocal dual bivector basis

$$\boldsymbol{\Phi}_1 = \frac{[\mathbf{A}_2\mathbf{A}_3]}{(\mathbf{A}_1\mathbf{A}_2\mathbf{A}_3)}, \quad \boldsymbol{\Phi}_2 = \frac{[\mathbf{A}_3\mathbf{A}_1]}{(\mathbf{A}_1\mathbf{A}_2\mathbf{A}_3)}, \quad \boldsymbol{\Phi}_3 = \frac{[\mathbf{A}_1\mathbf{A}_2]}{(\mathbf{A}_1\mathbf{A}_2\mathbf{A}_3)},$$

after defining the antisymmetric dual bivector function

$$[\mathbf{BC}] = (\mathbf{B}\rfloor\boldsymbol{\kappa}) \wedge (\mathbf{C}\rfloor\boldsymbol{\kappa}) \in \mathbb{F}_2$$

and the scalar function of three bivectors

$$(\mathbf{ABC}) = \mathbf{A}|[\mathbf{BC}] = \mathbf{k}|((\mathbf{A}\rfloor\boldsymbol{\kappa}) \wedge (\mathbf{B}\rfloor\boldsymbol{\kappa}) \wedge (\mathbf{C}\rfloor\boldsymbol{\kappa})).$$

The trivector $\mathbf{k}$ and the dual trivector $\boldsymbol{\kappa} \in \mathbb{F}_3$ satisfy $\mathbf{k}|\boldsymbol{\kappa} = 1$. Expand the general bivector dyadic as $\overline{\overline{\mathsf{A}}} = \mathbf{A}_1\boldsymbol{\Psi}_1 + \mathbf{A}_2\boldsymbol{\Psi}_2 + \mathbf{A}_3\boldsymbol{\Psi}_3$, where $\{\boldsymbol{\Psi}_i\}$ is another basis of dual bivectors. Finally show that

$$\text{Det}\,\overline{\overline{\mathsf{A}}} = (\mathbf{A}_1\mathbf{A}_2\mathbf{A}_3)(\boldsymbol{\Psi}_1\boldsymbol{\Psi}_2\boldsymbol{\Psi}_2).$$

2.4.12 Applying the formula for the inverse of a bivector dyadic given in the previous problem prove that, for $n = 3$, any bivector dyadic $\overline{\overline{\mathsf{A}}} \in \mathbb{E}_2\mathbb{F}_2$ satisfying $\text{Det}\,\overline{\overline{\mathsf{A}}} \neq 0$ can be expressed in the form $\overline{\overline{\mathsf{A}}} = \overline{\overline{\mathsf{Q}}}^{(2)}$ in terms of some dyadic $\overline{\overline{\mathsf{Q}}} \in \mathbb{E}_1\mathbb{F}_1$. Similar property is valid also for metric dyadics $\overline{\overline{\mathsf{A}}} \in \mathbb{E}_2\mathbb{E}_2$.

2.4.13 Show that the determinant of a Euclidean ($n = 3$) bivector dyadic $\overline{\overline{\mathsf{A}}} \in \mathbb{E}_2\mathbb{F}_2$ defined in Problem 2.4.11 satisfies

$$\text{Det}\,\overline{\overline{\mathsf{A}}}^{-1} = 1/\text{Det}\,\overline{\overline{\mathsf{A}}}$$

when $\text{Det}\overline{\overline{\mathsf{A}}} \neq 0$.

2.4.14 Assuming a Minkowskian ($n = 4$) bivector dyadic $\overline{\overline{\mathsf{A}}} \in \mathbb{E}_2\mathbb{E}_2$ of the form

$$\overline{\overline{\mathsf{A}}} = \sum_{i<j}(\mathbf{a}_i \wedge \mathbf{a}_j)(\mathbf{b}_i \wedge \mathbf{b}_j),$$

and assuming that $\{\mathbf{a}_i\}$ and $\{\mathbf{b}_j\}$, $i, j = 1\cdots 4$, make two systems of basis vectors, show that its inverse can be expressed in the form

$$\overline{\overline{\mathsf{A}}}^{-1} = 3\frac{\boldsymbol{\kappa}_N\boldsymbol{\kappa}_N \lfloor\lfloor \overline{\overline{\mathsf{A}}}^T}{\boldsymbol{\kappa}_N\boldsymbol{\kappa}_N || \overline{\overline{\mathsf{A}}}^{(2)}}$$

where $\boldsymbol{\kappa}_N$ is some dual quadrivector.

2.4.15 Expanding the general Minkowskian ($n = 4$) bivector dyadic $\overline{\overline{\mathsf{M}}} \in \mathbb{E}_2\mathbb{F}_2$ as

$$\overline{\overline{\mathsf{M}}} = \overline{\overline{\mathsf{A}}} + \mathbf{e}_4 \wedge \overline{\overline{\mathsf{B}}} + \overline{\overline{\mathsf{C}}} \wedge \boldsymbol{\varepsilon}_4 + \mathbf{e}_4 \wedge \overline{\overline{\mathsf{D}}} \wedge \boldsymbol{\varepsilon}_4,$$

where $\overline{\overline{\mathsf{A}}} \in \mathbb{E}_2\mathbb{F}_2$, $\overline{\overline{\mathsf{B}}} \in \mathbb{E}_1\mathbb{F}_2$, $\overline{\overline{\mathsf{C}}} \in \mathbb{E}_2\mathbb{F}_1$ and $\overline{\overline{\mathsf{D}}} \in \mathbb{E}_1\mathbb{F}_1$ are Euclidean ($n = 3$) dyadics, show that its inverse can be expressed as

$$\begin{aligned}\overline{\overline{\mathsf{M}}}^{-1} &= (\overline{\overline{\mathsf{A}}} - \overline{\overline{\mathsf{C}}}|\overline{\overline{\mathsf{D}}}^{-1}|\overline{\overline{\mathsf{B}}})^{-1} - (\overline{\overline{\mathsf{B}}} - \overline{\overline{\mathsf{D}}}|\overline{\overline{\mathsf{C}}}^{-1}|\overline{\overline{\mathsf{A}}})^{-1} \wedge \boldsymbol{\varepsilon}_4 \\ &\quad -\mathbf{e}_4 \wedge (\overline{\overline{\mathsf{C}}} - \overline{\overline{\mathsf{A}}}|\overline{\overline{\mathsf{B}}}^{-1}|\overline{\overline{\mathsf{D}}})^{-1} + \mathbf{e}_e \wedge (\overline{\overline{\mathsf{D}}} - \overline{\overline{\mathsf{B}}}|\overline{\overline{\mathsf{A}}}^{-1}|\overline{\overline{\mathsf{C}}})^{-1} \wedge \boldsymbol{\varepsilon}_4.\end{aligned}$$

Here we assume that the inverses of the dyadics $\overline{\overline{\mathsf{A}}}\cdots\overline{\overline{\mathsf{D}}}$ exist.

2.4.16 Check the result of the previous problem by forming the product $\overline{\overline{\mathsf{M}}}|\overline{\overline{\mathsf{M}}}^{-1}$.

2.5 METRIC DYADICS

Metric dyadics represent mappings between multivectors and dual multivectors of the same grade ($\mathbb{E}_p \to \mathbb{F}_p$ or $\mathbb{F}_p \to \mathbb{E}_p$). They are elements of the respective dyadic spaces $\mathbb{F}_p\mathbb{F}_p$ and $\mathbb{E}_p\mathbb{E}_p$. Symmetric metric dyadics allow one to define the dot product (scalar product) between two multivectors or two dual multivectors of the same grade, which gives a means to measure magnitudes of multivectors and dual multivectors.

2.5.1 Dot product

The dot product of two vectors $\mathbf{a}, \mathbf{b}$ or two dual vectors $\boldsymbol{\alpha}, \boldsymbol{\beta}$ is a bilinear scalar symmetric function of its arguments,

$$f(\mathbf{a}, \mathbf{b}) = \mathbf{a} \cdot \mathbf{b} = \mathbf{b} \cdot \mathbf{a}, \quad f(\boldsymbol{\alpha}, \boldsymbol{\beta}) = \boldsymbol{\alpha} \cdot \boldsymbol{\beta} = \boldsymbol{\beta} \cdot \boldsymbol{\alpha}, \tag{2.193}$$

and the definition is straightforwardly extended to multivectors and dual multivectors. Because $\boldsymbol{\alpha} \cdot \boldsymbol{\beta}$ is a linear function of $\boldsymbol{\beta}$, it can be expressed in the form $\boldsymbol{\alpha} \cdot \boldsymbol{\beta} = \mathbf{f}(\boldsymbol{\alpha})|\boldsymbol{\beta}$, in terms of a linear vector function $\mathbf{f}(\boldsymbol{\alpha})$. Because a linear vector function $\mathbf{f}(\boldsymbol{\alpha})$ can be expressed in terms of a dyadic $\overline{\overline{\mathsf{G}}} \in \mathbb{E}_1\mathbb{E}_1$ as $\mathbf{f}(\boldsymbol{\alpha}) = \overline{\overline{\mathsf{G}}}|\boldsymbol{\alpha}$, the dot product between two dual vectors depends on the chosen dyadic $\overline{\overline{\mathsf{G}}}$. Similarly, the dot product between two vectors depends on a dyadic $\overline{\overline{\Gamma}} \in \mathbb{F}_1\mathbb{F}_1$ (Figure 2.5). Symmetry of the dot products requires symmetry of the dyadics $\overline{\overline{\mathsf{G}}}, \overline{\overline{\Gamma}}$:

$$\mathbf{a} \cdot \mathbf{b} = \mathbf{a}|\overline{\overline{\Gamma}}|\mathbf{b} = \mathbf{b}|\overline{\overline{\Gamma}}|\mathbf{a}, \quad \overline{\overline{\Gamma}}^T = \overline{\overline{\Gamma}}, \tag{2.194}$$

$$\boldsymbol{\alpha} \cdot \boldsymbol{\beta} = \boldsymbol{\alpha}|\overline{\overline{\mathsf{G}}}|\boldsymbol{\beta} = \boldsymbol{\beta}|\overline{\overline{\mathsf{G}}}|\boldsymbol{\alpha}, \quad \overline{\overline{\mathsf{G}}}^T = \overline{\overline{\mathsf{G}}}. \tag{2.195}$$

Because of the definition above, nonsymmetric dyadics in the spaces $\mathbb{E}_1\mathbb{E}_1$, $\mathbb{F}_1\mathbb{F}_1$ cannot strictly be called metric dyadics and should rather be called generalized metric dyadics. However, for simplicity, all dyadics in these spaces, symmetric or nonsymmetric, will be called metric dyadics.

Because $\mathbf{a} \cdot \mathbf{a}$ produces a scalar, its value gives an idea of the magnitude of the vector $\mathbf{a}$. However, this scalar may be zero for $\mathbf{a} \neq 0$ in which case the metric dyadic is called indefinite. Or it may be a complex number for a complex vector $\mathbf{a}$. In these cases the magnitude concept is not so simple.

2.5.2 Metric dyadics

Two metric dyadics $\overline{\overline{\mathsf{G}}} \in \mathbb{E}_1\mathbb{E}_1$ and $\overline{\overline{\Gamma}} \in \mathbb{F}_1\mathbb{F}_1$ could basically be chosen independently as any real and symmetric dyadics with finite inverses. However, let us assume that the two metric dyadics are related and of the form

$$\overline{\overline{\mathsf{G}}} = \sum_{i=1}^{n} \sigma_i \mathbf{e}_i \mathbf{e}_i, \quad \overline{\overline{\Gamma}} = \sum_{i=1}^{n} \frac{1}{\sigma_i} \boldsymbol{\varepsilon}_i \boldsymbol{\varepsilon}_i, \tag{2.196}$$

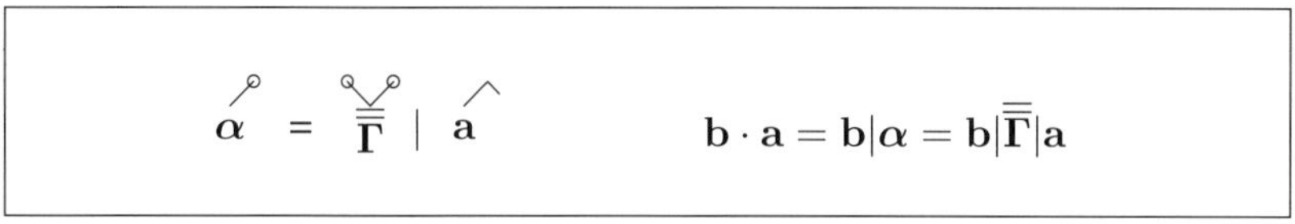

Fig. 2.5 *Dot product of two vectors is defined through a metric dyadic* $\overline{\overline{\Gamma}}$.

where $\{\mathbf{e}_i\}$ and $\{\boldsymbol{\varepsilon}_i\}$ are reciprocal vector and dual vector bases. In this case we call the two metric dyadics reciprocal to one another. The signature coefficients $\{\sigma_i\}$ are assumed real and either $+1$ or -1. Initially any finite values could have been chosen for the σ_i, but they can be normalized to ± 1 by changing the basis vectors accordingly. When all basis vectors are real, the case $\sigma_i = +1$ or $\sigma_i = -1$ for all i is called definite metric and other cases indefinite. Allowing complex basis vectors this definition does not have a meaning since any $\sigma_i = -1$ can be replaced by $\sigma_i = +1$ when replacing the corresponding basis vector $\mathbf{e}_i$ by $j\mathbf{e}_i$.

Let us concentrate on two special cases with real basis vectors: (1) the definite Euclidean metric: $\sigma_i = 1$ for all $i = 1 \cdots n$; and (2) the indefinite Minkowskian metric in four dimensions: $\sigma_1 = \sigma_2 = \sigma_3 = 1$, $\sigma_4 = -1$. For $n = 3$ the Euclidean metric dyadics and for $n = 4$ the Minkowskian metric dyadics are denoted by

$$\overline{\overline{\mathsf{G}}}_{\mathrm{E}} = \mathbf{e}_1\mathbf{e}_1 + \mathbf{e}_2\mathbf{e}_2 + \mathbf{e}_3\mathbf{e}_3, \tag{2.197}$$

$$\overline{\overline{\Gamma}}_{\mathrm{E}} = \boldsymbol{\varepsilon}_1\boldsymbol{\varepsilon}_1 + \boldsymbol{\varepsilon}_2\boldsymbol{\varepsilon}_2 + \boldsymbol{\varepsilon}_3\boldsymbol{\varepsilon}_3, \tag{2.198}$$

$$\overline{\overline{\mathsf{G}}}_{\mathrm{M}} = \mathbf{e}_1\mathbf{e}_1 + \mathbf{e}_2\mathbf{e}_2 + \mathbf{e}_3\mathbf{e}_3 - \mathbf{e}_4\mathbf{e}_4 = \overline{\overline{\mathsf{G}}}_{\mathrm{E}} - \mathbf{e}_4\mathbf{e}_4, \tag{2.199}$$

$$\overline{\overline{\Gamma}}_{\mathrm{M}} = \boldsymbol{\varepsilon}_1\boldsymbol{\varepsilon}_1 + \boldsymbol{\varepsilon}_2\boldsymbol{\varepsilon}_2 + \boldsymbol{\varepsilon}_3\boldsymbol{\varepsilon}_3 - \boldsymbol{\varepsilon}_4\boldsymbol{\varepsilon}_4 = \overline{\overline{\Gamma}}_{\mathrm{E}} - \boldsymbol{\varepsilon}_4\boldsymbol{\varepsilon}_4. \tag{2.200}$$

The subscripts E and M will be suppressed when there is no fear of misunderstanding.

The choice of basis $\{\mathbf{e}_i\}$ and signature σ_i defines the metric dyadic $\overline{\overline{\mathsf{G}}}$ uniquely. It also defines the dual metric dyadic $\overline{\overline{\Gamma}}$ uniquely, because $\overline{\overline{\Gamma}}$ happens to be the inverse of the dyadic $\overline{\overline{\mathsf{G}}}$. To see this, we can write from the orthogonality of the reciprocal bases

$$\overline{\overline{\mathsf{G}}}|\overline{\overline{\Gamma}} = \sum_{i,j} \frac{\sigma_i}{\sigma_j}\mathbf{e}_i\mathbf{e}_i|\boldsymbol{\varepsilon}_j\boldsymbol{\varepsilon}_j = \sum_i \mathbf{e}_i\boldsymbol{\varepsilon}_i = \overline{\overline{\mathsf{I}}}, \tag{2.201}$$

$$\overline{\overline{\Gamma}}|\overline{\overline{\mathsf{G}}} = \sum_{i,j} \frac{\sigma_i}{\sigma_j}\boldsymbol{\varepsilon}_j\boldsymbol{\varepsilon}_j|\mathbf{e}_i\mathbf{e}_i = \sum_i \boldsymbol{\varepsilon}_i\mathbf{e}_i = \overline{\overline{\mathsf{I}}}^T, \tag{2.202}$$

whence

$$\overline{\overline{\Gamma}} = \overline{\overline{\mathsf{G}}}^{-1}, \quad \overline{\overline{\mathsf{G}}} = \overline{\overline{\Gamma}}^{-1}. \tag{2.203}$$

2.5.3 Properties of the dot product

The basis vectors and dual vectors defining the metric dyadics are orthogonal in the dot product:

$$\mathbf{e}_i \cdot \mathbf{e}_j = \mathbf{e}_i|\overline{\overline{\Gamma}}|\mathbf{e}_j = \sigma_i\delta_{ij}, \quad \boldsymbol{\varepsilon}_i \cdot \boldsymbol{\varepsilon}_j = \boldsymbol{\varepsilon}_i|\overline{\overline{\mathsf{G}}}|\boldsymbol{\varepsilon}_j = \sigma_i\delta_{ij}. \tag{2.204}$$

As an example we write the dot product of two Minkowskian vectors as

$$\begin{aligned}\mathbf{a}_\mathrm{M}\cdot\mathbf{b}_\mathrm{M} &= \sum_{i,j=1}^{4} a_i\mathbf{e}_i|\overline{\overline{\Gamma}}_\mathrm{M}|b_j\mathbf{e}_j = \sum_{i=1}^{4} a_i b_i\sigma_i \\ &= a_1b_1 + a_2b_2 + a_3b_3 - a_4b_4. \qquad (2.205)\end{aligned}$$

In the Minkowskian metric one should note that the special basis vector $\mathbf{e}_4$ satisfying $\mathbf{e}_4\cdot\mathbf{e}_4 = -1$ is not an imaginary unit vector, but that the dot product involves a minus sign. This difference is essential if we consider vectors with complex-valued coefficients a_i. The metric in a Euclidean space is definite in the sense that $\mathbf{a}\cdot\mathbf{a}\geq 0$ for real vectors $\mathbf{a}$ and $\mathbf{a}\cdot\mathbf{a}=0$ only for $\mathbf{a}=0$. In the Minkowskian space the metric is indefinite because we can have $\mathbf{a}_\mathrm{M}\cdot\mathbf{a}_\mathrm{M} < 0$ for real vectors $\mathbf{a}_\mathrm{M}$ and $\mathbf{a}_\mathrm{M}\cdot\mathbf{a}_\mathrm{M} = 0$ can be valid for some vectors, e.g., for $\mathbf{a}_\mathrm{M} = \mathbf{e}_1 + \mathbf{e}_4 \neq 0$.

2.5.4 Metric in multivector spaces

The metric dyadic $\overline{\overline{\Gamma}}$ induces a metric dyadic in the p-vector space as the pth double-wedge power:

$$(\overline{\overline{\Gamma}}|\mathbf{a}_1)\wedge(\overline{\overline{\Gamma}}|\mathbf{a}_2)\wedge\cdots\wedge(\overline{\overline{\Gamma}}|\mathbf{a}_p) = \overline{\overline{\Gamma}}^{(p)}|(\mathbf{a}_1\wedge\mathbf{a}_2\wedge\cdots\wedge\mathbf{a}_p). \qquad (2.206)$$

Thus, $\overline{\overline{\Gamma}}^{(p)}$ maps a p-vector to a dual p-vector. The dot product of two basis p-vectors $\mathbf{e}_J^p$ with the ordered multi-index $J = j_1 j_2\cdots j_p$, can be defined as

$$\mathbf{e}_J^p\cdot\mathbf{e}_J^p = \mathbf{e}_J^p|\overline{\overline{\Gamma}}^{(p)}|\mathbf{e}_J^p = \sigma_J, \qquad (2.207)$$

with the multi-index signature defined as the product

$$\sigma_J = \sigma_{j_1}\sigma_{j_2}\cdots\sigma_{j_p}. \qquad (2.208)$$

The basis multivectors satisfy orthogonality with respect to the dot product as

$$\mathbf{e}_J^p\cdot\mathbf{e}_K^p = \sigma_J\delta_{JK} = \sigma_J\delta_{j_1k_1}\delta_{j_2k_2}\cdots\delta_{j_pk_p}. \qquad (2.209)$$

For example, the double-wedge square of a metric dyadic $\overline{\overline{\Gamma}}$ is

$$\overline{\overline{\Gamma}}^{(2)} = \frac{1}{2}\overline{\overline{\Gamma}}{}^\wedge_\wedge\overline{\overline{\Gamma}} = \sum_{i<j}\sigma_i\sigma_j\boldsymbol{\varepsilon}_{ij}\boldsymbol{\varepsilon}_{ij}, \qquad (2.210)$$

which can be generalized to any power p.

Instead of continuing with the general case, let us again consider the Euclidean and Minkowskian spaces which are of particular interest in the electromagnetic theory. The different multi-wedge powers of metric dyadics can be expressed as

$$\overline{\overline{\mathsf{G}}}_\mathrm{E}^{(2)} = \mathbf{e}_{12}\mathbf{e}_{12} + \mathbf{e}_{23}\mathbf{e}_{23} + \mathbf{e}_{31}\mathbf{e}_{31}, \qquad (2.211)$$

$$\overline{\overline{\mathsf{G}}}_{\mathbf{E}}^{(3)} = \mathbf{e}_{123}\mathbf{e}_{123}, \tag{2.212}$$

$$\overline{\overline{\mathsf{G}}}_{\mathbf{M}}^{(2)} = \mathbf{e}_{12}\mathbf{e}_{12} + \mathbf{e}_{23}\mathbf{e}_{23} + \mathbf{e}_{31}\mathbf{e}_{31} - \mathbf{e}_{14}\mathbf{e}_{14} - \mathbf{e}_{24}\mathbf{e}_{24} - \mathbf{e}_{34}\mathbf{e}_{34}, \tag{2.213}$$

$$\overline{\overline{\mathsf{G}}}_{\mathbf{M}}^{(3)} = \mathbf{e}_{123}\mathbf{e}_{123} - \mathbf{e}_{124}\mathbf{e}_{124} - \mathbf{e}_{234}\mathbf{e}_{234} - \mathbf{e}_{314}\mathbf{e}_{314}, \tag{2.214}$$

$$\overline{\overline{\mathsf{G}}}_{\mathbf{M}}^{(4)} = -\mathbf{e}_{1234}\mathbf{e}_{1234}. \tag{2.215}$$

Similar dual expansions can be written for the powers of $\overline{\overline{\Gamma}}_{\mathbf{E}}$ and $\overline{\overline{\Gamma}}_{\mathbf{M}}$ by changing $\mathbf{e}$ to $\boldsymbol{\varepsilon}$. The relations between these metric dyadics are

$$\overline{\overline{\mathsf{G}}}_{\mathbf{M}}^{(2)} = \frac{1}{2}(\overline{\overline{\mathsf{G}}}_{\mathbf{E}} - \mathbf{e}_4\mathbf{e}_4)^{\wedge}_{\wedge}(\overline{\overline{\mathsf{G}}}_{\mathbf{E}} - \mathbf{e}_4\mathbf{e}_4) = \overline{\overline{\mathsf{G}}}_{\mathbf{E}}^{(2)} - \overline{\overline{\mathsf{G}}}_{\mathbf{E}}{}^{\wedge}_{\wedge}\mathbf{e}_4\mathbf{e}_4. \tag{2.216}$$

$$\overline{\overline{\mathsf{G}}}_{\mathbf{M}}^{(3)} = \frac{1}{6}(\overline{\overline{\mathsf{G}}}_{\mathbf{E}} - \mathbf{e}_4\mathbf{e}_4)^{\wedge}_{\wedge}(\overline{\overline{\mathsf{G}}}_{\mathbf{E}} - \mathbf{e}_4\mathbf{e}_4)^{\wedge}_{\wedge}(\overline{\overline{\mathsf{G}}}_{\mathbf{E}} - \mathbf{e}_4\mathbf{e}_4) = \overline{\overline{\mathsf{G}}}_{\mathbf{E}}^{(3)} - \overline{\overline{\mathsf{G}}}_{\mathbf{E}}^{(2)}{}^{\wedge}_{\wedge}\mathbf{e}_4\mathbf{e}_4. \tag{2.217}$$

$$\overline{\overline{\mathsf{G}}}_{\mathbf{M}}^{(4)} = -\overline{\overline{\mathsf{G}}}_{\mathbf{E}}^{(3)}{}^{\wedge}_{\wedge}\mathbf{e}_4\mathbf{e}_4. \tag{2.218}$$

The dot product can now be extended to multivectors and dual multivectors. For example, for two bivectors in the four-dimensional Minkowskian space we have

$$\begin{aligned}\mathbf{A}_{\mathbf{M}} \cdot \mathbf{B}_{\mathbf{M}} &= \sum_{i,j=1}^{4} A_{ij}\mathbf{e}_{ij}|\overline{\overline{\Gamma}}_{\mathbf{M}}{}^{(2)}|\sum_{k,\ell=1}^{4} B_{k\ell}\mathbf{e}_{k\ell} \\ &= A_{12}B_{12} + A_{13}B_{13} + A_{23}B_{23} - A_{14}B_{14} - A_{24}B_{24} - A_{34}B_{34}\end{aligned} \tag{2.219}$$

Another form is obtained when expanding the bivectors as

$$\mathbf{A}_{\mathbf{M}} = \mathbf{A}_{\mathbf{E}} + \mathbf{a}_{\mathbf{E}} \wedge \mathbf{e}_4, \quad \mathbf{B}_{\mathbf{M}} = \mathbf{B}_{\mathbf{E}} + \mathbf{b}_{\mathbf{E}} \wedge \mathbf{e}_4, \tag{2.220}$$

where $\mathbf{a}_{\mathbf{E}}, \mathbf{b}_{\mathbf{E}}$ and $\mathbf{A}_{\mathbf{E}}, \mathbf{B}_{\mathbf{E}}$ are, respectively, Euclidean vectors and bivectors. In this notation the result is

$$\begin{aligned}\mathbf{A}_{\mathbf{M}} \cdot \mathbf{B}_{\mathbf{M}} &= \mathbf{A}_{\mathbf{E}}|\overline{\overline{\Gamma}}_{\mathbf{E}}^{(2)}|\mathbf{B}_{\mathbf{E}} - (\mathbf{a}_{\mathbf{E}} \wedge \mathbf{e}_4)|(\overline{\overline{\Gamma}}_{\mathbf{E}}{}^{\wedge}_{\wedge}\boldsymbol{\varepsilon}_4\boldsymbol{\varepsilon}_4)|(\mathbf{b}_{\mathbf{E}} \wedge \mathbf{e}_4) \\ &= \mathbf{A}_{\mathbf{E}} \cdot \mathbf{B}_{\mathbf{E}} - \mathbf{a}_{\mathbf{E}} \cdot \mathbf{b}_{\mathbf{E}}.\end{aligned} \tag{2.221}$$

The indefinite property of the Minkowskian dot product is clearly seen.

Problems

2.5.1 Study the symmetric $\mathbb{E}_1\mathbb{E}_1$ dyadic $\overline{\overline{\mathsf{D}}} = \mathbf{aa} + \mathbf{bc} + \mathbf{cb}$. Can it be expressed in the form of a three-dimensional metric dyadic $\overline{\overline{\mathsf{G}}} = \sum \sigma_i \mathbf{e}_i\mathbf{e}_i$, $i = 1, 2, 3$, $\sigma_i = \pm 1$? If yes, assuming $\mathbf{a}\wedge\mathbf{b}\wedge\mathbf{c} \neq 0$ check the condition $\mathbf{e}_1 \wedge \mathbf{e}_2 \wedge \mathbf{e}_3 \neq 0$.

2.5.2 Find the inverse of the dyadic in the previous problem.

2.5.3 Show that the dual multivector metric dyadics satisfy $\overline{\overline{\mathsf{G}}}^{(p)} = (\overline{\overline{\Gamma}}^{(p)})^{-1}$.

2.5.4 Find the inverse of the metric dyadic

$$\overline{\overline{\mathsf{G}}} = \overline{\overline{\mathsf{S}}} + \mathbf{aa},$$

where $\overline{\overline{\mathsf{S}}}$ is a metric dyadic whose inverse is assumed known.

2.5.5 Show that the inverse of a bivector metric dyadic $\overline{\overline{\mathsf{A}}} \in \mathbb{E}_2\mathbb{E}_2$ in the three-dimensional space ($n = 3$) can be written as

$$\overline{\overline{\mathsf{A}}}^{-1} = \frac{1}{(\overline{\overline{\mathsf{A}}}\rfloor\rfloor\boldsymbol{\kappa\kappa})^{(3)}||\mathbf{kk}}(\overline{\overline{\mathsf{A}}}^T\rfloor\rfloor\boldsymbol{\kappa\kappa})^{(2)},$$

where $\mathbf{k}$ is a trivector and $\boldsymbol{\kappa} \in \mathbb{F}_3$ a dual trivector satisfying $\mathbf{k}|\boldsymbol{\kappa} = 1$.

2.5.6 Show that in the case $n = 3$ any bivector dyadic $\overline{\overline{\mathsf{A}}} \in \mathbb{E}_2\mathbb{E}_2$ satisfying $(\overline{\overline{\mathsf{A}}}\rfloor\rfloor\boldsymbol{\kappa\kappa})^{(3)} \neq 0$ can be expressed in the form $\overline{\overline{\mathsf{A}}} = \overline{\overline{\mathsf{Q}}}^{(2)}$ for some dyadic $\overline{\overline{\mathsf{Q}}} \in \mathbb{E}_1\mathbb{E}_1$.

2.5.7 Expanding the general Minkowskian ($n = 4$) metric bivector dyadic $\overline{\overline{\mathsf{M}}} \in \mathbb{E}_2\mathbb{E}_2$ as

$$\overline{\overline{\mathsf{M}}} = \overline{\overline{\mathsf{A}}} + \mathbf{e}_4 \wedge \overline{\overline{\mathsf{B}}} + \overline{\overline{\mathsf{C}}} \wedge \mathbf{e}_4 + \mathbf{e}_4 \wedge \overline{\overline{\mathsf{D}}} \wedge \mathbf{e}_4,$$

where $\overline{\overline{\mathsf{A}}} \in \mathbb{E}_2\mathbb{E}_2$, $\overline{\overline{\mathsf{B}}} \in \mathbb{E}_1\mathbb{E}_2$, $\overline{\overline{\mathsf{C}}} \in \mathbb{E}_2\mathbb{E}_1$ and $\overline{\overline{\mathsf{D}}} \in \mathbb{E}_1\mathbb{E}_1$ are Euclidean ($n = 3$) dyadics, show that its inverse $\overline{\overline{\mathsf{M}}}^{-1} \in \mathbb{F}_2\mathbb{F}_2$ can be expressed as

$$\begin{aligned}\overline{\overline{\mathsf{M}}}^{-1} &= (\overline{\overline{\mathsf{A}}} - \overline{\overline{\mathsf{C}}}|\overline{\overline{\mathsf{D}}}^{-1}|\overline{\overline{\mathsf{B}}})^{-1} + (\overline{\overline{\mathsf{A}}} - \overline{\overline{\mathsf{C}}}|\overline{\overline{\mathsf{D}}}^{-1}|\overline{\overline{\mathsf{B}}})^{-1}|\overline{\overline{\mathsf{C}}}|\overline{\overline{\mathsf{D}}}^{-1} \wedge \boldsymbol{\varepsilon}_4 \\ &\quad + \boldsymbol{\varepsilon}_4 \wedge (\overline{\overline{\mathsf{D}}} - \overline{\overline{\mathsf{B}}}|\overline{\overline{\mathsf{A}}}^{-1}|\overline{\overline{\mathsf{C}}})^{-1}|\overline{\overline{\mathsf{B}}}|\overline{\overline{\mathsf{A}}}^{-1} + \boldsymbol{\varepsilon}_4 \wedge (\overline{\overline{\mathsf{D}}} - \overline{\overline{\mathsf{B}}}|\overline{\overline{\mathsf{A}}}^{-1}|\overline{\overline{\mathsf{C}}})^{-1} \wedge \boldsymbol{\varepsilon}_4.\end{aligned}$$

Here we assume the existence of the inverse dyadics $\overline{\overline{\mathsf{A}}}$ and $\overline{\overline{\mathsf{D}}}$.

2.5.8 Check the result of the previous problem by forming the dyadic $\overline{\overline{\mathsf{M}}}|\overline{\overline{\mathsf{M}}}^{-1}$.

2.5.9 Show that the inverse of Problem 2.5.7 can also be written as

$$\overline{\overline{\mathsf{M}}}^{-1} = \overline{\overline{\mathsf{D}}}^{-1}{}^{\wedge}_{\wedge}\boldsymbol{\varepsilon}_4\boldsymbol{\varepsilon}_4 + (\overline{\overline{\mathsf{I}}}_{\mathbf{E}}^{(2)T} - \boldsymbol{\varepsilon}_4\wedge\overline{\overline{\mathsf{D}}}^{-1}|\overline{\overline{\mathsf{B}}})|(\overline{\overline{\mathsf{A}}} + \overline{\overline{\mathsf{C}}}|\overline{\overline{\mathsf{D}}}^{-1}|\overline{\overline{\mathsf{B}}})^{-1}|(\overline{\overline{\mathsf{I}}}_{\mathbf{E}}^{(2)} - \overline{\overline{\mathsf{C}}}|\overline{\overline{\mathsf{D}}}^{-1}\wedge\boldsymbol{\varepsilon}_4).$$

2.5.10 Expanding the general Minkowskian ($n = 4$) metric dyadic $\overline{\overline{\mathsf{Q}}} \in \mathbb{E}_1\mathbb{E}_1$ as

$$\overline{\overline{\mathsf{Q}}} = \overline{\overline{\mathsf{A}}} + \mathbf{e}_4\mathbf{a} + \mathbf{b}\mathbf{e}_4 + c\mathbf{e}_4\mathbf{e}_4,$$

where $\overline{\overline{\mathsf{A}}} \in \mathbb{E}_1\mathbb{E}_1$ is a Euclidean ($n = 3$) dyadic, $\mathbf{a}$ and $\mathbf{b} \in \mathbb{E}_1$ are two Euclidean vectors, and c is a scalar, show that its inverse can be expressed as

$$\overline{\overline{\mathsf{Q}}}^{-1} = \overline{\overline{\mathsf{A}}}^{-1} + \frac{(\overline{\overline{\mathsf{A}}}^{-1}|\mathbf{b} - \boldsymbol{\varepsilon}_4)(\mathbf{a}|\overline{\overline{\mathsf{A}}}^{-1} - \boldsymbol{\varepsilon}_4)}{c - \mathbf{a}|\overline{\overline{\mathsf{A}}}^{-1}|\mathbf{b}},$$

where $\overline{\overline{\mathsf{A}}}^{-1}$ is a Euclidean three-dimensional inverse. *Hint*: Show that $\overline{\overline{\mathsf{Q}}}^{-1}|\overline{\overline{\mathsf{Q}}} = \overline{\overline{\mathsf{I}}}_{\mathbf{M}}^T$, where $\overline{\overline{\mathsf{I}}}_{\mathbf{M}}$ is the Minkowskian four-dimensional unit dyadic.

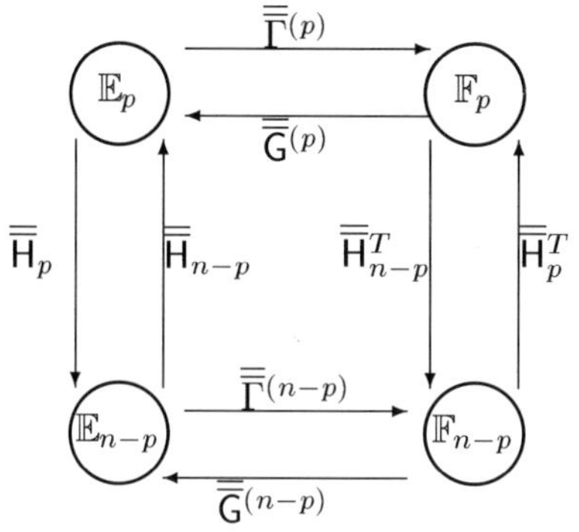

Fig. 2.6 *Mappings between different vector and dual vector spaces of the same dimension* $C_p^n = n!/p!(n-p)! = C_{n-p}^n$.

2.6 HODGE DYADICS

2.6.1 Complementary spaces

There are four spaces of multivectors and dual multivectors which have the same dimension $C_p^n = C_{n-p}^n = n!/p!(n-p)!$, namely $\mathbb{E}_p$ (p-vector space), $\mathbb{F}_p$ (dual p-vector space), $\mathbb{E}_{n-p}$ (complementary p-vector space) and $\mathbb{F}_{n-p}$ (complementary dual p-vector space). However, in the case when n is even and p equals $n/2$ the complementary spaces coincide with the original spaces and there are only two such spaces. For example, in the four-dimensional space, the multivectors complementary to bivectors are also bivectors.

It was seen above that the metric dyadics $\overline{\overline{\mathsf{G}}}^{(p)}, \overline{\overline{\Gamma}}^{(p)}$ define mappings between the dual spaces $\mathbb{E}_p$ and $\mathbb{F}_p$. We now define another set of dyadics mapping $\mathbb{E}_p$ and $\mathbb{F}_p$ to their respective complementary spaces $\mathbb{E}_{n-p}$ and $\mathbb{F}_{n-p}$ and conversely. This kind of dyadics are called Hodge dyadics $\overline{\overline{\mathsf{H}}}_p$ (Figure 2.6), and they are closely related to the metric dyadics.

In the mathematical literature a mapping $\mathbb{E}_p \rightarrow \mathbb{E}_{n-p}$ for any $0 \leq p \leq n$ is usually denoted in shorthand by $*$, which is called the Hodge star operator.[4] Thus, if $\mathbf{a}^p$ is a p-vector and $\mathbf{b}^{n-p}$ is an $(n-p)$-vector, $*\mathbf{a}^p$ is an $(n-p)$-vector and $*\mathbf{b}^{n-p}$ is a p-vector.

The Hodge star operator is shorthand for what we call the Hodge dyadic. The star operator may appear somewhat mystical because its rules are not easily memorizable (see reference 18, Appendix F). While it is not easy to associate any algebraic structure to the star sign, Hodge dyadics can be manipulated like any dyadics. Also, an explicit relation to metric dyadics helps in the analysis. In many cases it may actually be easier to work with metric dyadics than with the corresponding Hodge dyadics.

[4]One has to be careful not to mistake the Hodge star for complex conjugation.

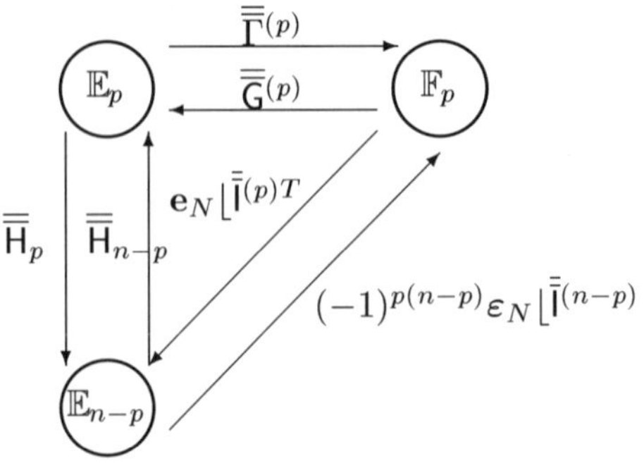

Fig. 2.7 *Diagonal mappings, corresponding to the relations (2.225) and (2.229), can be added to the diagram of Figure 2.6.*

2.6.2 Hodge dyadics

A Hodge dyadic performs a mapping between two complementary multivector spaces. Denoting again for clarity the grade of the multivector by a superscript, mapping of a p-vector $\mathbf{b}^p$ to an $(n-p)$-vector $\mathbf{c}^{n-p}$ can be represented through the Hodge dyadic $\overline{\overline{\mathsf{H}}}_p \in \mathbb{E}_{n-p}\mathbb{F}_p$ as

$$\mathbf{c}^{n-p} = \overline{\overline{\mathsf{H}}}_p|\mathbf{b}^p. \tag{2.222}$$

Let us now define Hodge dyadics $\overline{\overline{\mathsf{H}}}_p$ through their relation to metric dyadics $\overline{\overline{\Gamma}}^{(p)} \in \mathbb{F}_p\mathbb{F}_p$. Because a symmetric metric dyadic defines a dot product, the corresponding Hodge dyadic can also be defined through the dot product. Let us assume that for any two p-vectors $\mathbf{a}^p, \mathbf{b}^p$ and an n-vector $\mathbf{e}_N$ the following relation holds:

$$\mathbf{a}^p \wedge \overline{\overline{\mathsf{H}}}_p|\mathbf{b}^p = \mathbf{e}_N(\mathbf{a}^p \cdot \mathbf{b}^p) = \mathbf{e}_N(\mathbf{a}^p|\overline{\overline{\Gamma}}^{(p)}|\mathbf{b}^p). \tag{2.223}$$

As an example, for $n = 3$, $\overline{\overline{\mathsf{H}}}_1$ maps a vector $\mathbf{b}$ to a bivector whence $\mathbf{a} \wedge \overline{\overline{\mathsf{H}}}_1|\mathbf{b}$ is required to give the trivector $\mathbf{e}_{123}(\mathbf{a}\cdot\mathbf{b})$.

Because (2.223) is valid for any p-vector $\mathbf{b}^p$, the same relation can be written in $\mathbb{E}_n\mathbb{F}_p$ dyadic form as

$$\mathbf{a}^p \wedge \overline{\overline{\mathsf{H}}}_p = \mathbf{e}_N(\mathbf{a}^p|\overline{\overline{\Gamma}}^{(p)}). \tag{2.224}$$

The n-vector $\mathbf{e}_N$ is related to the metric dyadic $\overline{\overline{\mathsf{G}}} = \overline{\overline{\Gamma}}^{-1}$ which is defined in terms of the vector basis $\{\mathbf{e}_i\}$. The relation (2.224) is linear and valid for any p-vector $\mathbf{a}^p$ if it is valid for all basis p-vectors $\mathbf{e}_J^p$. Now the following simple relation between a metric dyadic and the Hodge dyadic can be derived:

$$\overline{\overline{\mathsf{H}}}_p = \mathbf{e}_N\lfloor\overline{\overline{\Gamma}}^{(p)} = \mathbf{e}_N\lfloor\overline{\overline{\mathsf{I}}}^{(p)T}|\overline{\overline{\Gamma}}^{(p)}, \tag{2.225}$$

which states that $\overline{\overline{\Gamma}}^{(p)}$ is mapped to $\overline{\overline{\mathsf{H}}}_p$ through the $\mathbb{E}_{n-p}\mathbb{E}_p$ dyadic $\mathbf{e}_N\lfloor\overline{\overline{\mathsf{I}}}^{(p)T}$. This can be verified through the following identity between different unit dyadics (proof left as an exercise)

$$\overline{\overline{\mathsf{I}}}^{(n-p)} = \overline{\overline{\mathsf{I}}}^{(n)}\lfloor\lfloor\overline{\overline{\mathsf{I}}}^{(p)T} = \mathbf{e}_N\lfloor\sum \boldsymbol{\varepsilon}_J^p(\boldsymbol{\varepsilon}_N\lfloor\mathbf{e}_J^p), \tag{2.226}$$

where summation is over the p-index $J = j_1 j_2 \cdots j_p$. From (2.224) with $\mathbf{a}^p = \mathbf{e}_J^p$ we can arrive at (2.225) through the following steps:

$$\begin{aligned}\overline{\overline{\mathsf{H}}}_p &= \overline{\overline{\mathsf{I}}}^{(n-p)}|\overline{\overline{\mathsf{H}}}_p = \mathbf{e}_N\lfloor\sum \boldsymbol{\varepsilon}_J^p(\boldsymbol{\varepsilon}_N\lfloor\mathbf{e}_J^p)|\overline{\overline{\mathsf{H}}}_p = \mathbf{e}_N\lfloor\sum \boldsymbol{\varepsilon}_J^p(\boldsymbol{\varepsilon}_N|(\mathbf{e}_J^p\wedge\overline{\overline{\mathsf{H}}}_p)) \\ &= \mathbf{e}_N\lfloor\sum \boldsymbol{\varepsilon}_J^p(\boldsymbol{\varepsilon}_N|\mathbf{e}_N)(\mathbf{e}_J^p|\overline{\overline{\Gamma}}^{(p)}) = (\mathbf{e}_N\lfloor\overline{\overline{\mathsf{I}}}^{(p)T})|\overline{\overline{\Gamma}}^{(p)} = \mathbf{e}_N\lfloor\overline{\overline{\Gamma}}^{(p)}. \end{aligned} \tag{2.227}$$

From (2.225) we see that the Hodge dyadic $\overline{\overline{\mathsf{H}}}_p$ is directly related to the metric dyadic $\overline{\overline{\Gamma}}^{(p)}$, which depends on the choice of the n-dimensional vector basis $\{\mathbf{e}_i\}$ and the grade p of multivectors upon which it operates. The relation can be visualized by adding a diagonal connection in Figure 2.6, as seen in Figure 2.7.

The inverse of (2.225) can be written in terms of the identity (1.122) which implies

$$\mathbf{e}_N\lfloor(\boldsymbol{\varepsilon}_N\lfloor\overline{\overline{\mathsf{H}}}_p) = (-1)^{p(n-p)}\overline{\overline{\mathsf{H}}}_p. \tag{2.228}$$

From this we obtain the relation

$$\overline{\overline{\Gamma}}^{(p)} = (-1)^{p(n-p)}\boldsymbol{\varepsilon}_N\lfloor\overline{\overline{\mathsf{H}}}_p = \left((-1)^{p(n-p)}\boldsymbol{\varepsilon}_N\lfloor\overline{\overline{\mathsf{I}}}^{(n-p)}\right)|\overline{\overline{\mathsf{H}}}_p. \tag{2.229}$$

The factor $(-1)^{p(n-p)}$ equals -1 only when n is even and p odd, otherwise it gives $+1$. Since the (proper) metric dyadic is symmetric, the transpose of (2.229) gives

$$\overline{\overline{\Gamma}}^{(p)} = \overline{\overline{\Gamma}}^{(p)T} = \overline{\overline{\mathsf{H}}}_p^T\rfloor\boldsymbol{\varepsilon}_N = \overline{\overline{\mathsf{H}}}_p^T|(\overline{\overline{\mathsf{I}}}^{(n-p)T}\rfloor\boldsymbol{\varepsilon}_N). \tag{2.230}$$

(2.229) can be visualized through another diagonal connection to Figure 2.6 as shown in Figure 2.8. Other formulas connecting the Hodge and metric dyadics can be obtained from the diagrams in Figures 2.7, 2.8.

Given a Hodge dyadic $\overline{\overline{\mathsf{H}}}_p$, from (2.229) we can obtain the corresponding metric dyadic as the inverse $\overline{\overline{\mathsf{G}}}^{(p)} = \overline{\overline{\Gamma}}^{-1(p)}$. Because a metric dyadic defining a dot product must be symmetric, (2.230) sets a condition for a corresponding Hodge dyadic $\overline{\overline{\mathsf{H}}}_e$ in the space $\mathbb{E}_{n-p}\mathbb{F}_p$ by requiring $\boldsymbol{\varepsilon}_N\lfloor\overline{\overline{\mathsf{H}}}_p$ to be symmetric. The same formulas (2.225), (2.229) serve to define relations between more general Hodge dyadics and nonsymmetric metric dyadics as well. After this quite general introduction let us consider three- and two-dimensional Euclidean and four-dimensional Minkowskian metrics and their Hodge dyadics in more detail.

2.6.3 Three-dimensional Euclidean Hodge dyadics

In the three-dimensional Euclidean space, vectors and bivectors are complementary because they have the same dimension $C_1^3 = C_2^3 = 3$. The Hodge dyadic $\overline{\overline{\mathsf{H}}}_1$ maps

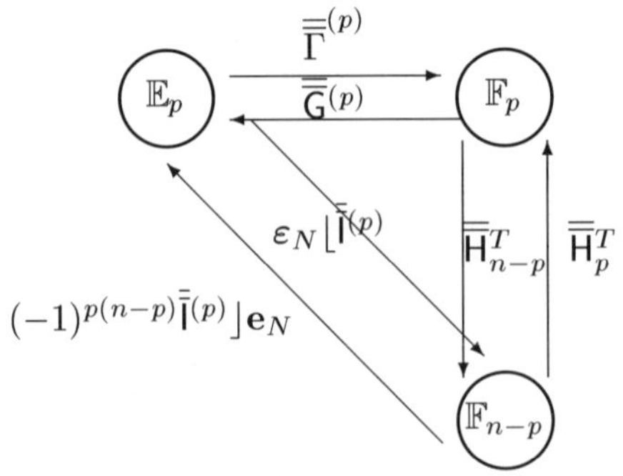

Fig. 2.8 *Adding a second set of diagonal mappings completes the diagram of Figure 2.6.*

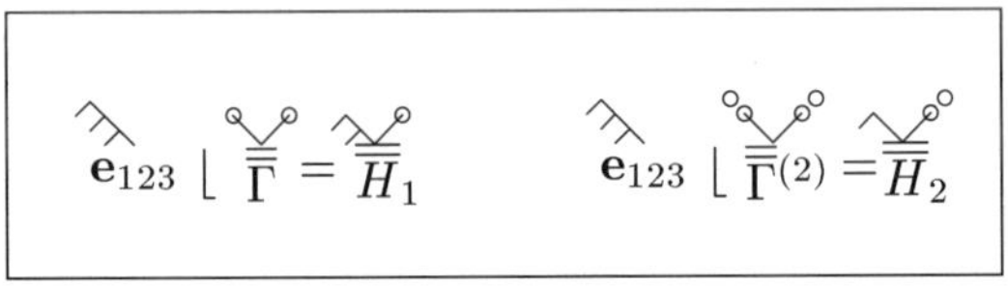

Fig. 2.9 *Visualization of relations between three-dimensional Hodge dyadics and metric dyadics.*

vectors to bivectors: $\mathbb{E}_1 \rightarrow \mathbb{E}_2$, while $\overline{\overline{\mathsf{H}}}_2$ maps bivectors to vectors: $\mathbb{E}_2 \rightarrow \mathbb{E}_1$. In the chosen basis they have the expansions

$$\overline{\overline{\mathsf{H}}}_1 = \mathbf{e}_{123} \lfloor \overline{\overline{\Gamma}} = \mathbf{e}_{123} \lfloor \sum \varepsilon_i \varepsilon_i = \mathbf{e}_{12}\varepsilon_3 + \mathbf{e}_{23}\varepsilon_1 + \mathbf{e}_{31}\varepsilon_2, \tag{2.231}$$

$$\overline{\overline{\mathsf{H}}}_2 = \mathbf{e}_{123} \lfloor \overline{\overline{\Gamma}}^{(2)} = \mathbf{e}_{123} \lfloor \sum \varepsilon_{ij} \varepsilon_{ij} = \mathbf{e}_1\varepsilon_{23} + \mathbf{e}_2\varepsilon_{31} + \mathbf{e}_3\varepsilon_{12}. \tag{2.232}$$

These expressions imply the conditions

$$\overline{\overline{\mathsf{H}}}_1 | \overline{\overline{\mathsf{H}}}_2 = \overline{\overline{\mathsf{I}}}^{(2)}, \quad \overline{\overline{\mathsf{H}}}_2 | \overline{\overline{\mathsf{H}}}_1 = \overline{\overline{\mathsf{I}}}, \tag{2.233}$$

whence we can write

$$\overline{\overline{\mathsf{H}}}_2 = \overline{\overline{\mathsf{H}}}_1^{-1}, \quad \overline{\overline{\mathsf{H}}}_1 = \overline{\overline{\mathsf{H}}}_2^{-1}. \tag{2.234}$$

Relations to the metric dyadic There are a lot of different relations between the Hodge dyadics and the metric dyadics (Figure 2.9):

$$\overline{\overline{\mathsf{H}}}_1 \rfloor \mathbf{e}_{123} = \mathbf{e}_{12}\mathbf{e}_{12} + \mathbf{e}_{23}\mathbf{e}_{23} + \mathbf{e}_{31}\mathbf{e}_{31} = \overline{\overline{\mathsf{G}}}^{(2)}, \tag{2.235}$$

$$\varepsilon_{123} \lfloor \overline{\overline{\mathsf{H}}}_1 = \varepsilon_1\varepsilon_1 + \varepsilon_2\varepsilon_2 + \varepsilon_3\varepsilon_3 = \overline{\overline{\Gamma}}. \tag{2.236}$$

$$\mathbf{e}_{123} \lfloor \overline{\overline{\Gamma}} = \overline{\overline{\mathsf{H}}}_1, \quad \overline{\overline{\mathsf{G}}}^{(2)} \rfloor \varepsilon_{123} = \overline{\overline{\mathsf{H}}}_1. \tag{2.237}$$

$$\overline{\overline{\mathsf{H}}}_2 \rfloor \mathbf{e}_{123} = \overline{\overline{\mathsf{G}}}, \qquad \boldsymbol{\varepsilon}_{123} \lfloor \overline{\overline{\mathsf{H}}}_2 = \overline{\overline{\Gamma}}^{(2)}, \tag{2.238}$$

$$\overline{\overline{\mathsf{G}}} \rfloor \boldsymbol{\varepsilon}_{123} = \overline{\overline{\mathsf{H}}}_2, \qquad \mathbf{e}_{123} \lfloor \overline{\overline{\Gamma}}^{(2)} = \overline{\overline{\mathsf{H}}}_2. \tag{2.239}$$

The transposed Hodge dyadic $\overline{\overline{\mathsf{H}}}_1^T$ maps dual bivectors to dual vectors: $\mathbb{F}_2 \rightarrow \mathbb{F}_1$. Thus, it is dual to $\overline{\overline{\mathsf{H}}}_2$. It satisfies

$$\overline{\overline{\mathsf{H}}}_1^T = \boldsymbol{\varepsilon}_1 \mathbf{e}_{23} + \boldsymbol{\varepsilon}_2 \mathbf{e}_{31} + \boldsymbol{\varepsilon}_3 \mathbf{e}_{12} = \overline{\overline{\mathsf{H}}}_2^{-1T}, \tag{2.240}$$

$$\mathbf{e}_{123} \lfloor \overline{\overline{\mathsf{H}}}_1^T = \overline{\overline{\mathsf{G}}}^{(2)}, \qquad \overline{\overline{\mathsf{H}}}_1^T \rfloor \boldsymbol{\varepsilon}_{123} = \overline{\overline{\Gamma}}, \tag{2.241}$$

$$\boldsymbol{\varepsilon}_{123} \lfloor \overline{\overline{\mathsf{G}}}^{(2)} = \overline{\overline{\mathsf{H}}}_1^T, \qquad \overline{\overline{\Gamma}} \rfloor \mathbf{e}_{123} = \overline{\overline{\mathsf{H}}}_1^T. \tag{2.242}$$

Correspondingly, the transpose of the Hodge dyadic $\overline{\overline{\mathsf{H}}}_2$ is dual to $\overline{\overline{\mathsf{H}}}_1$:

$$\overline{\overline{\mathsf{H}}}_2^T = \boldsymbol{\varepsilon}_{12} \mathbf{e}_3 + \boldsymbol{\varepsilon}_{23} \mathbf{e}_1 + \boldsymbol{\varepsilon}_{31} \mathbf{e}_2 = \overline{\overline{\mathsf{H}}}_1^{-1T}. \tag{2.243}$$

Also, it performs the mapping $\mathbb{F}_2 \rightarrow \mathbb{F}_1$. The transposed Hodge dyadics satisfy

$$\overline{\overline{\mathsf{H}}}_2^T | \overline{\overline{\mathsf{H}}}_1^T = \overline{\overline{\mathsf{I}}}^{(2)T}, \qquad \overline{\overline{\mathsf{H}}}_1^T | \overline{\overline{\mathsf{H}}}_2^T = \overline{\overline{\mathsf{I}}}^T. \tag{2.244}$$

Properties of Hodge dyadics These relations imply the property known in mathematics that the inverse of the Hodge star operator equals the operator itself in the Euclidean space [18]. Other important results are

$$\mathbf{e}_i \wedge \overline{\overline{\mathsf{H}}}_2 = \overline{\overline{\mathsf{H}}}_1 \wedge \boldsymbol{\varepsilon}_i, \tag{2.245}$$

$$\overline{\overline{\mathsf{H}}}_2 \wedge \boldsymbol{\varepsilon}_i = \mathbf{e}_i \boldsymbol{\varepsilon}_{123}, \qquad \mathbf{e}_i \wedge \overline{\overline{\mathsf{H}}}_1 = \mathbf{e}_{123} \boldsymbol{\varepsilon}_i, \tag{2.246}$$

$$\mathbf{e}_{ij} \wedge \overline{\overline{\mathsf{H}}}_2 = \mathbf{e}_{123} \boldsymbol{\varepsilon}_{ij}, \qquad \overline{\overline{\mathsf{H}}}_1 \wedge \boldsymbol{\varepsilon}_{ij} = \mathbf{e}_{ij} \boldsymbol{\varepsilon}_{123}. \tag{2.247}$$

As an example, let us check the condition (2.223) for $\overline{\overline{\mathsf{H}}}_1$ and any two vectors $\mathbf{a} = \sum a_i \mathbf{e}_i$, $\mathbf{b} = \sum b_i \mathbf{e}_i$, applying the above relations:

$$\begin{aligned} \mathbf{a} \wedge (\overline{\overline{\mathsf{H}}}_1 | \mathbf{b}) &= \sum_{i,j} a_i b_j \mathbf{e}_i \wedge \overline{\overline{\mathsf{H}}}_1 | \mathbf{e}_j = \mathbf{e}_{123} \sum_{i,j} a_i b_j \boldsymbol{\varepsilon}_i | \mathbf{e}_j \\ &= \mathbf{e}_{123} \sum_i a_i b_i = \mathbf{e}_{123} (\mathbf{a} | \overline{\overline{\Gamma}} | \mathbf{b}). \end{aligned} \tag{2.248}$$

As another example, for $\overline{\overline{\mathsf{H}}}_2$ and any two bivectors $\mathbf{A} = \sum A_{ij} \mathbf{e}_{ij}$, $\mathbf{B} = \sum B_{k\ell} \mathbf{e}_{k\ell}$ we have

$$\begin{aligned} \mathbf{A} \wedge (\overline{\overline{\mathsf{H}}}_2 | \mathbf{B}) &= \sum_{i,j,k,\ell} A_{ij} B_{k\ell} \mathbf{e}_{ij} \wedge \overline{\overline{\mathsf{H}}}_2 | \mathbf{e}_{k\ell} = \mathbf{e}_{123} \sum_{i,j,k,\ell} A_{ij} B_{k\ell} \boldsymbol{\varepsilon}_{ij} | \mathbf{e}_{k\ell} \\ &= \mathbf{e}_{123} \sum_{i,j} A_{ij} B_{ij} = \mathbf{e}_{123} (\mathbf{A} | \overline{\overline{\Gamma}}^{(2)} | \mathbf{B}). \end{aligned} \tag{2.249}$$

The corresponding conditions for the dual and transposed Hodge dyadics follow in the same manner.

2.6.4 Two-dimensional Euclidean Hodge dyadic

The two-dimensional Euclidean space can be understood as a subspace in the three-dimensional Euclidean space where all vectors $\mathbf{a}$ satisfy $\mathbf{a}|\boldsymbol{\varepsilon}_3 = 0$ and all dual vectors $\boldsymbol{\alpha}$ satisfy $\boldsymbol{\alpha}|\mathbf{e}_3 = 0$. Since bivectors have the dimension of scalars, 1, the only existing Hodge dyadic in the 2D space maps vectors to vectors.

The Hodge dyadic can be obtained from (2.225) as

$$\overline{\overline{\mathsf{H}}}_1 = \mathbf{e}_N \lfloor \overline{\overline{\Gamma}} = \mathbf{e}_{12} \lfloor (\boldsymbol{\varepsilon}_1\boldsymbol{\varepsilon}_1 + \boldsymbol{\varepsilon}_2\boldsymbol{\varepsilon}_2) = \mathbf{e}_2\boldsymbol{\varepsilon}_1 - \mathbf{e}_1\boldsymbol{\varepsilon}_2. \tag{2.250}$$

To check that the definition (2.250) is in accord with the basic definition (2.223), we write

$$\begin{aligned} \mathbf{a} \wedge (\overline{\overline{\mathsf{H}}}_1|\mathbf{b}) &= (a_1\mathbf{e}_1 + a_2\mathbf{e}_2) \wedge (\mathbf{e}_2 b_1 - \mathbf{e}_1 b_2) \\ &= \mathbf{e}_{12}(a_1 b_1 + a_2 b_2) = \mathbf{e}_{12}(\mathbf{a}|\overline{\overline{\Gamma}}|\mathbf{b}). \end{aligned} \tag{2.251}$$

The two-dimensional Hodge dyadic satisfies

$$\overline{\overline{\mathsf{H}}}_1^2 = \overline{\overline{\mathsf{H}}}_1|\overline{\overline{\mathsf{H}}}_1 = -\overline{\overline{\mathsf{I}}}, \quad \Rightarrow \quad \overline{\overline{\mathsf{H}}}_1^{-1} = -\overline{\overline{\mathsf{H}}}_1, \tag{2.252}$$

which means that $\overline{\overline{\mathsf{H}}}_1$ acts like the imaginary unit. Hodge dyadics of this kind can only be found in spaces with even dimension n for $p = n/2$ where they map p-vectors to p-vectors.

Two-dimensional rotation dyadic Combining $\overline{\overline{\mathsf{H}}}_1$ with the unit dyadic as

$$\begin{aligned} \overline{\overline{\mathsf{R}}}(\theta) &= \overline{\overline{\mathsf{I}}}\cos\theta + \overline{\overline{\mathsf{H}}}_1 \sin\theta \\ &= \mathbf{e}_1\boldsymbol{\varepsilon}_1 \cos\theta - \mathbf{e}_1\boldsymbol{\varepsilon}_2 \sin\theta + \mathbf{e}_2\boldsymbol{\varepsilon}_1 \sin\theta + \mathbf{e}_2\boldsymbol{\varepsilon}_2 \cos\theta, \end{aligned} \tag{2.253}$$

which can also be symbolically expressed as

$$\overline{\overline{\mathsf{R}}}(\theta) = e^{\overline{\overline{\mathsf{H}}}_1\theta} = \overline{\overline{\mathsf{I}}} + \overline{\overline{\mathsf{H}}}_1\theta - \frac{1}{2!}\overline{\overline{\mathsf{I}}}\theta^2 - \frac{1}{3!}\overline{\overline{\mathsf{H}}}_1\theta^3 + \cdots, \tag{2.254}$$

gives a dyadic which rotates a given vector by the angle θ:

$$\overline{\overline{\mathsf{R}}}(\theta)|(\mathbf{e}_1 a_1 + \mathbf{e}_2 a_2) = (\mathbf{e}_1\cos\theta + \mathbf{e}_2\sin\theta)a_1 + (\mathbf{e}_2\cos\theta - \mathbf{e}_1\sin\theta)a_2. \tag{2.255}$$

The rotation dyadic obeys the obvious properties $\overline{\overline{\mathsf{R}}}(0) = \overline{\overline{\mathsf{I}}}$, $\overline{\overline{\mathsf{R}}}(\pi/2) = \overline{\overline{\mathsf{H}}}_1$ and

$$\overline{\overline{\mathsf{R}}}(\theta_1)|\overline{\overline{\mathsf{R}}}(\theta_2) = \overline{\overline{\mathsf{R}}}(\theta_2)|\overline{\overline{\mathsf{R}}}(\theta_1) = \overline{\overline{\mathsf{R}}}(\theta_1 + \theta_2). \tag{2.256}$$

The duality product of a vector $\mathbf{a}$ and a dual vector $\boldsymbol{\alpha}$ is invariant if we define the reciprocal rotation dyadic for dual vectors as

$$\overline{\overline{\mathsf{R}}}^T(-\theta) = \overline{\overline{\mathsf{I}}}^T\cos\theta - \overline{\overline{\mathsf{H}}}_1^T \sin\theta, \tag{2.257}$$

because from (2.256) we then have

$$(\overline{\overline{\mathsf{R}}}^T(-\theta)|\boldsymbol{\alpha})|(\overline{\overline{\mathsf{R}}}(\theta)|\mathbf{a}) = \boldsymbol{\alpha}|\overline{\overline{\mathsf{R}}}(-\theta)|\overline{\overline{\mathsf{R}}}(\theta)|\mathbf{a} = \boldsymbol{\alpha}|\mathbf{a}. \tag{2.258}$$

Since the Hodge dyadic $\overline{\overline{\mathsf{H}}}_1$ depends on the chosen basis $\mathbf{e}_1, \mathbf{e}_2$, change of the basis also changes the rotation dyadic.

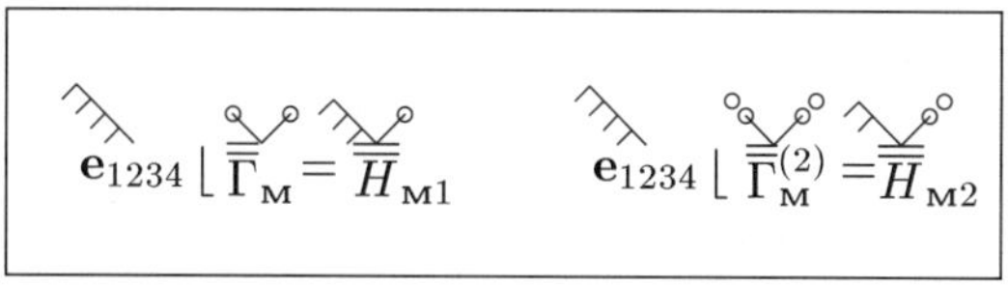

Fig. 2.10 *Visualization of relations between four-dimensional Hodge dyadics and metric dyadics.*

2.6.5 Four-dimensional Minkowskian Hodge dyadics

Let us denote the Hodge dyadics corresponding to the four-dimensional Minkowskian metric by the subscript M. Because in this case vectors and trivectors have the same dimension 4 and bivectors have the dimension 6, there exist three types of Hodge dyadics: $\overline{\overline{\mathsf{H}}}_{\mathrm{M1}}$ mapping vectors to trivectors, $\overline{\overline{\mathsf{H}}}_{\mathrm{M2}}$ mapping bivectors to bivectors, and $\overline{\overline{\mathsf{H}}}_{\mathrm{M3}}$ mapping trivectors to vectors. For dual multivectors the corresponding Hodge dyadics appear in respective form $\overline{\overline{\mathsf{H}}}{}^T_{\mathrm{M3}}$, $\overline{\overline{\mathsf{H}}}{}^T_{\mathrm{M2}}$, $\overline{\overline{\mathsf{H}}}{}^T_{\mathrm{M1}}$.

From (2.225) we can now write expansions for the different Hodge dyadics. They can be best memorized in a form where the fourth dimension is separated from the three Euclidean ones as in (1.28). An expansion for each Hodge dyadic in terms of the corresponding three-dimensional Euclidean Hodge dyadics $\overline{\overline{\mathsf{H}}}_{\mathrm{E1}}$, $\overline{\overline{\mathsf{H}}}_{\mathrm{E2}}$ appears often useful and is given below:

$$\begin{aligned}
\overline{\overline{\mathsf{H}}}_{\mathrm{M1}} &= \mathbf{e}_{1234}\lfloor\overline{\overline{\Gamma}}_{\mathrm{M}} \\
&= \mathbf{e}_{1234}\lfloor(\boldsymbol{\varepsilon}_1\boldsymbol{\varepsilon}_1 + \boldsymbol{\varepsilon}_2\boldsymbol{\varepsilon}_2 + \boldsymbol{\varepsilon}_3\boldsymbol{\varepsilon}_3 - \boldsymbol{\varepsilon}_4\boldsymbol{\varepsilon}_4) \\
&= \mathbf{e}_{234}\boldsymbol{\varepsilon}_1 + \mathbf{e}_{314}\boldsymbol{\varepsilon}_2 + \mathbf{e}_{124}\boldsymbol{\varepsilon}_3 + \mathbf{e}_{123}\boldsymbol{\varepsilon}_4 \\
&= \mathbf{e}_4 \wedge \overline{\overline{\mathsf{H}}}_{\mathrm{E1}} + \mathbf{e}_{123}\boldsymbol{\varepsilon}_4, \qquad (2.259)\\
\overline{\overline{\mathsf{H}}}_{\mathrm{M2}} &= \mathbf{e}_{1234}\lfloor\overline{\overline{\Gamma}}{}^{(2)}_{\mathrm{M}} \\
&= \mathbf{e}_{1234}\lfloor(\boldsymbol{\varepsilon}_{12}\boldsymbol{\varepsilon}_{12} + \boldsymbol{\varepsilon}_{23}\boldsymbol{\varepsilon}_{23} + \boldsymbol{\varepsilon}_{31}\boldsymbol{\varepsilon}_{31} - \boldsymbol{\varepsilon}_{14}\boldsymbol{\varepsilon}_{14} - \boldsymbol{\varepsilon}_{24}\boldsymbol{\varepsilon}_{24} - \boldsymbol{\varepsilon}_{34}\boldsymbol{\varepsilon}_{34}) \\
&= -(\mathbf{e}_{12}\boldsymbol{\varepsilon}_{34} + \mathbf{e}_{23}\boldsymbol{\varepsilon}_{14} + \mathbf{e}_{31}\boldsymbol{\varepsilon}_{24}) + (\mathbf{e}_{14}\boldsymbol{\varepsilon}_{23} + \mathbf{e}_{24}\boldsymbol{\varepsilon}_{31} + \mathbf{e}_{34}\boldsymbol{\varepsilon}_{12}) \\
&= -(\mathbf{e}_{12}\boldsymbol{\varepsilon}_3 + \mathbf{e}_{23}\boldsymbol{\varepsilon}_1 + \mathbf{e}_{31}\boldsymbol{\varepsilon}_2)\wedge\boldsymbol{\varepsilon}_4 - \mathbf{e}_4\wedge(\mathbf{e}_1\boldsymbol{\varepsilon}_{23} + \mathbf{e}_2\boldsymbol{\varepsilon}_{31} + \mathbf{e}_3\boldsymbol{\varepsilon}_{12}) \\
&= -\overline{\overline{\mathsf{H}}}_{\mathrm{E1}}\wedge\boldsymbol{\varepsilon}_4 - \mathbf{e}_4\wedge\overline{\overline{\mathsf{H}}}_{\mathrm{E2}} = -\overline{\overline{\mathsf{H}}}{}^{-1}_{\mathrm{M2}}, \qquad (2.260)\\
\overline{\overline{\mathsf{H}}}_{\mathrm{M3}} &= \mathbf{e}_{1234}\lfloor\overline{\overline{\Gamma}}{}^{(3)}_{\mathrm{M}} \\
&= \mathbf{e}_{1234}\lfloor(\boldsymbol{\varepsilon}_{123}\boldsymbol{\varepsilon}_{123} - \boldsymbol{\varepsilon}_{124}\boldsymbol{\varepsilon}_{124} - \boldsymbol{\varepsilon}_{234}\boldsymbol{\varepsilon}_{234} - \boldsymbol{\varepsilon}_{314}\boldsymbol{\varepsilon}_{314}) \\
&= \mathbf{e}_1\boldsymbol{\varepsilon}_{234} + \mathbf{e}_2\boldsymbol{\varepsilon}_{314} + \mathbf{e}_3\boldsymbol{\varepsilon}_{124} + \mathbf{e}_4\boldsymbol{\varepsilon}_{123} \\
&= \overline{\overline{\mathsf{H}}}_{\mathrm{E2}}\wedge\boldsymbol{\varepsilon}_4 + \mathbf{e}_4\boldsymbol{\varepsilon}_{123} = \overline{\overline{\mathsf{H}}}{}^{-1}_{\mathrm{M1}}. \qquad (2.261)
\end{aligned}$$

Checking the expressions One can easily check that these dyadics satisfy (2.223) by making decompositions for the multivectors and Hodge dyadics. For $\overline{\overline{\mathsf{H}}}_{\mathrm{M1}}$ we expand

$$\mathbf{a}_{\mathrm{M}} \wedge \overline{\overline{\mathsf{H}}}_{\mathrm{M1}}|\mathbf{b}_{\mathrm{M}}$$

$$\begin{aligned}
&= (\mathbf{a}_\mathrm{E} + \mathbf{e}_4 a_4) \wedge (\mathbf{e}_{123}\varepsilon_4 + \mathbf{e}_4 \wedge \overline{\overline{\mathsf{H}}}_{\mathrm{E}1})|(\mathbf{b}_\mathrm{E} + \mathbf{e}_4 b_4) \\
&= -\mathbf{e}_4 \wedge \mathbf{a}_\mathrm{E} \wedge \overline{\overline{\mathsf{H}}}_{\mathrm{E}1}|\mathbf{b}_\mathrm{E} - a_4 \mathbf{e}_{1234} b_4 \\
&= \mathbf{e}_{1234}(\mathbf{a}_\mathrm{E}|\overline{\overline{\Gamma}}_\mathrm{E}|\mathbf{b}_\mathrm{E} - a_4 b_4) \\
&= \mathbf{e}_{1234}(\mathbf{a}_\mathrm{E}|\overline{\overline{\Gamma}}_\mathrm{M}|\mathbf{b}_\mathrm{E}). \qquad (2.262)
\end{aligned}$$

Similarly, we write for two bivectors $\mathbf{A}_\mathrm{M}, \mathbf{B}_\mathrm{M}$

$$\begin{aligned}
&\mathbf{A}_\mathrm{M} \wedge \overline{\overline{\mathsf{H}}}_{\mathrm{M}2}|\mathbf{B}_\mathrm{M} \\
&= -(\mathbf{A}_\mathrm{E} + \mathbf{a}_\mathrm{E} \wedge \mathbf{e}_4) \wedge (\overline{\overline{\mathsf{H}}}_{\mathrm{E}1} \wedge \varepsilon_4 + \mathbf{e}_4 \wedge \overline{\overline{\mathsf{H}}}_{\mathrm{E}2})|(\mathbf{B}_\mathrm{E} + \mathbf{b}_\mathrm{E} \wedge \mathbf{e}_4) \\
&= -\mathbf{A}_\mathrm{E} \wedge \mathbf{e}_4 \wedge \overline{\overline{\mathsf{H}}}_{\mathrm{E}2}|\mathbf{B}_\mathrm{E} - \mathbf{a}_\mathrm{E} \wedge \mathbf{e}_4 \wedge (\overline{\overline{\mathsf{H}}}_{\mathrm{E}1} \wedge \varepsilon_4)|(\mathbf{b}_\mathrm{E} \wedge \mathbf{e}_4) \\
&= -\mathbf{e}_4 \wedge (\mathbf{A}_\mathrm{E} \wedge \overline{\overline{\mathsf{H}}}_{\mathrm{E}2}|\mathbf{B}_\mathrm{E}) + \mathbf{e}_4 \wedge (\mathbf{a}_\mathrm{E} \wedge \overline{\overline{\mathsf{H}}}_{\mathrm{E}1}|\mathbf{b}_\mathrm{E}) \\
&= \mathbf{e}_4 \wedge \mathbf{e}_{123}(-\mathbf{A}_\mathrm{E}|\overline{\overline{\Gamma}}{}^{(2)}_\mathrm{E}|\mathbf{B}_\mathrm{E}) + \mathbf{a}_\mathrm{E}|\overline{\overline{\Gamma}}_\mathrm{E}|\mathbf{b}_\mathrm{E}) \\
&= \mathbf{e}_{1234}(\mathbf{A}_\mathrm{M}|\overline{\overline{\Gamma}}_\mathrm{M}{}^{(2)}|\mathbf{B}_\mathrm{M}). \qquad (2.263)
\end{aligned}$$

Finally, for two trivectors $\mathbf{a}_\mathrm{M}, \mathbf{b}_\mathrm{M}$ which can be expressed in terms of Euclidean trivectors $\mathbf{a}_\mathrm{E} = a_{123}\mathbf{e}_{123}$, $\mathbf{b}_\mathrm{E} = b_{123}\mathbf{e}_{123}$ and bivectors $\mathbf{A}_\mathrm{E}, \mathbf{B}_\mathrm{E}$, we have

$$\begin{aligned}
&\mathbf{a}_\mathrm{M} \wedge \overline{\overline{\mathsf{H}}}_{\mathrm{M}3}|\mathbf{b}_\mathrm{M} \\
&= (a_{123}\mathbf{e}_{123} + \mathbf{A}_\mathrm{E} \wedge \mathbf{e}_4) \wedge (\overline{\overline{\mathsf{H}}}_{\mathrm{E}2} \wedge \varepsilon_4 + \mathbf{e}_4 \varepsilon_{123})|(b^{123}\mathbf{e}_{123} + \mathbf{B} \wedge \mathbf{e}_4) \\
&= a_{123}b_{123}\mathbf{e}_{1234} + \mathbf{A}_\mathrm{E} \wedge \mathbf{e}_4 \wedge (\overline{\overline{\mathsf{H}}}_{\mathrm{E}2} \wedge \varepsilon_4)|(\mathbf{B}_\mathrm{E} \wedge \mathbf{e}_4) \\
&= a_{123}b_{123}\mathbf{e}_{1234} + \mathbf{e}_4 \wedge \mathbf{A}_\mathrm{E} \wedge \overline{\overline{\mathsf{H}}}_{\mathrm{E}2}|\mathbf{B} \\
&= \mathbf{e}_{1234}(\mathbf{a}_\mathrm{E}|\varepsilon_{123}\varepsilon_{123}|\mathbf{b}_\mathrm{E}) - \mathbf{e}_{1234}(\mathbf{A}_\mathrm{E}|\overline{\overline{\Gamma}}{}^{(2)}_\mathrm{E}|\mathbf{B}_\mathrm{E}) \\
&= \mathbf{e}_{1234}(\mathbf{a}_\mathrm{M}|\overline{\overline{\Gamma}}_\mathrm{M}{}^{(3)}|\mathbf{b}_\mathrm{M}). \qquad (2.264)
\end{aligned}$$

Relations to the metric dyadic Various relations of Hodge dyadics and the Minkowskian metric dyadics can be easily obtained:

$$\overline{\overline{\mathsf{H}}}_{\mathrm{M}1}\rfloor\mathbf{e}_{1234} = \overline{\overline{\mathsf{G}}}{}^{(3)}_\mathrm{M}, \quad \varepsilon_{1234}\lfloor\overline{\overline{\mathsf{H}}}_{\mathrm{M}1} = -\overline{\overline{\Gamma}}_\mathrm{M}, \qquad (2.265)$$

$$\overline{\overline{\mathsf{H}}}_{\mathrm{M}2}\rfloor\mathbf{e}_{1234} = -\overline{\overline{\mathsf{G}}}{}^{(2)}_\mathrm{M}, \quad \varepsilon_{1234}\lfloor\overline{\overline{\mathsf{H}}}_{\mathrm{M}2} = \overline{\overline{\Gamma}}{}^{(2)}_\mathrm{M}, \qquad (2.266)$$

$$\overline{\overline{\mathsf{H}}}_{\mathrm{M}3}\rfloor\mathbf{e}_{1234} = \overline{\overline{\mathsf{G}}}_\mathrm{M}, \quad \varepsilon_{1234}\lfloor\overline{\overline{\mathsf{H}}}_{\mathrm{M}3} = -\overline{\overline{\Gamma}}{}^{(3)}_\mathrm{M}. \qquad (2.267)$$

Converse relations are

$$\overline{\overline{\mathsf{H}}}_{\mathrm{M}1} = \mathbf{e}_{1234}\lfloor\overline{\overline{\Gamma}}_\mathrm{M}, \quad \overline{\overline{\mathsf{H}}}{}^T_{\mathrm{M}1} = \varepsilon_{1234}\lfloor\overline{\overline{\mathsf{G}}}{}^{(3)}_\mathrm{M}, \qquad (2.268)$$

$$\overline{\overline{\mathsf{H}}}_{\mathrm{M}2} = \mathbf{e}_{1234}\lfloor\overline{\overline{\Gamma}}{}^{(2)}_\mathrm{M}, \quad \overline{\overline{\mathsf{H}}}{}^T_{\mathrm{M}2} = -\varepsilon_{1234}\lfloor\overline{\overline{\mathsf{G}}}{}^{(2)}_\mathrm{M}, \qquad (2.269)$$

$$\overline{\overline{\mathsf{H}}}_{\mathrm{M}3} = \mathbf{e}_{1234}\lfloor\overline{\overline{\Gamma}}{}^{(3)}_\mathrm{M}, \quad \overline{\overline{\mathsf{H}}}{}^T_{\mathrm{M}3} = \varepsilon_{1234}\lfloor\overline{\overline{\mathsf{G}}}_\mathrm{M}. \qquad (2.270)$$

The transposed Hodge dyadics which operate on dual multivectors can be easily derived and are given here for convenience:

$$\begin{aligned}\overline{\overline{\mathsf{H}}}^T_{\mathrm{M}1} &= \varepsilon_1\mathbf{e}_{234} + \varepsilon_2\mathbf{e}_{314} + \varepsilon_3\mathbf{e}_{124} + \varepsilon_4\mathbf{e}_{123} \\ &= \overline{\overline{\mathsf{H}}}^T_{\mathrm{E}1} \wedge \mathbf{e}_4 + \varepsilon_4\mathbf{e}_{123},\end{aligned} \tag{2.271}$$

$$\begin{aligned}\overline{\overline{\mathsf{H}}}^T_{\mathrm{M}2} &= (\varepsilon_{41}\mathbf{e}_{23} + \varepsilon_{42}\mathbf{e}_{31} + \varepsilon_{43}\mathbf{e}_{12}) + (\varepsilon_{23}\mathbf{e}_{14} + \varepsilon_{31}\mathbf{e}_{24} + \varepsilon_{12}\mathbf{e}_{34}) \\ &= \varepsilon_4 \wedge \overline{\overline{\mathsf{H}}}^T_{\mathrm{E}1} + \overline{\overline{\mathsf{H}}}^T_{\mathrm{E}2} \wedge \mathbf{e}_4,\end{aligned} \tag{2.272}$$

$$\begin{aligned}\overline{\overline{\mathsf{H}}}^T_{\mathrm{M}3} &= \overline{\overline{\mathsf{H}}}^{-1}_{\mathrm{M}1} = \varepsilon_{123}\mathbf{e}_4 + \varepsilon_{234}\mathbf{e}_1 + \varepsilon_{314}\mathbf{e}_2 + \varepsilon_{124}\mathbf{e}_3 \\ &= \varepsilon_{123}\mathbf{e}_4 + \varepsilon_4 \wedge \overline{\overline{\mathsf{H}}}^T_{\mathrm{E}2}.\end{aligned} \tag{2.273}$$

Relations to the unit dyadic The different Minkowskian Hodge dyadics have also connections to the four-dimensional unit dyadic $\overline{\overline{\mathsf{I}}}_{\mathrm{M}}$ and its double-wedge powers. Applying the relations between the Euclidean and Minkowskian unit dyadics,

$$\overline{\overline{\mathsf{I}}}_{\mathrm{M}} = \overline{\overline{\mathsf{I}}}_{\mathrm{E}} + \mathbf{e}_4\varepsilon_4, \tag{2.274}$$

$$\overline{\overline{\mathsf{I}}}^{(2)}_{\mathrm{M}} = \overline{\overline{\mathsf{I}}}^{(2)}_{\mathrm{E}} + \overline{\overline{\mathsf{I}}}_{\mathrm{E}} {}^{\wedge}_{\wedge} \mathbf{e}_4\varepsilon_4, \tag{2.275}$$

$$\overline{\overline{\mathsf{I}}}^{(3)}_{\mathrm{M}} = \overline{\overline{\mathsf{I}}}^{(3)}_{\mathrm{E}} + \overline{\overline{\mathsf{I}}}^{(2)}_{\mathrm{E}} {}^{\wedge}_{\wedge} \mathbf{e}_4\varepsilon_4. \tag{2.276}$$

we can write the following relations between the different Minkowskian Hodge dyadics and unit dyadics:

$$\overline{\overline{\mathsf{H}}}_{\mathrm{M}3}|\overline{\overline{\mathsf{H}}}_{\mathrm{M}1} = \overline{\overline{\mathsf{I}}}_{\mathrm{M}}, \tag{2.277}$$

$$\overline{\overline{\mathsf{H}}}_{\mathrm{M}1}|\overline{\overline{\mathsf{H}}}_{\mathrm{M}3} = \overline{\overline{\mathsf{I}}}^{(3)}_{\mathrm{M}}, \tag{2.278}$$

$$\overline{\overline{\mathsf{H}}}^2_{\mathrm{M}2} = \overline{\overline{\mathsf{H}}}_{\mathrm{M}2}|\overline{\overline{\mathsf{H}}}_{\mathrm{M}2} = -\overline{\overline{\mathsf{I}}}^{(2)}_{\mathrm{M}}, \tag{2.279}$$

$$\overline{\overline{\mathsf{H}}}^{-1}_{\mathrm{M}2} = -\overline{\overline{\mathsf{H}}}_{\mathrm{M}2}. \tag{2.280}$$

These formulas show properties for the Hodge star operator in the Minkowskian space known from the literature [18]: $(*)^2 = 1$ when p is odd and $(*)^2 = -1$ when p is even.

Four-dimensional rotation dyadic The Hodge dyadic $\overline{\overline{\mathsf{H}}}_{\mathrm{M}2}$ resembles again the imaginary unit because its square equals the negative of the unit dyadic of the bivector space, as is seen from (2.279). Similarly to the two-dimensional case, we can form a rotation dyadic by combining $\overline{\overline{\mathsf{H}}}_{\mathrm{M}2}$ and the four-dimensional bivector unit dyadic $\overline{\overline{\mathsf{I}}}^{(2)}_{\mathrm{M}}$ as

$$\begin{aligned}\overline{\overline{\mathsf{R}}}_{\mathrm{M}}(\theta) &= \overline{\overline{\mathsf{I}}}^{(2)}_{\mathrm{M}} \cos\theta + \overline{\overline{\mathsf{H}}}_{\mathrm{M}2} \sin\theta \\ &= (\overline{\overline{\mathsf{I}}}^{(2)}_{\mathrm{E}} + \overline{\overline{\mathsf{I}}}_{\mathrm{E}} {}^{\wedge}_{\wedge} \mathbf{e}_4\varepsilon_4)\cos\theta - (\overline{\overline{\mathsf{H}}}_{\mathrm{E}1} \wedge \varepsilon_4 + \mathbf{e}_4 \wedge \overline{\overline{\mathsf{H}}}_{\mathrm{E}2})\sin\theta.\end{aligned} \tag{2.281}$$

The dyadic satisfies

$$\overline{\overline{\mathsf{R}}}_{\mathbf{M}}(0) = \overline{\overline{\mathsf{I}}}^{(2)}_{\mathbf{M}}, \quad \overline{\overline{\mathsf{R}}}_{\mathbf{M}}(\pm\pi/2) = \pm\overline{\overline{\mathsf{H}}}_{\mathbf{M}2}, \quad \overline{\overline{\mathsf{R}}}_{\mathbf{M}}(\pi) = -\overline{\overline{\mathsf{I}}}^{(2)}_{\mathbf{M}} \tag{2.282}$$

and the product rule

$$\overline{\overline{\mathsf{R}}}_{\mathbf{M}}(\theta_1)|\overline{\overline{\mathsf{R}}}_{\mathbf{M}}(\theta_2) = \overline{\overline{\mathsf{R}}}_{\mathbf{M}}(\theta_2)|\overline{\overline{\mathsf{R}}}_{\mathbf{M}}(\theta_1) = \overline{\overline{\mathsf{R}}}_{\mathbf{M}}(\theta_1 + \theta_2). \tag{2.283}$$

The rotation dyadic $\overline{\overline{\mathsf{R}}}_{\mathbf{M}}(\theta)$ maps a bivector $\mathbf{A}_{\mathbf{M}} = \mathbf{A}_{\mathbf{E}} + \mathbf{a}_{\mathbf{E}} \wedge \mathbf{e}_4$ to another bivector $\mathbf{A}'_{\mathbf{M}} = \mathbf{A}'_{\mathbf{E}} + \mathbf{a}'_{\mathbf{E}} \wedge \mathbf{e}_4$ with

$$\mathbf{A}'_{\mathbf{E}} = \mathbf{A}_{\mathbf{E}} \cos\theta - (\overline{\overline{\mathsf{H}}}_{\mathbf{E}1}|\mathbf{a}_{\mathbf{E}}) \sin\theta, \tag{2.284}$$

$$\mathbf{a}'_{\mathbf{E}} = \mathbf{a}_{\mathbf{E}} \cos\theta + (\overline{\overline{\mathsf{H}}}_{\mathbf{E}2}|\mathbf{A}_{\mathbf{E}}) \sin\theta. \tag{2.285}$$

Since the Hodge dyadic $\overline{\overline{\mathsf{H}}}_{\mathbf{M}2}$ depends on the chosen basis $\{\mathbf{e}_i\}$, change of the basis also changes the rotation dyadic. This can be visualized as a change of the axis of rotation. The dual form of the rotation dyadic is often called the duality rotation when applied to the electromagnetic dual bivector, in which case it has the effect of changing electricity to magnetism and conversely.

Problems

2.6.1 Prove that the relation $\overline{\overline{\mathsf{H}}}_p = \mathbf{e}_N \lfloor \overline{\overline{\Gamma}}^{(p)}$ follows from (2.224). *Hint*: Applying the identity

$$\overline{\overline{\mathsf{I}}}^{(n-p)} = \overline{\overline{\mathsf{I}}}^{(n)} \lfloor\lfloor \overline{\overline{\mathsf{I}}}^{(p)T} = \mathbf{e}_N \lfloor \sum \boldsymbol{\varepsilon}_J (\boldsymbol{\varepsilon}_N \lfloor \mathbf{e}_J),$$

where J is a p-index, expand $\overline{\overline{\mathsf{H}}}_p = \overline{\overline{\mathsf{I}}}^{(n-p)}|\overline{\overline{\mathsf{H}}}_p$ and use (2.224).

2.6.2 Check (2.256).

2.6.3 Verify (2.245).

2.6.4 Derive (2.265) – (2.267).

2.6.5 Derive (2.268) – (2.270).

2.6.6 Prove that $\overline{\overline{\mathsf{H}}}^2_{\mathbf{M}2} = -\overline{\overline{\mathsf{I}}}^{(2)}_{\mathbf{M}}$.

3

Differential Forms

Dual multivector fields, dual multivector functions of the space position vector or space-time vector, are called differential forms. Because physical quantities like electromagnetic fields, currents, and charge densities are integrable over lines, areas, volumes, and time, they can be represented by differential forms of various grades.

3.1 DIFFERENTIATION

In the theory of differential forms, the central role is played by the differential operator $\mathbf{d}$. Its counterpart in the three-dimensional Gibbsian vector analysis is ∇, the Hamiltonian nabla operator.[1] In the four-dimensional Minkowskian space, $\mathbf{d}$ also involves differentiation with respect to time.

3.1.1 Three-dimensional space

Denoting the three-dimensional position vector by $\mathbf{r}$, coordinate functions are denoted either as $x_i(\mathbf{r})$, $i = 1, 2, 3$, or, with an emphasis on the Cartesian coordinates, as $x(\mathbf{r})$, $y(\mathbf{r})$, and $z(\mathbf{r})$. The differential operator $\mathbf{d}$ operating on a scalar function $f(\mathbf{r})$ gives a dual vector $\mathbf{d}f(\mathbf{r})$ (dimension of $\mathbf{d}$ is 1/m). A dual-vector basis can be formed by operating on three coordinate functions, for example, $\{\boldsymbol{\varepsilon}_i\} = \{\mathbf{d}x_i\}$ or $\{\mathbf{d}x, \mathbf{d}y, \mathbf{d}z\}$. The latter three dual vectors correspond to the Gibbsian unit vectors $\mathbf{u}_x = \nabla x$, $\mathbf{u}_y = \nabla y$, $\mathbf{u}_z = \nabla z$. The vector basis reciprocal to $\{\boldsymbol{\varepsilon}_i\}$ is denoted by $\{\mathbf{e}_i\}$ and, in Cartesian form, by $\{\mathbf{e}_x, \mathbf{e}_y, \mathbf{e}_z\}$.

[1]Some authors like Meetz and Engl [57] write ∇ instead of $\mathbf{d}$.

From the differential of a scalar function of spatial variables $\phi(x, y, z)$

$$\mathbf{d}\phi = \mathbf{d}x\,\partial_x\phi + \mathbf{d}y\,\partial_y\phi + \mathbf{d}z\,\partial_z\phi \tag{3.1}$$

we can extract the differential operator $\mathbf{d}$ in the Cartesian basis as

$$\mathbf{d} = \mathbf{d}x\partial_x + \mathbf{d}y\partial_y + \mathbf{d}z\partial_z, \tag{3.2}$$

which is similar to the expansion of the gradient operator in the Gibbsian vector analysis. The similarity can be given a concrete form through the Euclidean metric dyadic

$$\overline{\overline{\mathsf{G}}} = \sum \mathbf{e}_i\mathbf{e}_i = \mathbf{e}_x\mathbf{e}_x + \mathbf{e}_y\mathbf{e}_y + \mathbf{e}_z\mathbf{e}_z, \tag{3.3}$$

as

$$\nabla = \overline{\overline{\mathsf{G}}}|\mathbf{d} = \sum \mathbf{e}_i\partial_{x_i} = \mathbf{e}_x\partial_x + \mathbf{e}_y\partial_y + \mathbf{e}_z\partial_z. \tag{3.4}$$

Thus, unlike $\mathbf{d}$, ∇ depends on the metric through some chosen basis system $\{\mathbf{e}_i\}$, which again depends on the chosen coordinate functions $x_i(\mathbf{r})$ through the dual basis $\{\mathbf{d}x_i(\mathbf{r})$.

One-forms Operating on a scalar function $\phi(\mathbf{r})$ by the operator $\mathbf{d}$ gives a dual vector field $\mathbf{d}\phi(\mathbf{r})$, whose Cartesian expansion is (3.1). The space formed by linear combinations of differentials of scalar functions of $\mathbf{r}$ is called the space of one-forms. Any one-form $\boldsymbol{\phi}(\mathbf{r})$ can be written in a basis of dual vectors as

$$\boldsymbol{\phi} = \sum \phi_i\mathbf{d}x_i = \phi_x\mathbf{d}x + \phi_y\mathbf{d}y + \phi_z\mathbf{d}z, \tag{3.5}$$

where the ϕ_i or ϕ_x, ϕ_y, ϕ_z are three scalar functions of $\mathbf{r}$.

Two-forms The differential $\mathbf{d}\wedge$ of the one-form (3.5) gives a two-form:

$$\mathbf{d}\wedge\boldsymbol{\phi} = \mathbf{d}\phi_x\wedge\mathbf{d}x + \mathbf{d}\phi_y\wedge\mathbf{d}y + \mathbf{d}\phi_z\wedge\mathbf{d}z. \tag{3.6}$$

Due to the antisymmetric property of the wedge product we have $\mathbf{d}\wedge\mathbf{d}\phi(\mathbf{r}) = 0$ for any scalar function $\phi(\mathbf{r})$, which means that $\mathbf{d}\wedge\mathbf{d}$ is a null operator. Substituting (3.2) and applying properties of the wedge product, one obtains for the resulting two-form in Cartesian coordinates the result

$$\begin{aligned}\boldsymbol{\Phi} &= \mathbf{d}\wedge\boldsymbol{\phi} = (\partial_x\phi_y - \partial_y\phi_x)\mathbf{d}x\wedge\mathbf{d}y \\ &\quad +(\partial_y\phi_z - \partial_z\phi_y)\mathbf{d}y\wedge\mathbf{d}z + (\partial_z\phi_x - \partial_x\phi_z)\mathbf{d}z\wedge\mathbf{d}x.\end{aligned} \tag{3.7}$$

The components of this two-form correspond to the components of the curl of a Gibbsian vector. The basis of three one-forms $\{\mathbf{d}x_i\}$ generates a basis of three two-forms $\{\mathbf{d}x_i\wedge\mathbf{d}x_j\}$ with reciprocal basis bivectors $\mathbf{e}_{ij} = \mathbf{e}_i\wedge\mathbf{e}_j$ satisfying

$$\begin{aligned}(\mathbf{e}_i\wedge\mathbf{e}_j)|(\mathbf{d}x_k\wedge\mathbf{d}x_\ell) &= (\mathbf{e}_i|\mathbf{d}x_k)(\mathbf{e}_j|\mathbf{d}x_\ell) - (\mathbf{e}_i|\mathbf{d}x_\ell)(\mathbf{e}_j|\mathbf{d}x_k) \\ &= \delta_{ik}\delta_{j\ell} - \delta_{i\ell}\delta_{jk}.\end{aligned} \tag{3.8}$$

Any two-form in the three-dimensional Euclidean space can be expanded in terms of the Cartesian basis two-forms as

$$\begin{aligned}\mathbf{\Phi} &= \sum_{i<j} \Phi_{ij}\mathbf{d}x_i \wedge \mathbf{d}x_j \\ &= \Phi_{xy}\mathbf{d}x \wedge \mathbf{d}y + \Phi_{yz}\mathbf{d}y \wedge \mathbf{d}z + \Phi_{zx}\mathbf{d}z \wedge \mathbf{d}x,\end{aligned} \tag{3.9}$$

where the coefficients Φ_{ij}, $i < j$ or Φ_{xy}, Φ_{yz}, Φ_{zx} in cyclic ordering are three scalar functions of the space variable $\mathbf{r}$.

Three-forms Operating the two-form (3.9) by $\mathbf{d}\wedge$ gives the three-form,

$$\gamma = \mathbf{d} \wedge \mathbf{\Phi} = (\partial_z \Phi_{xy} + \partial_x \Phi_{yz} + \partial_y \Phi_{zx})\mathbf{d}x \wedge \mathbf{d}y \wedge \mathbf{d}z. \tag{3.10}$$

Any three-form in the three-dimensional space can be written as a scalar multiple of the basis three-form, as $\gamma = \gamma_{xyz}\mathbf{d}x \wedge \mathbf{d}y \wedge \mathbf{d}z$ or, more generally, $\gamma = \gamma_{123}\mathbf{d}x_1 \wedge \mathbf{d}x_2 \wedge \mathbf{d}x_3$, where γ_{xyz} or γ_{123} is the component of the three-form in the corresponding basis. The scalar component in (3.10) resembles the Gibbsian divergence of a vector.

From the previous definitions, it is seen that the differential operation $\mathbf{d}\wedge$ in the three-dimensional space may correspond to either gradient, curl, or divergence of the Gibbsian vector analysis, depending on the multiform quantity under operation:

- ϕ scalar $\rightarrow$ $\mathbf{d}\phi$ $\sim$ gradient
- $\boldsymbol{\phi}$ one-form $\rightarrow$ $\mathbf{d} \wedge \boldsymbol{\phi}$ $\sim$ curl
- $\mathbf{\Phi}$ two-form $\rightarrow$ $\mathbf{d} \wedge \mathbf{\Phi}$ $\sim$ divergence

Leibnitz rule The Leibnitz rule for the differentiation of scalar functions,

$$\partial_x(f(x)g(x)) = g(x)\partial_x f(x) + f(x)\partial_x g(x), \tag{3.11}$$

can now be extended to the product of a p-form $\boldsymbol{\phi}^p(\mathbf{r})$ and a q-form $\boldsymbol{\psi}^q(\mathbf{r})$ as

$$\mathbf{d} \wedge (\boldsymbol{\phi}^p \wedge \boldsymbol{\psi}^q) = (\mathbf{d} \wedge \boldsymbol{\phi}^p) \wedge \boldsymbol{\psi}^q_c + (-1)^p \boldsymbol{\phi}^p_c \wedge (\mathbf{d} \wedge \boldsymbol{\psi}^q). \tag{3.12}$$

For clarity, quantities constant in differentiation have been given the subscript c. If $p = 0$ or $q = 0$, the corresponding $\wedge$ product sign must be omitted. The Leibnitz rule covers the known Gibbsian vector differentiation formulas for the expressions of the form $\nabla(fg)$, $\nabla \times (f\mathbf{g})$, $\nabla \cdot (f\mathbf{g})$, and $\nabla \cdot (\mathbf{f} \times \mathbf{g})$. The rule (3.12) is valid in the general n-dimensional space.

Second-order differentiation The differential-form operator $\mathbf{d}\wedge$ is independent of any metric. Because $\mathbf{d} \wedge \mathbf{d} = 0$, second-order differentiation requires some dyadic which prefers certain directions in space, say $\overline{\overline{\mathsf{A}}} \in \mathbb{F}_1\mathbb{E}_1$. For example, $\mathbf{d} \wedge (\overline{\overline{\mathsf{A}}}|\mathbf{d})$ makes a second-order two-form operator when $\overline{\overline{\mathsf{A}}}$ is not a multiple of $\overline{\overline{\mathsf{I}}}$. As another example, the Euclidean metric dyadic $\overline{\overline{\mathsf{G}}} = \sum \mathbf{e}_i\mathbf{e}_i$ defines the second-order scalar operator

$$\mathbf{d} \cdot \mathbf{d} = \mathbf{d}|\overline{\overline{\mathsf{G}}}|\mathbf{d} = \mathbf{d}|\nabla = \partial_x^2 + \partial_y^2 + \partial_z^2 = \nabla^2. \tag{3.13}$$

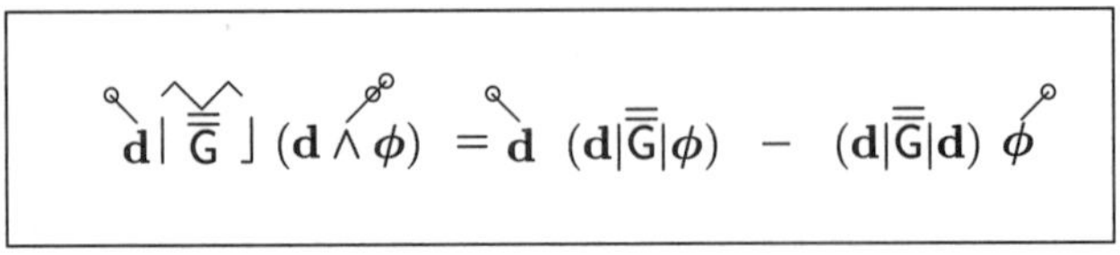

Fig. 3.1 *Second-order differentiation rule (3.15) for a one-form $\phi(\mathbf{r})$ corresponds to the curlcurl formula of Gibbsian vector analysis.*

∇^2 is the familiar Laplacian operator which can also be obtained as $\nabla^2 = \nabla \cdot \nabla = \nabla|\overline{\overline{\Gamma}}|\nabla$. It must once again be noted that ∇^2 depends on the metric through the chosen basis $\{\mathbf{e}_i\}$. With the help of the "bac cab formula" (1.50)

$$\mathbf{a}\rfloor(\beta \wedge \gamma) = \beta(\mathbf{a}|\gamma) - \gamma(\mathbf{a}|\beta), \tag{3.14}$$

we can write a useful second-order operation rule for one-forms $\phi(\mathbf{r})$,

$$(\mathbf{d}|\overline{\overline{\mathsf{G}}})\rfloor(\mathbf{d} \wedge \phi) = \mathbf{d}(\mathbf{d}|\overline{\overline{\mathsf{G}}}|\phi) - (\mathbf{d}|\overline{\overline{\mathsf{G}}}|\mathbf{d})\phi, \tag{3.15}$$

which is also valid when $\overline{\overline{\mathsf{G}}}$ is replaced by any dyadic $\overline{\overline{\mathsf{A}}} \in \mathbb{E}_1\mathbb{E}_1$, not necessarily symmetric. For a symmetric metric dyadic $\overline{\overline{\mathsf{G}}}$ (3.15) can also be written as

$$\begin{aligned} \nabla\rfloor(\mathbf{d} \wedge \phi) &= \mathbf{d}(\nabla|\phi) - \nabla^2\phi \\ &= \mathbf{d}(\partial_x\phi_x + \partial_y\phi_y + \partial_z\phi_z) - \nabla^2\phi, \end{aligned} \tag{3.16}$$

whence it corresponds to the familiar Gibbsian "curlcurl" formula

$$\nabla \times (\nabla \times \mathbf{f}) = \nabla(\nabla \cdot \mathbf{f}) - \nabla^2\mathbf{f}, \tag{3.17}$$

where $\mathbf{f}$ is a vector function.

3.1.2 Four-dimensional space

The Minkowskian four-dimensional spacetime can be studied in terms of the spacetime coordinates denoted by $x_1 = x$, $x_2 = y$, $x_3 = z$, $x_4 = \tau$ with τ defined as

$$\tau = ct, \quad c = 1/\sqrt{\mu_o\epsilon_o}. \tag{3.18}$$

Because τ is time normalized by the velocity of light c, it has the same physical dimension (meters) as the three spatial coordinates. The basis of one-forms is now $\varepsilon_x = \mathbf{d}x$, $\varepsilon_y = \mathbf{d}y$, $\varepsilon_z = \mathbf{d}z$, and $\varepsilon_\tau = \mathbf{d}\tau$. It is often advantageous to separate the three-dimensional Euclidean space components and normalized time components in multiform expressions. For clarity we occasionally denote the differential operator $\mathbf{d}$ by $\mathbf{d}_{\mathbb{E}}$ when it operates on the three-dimensional Euclidean space vector $\mathbf{r}$ only. To emphasize operation on all four coordinates in the Minkowskian space, we occasionally denote $\mathbf{d}$ by $\mathbf{d}_{\mathbf{M}}$.

One-forms The differential of a scalar function $\phi(\mathbf{r}, t)$ can be represented as

$$\begin{aligned}\mathbf{d}\phi = \sum_{i=1}^{4} \mathbf{d}x_i \partial_{x_i}\phi &= \mathbf{d}x\partial_x\phi + \mathbf{d}y\partial_y\phi + \mathbf{d}z\partial_z\phi + \mathbf{d}\tau\partial_\tau\phi \\ &= \mathbf{d}_{\mathbf{E}}\phi + \mathbf{d}\tau\partial_\tau\phi,\end{aligned} \tag{3.19}$$

whence the Minkowskian differential operator can be identified as

$$\mathbf{d}_{\mathbf{M}} = \sum_{i=1}^{4} \mathbf{d}x_i \partial_{x_i} = \mathbf{d}x\partial_x + \mathbf{d}y\partial_y + \mathbf{d}z\partial_z + \mathbf{d}\tau\partial_\tau = \mathbf{d}_{\mathbf{E}} + \mathbf{d}\tau\partial_\tau. \tag{3.20}$$

The general one-form has the form

$$\boldsymbol{\phi} = \sum \phi_i \mathbf{d}x_i = \boldsymbol{\phi}_{\mathbf{E}} + \phi_\tau \mathbf{d}\tau, \quad \boldsymbol{\phi}_{\mathbf{E}} = \phi_x \mathbf{d}x + \phi_y \mathbf{d}y + \phi_z \mathbf{d}z. \tag{3.21}$$

Applying the Minkowskian metric dyadic

$$\overline{\overline{\mathsf{G}}}_{\mathrm{M}} = \mathbf{e}_1\mathbf{e}_1 + \mathbf{e}_2\mathbf{e}_2 + \mathbf{e}_3\mathbf{e}_3 - \mathbf{e}_4\mathbf{e}_4 = \overline{\overline{\mathsf{G}}}_{\mathrm{E}} - \mathbf{e}_4\mathbf{e}_4, \tag{3.22}$$

where the basis vector $\mathbf{e}_4$ satisfies

$$\mathbf{e}_4|\mathbf{d}x_i = \delta_{i4}, \tag{3.23}$$

the vector four-gradient operator can be expressed as

$$\Box = \overline{\overline{\mathsf{G}}}_{\mathrm{M}}|\mathbf{d} = \nabla - \mathbf{e}_4\partial_\tau = \nabla - \mathbf{e}_4\frac{1}{c}\partial_t. \tag{3.24}$$

The second-order d'Alembertian scalar operator is obtained as

$$\Box^2 = \mathbf{d}|\overline{\overline{\mathsf{G}}}_{\mathrm{M}}|\mathbf{d} = \mathbf{d}|\Box = \nabla^2 - \partial_\tau^2 = \nabla^2 - \frac{1}{c^2}\partial_t^2. \tag{3.25}$$

Defining the Minkowskian dot operator between dual vectors we could also write $\Box^2 = \mathbf{d}_{\mathbf{M}} \cdot \mathbf{d}_{\mathbf{M}}$.

Two-forms The two-form obtained from the Minkowskian one-form (3.21) through differential operation can be decomposed as

$$\begin{aligned}\mathbf{d} \wedge \boldsymbol{\phi} &= (\mathbf{d}_{\mathbf{E}} + \mathbf{d}\tau\partial_\tau) \wedge (\boldsymbol{\phi}_{\mathbf{E}} + \phi_\tau \mathbf{d}\tau) \\ &= \mathbf{d}_{\mathbf{E}} \wedge \boldsymbol{\phi}_{\mathbf{E}} + (\mathbf{d}_{\mathbf{E}}\phi_\tau - \partial_\tau\boldsymbol{\phi}_{\mathbf{E}}) \wedge \mathbf{d}\tau.\end{aligned} \tag{3.26}$$

To expand this, the first term can be directly copied from (3.7),

$$\begin{aligned}\mathbf{d}_{\mathbf{E}} \wedge \boldsymbol{\phi}_{\mathbf{E}} &= (\partial_x\phi_y - \partial_y\phi_x)\mathbf{d}x \wedge \mathbf{d}y + (\partial_y\phi_z - \partial_z\phi_y)\mathbf{d}y \wedge \mathbf{d}z \\ &\quad + (\partial_z\phi_x - \partial_x\phi_z)\mathbf{d}z \wedge \mathbf{d}x,\end{aligned} \tag{3.27}$$

while the last term in (3.26) contains the one-form

$$\mathbf{d}_{\mathbb{E}}\phi_\tau - \partial_\tau\boldsymbol{\phi}_{\mathbb{E}} = (\partial_x\phi_\tau - \partial_\tau\phi_x)\mathbf{d}x + (\partial_y\phi_\tau - \partial_\tau\phi_y)\mathbf{d}y + (\partial_z\phi_\tau - \partial_\tau\phi_z)\mathbf{d}z. \tag{3.28}$$

$\mathbf{d}\wedge\boldsymbol{\phi}$ in (3.26) can be interpreted as a generalization of the curl operation to the four-dimensional Minkowskian space. Being independent of the dimension, the "curlcurl" rule (3.15) can then be generalized to

$$(\mathbf{d}|\overline{\overline{\mathsf{G}}}_{\mathrm{M}})\rfloor(\mathbf{d}\wedge\boldsymbol{\phi}) = \mathbf{d}(\mathbf{d}|\overline{\overline{\mathsf{G}}}_{\mathrm{M}}|\boldsymbol{\phi}) - (\mathbf{d}|\overline{\overline{\mathsf{G}}}_{\mathrm{M}}|\mathbf{d})\boldsymbol{\phi} \tag{3.29}$$

or, in terms of the four-gradient vector operator $\Box = \mathbf{d}|\overline{\overline{\mathsf{G}}}_{\mathrm{M}}$, to

$$\Box\rfloor(\mathbf{d}\wedge\boldsymbol{\phi}) = \mathbf{d}(\Box|\boldsymbol{\phi}) - \Box^2\boldsymbol{\phi}. \tag{3.30}$$

This rule actually generalizes the classical Helmholtz decomposition theorem to four dimensions, as will be seen. The rule (3.29) is also valid more generally because the symmetric metric dyadic $\overline{\overline{\mathsf{G}}}_{\mathrm{M}}$ can be replaced by any Minkowskian $\mathbb{E}_1\mathbb{E}_1$ dyadic, not necessarily symmetric.

The two-form $\boldsymbol{\Phi}$ is an element of the space $\mathbb{F}_2$ of six dimensions, and its general form is

$$\boldsymbol{\Phi} = \sum_{i<j}\Phi_{ij}\mathbf{d}x_i\wedge\mathbf{d}x_j \tag{3.31}$$

and, in Cartesian coordinates,

$$\boldsymbol{\Phi} = \Phi_{xy}\mathbf{d}x\wedge\mathbf{d}y + \Phi_{yz}\mathbf{d}y\wedge\mathbf{d}z + \Phi_{zx}\mathbf{d}z\wedge\mathbf{d}x + \Phi_{x\tau}\mathbf{d}x\wedge\mathbf{d}\tau + \Phi_{y\tau}\mathbf{d}y\wedge\mathbf{d}\tau + \Phi_{z\tau}\mathbf{d}z\wedge\mathbf{d}\tau. \tag{3.32}$$

This can be split in its spatial and temporal components as

$$\boldsymbol{\Phi} = \boldsymbol{\Phi}_{\mathbb{E}} + \boldsymbol{\phi}_{\mathbb{E}}\wedge\mathbf{d}\tau, \tag{3.33}$$

where $\boldsymbol{\Phi}_{\mathbb{E}}$ is a two-form, and $\boldsymbol{\phi}_{\mathbb{E}}$ a one-form, in the Euclidean space.

Three-forms Differentiating the general Minkowskian two-form $\boldsymbol{\Phi}$ written as (3.33), the following three-form is obtained:

$$\boldsymbol{\gamma} = \mathbf{d}\wedge\boldsymbol{\Phi} = \mathbf{d}_{\mathbb{E}}\wedge\boldsymbol{\Phi}_{\mathbb{E}} + (\mathbf{d}_{\mathbb{E}}\wedge\boldsymbol{\phi}_{\mathbb{E}} + \partial_\tau\boldsymbol{\Phi}_{\mathbb{E}})\wedge\mathbf{d}\tau, \tag{3.34}$$

with

$$\mathbf{d}_{\mathbb{E}}\wedge\boldsymbol{\Phi}_{\mathbb{E}} = (\partial_x\Phi_{yz} + \partial_y\Phi_{zx} + \partial_z\Phi_{xy})\mathbf{d}x\wedge\mathbf{d}y\wedge\mathbf{d}z \tag{3.35}$$

$$\mathbf{d}_{\mathbb{E}}\wedge\boldsymbol{\phi}_{\mathbb{E}} = (\partial_x\Phi_{y\tau} - \partial_y\Phi_{x\tau})\mathbf{d}x\wedge\mathbf{d}y + (\partial_y\Phi_{z\tau} - \partial_z\Phi_{y\tau})\mathbf{d}y\wedge\mathbf{d}z + (\partial_z\Phi_{x\tau} - \partial_x\Phi_{z\tau})\mathbf{d}z\wedge\mathbf{d}x, \tag{3.36}$$

$$\partial_\tau\boldsymbol{\Phi}_{\mathbb{E}} = \partial_\tau(\Phi_{xy}\mathbf{d}x\wedge\mathbf{d}y + \Phi_{yz}\mathbf{d}y\wedge\mathbf{d}z + \Phi_{zx}\mathbf{d}z\wedge\mathbf{d}x). \tag{3.37}$$

It is seen that this expression contains terms corresponding to divergence and curl expressions in three dimensions.

Invoking the bac cab rule for dual bivectors $\mathbf{\Gamma}$, (1.91), reproduced here as

$$\mathbf{a}\rfloor(\boldsymbol{\beta}\wedge\mathbf{\Gamma}) = \boldsymbol{\beta}\wedge(\mathbf{a}\rfloor\mathbf{\Gamma}) + \mathbf{\Gamma}(\mathbf{a}|\boldsymbol{\beta}), \tag{3.38}$$

corresponding to (3.15) and (3.29), we can write a second-order differentiation rule for the two-form $\mathbf{\Phi}$

$$(\overline{\overline{\mathsf{G}}}_{\mathrm{M}}|\mathbf{d})\rfloor(\mathbf{d}\wedge\mathbf{\Phi}) = \mathbf{d}\wedge(\mathbf{d}|\overline{\overline{\mathsf{G}}}_{\mathrm{M}}\rfloor\mathbf{\Phi}) + (\mathbf{d}|\overline{\overline{\mathsf{G}}}_{\mathrm{M}}|\mathbf{d})\mathbf{\Phi}, \tag{3.39}$$

where $\overline{\overline{\mathsf{G}}}_{\mathrm{M}}$ is the Minkowskian metric dyadic (3.22). This rule differs from that given for one-forms in (3.29) by a sign. An alternative form for (3.39),

$$\Box\rfloor(\mathbf{d}\wedge\mathbf{\Phi}) = \mathbf{d}\wedge(\Box\rfloor\mathbf{\Phi}) + \Box^2\mathbf{\Phi}, \tag{3.40}$$

gives another generalization of the "curlcurl" rule in four dimensions.

Four-forms Finally, differentiation of the general three-form

$$\begin{aligned}\gamma &= (\gamma_{xy\tau}\mathbf{d}x\wedge\mathbf{d}y + \gamma_{zx\tau}\mathbf{d}z\wedge\mathbf{d}x + \gamma_{yz\tau}\mathbf{d}y\wedge\mathbf{d}z)\wedge\mathbf{d}\tau \\ &\quad +\gamma_{xyz}\mathbf{d}x\wedge\mathbf{d}y\wedge\mathbf{d}z\end{aligned} \tag{3.41}$$

produces the four-form

$$\mathbf{d}\wedge\gamma = (\partial_z\gamma_{xy\tau} + \partial_y\gamma_{zx\tau} + \partial_x\gamma_{yz\tau} - \partial_\tau\gamma_{xyz})\mathbf{d}x\wedge\mathbf{d}y\wedge\mathbf{d}z\wedge\mathbf{d}\tau. \tag{3.42}$$

The scalar coefficient can be interpreted as the generalization of divergence to four dimensions. Note that the minus sign is not due to any metric but to the chosen ordering of the indices. Any four-form is a multiple of the basis four-form $\varepsilon_{1234} = \mathbf{d}x\wedge\mathbf{d}y\wedge\mathbf{d}z\wedge\mathbf{d}\tau$.

3.1.3 Spatial and temporal components

In the four-dimensional Minkowskian space, any p-form can be expanded in two terms which are called its spatial and temporal components. Because of orthogonality, this allows splitting equations in their spatial and temporal components. Thus, any one-form equation in four-dimensional space,

$$\phi = \phi_{\mathbf{E}} + \phi_\tau\mathbf{d}\tau = 0, \tag{3.43}$$

can be split in its components as

$$\phi_{\mathbf{E}} = 0, \qquad \phi_\tau = 0. \tag{3.44}$$

Similarly, any two-form equation of the form

$$\mathbf{\Phi} = \mathbf{\Phi}_{\mathbf{E}} + \phi_{\mathbf{E}}\wedge\mathbf{d}\tau = 0, \tag{3.45}$$

can be split as

$$\mathbf{\Phi}_{\mathrm{E}} = 0, \quad \boldsymbol{\phi}_{\mathrm{E}} = 0. \tag{3.46}$$

For example, consider a two-form equation for the four-dimensional one-form $\boldsymbol{\phi} = \boldsymbol{\phi}_{\mathrm{E}} + \phi_\tau \mathbf{d}\tau$:

$$\mathbf{d} \wedge \boldsymbol{\phi} = \mathbf{d}_{\mathrm{E}} \wedge \boldsymbol{\phi}_{\mathrm{E}} + \mathbf{d}\tau \wedge \partial_\tau \boldsymbol{\phi}_{\mathrm{E}} + \mathbf{d}_{\mathrm{E}}\phi_\tau \wedge \mathbf{d}\tau = 0. \tag{3.47}$$

This can be split in two equations as

$$\mathbf{d}_{\mathrm{E}} \wedge \boldsymbol{\phi}_{\mathrm{E}} = 0, \quad \mathbf{d}_{\mathrm{E}}\phi_\tau - \partial_\tau \boldsymbol{\phi}_{\mathrm{E}} = 0. \tag{3.48}$$

Finally, the general three-form equation

$$\boldsymbol{\gamma} = \boldsymbol{\gamma}_{\mathrm{E}} + \mathbf{\Psi}_{\mathrm{E}} \wedge \mathbf{d}\tau = 0 \tag{3.49}$$

can be split into $\boldsymbol{\gamma}_{\mathrm{E}} = 0$ and $\mathbf{\Psi}_{\mathrm{E}} = 0$. For example, the three-form equation for a two-form $\mathbf{\Phi} = \mathbf{\Phi}_{\mathrm{E}} + \boldsymbol{\phi}_{\mathrm{E}} \wedge \mathbf{d}\tau$,

$$\begin{aligned} \mathbf{d} \wedge \mathbf{\Phi} &= (\mathbf{d}_{\mathrm{E}} + \mathbf{d}\tau\partial_\tau) \wedge (\mathbf{\Phi}_{\mathrm{E}} + \boldsymbol{\phi}_{\mathrm{E}} \wedge \mathbf{d}\tau) \\ &= \mathbf{d}_{\mathrm{E}} \wedge \mathbf{\Phi}_{\mathrm{E}} + (\partial_\tau \mathbf{\Phi}_{\mathrm{E}} + \mathbf{d}_{\mathrm{E}} \wedge \boldsymbol{\phi}_{\mathrm{E}}) \wedge \mathbf{d}\tau = 0, \end{aligned} \tag{3.50}$$

is split in its spatial and temporal equations as

$$\mathbf{d}_{\mathrm{E}} \wedge \mathbf{\Phi}_{\mathrm{E}} = 0, \quad \mathbf{d}_{\mathrm{E}} \wedge \boldsymbol{\phi}_{\mathrm{E}} + \partial_\tau \mathbf{\Phi}_{\mathrm{E}} = 0. \tag{3.51}$$

Actually these two equations make one set of the Maxwell equations in three-dimensional representation, as will be seen in Chapter 4. In terms of differential forms, the Maxwell equations are represented by the simplest possible differential relations in the Minkowskian four-dimensional space. In fact, it is hard to imagine a simpler form for a differential equation than $\mathbf{d} \wedge \mathbf{\Phi} = 0$.

Problems

3.1.1 Verify by expanding in components that the two-form $\mathbf{d} \wedge (\mathbf{d}\phi)$ vanishes for any one-form $\boldsymbol{\phi} = \sum \phi_i \mathbf{d}x_i$ in n-dimensional space.

3.1.2 Derive through basis expansion of different terms the rule

$$(\mathbf{d}|\overline{\overline{\mathsf{G}}})\rfloor(\mathbf{d} \wedge \boldsymbol{\phi}) = \mathbf{d}(\mathbf{d}|\overline{\overline{\mathsf{G}}}|\boldsymbol{\phi}) - (\mathbf{d}|\overline{\overline{\mathsf{G}}}|\mathbf{d})\boldsymbol{\phi}$$

where $\boldsymbol{\phi}$ is a one-form and $\overline{\overline{\mathsf{G}}}$ a metric dyadic.

3.1.3 Prove the differentiation rule

$$\mathbf{d} \wedge (\boldsymbol{\phi}^p \wedge \boldsymbol{\psi}^q) = (\mathbf{d} \wedge \boldsymbol{\phi}^p) \wedge \boldsymbol{\psi}^q_c + (-1)^p \boldsymbol{\phi}^p_c \wedge (\mathbf{d} \wedge \boldsymbol{\psi}^q)$$

when $\boldsymbol{\phi}^p$ is a p-form and $\boldsymbol{\psi}^q$ is a q-form, and the subscript c denotes "constant" (no differentiation on this quantity).

3.2 DIFFERENTIATION THEOREMS

3.2.1 Poincaré's lemma and de Rham's theorem

From the antisymmetry of the wedge product, it follows that the second-order operator $\mathbf{d} \wedge \mathbf{d}$ vanishes. This simple fact is valid in any space of n dimensions and is called Poincaré's lemma. As special cases we obtain a number of theorems known from Gibbsian vector analysis. When ϕ is a scalar function, we have

$$\mathbf{d} \wedge \mathbf{d}\phi = 0, \quad \rightarrow \quad \nabla \times (\nabla f) = 0. \tag{3.52}$$

When $\boldsymbol{\phi}$ is a one-form, we have

$$\mathbf{d} \wedge \mathbf{d} \wedge \boldsymbol{\phi} = 0, \quad \rightarrow \quad \nabla \cdot (\nabla \times \mathbf{f}) = 0, \tag{3.53}$$

where the Gibbsian rule is valid for a vector $\mathbf{f}$. The inverse of Poincaré's lemma is called de Rham's theorem [24]:

> If $\boldsymbol{\chi}^p$ is a p-form $(1 \leq p < n)$ and $\mathbf{d} \wedge \boldsymbol{\chi}^p = 0$, there exists a $(p-1)$-form $\boldsymbol{\xi}^{p-1}$ so that one can write $\boldsymbol{\chi}^p = \mathbf{d} \wedge \boldsymbol{\xi}^{p-1}$.

This is a local property and valid in a star-like region which is topologically similar to a ball [24]. In a donut-like region there does not necessarily exist a unique $(p-1)$-form $\boldsymbol{\xi}^{p-1}$ for a given p-form $\boldsymbol{\chi}^p$. It may change its value along a closed path in the donut. By making a cut in the donut, a unique value can be, however, ascertained.

Closed and exact forms A p-form $\boldsymbol{\chi}^p$ satisfying $\mathbf{d} \wedge \boldsymbol{\chi}^p = 0$ is called a closed p-form. On the other hand, a p-form $\boldsymbol{\chi}^p = \mathbf{d} \wedge \boldsymbol{\xi}^{p-1}$ is called an exact form. An exact p-form is always closed due to Poincaré's lemma. A closed form is locally exact due to De Rham's theorem, but globally only in a star-like region.

Second-order operators As was seen earlier, double differentiation $\mathbf{d} \wedge \mathbf{d} \wedge$ does not produce a second-order differential operator. In the Euclidean three-dimensional space the Hodge dyadic transforming one-forms to two-forms

$$\overline{\overline{\mathsf{H}}}_2^T = (\mathbf{d}x \wedge \mathbf{d}y)\mathbf{e}_z + (\mathbf{d}y \wedge \mathbf{d}z)\mathbf{e}_x + (\mathbf{d}z \wedge \mathbf{d}x)\mathbf{e}_y \tag{3.54}$$

can be applied to define a two-form differential operator which is sometimes called the co-differential operator [24],

$$\boldsymbol{\delta} = \overline{\overline{\mathsf{H}}}_2^T|\mathbf{d} = (\mathbf{d}x \wedge \mathbf{d}y)\partial_z + (\mathbf{d}y \wedge \mathbf{d}z)\partial_x + (\mathbf{d}z \wedge \mathbf{d}x)\partial_y. \tag{3.55}$$

This allows one to form a second-order three-form differential operator as

$$\mathbf{d} \wedge \boldsymbol{\delta} = (\partial_x^2 + \partial_y^2 + \partial_z^2)\mathbf{d}x \wedge \mathbf{d}y \wedge \mathbf{d}z = \nabla^2 \mathbf{d}x \wedge \mathbf{d}y \wedge \mathbf{d}z. \tag{3.56}$$

From this, the scalar Laplace operator can be introduced as

$$\nabla^2 = \mathbf{e}_{xyz}|(\mathbf{d} \wedge \boldsymbol{\delta}) = \mathbf{e}_{xyz}|(\mathbf{d} \wedge (\overline{\overline{\mathsf{H}}}_2^T|\mathbf{d})). \tag{3.57}$$

Applying the dual form of (2.223), we see that this definition actually coincides with the definition $\nabla^2 = \mathbf{d}|\overline{\overline{\mathsf{G}}}|\mathbf{d}$ of (3.13) when the Euclidean metric dyadic $\overline{\overline{\mathsf{G}}}$ corresponds to the Hodge dyadic $\overline{\overline{\mathsf{H}}}_2$.

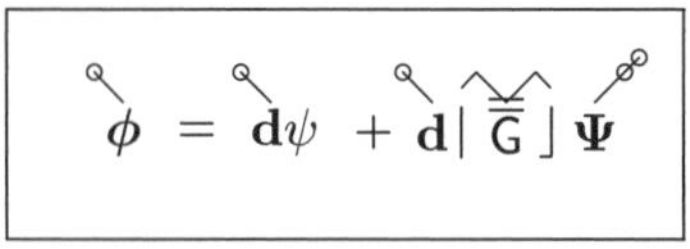

Fig. 3.2 *Visualization of the Helmholtz decomposition (3.61) of one-form ϕ in terms of scalar and two-form potentials ψ, $\boldsymbol{\Psi}$.*

3.2.2 Helmholtz decomposition

In Gibbsian vector analysis the classical Helmholtz theorem allows one to represent a given vector field $\mathbf{f}(\mathbf{r})$ in terms of a scalar potential $\phi(\mathbf{r})$ and a vector potential $\mathbf{A}(\mathbf{r})$ as a sum of irrotational and solenoidal components,

$$\mathbf{f}(\mathbf{r}) = \nabla\phi(\mathbf{r}) + \nabla \times \mathbf{A}(\mathbf{r}). \tag{3.58}$$

This decomposition can be extended to the space of n dimensions in differential-form notation. Let $\overline{\overline{\mathsf{G}}} \in \mathbb{E}_1\mathbb{E}_1$ be a metric dyadic (not necessarily symmetric), and let $\eta(\mathbf{r})$ be a one-form. Then, from the identity (3.15) we can write the relation

$$(\mathbf{d}|\overline{\overline{\mathsf{G}}}|\mathbf{d})\eta = \mathbf{d}(\mathbf{d}|\overline{\overline{\mathsf{G}}}|\eta) - \mathbf{d}|\overline{\overline{\mathsf{G}}}\rfloor(\mathbf{d}\wedge\eta). \tag{3.59}$$

Now, if the one-form $\eta(\mathbf{r})$ is a solution of the generalized Poisson equation (wave equation in the four-dimensional Minkowski space),

$$(\mathbf{d}|\overline{\overline{\mathsf{G}}}|\mathbf{d})\eta(\mathbf{r}) = \phi(\mathbf{r}), \tag{3.60}$$

from the differentiation rule (3.59) we see that any given one-form $\phi(\mathbf{r})$ can be decomposed as

$$\phi(\mathbf{r}) = \mathbf{d}\psi(\mathbf{r}) + (\mathbf{d}|\overline{\overline{\mathsf{G}}})\rfloor\boldsymbol{\Psi}(\mathbf{r}), \tag{3.61}$$

in terms of a scalar potential ψ and a two-form potential $\boldsymbol{\Psi}$, Figure 3.2. These potentials can be given in terms of the one-form η as

$$\psi(\mathbf{r}) = \mathbf{d}|\overline{\overline{\mathsf{G}}}|\eta(\mathbf{r}), \tag{3.62}$$

$$\boldsymbol{\Psi}(\mathbf{r}) = -\mathbf{d}\wedge\eta(\mathbf{r}). \tag{3.63}$$

From the expression (3.63) of the two-form potential we see that it must satisfy

$$\mathbf{d}\wedge\boldsymbol{\Psi}(\mathbf{r}) = 0. \tag{3.64}$$

Actually, $\boldsymbol{\Psi}$ is not uniquely defined by the one-form ϕ. Equation (3.64) can be taken as an extra condition for the two-form potential and it is called the gauge condition.

Equation (3.61) gives the Helmholtz decomposition theorem for a given one-form ϕ. The term $\mathbf{d}\psi(\mathbf{r})$ in (3.61) corresponds to the irrotational (gradient) part in the

Gibbsian form of the decomposition theorem (3.58), because it satisfies $\mathbf{d}\wedge\mathbf{d}\psi(\mathbf{r}) = 0$. Also, $(\mathbf{d}|\overline{\overline{\mathsf{G}}})\rfloor\mathbf{\Psi}(\mathbf{r})$ corresponds to the divergenceless (curl) part because it satisfies

$$\begin{aligned}(\mathbf{d}|\overline{\overline{\mathsf{G}}})|((\mathbf{d}|\overline{\overline{\mathsf{G}}})\rfloor\mathbf{\Psi}(\mathbf{r})) &= ((\mathbf{d}|\overline{\overline{\mathsf{G}}})\wedge(\mathbf{d}|\overline{\overline{\mathsf{G}}}))|\mathbf{\Psi}(\mathbf{r}) \\ &= (\mathbf{d}\wedge\mathbf{d})|\overline{\overline{\mathsf{G}}}^{(2)}|\mathbf{\Psi}(\mathbf{r}) = 0. \end{aligned} \tag{3.65}$$

$(\mathbf{d}|\overline{\overline{\mathsf{G}}})|$ corresponds to a divergence operation of Gibbsian analysis when operating on the one-form term.

Potential equations Given the one-form function $\boldsymbol{\phi}(\mathbf{r})$, its potentials $\psi(\mathbf{r}), \mathbf{\Psi}(\mathbf{r})$ can be expressed in terms of the one-form $\boldsymbol{\eta}(\mathbf{r})$, which is a solution of the generalized Poisson equation (3.60). However, we can also form individual equations for the scalar and two-form potentials. In fact, operating (3.61) by $(\mathbf{d}|\overline{\overline{\mathsf{G}}})|$ we obtain a second-order equation for the scalar potential:

$$(\mathbf{d}|\overline{\overline{\mathsf{G}}}|\mathbf{d})\psi(\mathbf{r}) = \mathbf{d}|\overline{\overline{\mathsf{G}}}|\boldsymbol{\phi}(\mathbf{r}), \tag{3.66}$$

and operating by $\mathbf{d}\wedge$ we have another equation for the two-form potential:

$$\mathbf{d}\wedge(\overline{\overline{\mathsf{G}}}|\mathbf{d})\rfloor\mathbf{\Psi}(\mathbf{r})) = \mathbf{d}\wedge\boldsymbol{\phi}(\mathbf{r}). \tag{3.67}$$

To simplify the latter we apply again the bac cab rule (1.90)

$$\boldsymbol{\alpha}\wedge(\mathbf{b}\rfloor\mathbf{\Gamma}) = \mathbf{b}\rfloor(\boldsymbol{\alpha}\wedge\mathbf{\Gamma}) - \mathbf{\Gamma}(\mathbf{a}|\boldsymbol{\beta}), \tag{3.68}$$

which is valid for any vector $\mathbf{b}$, dual vector $\boldsymbol{\beta}$, and dual bivector $\mathbf{\Gamma}$. Substituting $\mathbf{d}$ for $\boldsymbol{\alpha}$, $\mathbf{d}|\overline{\overline{\mathsf{G}}}$ for $\mathbf{b}$ and $\mathbf{\Psi}$ for $\mathbf{\Gamma}$ and using the gauge condition (3.64), (3.67) becomes

$$(\mathbf{d}|\overline{\overline{\mathsf{G}}}|\mathbf{d})\mathbf{\Psi}(\mathbf{r}) = -\mathbf{d}\wedge\boldsymbol{\phi}(\mathbf{r}). \tag{3.69}$$

Uniqueness for the decomposition (3.61) requires additional conditions on the boundary of the region or infinity for the two potentials.

Helmholtz decomposition in four dimensions Gibbsian form (3.58) for the Helmholtz decomposition theorem is only valid in the three-dimensional space. However, the differential-form formulation (3.61) is valid for any dimension n. Considering the four-dimensional Minkowskian space, let us interprete the decomposition formula by expanding all quantities in their spatial and temporal components:

$$\boldsymbol{\phi}_{\mathrm{M}}(\mathbf{r},\tau) = \boldsymbol{\phi}_{\mathrm{E}}(\mathbf{r},\tau) + \phi(\mathbf{r},\tau)\mathbf{d}\tau, \tag{3.70}$$

$$\mathbf{\Psi}_{\mathrm{M}}(\mathbf{r},\tau) = \mathbf{\Psi}_{\mathrm{E}}(\mathbf{r},\tau) + \boldsymbol{\psi}_{\mathrm{E}}(\mathbf{r},\tau)\wedge\mathbf{d}\tau, \tag{3.71}$$

$$\overline{\overline{\mathsf{G}}}_{\mathrm{M}} = \mathbf{e}_x\mathbf{e}_x + \mathbf{e}_y\mathbf{e}_y + \mathbf{e}_z\mathbf{e}_z - \mathbf{e}_\tau\mathbf{e}_\tau. \tag{3.72}$$

Here, again, $\mathbf{\Psi}_{\mathrm{E}}$ is a Euclidean three-dimensional two-form, $\boldsymbol{\phi}_{\mathrm{E}}, \boldsymbol{\psi}_{\mathrm{E}}$ are Euclidean three-dimensional one-forms, and ϕ, ψ are scalars. Now (3.61) becomes

$$\begin{aligned}\boldsymbol{\phi}_{\mathrm{M}} &= \boldsymbol{\phi}_{\mathrm{E}} + \phi\mathbf{d}\tau \\ &= (\mathbf{d}_{\mathrm{E}} + \mathbf{d}\tau\partial_\tau)\psi + (\mathbf{d}_{\mathrm{E}}|\overline{\overline{\mathsf{G}}}_{\mathrm{E}} - \mathbf{e}_\tau\partial_\tau)\rfloor(\mathbf{\Psi}_{\mathrm{E}} + \boldsymbol{\psi}_{\mathrm{E}}\wedge\mathbf{d}\tau) \\ &= \mathbf{d}_{\mathrm{E}}\psi + \mathbf{d}\tau\partial_\tau\psi + (\mathbf{d}_{\mathrm{E}}|\overline{\overline{\mathsf{G}}}_{\mathrm{E}})\rfloor\mathbf{\Psi}_{\mathrm{E}} - (\mathbf{d}_{\mathrm{E}}|\overline{\overline{\mathsf{G}}}_{\mathrm{E}})|\boldsymbol{\psi}_{\mathrm{E}})\mathbf{d}\tau - \partial_\tau\boldsymbol{\psi}_{\mathrm{E}}. \end{aligned} \tag{3.73}$$

In the expansion we have twice applied the bac cab rule (1.50). Separating finally the spatial and temporal components, the equation is split in two:

$$\phi_{\mathbf{E}}(\mathbf{r},\tau) = \mathbf{d}_{\mathbf{E}}\psi(\mathbf{r},\tau) + (\mathbf{d}_{\mathbf{E}}|\overline{\overline{\mathsf{G}}}_{\mathbf{E}})\rfloor\boldsymbol{\Psi}_{\mathbf{E}}(\mathbf{r},\tau) - \partial_\tau\boldsymbol{\psi}_{\mathbf{E}}(\mathbf{r},\tau), \quad (3.74)$$

$$\phi(\mathbf{r},\tau) = \partial_\tau\psi(\mathbf{r},\tau) - (\mathbf{d}_{\mathbf{E}}|\overline{\overline{\mathsf{G}}}_{\mathbf{E}})|\boldsymbol{\psi}_{\mathbf{E}}(\mathbf{r},\tau). \quad (3.75)$$

The gauge condition (3.64) gives

$$(\mathbf{d}_{\mathbf{E}} + \mathbf{d}\tau\partial_\tau)\wedge(\boldsymbol{\Psi}_{\mathbf{E}}(\mathbf{r},\tau) + \boldsymbol{\psi}_{\mathbf{E}}(\mathbf{r},\tau)\wedge\mathbf{d}\tau) \quad (3.76)$$

$$= \mathbf{d}_{\mathbf{E}}\wedge\boldsymbol{\Psi}_{\mathbf{E}}(\mathbf{r},\tau) + (\partial_\tau\boldsymbol{\Psi}_{\mathbf{E}}(\mathbf{r},\tau) + \mathbf{d}_{\mathbf{E}}\wedge\boldsymbol{\psi}_{\mathbf{E}}(\mathbf{r},\tau))\wedge\mathbf{d}\tau = 0, \quad (3.77)$$

and it can be split into

$$\mathbf{d}_{\mathbf{E}}\wedge\boldsymbol{\Psi}_{\mathbf{E}}(\mathbf{r},\tau) = 0, \quad \partial_\tau\boldsymbol{\Psi}_{\mathbf{E}}(\mathbf{r},\tau) + \mathbf{d}_{\mathbf{E}}\wedge\boldsymbol{\psi}_{\mathbf{E}}(\mathbf{r},\tau) = 0. \quad (3.78)$$

These conditions are again similar to one half of the Maxwell equations.

Problems

3.2.1 Replacing the Hodge dyadic (3.54) by that of the four-dimensional Minkowski space, define the co-differential operator and the D'Alembertian operator $\Box^2$.

3.2.2 Derive (3.74), (3.75) by doing the intemediate steps.

3.3.3 Defining the metric dyadic $\overline{\overline{\mathsf{G}}}$ as (3.3) and the Hodge dyadic $\overline{\overline{\mathsf{H}}}_2$ as (3.54), show that the following identity holds:

$$\mathbf{e}_{123}|(\boldsymbol{\alpha}\wedge(\overline{\overline{\mathsf{H}}}{}_2^T|\boldsymbol{\beta})) = \boldsymbol{\alpha}|\overline{\overline{\mathsf{G}}}|\boldsymbol{\beta}$$

for any three-dimensional dual vectors $\boldsymbol{\alpha},\boldsymbol{\beta}$. Thus, it is valid for $\boldsymbol{\alpha}=\boldsymbol{\beta}=\mathbf{d}$, the differential operator.

3.3 INTEGRATION

3.3.1 Manifolds

A manifold is a topological space which locally resembles a Euclidean space. An example of a manifold is the surface of a three-dimensional region, for example, the surface of a sphere. On every point of the surface, one can attach a tangent plane which forms a local two-dimensional Euclidean space. A curved line is a manifold, and the tangent line at each of its points forms a one-dimensional Euclidean space. A manifold can be represented through coordinates x_i which are functions of the position on the manifold. In the following we only consider manifolds in the three-dimensional space. One-, two-, and three-dimensional manifolds (curves, surfaces, and volumes) will be denoted by $\mathcal{C}$, $\mathcal{S}$, and $\mathcal{V}$, respectively.

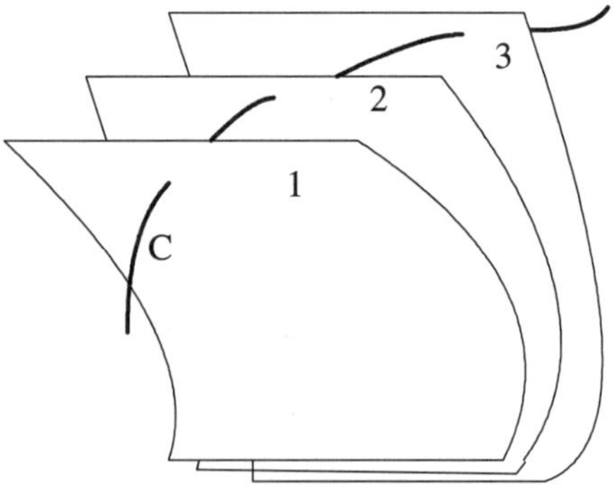

Fig. 3.3 *Integration of a one-form $\phi(\mathbf{r})$ along a curve C is visualized through piercing surfaces making the one-form.*

Differential forms are integrable over manifolds of the same dimension. For example, a one-form in the three-dimensional space

$$\phi = \phi_1 \mathbf{d}x_1 + \phi_2 \mathbf{d}x_2 + \phi_3 \mathbf{d}x_3 \tag{3.79}$$

can be integrated along a curve C. If the curve is approximated by a broken line of consecutive line elements represented by a set of vectors $\Delta_i \mathbf{r}$, the integral I can be approximated by the sum

$$I \approx \sum \phi(\mathbf{r}_i)|\Delta_i \mathbf{r}. \tag{3.80}$$

When the broken line approaches the curve C, the limit can be written symbolically as the duality product of the one-form and the manifold [18]

$$I = \phi|C. \tag{3.81}$$

Because a dual vector can be visualized through a set of parallel planes whose density corresponds to the magnitude of the dual vector, any one-form can be visualized through a set of surfaces filling the whole space. At any particular point of the space, the local dual vector consists of a set of parallel planes tangent to the surface whose density equals that of the surfaces at that point. Thus, the integral $\phi|C$ gives the number of surfaces pierced by curve C in Figure 3.3, and the sign depends on the direction of integration. A suitable physical analogue is the integral of the static electric field, which can be represented through a set of constant-potential surfaces. The result is the voltage. It can be measured by counting the number of one-volt surfaces pierced by the curve in the positive direction minus number of surfaces pierced in the negative direction.

Similarly, a two-form can be integrated over a surface S. Writing $\Delta_{ij}\mathbf{S} = \Delta_i \mathbf{r} \wedge \Delta_j \mathbf{r}$ for a surface element, one can interpret the limit as the surface integral

$$I \approx \sum \sum \mathbf{\Phi}|\Delta_{ij}\mathbf{S} \rightarrow I = \mathbf{\Phi}|S. \tag{3.82}$$

Finally, a three-form $\boldsymbol{\alpha}$ can be integrated over a volume $\mathcal{V}$ as

$$I \approx \sum\sum\sum \boldsymbol{\alpha}|\Delta_{ijk}\mathbf{V} \rightarrow I = \boldsymbol{\alpha}|\mathcal{V}, \tag{3.83}$$

when the volume element is the trivector $\Delta_{ijk}\mathbf{V} = \Delta_i\mathbf{r} \wedge \Delta_j\mathbf{r} \wedge \Delta_k\mathbf{r}$.

3.3.2 Stokes' theorem

The integral theorems of Gauss and Stokes, known in the Gibbsian vector analysis, are two aspects of a single theorem which in the differential-form language is called Stokes' theorem. If $\mathcal{D}$ is the domain of integration and its boundary is denoted by $\partial\mathcal{D}$, Stokes' theorem becomes [24]

$$(\mathbf{d} \wedge \boldsymbol{\phi})|\mathcal{D} = \boldsymbol{\phi}|\partial\mathcal{D}. \tag{3.84}$$

As a memory rule, the differential operator $\mathbf{d}$ can be thought of as moving through the duality product sign and transforming to ∂, which can be interpreted as the boundary operator.

Let us consider examples in the three-dimensional space. If ϕ is a zero-form (scalar field), $\mathbf{d}\phi$ is a one-form which can be integrated along a curve $\mathcal{C}$. If the two end points of the curve are denoted by $\mathcal{A}$ and $\mathcal{B}$, the boundary operator gives $\partial\mathcal{C} = \mathcal{B} - \mathcal{A}$ and (3.84) takes the form

$$\mathbf{d}\phi|\mathcal{C} = \phi|\partial\mathcal{C} = \phi|\mathcal{B} - \phi|\mathcal{A}. \tag{3.85}$$

In the Gibbsian vector analysis, this corresponds to integration of a gradient of a scalar function which leads to the difference of the values of the function at the endpoints.

The boundary $\partial\mathcal{S}$ of a surface $\mathcal{S}$ in the three-dimensional space is a closed curve. If $\boldsymbol{\phi}$ is a one-form, $\mathbf{d} \wedge \boldsymbol{\phi}$ is a two-form which can be integrated over the surface. Stokes' theorem (3.84) now gives

$$(\mathbf{d} \wedge \boldsymbol{\phi})|\mathcal{S} = \boldsymbol{\phi}|\partial\mathcal{S}, \tag{3.86}$$

which corresponds to Stokes' theorem of the Gibbsian vector analysis: The curl of a vector function integrated over a surface equals the path integral of the vector function round the boundary curve.

On the surface $\mathcal{S}$ one can write a theorem which corresponds to Poincaré's lemma, for multiforms:

$$\partial\partial\mathcal{S} = 0, \tag{3.87}$$

meaning that a closed curve does not have end points. More generally, it states that a boundary does not have a boundary. The theorem (3.87) also matches the following integral theorem:

$$(\mathbf{d} \wedge \mathbf{d}\phi)|\mathcal{S} = \mathbf{d}\phi|\partial\mathcal{S} = \phi|\partial\partial\mathcal{S} = 0, \tag{3.88}$$

valid for any scalar function ϕ and surface $\mathcal{S}$. The first term vanishes due to Poincaré's lemma and the last term vanishes due to (3.87). The middle term in Gibbsian notation is known to vanish because the integral around a closed loop of a gradient vanishes.

If $\mathbf{\Phi}$ is a two-form, $\mathbf{d} \wedge \mathbf{\Phi}$ is a three-form which can be integrated over a volume in a three-dimensional space. If the volume $\mathcal{V}$ is bounded by the surface $\partial\mathcal{V}$, Stokes' theorem reads

$$(\mathbf{d} \wedge \mathbf{\Phi})|\mathcal{V} = \mathbf{\Phi}|\partial\mathcal{V}. \tag{3.89}$$

In the language of Gibbsian vector analysis, this is the well-known Gauss' theorem: The volume integral of the divergence of a vector function equals the integral of the vector function over the boundary surface, the flow of the vector through the boundary surface. Again we have $\partial\partial\mathcal{V} = 0$, which means that a closed surface has no boundary curve. These mental pictures can be transferred to the space of n dimensions.

3.3.3 Euclidean simplexes

Integration over a manifold can be understood as the limit over small Euclidean elements, as was seen above. For a curve and a surface this corresponds to a broken-line and a plane-element approximation, respectively. Such elements of a manifold are called simplexes [24]. Simplexes play an important role in numerical electromagnetics when physical regions are approximated through finite elements [8]. Different orders of simplexes can be classified as follows.

- 0-simplex is a point (P_0)
- 1-simplex is a straight line element. It is defined through an ordered pair of points (P_0, P_1), provided these do not coincide.
- 2-simplex is a triangular surface element. It is defined through an ordered triple of points (P_0, P_1, P_2), provided these are not on the same line.
- 3-simplex is a tetrahedral volume element. It is defined through an ordered quadruple of points (P_0, P_1, P_2, P_3), provided they are not on the same plane.
- n-simplex is defined correspondingly through an ordered $(n+1)$-tuple of points $(P_0, P_1, \cdots, P_n)$.

All points within the boundary of the simplex belong to the simplex. The boundary of a p-simplex $\mathbf{s}$ is denoted by $\partial\mathbf{s}$, and it is a sum of $(p-1)$-simplexes

$$\partial(P_0, P_1, \cdots, P_n) = \sum_{i=1}^{n}(-1)^i(P_0, \cdots, P_{i-1}, P_{i+1}, \cdots, P_n). \tag{3.90}$$

Because every simplex is oriented through the order of the points, minus sign reverses that order. The orientation of a line segment is its direction (order of end points), that of a triangle is its sense of rotation (order of two sides) and that of a tetrahedron is its handedness (order of three sides joining all four points).

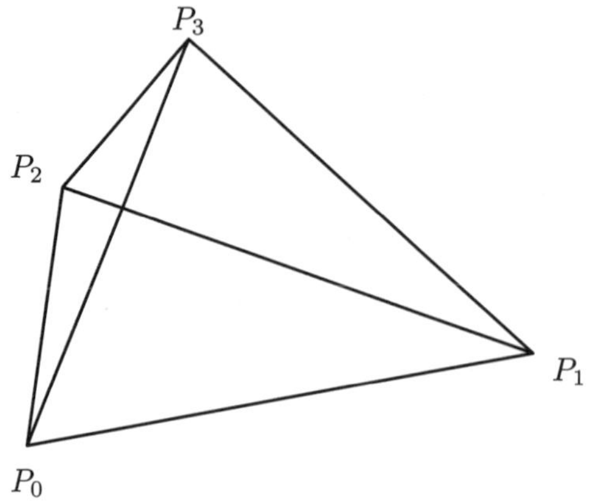

Fig. 3.4 *The boundary of the simplex* (P_0, P_1, P_2, P_3) *is* $(P_1, P_2, P_3) - (P_0, P_2, P_3) + (P_0, P_1, P_3) - (P_0, P_1, P_2)$*, defined by four triangles with orientation corresponding to the same handedness of rotation when looking out from the simplex.*

Chain and cycle A chain of p-simplexes $\mathbf{s}_i$ is a sum of the form

$$\mathbf{c} = \sum a^i \mathbf{s}_i, \tag{3.91}$$

where the a_i are real numbers. A chain of 1-simplexes is a broken line of consecutive line elements, and a chain of 2-simplexes is a surface of triangular elements. Multiplying a simplex $\mathbf{s}_i$ by a scalar a^i means changing its size.

The boundary of a chain of simplexes is

$$\partial \mathbf{c} = \sum a^i \partial \mathbf{s}_i. \tag{3.92}$$

For example, the boundary of a chain of 2-simplexes is a broken line bounding the surface of triangular elements. It has no boundary because of the rule (3.90). For example, for the 2-simplex (P_0, P_1, P_2) we can write for its boundary the 1-simplex

$$\partial(P_0, P_1, P_2) = (P_1, P_2) - (P_0, P_2) + (P_0, P_1), \tag{3.93}$$

whose boundary vanishes:

$$\begin{aligned} &\partial\partial(P_0, P_1, P_2) \\ &= \partial(P_1, P_2) - \partial(P_0, P_2) + \partial(P_0, P_1) \\ &= [(P_2) - (P_1)] - [(P_2) - (P_0)] + [(P_1) - (P_0)] = 0. \end{aligned} \tag{3.94}$$

A closed chain of simplexes is called a cycle. Because the boundary of a chain of simplexes is closed, it forms a cycle. A cycle is not, however, necessarily a boundary of a chain of simplexes. For example, a triangular net covering a Möbius strip has a boundary which is a cycle, but the net is not a chain of simplexes because it does not have a unique orientation.

By approximating a manifold through a chain of simplexes, any differential form can be integrated approximately. The finite element method is based on replacing the manifold through a chain of simplexes. Because there is a one-to-one correspondence between points of any p-simplex and a basic p-simplex, all p-simplexes can be mapped onto a basic p-simplex of unit size.

Problems

3.3.1 Study the one-form

$$\phi = \frac{1}{x^2 + y^2}(x\mathbf{d}y - y\mathbf{d}x).$$

Prove that it is closed by showing that it satisfies $\mathbf{d} \wedge \phi = 0$. Study whether it is also exact, by integrating along a loop $\mathcal{C}$ containing the origin. For an exact one-form, this should give $\phi|\mathcal{C} = 0$.

3.3.2 Derive a condition for the coefficient functions Φ_{ij} of the two-form

$$\mathbf{\Phi}(\mathbf{r}) = \sum_{i<j} \Phi_{ij}(\mathbf{r})\mathbf{d}x_i \wedge \mathbf{d}x_j$$

which makes it exact. Find the one-form $\phi(\mathbf{r}) = \sum_i \phi_i(\mathbf{r})\mathbf{d}x_i$ such that $\mathbf{\Phi}(\mathbf{r}) = \mathbf{d} \wedge \phi(\mathbf{r})$.

3.4 AFFINE TRANSFORMATIONS

Affine transformation is a linear transformation of space and time variables. Let us first consider the general n-dimensional space and later apply the transformation to three-dimensional Euclidean and four-dimensional Minkowskian spaces.

3.4.1 Transformation of differential forms

Ignoring the shift of origin, the affine transformation from the vector variable $\mathbf{x}$ to $\mathbf{x}_a$ can be expressed in terms of a dyadic $\overline{\overline{\mathsf{A}}} \in \mathbb{E}_1\mathbb{F}_1$, possessing a finite inverse $\overline{\overline{\mathsf{A}}}^{-1}$, as

$$\mathbf{x}_a = \overline{\overline{\mathsf{A}}}|\mathbf{x}, \quad \mathbf{x} = \overline{\overline{\mathsf{A}}}^{-1}|\mathbf{x}_a. \tag{3.95}$$

Choosing a base of vectors $\{\mathbf{a}_i\}$ or a base of dual vectors $\{\boldsymbol{\alpha}_i\}$ in n dimensions, any such dyadic $\overline{\overline{\mathsf{A}}}$ can be expressed as

$$\overline{\overline{\mathsf{A}}} = \sum \mathbf{a}_i\boldsymbol{\alpha}_i = \mathbf{a}_1\boldsymbol{\alpha}_1 + \mathbf{a}_2\boldsymbol{\alpha}_2 + \cdots \mathbf{a}_n\boldsymbol{\alpha}_n, \tag{3.96}$$

satisfying

$$\begin{aligned} \overline{\overline{\mathsf{A}}}^{(n)} &= \overline{\overline{\mathsf{I}}}^{(n)}\mathrm{det}\overline{\overline{\mathsf{A}}} \\ &= (\mathbf{a}_1 \wedge \mathbf{a}_2 \wedge \cdots \wedge \mathbf{a}_n)(\boldsymbol{\alpha}_1 \wedge \boldsymbol{\alpha}_2 \wedge \cdots \wedge \boldsymbol{\alpha}_n) \neq 0. \end{aligned} \tag{3.97}$$

The affine transformation can be pictured as a deformation made by stretching, mirroring, and rotating the space.

When the position vector $\mathbf{x} = \sum x_i \mathbf{e}_i$ is operated dyadically by $\mathbf{d}$, what results is the unit dyadic for one-forms:

$$\mathbf{dx} = \sum \mathbf{d}x_i \mathbf{e}_i = \sum \boldsymbol{\varepsilon}_i \mathbf{e}_i = \overline{\overline{\mathsf{I}}}^T. \tag{3.98}$$

Similar operation on the transformed position vector gives

$$\mathbf{dx}_a = \mathbf{d}(\overline{\overline{\mathsf{A}}}|\mathbf{x}) = \mathbf{dx}|\overline{\overline{\mathsf{A}}}^T = \overline{\overline{\mathsf{A}}}^T, \tag{3.99}$$

Requiring

$$\mathbf{d}_a \mathbf{x}_a = \overline{\overline{\mathsf{I}}}^T = \overline{\overline{\mathsf{A}}}^{-1T}|\mathbf{dx}_a, \tag{3.100}$$

we can define the transformed operator as

$$\mathbf{d}_a = \overline{\overline{\mathsf{A}}}^{-1T}|\mathbf{d}, \quad \mathbf{d} = \overline{\overline{\mathsf{A}}}^T|\mathbf{d}_a. \tag{3.101}$$

All dual vectors are transformed similarly. For example, a scalar function $\phi(\mathbf{x})$ is transformed to

$$\phi(\mathbf{x}) = \phi(\overline{\overline{\mathsf{A}}}^{-1}|\mathbf{x}_a) = \phi_a(\mathbf{x}_a), \tag{3.102}$$

whence its gradient obeys the rule

$$\mathbf{d}\phi(\mathbf{x}) = \mathbf{d}\phi_a(\mathbf{x}_a) = \overline{\overline{\mathsf{A}}}^T|\mathbf{d}_a\phi_a(\mathbf{x}_a). \tag{3.103}$$

Now since this is a one-form, all one-forms $\boldsymbol{\phi}(\mathbf{x})$ transform similarly as

$$\boldsymbol{\phi}(\mathbf{x}) = \overline{\overline{\mathsf{A}}}^T|\boldsymbol{\phi}_a(\mathbf{x}_a), \quad \boldsymbol{\phi}_a(\mathbf{x}_a) = \overline{\overline{\mathsf{A}}}^{-1T}|\boldsymbol{\phi}(\mathbf{x}). \tag{3.104}$$

For two-forms we apply the identity

$$(\overline{\overline{\mathsf{A}}}|\mathbf{a}) \wedge (\overline{\overline{\mathsf{A}}}|\mathbf{b}) = \overline{\overline{\mathsf{A}}}^{(2)}|(\mathbf{a} \wedge \mathbf{b}), \tag{3.105}$$

to obtain, for example,

$$\mathbf{d} \wedge \boldsymbol{\phi}(\mathbf{x}) = (\overline{\overline{\mathsf{A}}}^T|\mathbf{d}_a) \wedge (\overline{\overline{\mathsf{A}}}^T|\boldsymbol{\phi}_a(\mathbf{x}_a)) = \overline{\overline{\mathsf{A}}}^{(2)T}|(\mathbf{d}_a \wedge \boldsymbol{\phi}_a(\mathbf{x}_a)). \tag{3.106}$$

Thus, any two-form $\boldsymbol{\Phi}(\mathbf{x})$ transforms as

$$\boldsymbol{\Phi}(\mathbf{x}) = \overline{\overline{\mathsf{A}}}^{(2)T}|\boldsymbol{\Phi}_a(\mathbf{x}_a), \quad \boldsymbol{\Phi}_a(\mathbf{x}_a) = \overline{\overline{\mathsf{A}}}^{(-2)T}|\boldsymbol{\Phi}(\mathbf{x}). \tag{3.107}$$

In analogy, the corresponding rules for a three-form $\boldsymbol{\gamma}(\mathbf{x})$ are

$$\boldsymbol{\gamma}(\mathbf{x}) = \overline{\overline{\mathsf{A}}}^{(3)T}|\boldsymbol{\gamma}_a(\mathbf{x}_a), \quad \boldsymbol{\gamma}_a(\mathbf{x}_a) = \overline{\overline{\mathsf{A}}}^{(-3)T}|\boldsymbol{\gamma}(\mathbf{x}). \tag{3.108}$$

This leads to the folloing conclusion. If a law of nature between, say, a two-form $\boldsymbol{\Phi}$ and a three-form $\boldsymbol{\gamma}$ is expressed as a relation involving the differential operator as

$$\mathbf{d} \wedge \boldsymbol{\Phi}(\mathbf{x}) = \boldsymbol{\gamma}(\mathbf{x}), \tag{3.109}$$

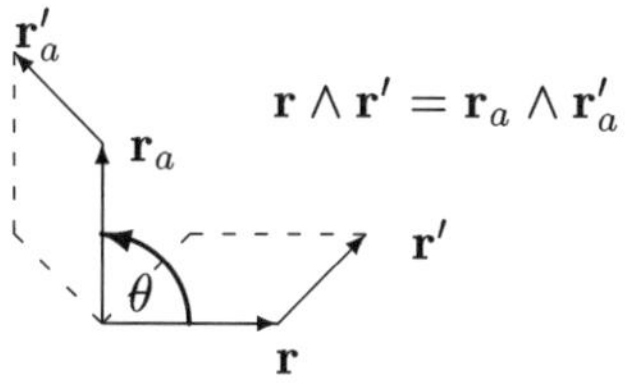

Fig. 3.5 *Rotation of a bivector in its plane by an angle θ does not change the bivector.*

applying the above rules we can see that it implies

$$(\overline{\overline{\mathsf{A}}}^T|\mathbf{d}_a)\wedge(\overline{\overline{\mathsf{A}}}^{(2)T}|\mathbf{\Phi}_a) = \overline{\overline{\mathsf{A}}}^{(3)T}|(\mathbf{d}_a\wedge\mathbf{\Phi}_a) = \overline{\overline{\mathsf{A}}}^{(3)T}|\gamma_a. \tag{3.110}$$

This means that the affine-transformed law has the same form

$$\mathbf{d}_a\wedge\mathbf{\Phi}_a(\mathbf{x}_a) = \gamma_a(\mathbf{x}_a), \tag{3.111}$$

which property is called covariance in physics.

3.4.2 Three-dimensional rotation

In the three-dimensional Euclidean space we denote the position vector $\mathbf{x}$ by $\mathbf{r}$. As an example of an affine transformation we consider the rotation dyadic

$$\overline{\overline{\mathsf{A}}} = \overline{\overline{\mathsf{R}}}(\theta) = \mathbf{e}_1\varepsilon_1 + (\cos\theta\mathbf{e}_2 + \sin\theta\mathbf{e}_3)\varepsilon_2 + (\cos\theta\mathbf{e}_3 - \sin\theta\mathbf{e}_2)\varepsilon_3, \tag{3.112}$$

where $\{\mathbf{e}_i\}$ and $\{\varepsilon_i\}$ are two reciprocal basis systems. The rotation dyadic transforms the position vector $\mathbf{r}$ to

$$\mathbf{r}_a = \overline{\overline{\mathsf{R}}}(\theta)|\mathbf{r} = x_1\mathbf{e}_1 + (x_2\cos\theta - x_3\sin\theta)\mathbf{e}_2 + (x_3\cos\theta + x_2\sin\theta)\mathbf{e}_3, \tag{3.113}$$

by rotating its x_1 and x_2 components by an angle θ. The rotation dyadic can be shown to satisfy

$$\overline{\overline{\mathsf{A}}}^{(2)} = (\cos\theta\mathbf{e}_{12} - \sin\theta\mathbf{e}_{31})\varepsilon_{12} + (\cos\theta\mathbf{e}_{31} + \sin\theta\mathbf{e}_{12})\varepsilon_{31} + \mathbf{e}_{23}\varepsilon_{23}, \tag{3.114}$$

$$\overline{\overline{\mathsf{A}}}^{(3)} = \mathbf{e}_{123}\varepsilon_{123} = \overline{\overline{\mathsf{I}}}^{(3)}. \tag{3.115}$$

The inverse dyadic becomes

$$\begin{aligned}\overline{\overline{\mathsf{A}}}^{-1} &= \mathbf{e}_1\varepsilon_1 + (\cos\theta\mathbf{e}_2 - \sin\theta\mathbf{e}_3)\varepsilon_2 + (\cos\theta\mathbf{e}_3 + \sin\theta\mathbf{e}_2)\varepsilon_3 \\ &= \overline{\overline{\mathsf{R}}}(-\theta),\end{aligned} \tag{3.116}$$

whence it corresponds to a rotation by the angle $-\theta$. Taking two vectors $\mathbf{r}$ and $\mathbf{r}'$, the bivector $\mathbf{r} \wedge \mathbf{r}'$ (oriented surface) is transformed to

$$\begin{aligned} \mathbf{r}_a \wedge \mathbf{r}_a' &= (\cos\theta \mathbf{e}_{12} - \sin\theta \mathbf{e}_{31})(x_1 x_2' - x_2 x_1') \\ &+(\cos\theta \mathbf{e}_{31} + \sin\theta \mathbf{e}_{12})(x_3 x_1' - x_1 x_3') \\ &+\mathbf{e}_{23}(x_2 x_3' - x_3 x_2'). \end{aligned} \tag{3.117}$$

If $x_1 = x_1' = 0$, rotation is in the plane defined by $\mathbf{r}$ and $\mathbf{r}'$, in which case $\mathbf{r}_a \wedge \mathbf{r}_a' = \mathbf{r} \wedge \mathbf{r}'$ for any θ and the rotation does not change the bivector. All trivectors (volumes with handedness) are invariant in the transformation. In the Euclidean metric, the rotation transformation implies

$$\begin{aligned} \mathbf{r}_a \cdot \mathbf{r}_a &= \mathbf{r}_a|\overline{\overline{\Gamma}}_E|\mathbf{r}_a = \sum x_{ai}^2 \\ &= x_1^2 + (x_2\cos\theta - x_3\sin\theta)^2 + (x_3\cos\theta + x_2\sin\theta)^2 \\ &= \sum x_i^2 = \mathbf{r}\cdot\mathbf{r}. \end{aligned} \tag{3.118}$$

Thus, the Euclidean length of the vector is invariant in the rotation, an obvious result from geometrical point of view. An affine transformation which preserves the quantity $\mathbf{r}\cdot\mathbf{r} = \mathbf{r}|\overline{\overline{\Gamma}}|\mathbf{r}$ in some metric is called an orthogonal transformation (with respect to that metric).

3.4.3 Four-dimensional rotation

In the four-dimensional Minkowskian space, the general affine dyadic $\overline{\overline{A}}$ involves transformation of the normalized time variable $x_4 = \tau = ct$ as well. The four-vector is written as

$$\mathbf{x} = \mathbf{r} + \mathbf{e}_4 x_4, \tag{3.119}$$

where $\mathbf{r}$ is the three-dimensional Euclidean position vector and x_4 is the normalized time variable.

As a special case of the affine transformation in the Minkowskian space, let us consider the orthogonal transformation by assuming that the length of a four-vector $\mathbf{x}$ in the Minkowskian metric is not changed:

$$\mathbf{x}_a\cdot\mathbf{x}_a = \mathbf{x}_a|\overline{\overline{\Gamma}}_M|\mathbf{x}_a = \mathbf{x}|\overline{\overline{A}}^T|\overline{\overline{\Gamma}}_M|\overline{\overline{A}}|\mathbf{x} = \mathbf{x}\cdot\mathbf{x} = \mathbf{x}|\overline{\overline{\Gamma}}_M|\mathbf{x}. \tag{3.120}$$

Because the dyadic $\overline{\overline{B}} = \overline{\overline{A}}^T|\overline{\overline{\Gamma}}_M|\overline{\overline{A}} - \overline{\overline{\Gamma}}_M \in \mathbb{F}_1\mathbb{F}_1$ is symmetric and satisfies $\mathbf{x}|\overline{\overline{B}}|\mathbf{x} = 0$ for all $\mathbf{x}$, it turns out that we must actually have $\overline{\overline{B}} = 0$, (proof left as a problem). This requires that transformation dyadic must satisfy

$$\overline{\overline{A}}^T|\overline{\overline{\Gamma}}_M|\overline{\overline{A}} = \overline{\overline{\Gamma}}_M. \tag{3.121}$$

More generally, suppressing the subscript, (3.121) describes the condition for a dyadic $\overline{\overline{A}}$ defining an orthogonal transformation with respect to a given metric dyadic $\overline{\overline{\Gamma}}$.

To simplify the problem, we assume that $\overline{\overline{\mathsf{A}}}|\mathbf{e}_i = \mathbf{e}_i$ for $i = 2, 3$, in which case the transformation does not change vectors in the subspace spanned by the basis vectors $\mathbf{e}_2, \mathbf{e}_3$ associated to the metric dyadic $\mathbf{\Gamma_M}$. Now the transformation dyadic can be written as

$$\overline{\overline{\mathsf{A}}} = \overline{\overline{\mathsf{A}}}' + \mathbf{e}_2\varepsilon_2 + \mathbf{e}_3\varepsilon_3, \tag{3.122}$$

$$\overline{\overline{\mathsf{A}}}' = A_{11}\mathbf{e}_1\varepsilon_1 + A_{14}\mathbf{e}_1\varepsilon_4 + A_{41}\mathbf{e}_4\varepsilon_1 + A_{44}\mathbf{e}_4\varepsilon_4. \tag{3.123}$$

The condition (3.121) is simplified to

$$\overline{\overline{\mathsf{A}}}'^T|(\varepsilon_1\varepsilon_1 - \varepsilon_4\varepsilon_4)|\overline{\overline{\mathsf{A}}}' = \varepsilon_1\varepsilon_1 - \varepsilon_4\varepsilon_4, \tag{3.124}$$

which can be split in four equations for the coefficients A_{ij}

$$A_{11}^2 - A_{41}^2 = 1, \quad A_{11}A_{14} - A_{41}A_{44} = 0, \tag{3.125}$$

$$A_{14}A_{11} - A_{44}A_{41} = 0, \quad A_{14}^2 - A_{44}^2 = -1, \tag{3.126}$$

of which two are the same. From these we obtain two possibilities (double signs correspond to one another)

$$A_{11} = \pm A_{44}, \qquad A_{14} = \pm A_{41}. \tag{3.127}$$

Expressing $A_{11} = \cosh\vartheta$, from the above relations we can solve all the other coefficients as

$$\begin{pmatrix} A_{11} & A_{14} \\ A_{41} & A_{44} \end{pmatrix} = \begin{pmatrix} \cosh\vartheta & \pm\sinh\vartheta \\ \sinh\vartheta & \pm\cosh\vartheta \end{pmatrix}. \tag{3.128}$$

Assuming real ϑ and restricting to transformations which do not reverse the direction of time, we must have $A_{44} > 0$. This requires that the double sign be taken as $\pm \to +$ and the affine transformation dyadic becomes

$$\overline{\overline{\mathsf{A}}} = (\cosh\vartheta\mathbf{e}_1 + \sinh\vartheta\mathbf{e}_4)\varepsilon_1 + \mathbf{e}_2\varepsilon_2 + \mathbf{e}_3\varepsilon_3 + (\sinh\vartheta\mathbf{e}_1 + \cosh\vartheta\mathbf{e}_4)\varepsilon_4, \tag{3.129}$$

also known as the Lorentz transformation. It maps the four-vector $\mathbf{x}$ to

$$\mathbf{x}_a = \overline{\overline{\mathsf{A}}}|\mathbf{x} = (\cosh\vartheta\mathbf{e}_1 + \sinh\vartheta\mathbf{e}_4)x_1 + \mathbf{e}_2x_2 + \mathbf{e}_3x_3 + (\sinh\vartheta\mathbf{e}_1 + \cosh\vartheta\mathbf{e}_4)x_4. \tag{3.130}$$

If we define

$$x_4 = ct, \quad v = c\tanh\vartheta, \tag{3.131}$$

the transformed quantities become

$$x_{1a} = \cosh\vartheta x_1 + \sinh\vartheta ct = \frac{c}{\sqrt{c^2 - v^2}}(x_1 + vt), \tag{3.132}$$

$$ct_a = \sinh\vartheta x_1 + \cosh\vartheta ct = \frac{c}{\sqrt{c^2 - v^2}}(\frac{v}{c}x_1 + ct), \tag{3.133}$$

which coincide with the Lorentz transformation formulas [38] corresponding to an observer moving with the velocity $-\mathbf{e}_1 v$. In fact, for $v \ll c$ we have $t_a \approx t$ and the x_1 coordinate changes as $x_{1a} \approx x_1 + vt$.

Problems

3.4.1 Prove that if $\mathbf{x}|\overline{\overline{\mathsf{B}}}|\mathbf{x} = 0$ for any vector $\mathbf{x}$, we must have $\overline{\overline{\mathsf{B}}} = 0$ when $\overline{\overline{\mathsf{B}}} \in \mathbb{F}_1\mathbb{F}_1$ is a symmetric dyadic. *Hint*: Set $\mathbf{x} = \mathbf{y} + \mathbf{z}$.

3.4.2 Show that any basis $\{\mathbf{a}_i\}$ orthogonal in Minkowskian sense $\mathbf{a}_i \cdot \mathbf{a}_j = \delta_{ij}$ remains orthogonal after an orthogonal transformation $\overline{\overline{\mathsf{A}}}$.

3.4.3 Find $\overline{\overline{\mathsf{A}}}^{(2)}, \overline{\overline{\mathsf{A}}}^{(3)}$, and $\overline{\overline{\mathsf{A}}}^{-1}$ for the dyadic $\overline{\overline{\mathsf{A}}}$ given in (3.129).

3.4.4 Find eigenvalues and eigenvectors of the Lorentz transformation dyadic

$$\overline{\overline{\mathsf{A}}} = c\mathbf{e}_1\boldsymbol{\varepsilon}_1 + s\mathbf{e}_4\boldsymbol{\varepsilon}_1 + s\mathbf{e}_1\boldsymbol{\varepsilon}_4 + c\mathbf{e}_4\boldsymbol{\varepsilon}_4 + \mathbf{e}_2\boldsymbol{\varepsilon}_2 + \mathbf{e}_3\boldsymbol{\varepsilon}_4,$$

where $c = \cosh\vartheta$, $s = \sinh\vartheta$. Express the dyadic in terms of the eigenvectors. Show that the inverse transformation $\overline{\overline{\mathsf{A}}}^{-1}$ corresponds to changing $\vartheta \to -\vartheta$.

3.4.5 Find the powers $\overline{\overline{\mathsf{A}}}^{(2)}, \overline{\overline{\mathsf{A}}}^{(3)}$, and $\operatorname{tr}\overline{\overline{\mathsf{A}}}$, $\det\overline{\overline{\mathsf{A}}}$ for the Lorentz-transformation dyadic of the previous problem.

4

Electromagnetic Fields and Sources

Because electromagnetic fields and sources are integrable over a space domain, they can be represented in terms of differential forms. In the Gibbsian three-dimensional formalism we usually distinguish between field intensities, integrable along lines in space, and flux densities, integrable over surfaces. The former can be represented in terms of one-forms like $\boldsymbol{\phi} \in \mathbb{F}_1$, the latter in terms of two-forms like $\boldsymbol{\Phi} \in \mathbb{F}_2$. In addition, there are volume density quantities and scalars, elements of the respective spaces $\mathbb{F}_3$ and $\mathbb{F}_0$. The differential operation $\mathbf{d}\wedge$ transforms a p-form quantity from the space $\mathbb{F}_p$ to a $p+1$-form in the space $\mathbb{F}_{p+1}$. Differential-form representation of electromagnetic field equations becomes very simple and elegant in the four-dimensional form.

4.1 BASIC ELECTROMAGNETIC QUANTITIES

The electric and magnetic field intensities $\mathbf{E}$, $\mathbf{H}$ are integrable along lines in space, whence they are one-forms. The path integral of the electric field gives the work done by the field when it moves a unit point charge along the path. For the charge Q, the work expressed in the Gibbsian notation is

$$W = Q\int_C \mathbf{E}\cdot d\mathbf{r} = Q\int_C (E_x\mathbf{u}_x + E_y\mathbf{u}_y + E_z\mathbf{u}_z)\cdot d\mathbf{r}. \tag{4.1}$$

In the differential-form notation the basis unit vectors $\mathbf{u}_x = \nabla x,\ \mathbf{u}_y = \nabla y,\ \mathbf{u}_z = \nabla z$ are replaced by the basis one-forms $\mathbf{d}x, \mathbf{d}y, \mathbf{d}z$. Inside the integral sign the bracketed term can be interpreted as a one-form which, instead of a new Greek letter, will be

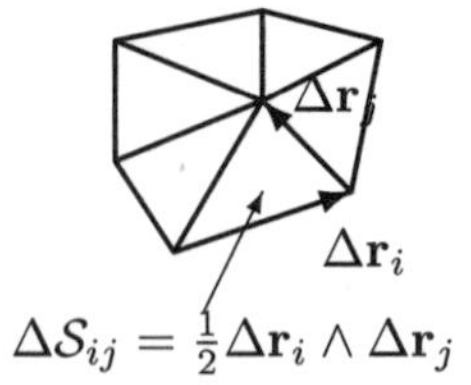

Fig. 4.1 *A surface $\mathcal{S}$ is a manifold which can be approximated by a set of simplexes (planar triangular elements represented by bivectors $\Delta\mathbf{S}_{ij}$) which results in a finite-element mesh.*

denoted by the traditional symbol $\mathbf{E}$:

$$\mathbf{E} = E_x\mathbf{d}x + E_y\mathbf{d}y + E_z\mathbf{d}z. \tag{4.2}$$

The integral (4.1) can be considered as a limit (Riemann sum) when the elements of work are summed over consequent path elements $\Delta_i\mathbf{r}$. By denoting the total path as a one-dimensional manifold $\mathcal{C}$, the total work is thus

$$W = Q\lim\sum \mathbf{E}|\Delta_i\mathbf{r} = Q\,\mathbf{E}|\mathcal{C}. \tag{4.3}$$

The electric field is a one-form because its duality product with a vector is a scalar. An example of a two-form is the density of electric current $\mathbf{J}$ which is integrable over a surface,

$$\mathbf{J} = J_{xy}\mathbf{d}x \wedge \mathbf{d}y + J_{yz}\mathbf{d}y \wedge \mathbf{d}z + J_{zx}\mathbf{d}z \wedge \mathbf{d}x. \tag{4.4}$$

The surface is defined by a two-dimensional manifold $\mathcal{S}$ which can be approximated by a chain of two-simplexes which are bivectors $\Delta\mathbf{S}_{ij} = (1/2)\Delta\mathbf{r}_i \wedge \Delta\mathbf{r}_j$ (Figure 4.1). The current I through the surface is then expressed as the integral

$$I = \lim\sum \mathbf{J}|\Delta\mathbf{S}_{ij} = \mathbf{J}|\mathcal{S}. \tag{4.5}$$

An example of a three-form, a quantity which is integrable over a volume in the three-dimensional space, is the charge density $\boldsymbol{\varrho}$:

$$\boldsymbol{\varrho} = \varrho\mathbf{d}x \wedge \mathbf{d}y \wedge \mathbf{d}z. \tag{4.6}$$

The total charge Q in a volume $\mathcal{V}$, which is the limit of a chain of three-simplexes $\Delta_{ijk}\mathbf{V} = \Delta\mathbf{r}_i \wedge \Delta\mathbf{r}_j \wedge \Delta\mathbf{r}_k$, equals

$$Q = \boldsymbol{\varrho}|\mathcal{V}. \tag{4.7}$$

Finally, as an example of a zero-form which is not integrable over a space region is the scalar potentialıpotential!scalar ϕ.

To summarize, various basic electromagnetic quantities can be expressed in 3D Euclidean differential forms as follows:

- **zero-forms**: scalar potential ϕ, magnetic scalar potential ϕ_m

- **one-forms**: electric field $\mathbf{E}$, magnetic field $\mathbf{H}$, vector potential $\mathbf{A}$, magnetic vector potential $\mathbf{A}_m$
- **two-forms**: electric flux density $\mathbf{D}$, magnetic flux density $\mathbf{B}$, electric current density $\mathbf{J}$, magnetic current density $\mathbf{J}_m$
- **three-forms**: electric charge density ϱ, magnetic charge density ϱ_m

For reasons of symmetry, electric quantities will also be denoted by $\phi_e, \mathbf{A}_e, \mathbf{J}_e, \varrho_e$.

In the Gibbsian three-dimensional vector analysis, both one-forms and two-forms are treated as vectors, sometimes distinguished as polar and axial (or pseudo) vectors. The electromagnetic vectors are coupled through the medium equations which for the isotropic medium are of the form $\mathbf{D} = \epsilon\mathbf{E}$, where ϵ is a scalar. In the differential-form formalism, ϵ cannot be a scalar. Instead, it is a Hodge dyadic $\overline{\overline{\epsilon}}$ which maps one-forms to two-forms.

4.2 MAXWELL EQUATIONS IN THREE DIMENSIONS

Let us now consider the Maxwell equations in terms of both three and four-dimensional differential-form formulations. The former case does not differ much from that of the Gibbsian representation.

4.2.1 Maxwell–Faraday equations

The first half of the Maxwell equations were called Maxwell–Faraday equations by Dechamps [18]. They can be written in Gibbsian vector formalism as

$$\nabla \times \mathbf{E} + \partial_t \mathbf{B} = 0, \tag{4.8}$$

$$\nabla \cdot \mathbf{B} = 0, \tag{4.9}$$

in the absence of magnetic sources. For the differential-form formulation we replace ∇ by $\mathbf{d}$. Recalling that curl of a vector is represented by $\mathbf{d}\wedge$ operating on a one-form and that divergence is represented by the same operation on a two-form, we again conclude that $\mathbf{E}$ must be a one-form and $\mathbf{B}$ a two-form. Thus, (4.8) becomes the two-form equation

$$\mathbf{d} \wedge \mathbf{E} + \partial_\tau \mathbf{B} = 0 \tag{4.10}$$

and (4.9) becomes the three-form equation

$$\mathbf{d} \wedge \mathbf{B} = 0. \tag{4.11}$$

Here, following [18], we have replaced the time variable t by normalized time $\tau = ct$. This gives the one-form $\mathbf{E}$ and the two-form $\mathbf{B}$ the same physical dimension, because the dimension of both $\mathbf{d}$ and ∂_τ is the same, 1/m.

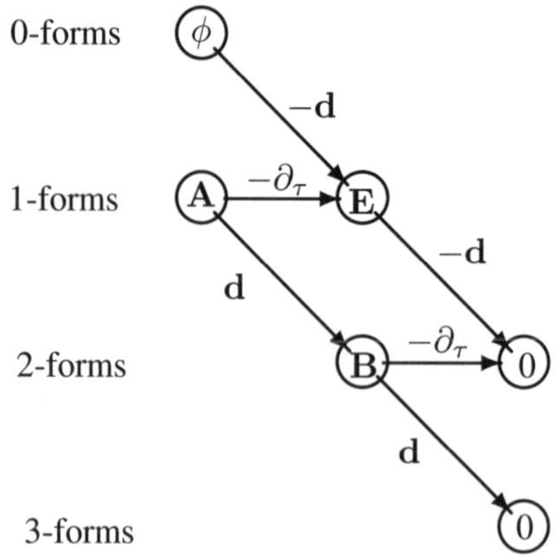

Fig. 4.2 *Maxwell-Faraday equations and potential relations as a Deschamps graph for electromagnetic differential-form quantities in three dimensions.*

Potentials From (4.11) and de Rham's theorem it now follows that $\mathbf{B}$ can locally be expressed in terms of a potential one-form $\mathbf{A}$,

$$\mathbf{B} = \mathbf{d} \wedge \mathbf{A}. \tag{4.12}$$

As noted before, the potential $\mathbf{A}$ is unique only in a star-like region. Conversely, $\mathbf{B}$ defined by (4.12) satisfies (4.11) identically. When substituted in (4.10), we have

$$\mathbf{d} \wedge (\mathbf{E} + \partial_\tau \mathbf{A}) = 0. \tag{4.13}$$

From de Rham's theorem again it follows that there locally exists a potential zero-form ϕ in terms of which the bracketed one-form in (4.13) can be expressed as $-\mathbf{d}\phi$. The electric field one-form can then be written as

$$\mathbf{E} = -\mathbf{d}\phi - \partial_\tau \mathbf{A}. \tag{4.14}$$

Equation (4.10) is now satisfied identically by $\mathbf{E}$ defined as (4.14).

Representation of fields in terms of the potentials $\mathbf{A}$ and ϕ is not unique, because the potentials can be changed as

$$\phi \rightarrow \phi + \partial_\tau \psi, \quad \mathbf{A} \rightarrow \mathbf{A} - \mathbf{d}\psi, \tag{4.15}$$

without changing $\mathbf{E}$ and $\mathbf{B}$. ψ can be any scalar function. A unique set of potentials can be obtained if an additional scalar condition is introduced which defines the function ψ. One such condition, called the Lorenz condition (to be discussed later), is valid in special media. The equations up to this point are independent of the nature of the electromagnetic medium.

The Maxwell–Faraday equations (4.10), (4.11) and the potential relations (4.12), (4.14) can be visualized by the flowgraph in Figure 4.2, as introduced by Deschamps [18]. Horizontal lines represent time differentiation $-\partial_\tau$, and oblique lines represent spatial differentiation $\mathbf{d}$ or $-\mathbf{d}$.

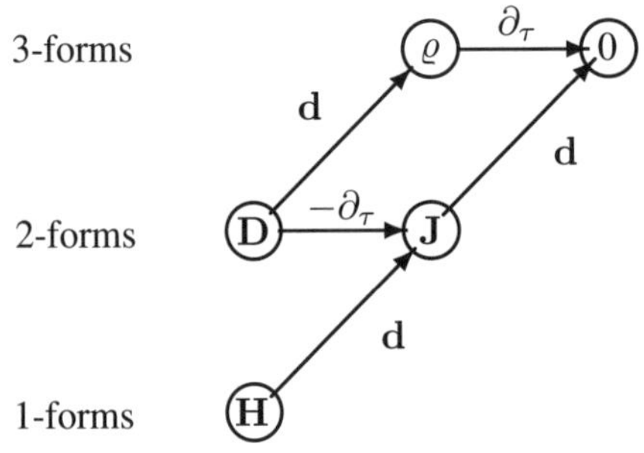

Fig. 4.3 *Maxwell-Ampère equations as a Deschamps graph for electromagnetic differential-form quantities in three dimensions.*

4.2.2 Maxwell–Ampère equations

The other half of the Maxwell equations was called the Maxwell–Ampère equations in reference 18. In Gibbsian vector formalism they are given as

$$\nabla \times \mathbf{H} - \partial_t \mathbf{D} = \mathbf{J}, \tag{4.16}$$

$$\nabla \cdot \mathbf{D} = \varrho. \tag{4.17}$$

From the analogy of the Maxwell–Faraday equations, the vector $\mathbf{H}$ is replaced by a one-form $\mathbf{H}$ while the vector $\mathbf{D}$ is replaced by a two-form $\mathbf{D}$. The differential equations then become

$$\mathbf{d} \wedge \mathbf{H} - \partial_\tau \mathbf{D} = \mathbf{J}, \tag{4.18}$$

$$\mathbf{d} \wedge \mathbf{D} = \boldsymbol{\varrho}. \tag{4.19}$$

Equation (4.18) is an equation between two-forms, and (4.19) is an equation between three-forms. The right-hand sides in both equations represent electromagnetic sources, the current density two-form $\mathbf{J}$ and the charge density three-form $\boldsymbol{\varrho}$. Together, they satisfy the condition of continuity obtained by taking the differential of (4.18), setting $\mathbf{d} \wedge \mathbf{d} \wedge \mathbf{H} = 0$ and substituting (4.19):

$$\mathbf{d} \wedge \mathbf{J} + \partial_\tau \boldsymbol{\varrho} = 0. \tag{4.20}$$

The Maxwell–Ampère equations can be visualized by another Deschamps flowgraph (Figure 4.3) [18].

Both Maxwell–Faraday and Maxwell–Ampère sets of equations are free from metric because they only involve the wedge product. The medium equations, to be defined later, combine these two sets of equations by introducing some metric dyadics.

4.2.3 Time-harmonic fields and sources

Time-harmonic sources corresponding to angular frequency ω can be represented through complex exponentials with time dependence $e^{j\omega t}$. In normalized time τ, this is expressed as $e^{jk_o\tau}$ with $k_o = \omega\sqrt{\mu_o\epsilon_o}$. Because all linear expressions carry the same exponential function, it is suppressed in equations. The Maxwell–Faraday equations (4.10), (4.11) for time-harmonic fields become

$$\mathbf{d} \wedge \mathbf{E} + jk_o\mathbf{B} = 0, \tag{4.21}$$

$$\mathbf{d} \wedge \mathbf{B} = 0, \tag{4.22}$$

while the Maxwell–Ampère equations (4.18), (4.19) are

$$\mathbf{d} \wedge \mathbf{H} - jk_o\mathbf{D} = \mathbf{J}, \tag{4.23}$$

$$\mathbf{d} \wedge \mathbf{D} = \boldsymbol{\varrho}. \tag{4.24}$$

From these the continuity condition (4.20) for time-harmonic sources can be derived as

$$\mathbf{d} \wedge \mathbf{J} + jk_o\boldsymbol{\varrho} = 0. \tag{4.25}$$

Because the charge three-form $\boldsymbol{\varrho}$ is uniquely determined through the time-harmonic current two-form $\mathbf{J}$, it is not needed as an independent source. Actually, the three-form equations can be derived from the two-form equations and the continuity condition for the source and, thus, can be omitted in the analysis of time-harmonic problems leaving $\mathbf{J}$ as the only source.

4.3 MAXWELL EQUATIONS IN FOUR DIMENSIONS

The Maxwell equations assume an especially simple form in the Minkowskian four-dimensional representation. In fact, one can say that they are the simplest possible differential equations for these field and source quantities. Let us again denote for clarity the differential operator operating on the Euclidean three-dimensional vector variable $\mathbf{r}$ by $\mathbf{d_E}$:

$$\mathbf{d} = \mathbf{d_E} + \mathbf{d}\tau\,\partial_\tau, \tag{4.26}$$

$$\mathbf{d_E} = \sum_{i=1}^{3} \mathbf{d}x_i\,\partial_{x_i} = \mathbf{d}x\,\partial_x + \mathbf{d}y\,\partial_y + \mathbf{d}z\,\partial_z. \tag{4.27}$$

Note that $\mathbf{d_E}$ is the same as $\mathbf{d}$ in the previous section where differentiation with respect to τ was taken separately.

4.3.1 The force field

In the four-dimensional representation of electromagnetic fields, the one-form $\mathbf{E}$ and two-form $\mathbf{B}$ are combined to a two-form $\boldsymbol{\Phi}$, which is called the force field by Deschamps [18] because $\mathbf{E}$ and $\mathbf{B}$ give rise to the physical force of the electromagnetic field (see Chapter 5),

$$\boldsymbol{\Phi} = \mathbf{B} + \mathbf{E} \wedge \mathbf{d}\tau. \tag{4.28}$$

In [58] $\mathbf{\Phi}$ is called the Faraday two-form. The Maxwell–Faraday set of equations (4.10), (4.11) can be combined to make a single equation:

$$\mathbf{d} \wedge \mathbf{\Phi} = 0. \tag{4.29}$$

Let us check this by substituting (4.28). Noting that $\mathbf{d}\tau \wedge \mathbf{E} = -\mathbf{E} \wedge \mathbf{d}\tau$ and $\mathbf{d}\tau \wedge \mathbf{B} = \mathbf{B} \wedge \mathbf{d}\tau$, we obtain

$$\begin{aligned} \mathbf{d} \wedge \mathbf{\Phi} &= (\mathbf{d}_{\mathrm{E}} + \mathbf{d}\tau\, \partial_\tau) \wedge (\mathbf{B} + \mathbf{E} \wedge \mathbf{d}\tau) \\ &= \mathbf{d}_{\mathrm{E}} \wedge \mathbf{B} + (\partial_\tau \mathbf{B} + \mathbf{d}_{\mathrm{E}} \wedge \mathbf{E}) \wedge \mathbf{d}\tau = 0. \end{aligned} \tag{4.30}$$

Because the spatial and temporal components of this equation must be satisfied separately, (4.29) leads to the equations (4.11) and (4.10).

Equation (4.29) is a kind of four-dimensional law for local electromagnetic flux conservation. In analogy to (4.20), which states that temporal change of the charge density three-form ϱ is associated to an electric current two-form $\mathbf{J}$, we could state that temporal change of the magnetic two-form "charge" $\mathbf{B}$ is associated to the electric one-form "current" $\mathbf{E}$ according to the Faraday law.

The four-potential Because the two-form $\mathbf{\Phi}$ satisfies (4.29), from de Rham's theorem it follows that there exists a one-form $\boldsymbol{\alpha}$, the four-potential, in terms of which we can locally write

$$\mathbf{\Phi} = \mathbf{d} \wedge \boldsymbol{\alpha}. \tag{4.31}$$

From Poincaré's lemma, (4.29) is then valid identically. Equation (4.31) can be interpreted so that the force field $\mathbf{\Phi}$ serves as the source of the potential field $\boldsymbol{\alpha}$ just like currents are the source of the electromagnetic field. This is not so easy to visualize when fields fill the whole space, but in some examples where the fields are restricted to a certain region, one can see that the potentials do not vanish outside this region even if the fields do. In physics this is associated with what is called the Bohm–Aharonov effect.

The four-potential $\boldsymbol{\alpha}$ can be expressed in its spatial and temporal components as

$$\boldsymbol{\alpha} = \mathbf{A} - \phi \mathbf{d}\tau, \tag{4.32}$$

whence we can expand

$$\begin{aligned} \mathbf{d} \wedge \boldsymbol{\alpha} &= (\mathbf{d}_{\mathrm{E}} + \mathbf{d}\tau \partial_\tau) \wedge (\mathbf{A} - \phi \mathbf{d}\tau) \\ &= \mathbf{d}_{\mathrm{E}} \wedge \mathbf{A} - (\partial_\tau \mathbf{A} + \mathbf{d}_{\mathrm{E}} \phi) \wedge \mathbf{d}\tau. \end{aligned} \tag{4.33}$$

By equating the right-hand sides of (4.28) and (4.31) with (4.33) inserted, one directly arrives at the three-dimensional equations (4.12) and (4.14). Again, the four-potential $\boldsymbol{\alpha}$ is not unique without some extra conditions. In fact, writing $\boldsymbol{\alpha} + \mathbf{d}\psi$ instead of $\boldsymbol{\alpha}$, where ψ is a zero-form (scalar function), and applying Poincaré's lemma, the same force field will result: $\mathbf{d} \wedge (\boldsymbol{\alpha} + \mathbf{d}\psi) = \mathbf{d} \wedge \boldsymbol{\alpha} = \mathbf{\Phi}$. Uniqueness can be attained through an additional scalar condition (gauge condition) for the four-potential $\boldsymbol{\alpha}$.

4.3.2 The source field

The source of the Maxwell–Ampère equations in the Minkowskian four-dimensional formalism is obtained by combining the current-density two-form **J** and the charge-density three-form ϱ to the four-current three-form γ

$$\gamma = \varrho - \mathbf{J} \wedge \mathbf{d}\tau. \tag{4.34}$$

This quantity satisfies the charge conservation law (4.20) in the simple form

$$\mathbf{d} \wedge \gamma = 0. \tag{4.35}$$

In fact, noting that $\mathbf{d}_{\mathbf{E}} \wedge \varrho$ vanishes identically because four-forms vanish in the three-dimensional space, one can show that the condition (4.35) is equal to (4.20). Details are left as an exercise.

Because the four-current satisfies (4.35), from de Rham's theorem there locally exists a two-form $\boldsymbol{\Psi}$ which is called the source field by Deschamps [18] and the Maxwell two-form in [58]. $\boldsymbol{\Psi}$ is a kind of potential field of the four-current γ. This two-form is governed by the equation

$$\mathbf{d} \wedge \boldsymbol{\Psi} = \gamma. \tag{4.36}$$

Due to Poincaré's lemma, the continuity condition (4.35) is now satisfied identically. Because any two-form in the Minkowskian four-dimensional space can be expressed in terms of a three-dimensional two-form (spatial part) and a three-dimensional one-form (temporal part), we can introduce two fields **D** and **H** by writing

$$\boldsymbol{\Psi} = \mathbf{D} - \mathbf{H} \wedge \mathbf{d}\tau. \tag{4.37}$$

Through this definition, (4.36) actually becomes the source field equation and it coincides with the set of Maxwell–Ampère equations (4.18), (4.19). This is seen by splitting (4.36) in its spatial and temporal components:

$$\gamma = \varrho - \mathbf{J} \wedge \mathbf{d}\tau = \mathbf{d}_{\mathbf{E}} \wedge \mathbf{D} + (\partial_\tau \mathbf{D} - \mathbf{d}_{\mathbf{E}} \wedge \mathbf{H}) \wedge \mathbf{d}\tau. \tag{4.38}$$

In conclusion, the second half of the Maxwell equations (4.36) is a direct consequence of the conservation of the four-current, (4.35).

4.3.3 Deschamps graphs

In the four-dimensional formulation the Deschamps graphs are extremely simple. The force field is associated with the sequence on the left, and the source field is associated with the sequence on the right, in Figure 4.4. The arrow stands for the operation $\mathbf{d}\wedge$.

The graphs in Figures 4.2 and 4.3 actually describe the inner structure of the respective two graphs in Figure 4.4. Similarly, Figures 4.2 and 4.3 could be expanded to display component equations where differentiations in terms of different spatial variables are depicted by lines of different directions. However, to draw the complete diagram in component form would need a sheet of four-dimensional paper.

$$\alpha \stackrel{\mathbf{d}}{\rightarrow} \mathbf{\Phi} \stackrel{\mathbf{d}}{\rightarrow} 0 \qquad \mathbf{\Psi} \stackrel{\mathbf{d}}{\rightarrow} \gamma \stackrel{\mathbf{d}}{\rightarrow} 0$$

Fig. 4.4 *Deschamps graphs for the two Maxwell equations in four dimensions have a very simple form.*

4.3.4 Medium equation

Additional equations are required to couple the force field and source field two-forms together. They are called constitutive equations, or more simply, medium equations because they depend on the macroscopic properties of the polarizable medium. A linear medium defines a linear relation between $\mathbf{\Phi}$ and $\mathbf{\Psi}$. It can be represented by a medium dyadic $\overline{\overline{\mathsf{M}}} \in \mathbb{F}_2\mathbb{E}_2$ in the four-dimensional formalism which maps a two-form to a two-form:

$$\mathbf{\Psi} = \overline{\overline{\mathsf{M}}}|\mathbf{\Phi}. \tag{4.39}$$

Medium equations will be discussed in more detail in Chapter 5.

4.3.5 Magnetic sources

Magnetic current two-form $\mathbf{J}_m$ and magnetic charge density three-form ϱ_m can be used as equivalent sources replacing some physical sources. In the four-dimensional formalism, they can be combined to the magnetic four-current density three-form as

$$\gamma_m = \varrho_m - \mathbf{J}_m \wedge \mathbf{d}\tau. \tag{4.40}$$

If the conservation equation similar to that of electric sources,

$$\mathbf{d} \wedge \gamma_m = 0, \tag{4.41}$$

is assumed to be valid, from de Rham's theorem the magnetic four-current γ_m can be represented in terms of a two-form which is now taken as the force field $\mathbf{\Phi}$. The relation is written as

$$\mathbf{d} \wedge \mathbf{\Phi} = \gamma_m. \tag{4.42}$$

This is the generalization of the force field equation (4.29) with the magnetic source inserted on the right-hand side. With the magnetic sources present the two-form $\mathbf{\Phi}$ could be called the magnetic source field, and $\mathbf{\Psi}$ the electric source field. Because these different names may actually cause some confusion, we mostly refer the field two-forms simply as $\mathbf{\Phi}$ and $\mathbf{\Psi}$ two-forms. With nonvanishing magnetic sources, $\mathbf{\Phi}$ cannot be derived from a four-potential α as in (4.31).

In case both electric and magnetic sources are present, the two Maxwell equations make a symmetric pair,

$$\mathbf{d} \wedge \mathbf{\Psi} = \gamma_e, \qquad \mathbf{d} \wedge \mathbf{\Phi} = \gamma_m, \tag{4.43}$$

$$\mathbf{d} \wedge \mathbf{E} = -\partial_\tau \mathbf{B} - \mathbf{J}_m \qquad \mathbf{d} \wedge \mathbf{B} = \varrho_m$$

$$\mathbf{d} \wedge \mathbf{H} = \partial_\tau \mathbf{D} - \mathbf{J}_e \qquad \mathbf{d} \wedge \mathbf{D} = \varrho_e$$

Fig. 4.5 *Adding magnetic sources gives a symmetric form to the Maxwell equations.*

where the electric sources have been added the extra subscript e (Figure 4.5). One can introduce a duality transformation $\mathbf{\Phi} \leftrightarrow \mathbf{\Psi}$, $\gamma_e \leftrightarrow \gamma_m$, through which electric and magnetic sources and fields are mapped onto one another. This transformation will be discussed in Chapter 6.

In case the electric sources vanish ($\gamma_e = 0$), the $\mathbf{\Psi}$ field can be derived from a magnetic four-potential $\boldsymbol{\alpha}_m$ as

$$\mathbf{\Psi} = \mathbf{d} \wedge \boldsymbol{\alpha}_m, \tag{4.44}$$

according to de Rham's theorem. In this case, $\mathbf{d} \wedge \mathbf{\Psi} = 0$ is automatically satisfied. When there are both electric and magnetic sources, the problem can be split into two components: Fields $\mathbf{\Phi}_e, \mathbf{\Psi}_e$ arising from the electric sources γ_e can be derived from an electric four-potential $\boldsymbol{\alpha}_e$, while the fields $\mathbf{\Phi}_m, \mathbf{\Psi}_m$ arising from the magnetic sources γ_m can be derived from a magnetic four-potential $\boldsymbol{\alpha}_m$.

Problems

4.3.1 Interpret in the language of Gibbsian vector analysis the following electromagnetic quantities when $\mathbf{\Phi}$ and $\mathbf{\Psi}$ are the two electromagnetic bivectors:

(a) $\mathbf{\Phi} \wedge \mathbf{\Phi}$

(b) $\mathbf{\Psi} \wedge \mathbf{\Psi}$

(c) $\mathbf{\Phi} \wedge \mathbf{\Psi}$

4.3.2 Prove in detail that the four-dimensional equation (4.35) equals the three-dimensional equation (4.20).

4.3.3 Interpret the conditions $\mathbf{\Phi} \wedge \mathbf{\Phi} = 0$ and $\mathbf{\Psi} \wedge \mathbf{\Psi} = 0$ in terms of Gibbsian field vectors. Show that they are satisfied by fields of a time-harmonic plane wave in any medium.

4.4 TRANSFORMATIONS

4.4.1 Coordinate transformations

The differential operator $\mathbf{d}$ operates on functions much in the same way as the ∇ operator and change of coordinates is a simple matter. As an example, let us consider two-dimensional functions f and g in terms of either Cartesian coordinates x, y or polar coordinates ρ, φ. The relation between the coordinates is

$$x = \rho \cos \varphi, \quad y = \rho \sin \varphi. \tag{4.45}$$

From these the relations between the basis one-forms are obtained as

$$\mathbf{d}x = \cos \varphi \mathbf{d}\rho - \rho \sin \varphi \mathbf{d}\varphi, \quad \mathbf{d}y = \sin \varphi \mathbf{d}\rho + \rho \cos \varphi \mathbf{d}\varphi, \tag{4.46}$$

and the relation between the basis two-forms are obtained as

$$\mathbf{d}x \wedge \mathbf{d}y = \rho(\cos^2 \varphi + \sin^2 \varphi)\mathbf{d}\rho \wedge \mathbf{d}\varphi = \rho \mathbf{d}\rho \wedge \mathbf{d}\varphi. \tag{4.47}$$

Considering the scalar functions f, g as functions of either x, y or ρ, φ, we can write

$$\mathbf{d}f = \partial_x f \mathbf{d}x + \partial_y f \mathbf{d}y = \partial_\rho f \mathbf{d}\rho + \partial_\varphi f \mathbf{d}\varphi, \tag{4.48}$$

$$\mathbf{d}g = \partial_x g \mathbf{d}x + \partial_y g \mathbf{d}y = \partial_\rho g \mathbf{d}\rho + \partial_\varphi g \mathbf{d}\varphi, \tag{4.49}$$

$$\begin{aligned} \mathbf{d}f \wedge \mathbf{d}g &= (\partial_x f \partial_y g - \partial_y f \partial_x g)\mathbf{d}x \wedge \mathbf{d}y \\ &= (\partial_\rho f \partial_\varphi g - \partial_\varphi f \partial_\rho g)\mathbf{d}\rho \wedge \mathbf{d}\varphi. \end{aligned} \tag{4.50}$$

From these we see that the expressions in the two coordinates are similar in form. Thus, there is no need to apply the relations (4.45) to change from one system of coordinates to the other one when forming expressions which do not depend on the metric.

As an electromagnetic example, let us consider the relation between the magnetic two-form $\mathbf{B}$ and the vector potential one-form $\mathbf{A}$

$$\mathbf{B} = \mathbf{d} \wedge \mathbf{A}. \tag{4.51}$$

Expressed in Cartesian coordinates we have

$$\mathbf{A} = A_x \mathbf{d}x + A_y \mathbf{d}y + A_z \mathbf{d}z, \tag{4.52}$$

$$\begin{aligned} \mathbf{B} &= (\partial_x A_y - \partial_y A_x)\mathbf{d}x \wedge \mathbf{d}y + (\partial_x A_z - \partial_z A_x)\mathbf{d}x \wedge \mathbf{d}z \\ &\quad + (\partial_y A_z - \partial_z A_y)\mathbf{d}y \wedge \mathbf{d}z. \end{aligned} \tag{4.53}$$

To transform these to polar coordinates, we don't have to recall any transformation formulas. The corresponding expressions can be written directly from those above by replacing $x \to r$, $y \to \theta$, and $z \to \varphi$. Thus, the potential in spherical coordinates is

$$\mathbf{A} = A_r \mathbf{d}r + A_\theta \mathbf{d}\theta + A_\varphi \mathbf{d}\varphi, \tag{4.54}$$

and the magnetic two-form becomes

$$\begin{aligned} \mathbf{B} &= (\partial_r A_\theta - \partial_\theta A_r)\mathbf{d}r \wedge \mathbf{d}\theta + (\partial_r A_\varphi - \partial_\varphi A_r)\mathbf{d}r \wedge \mathbf{d}\varphi \\ &\quad + (\partial_\theta A_\varphi - \partial_\varphi A_\theta)\mathbf{d}\theta \wedge \mathbf{d}\varphi. \end{aligned} \tag{4.55}$$

4.4.2 Affine transformation

Affine transformation in electromagnetics transforms sources, fields and media. Assuming a dyadic $\overline{\overline{\mathsf{A}}} \in \mathbb{E}_1\mathbb{F}_1$ in Minkowskian four-dimensional space, the four-vector $\mathbf{x}$ is transformed by $\mathbf{x}_a = \overline{\overline{\mathsf{A}}}|\mathbf{x}$ as in (3.95). Transformation of the field two-forms follow the rule (3.107):

$$\mathbf{\Phi}(\mathbf{x}) \quad \rightarrow \quad \mathbf{\Phi}_a(\mathbf{x}_a) = \overline{\overline{\mathsf{A}}}^{(-2)T}|\mathbf{\Phi}(\mathbf{x}), \tag{4.56}$$

$$\mathbf{\Psi}(\mathbf{x}) \quad \rightarrow \quad \mathbf{\Psi}_a(\mathbf{x}_a) = \overline{\overline{\mathsf{A}}}^{(-2)T}|\mathbf{\Psi}(\mathbf{x}). \tag{4.57}$$

The corresponding rules for the electric and magnetic source three-forms are

$$\gamma_e(\mathbf{x}) \quad \rightarrow \quad \gamma_{ea}(\mathbf{x}_a) = \overline{\overline{\mathsf{A}}}^{(-3)T}|\gamma_e(\mathbf{x}), \tag{4.58}$$

$$\gamma_m(\mathbf{x}) \quad \rightarrow \quad \gamma_{ma}(\mathbf{x}_a) = \overline{\overline{\mathsf{A}}}^{(-3)T}|\gamma_m(\mathbf{x}). \tag{4.59}$$

Thus, the transformed Maxwell equations have the same form as before the transformation,

$$\mathbf{d}_a \wedge \mathbf{\Phi}_a(\mathbf{x}_a) = \gamma_{ma}(\mathbf{x}_a), \tag{4.60}$$

$$\mathbf{d}_a \wedge \mathbf{\Psi}_a(\mathbf{x}_a) = \gamma_{ea}(\mathbf{x}_a). \tag{4.61}$$

Transformation of the medium equation (4.39) is found quite easily by writing

$$\begin{aligned} \mathbf{\Psi}_a(\mathbf{x}_a) &= \overline{\overline{\mathsf{A}}}^{(-2)T}|\mathbf{\Psi}(\mathbf{x}) = \overline{\overline{\mathsf{A}}}^{(-2)T}|\overline{\overline{\mathsf{M}}}|\mathbf{\Phi}(\mathbf{x}) \\ &= \overline{\overline{\mathsf{A}}}^{(-2)T}|\overline{\overline{\mathsf{M}}}|\overline{\overline{\mathsf{A}}}^{(2)T}|\mathbf{\Phi}_a(\mathbf{x}_a), \end{aligned} \tag{4.62}$$

from which we can identify the transformed medium dyadic as

$$\overline{\overline{\mathsf{M}}}_a = \overline{\overline{\mathsf{A}}}^{(-2)T}|\overline{\overline{\mathsf{M}}}|\overline{\overline{\mathsf{A}}}^{(2)T}. \tag{4.63}$$

Here $\overline{\overline{\mathsf{A}}}^{(-2)}$ denotes the double-cross square of $\overline{\overline{\mathsf{A}}}^{-1}$.

Example As an example, let us consider transformation of a medium with medium equations of the form

$$\mathbf{D} = \overline{\overline{\epsilon}}|\mathbf{E}, \quad \mathbf{H} = \overline{\overline{\mu}}^{-1}|\mathbf{B}, \tag{4.64}$$

where the permittivity $\overline{\overline{\epsilon}} \in \mathbb{F}_2\mathbb{E}_1$ and the inverse permeability $\overline{\overline{\mu}}^{-1} \in \mathbb{F}_1\mathbb{E}_2$ are defined by two Euclidean three-dimensional dyadics. In this case the four-dimensional medium dyadic has the special form

$$\overline{\overline{\mathsf{M}}} = \overline{\overline{\epsilon}} \wedge \mathbf{e}_\tau + \mathbf{d}\tau \wedge \overline{\overline{\mu}}^{-1} \tag{4.65}$$

(proof left as an exercize). Now consider an affine transformation defined by the dyadic

$$\overline{\overline{\mathsf{A}}} = A\overline{\overline{\mathsf{I}}}_{\mathbf{E}} + B\mathbf{e}_\tau\mathbf{d}\tau, \tag{4.66}$$

where $\overline{\overline{\mathsf{I}}}_{\mathbf{E}} \sum_{i=1}^{3} \mathbf{e}_i \boldsymbol{\varepsilon}_i$ is the three-dimensional unit dyadic. Let us find the transformed medium dyadic $\overline{\overline{\mathsf{M}}}_a$ from (4.63) to see if we can interpret it in terms of transformed permittivity and permeability dyadics.

To be able to use (4.63), we need to find the double-wedge squares of the transformation dyadic $\overline{\overline{\mathsf{A}}}$ and its inverse. The inverse of the transformation dyadic can be simply found to be

$$\overline{\overline{\mathsf{A}}}^{-1} = A^{-1}\overline{\overline{\mathsf{I}}}_{\mathbf{E}} + B^{-1}\mathbf{e}_\tau \mathbf{d}\tau, \tag{4.67}$$

as is easily verified through multiplication. After some algebra, the double-wedge squares are obtained as

$$\overline{\overline{\mathsf{A}}}^{(2)} = A^2\overline{\overline{\mathsf{I}}}_{\mathbf{E}}^{(2)} + AB\overline{\overline{\mathsf{I}}}_{\mathbf{E}} {}^{\wedge}_{\wedge} \mathbf{e}_\tau \mathbf{d}\tau, \tag{4.68}$$

$$\overline{\overline{\mathsf{A}}}^{(-2)} = A^{-2}\overline{\overline{\mathsf{I}}}_{\mathbf{E}}^{(2)} + (AB)^{-1}\overline{\overline{\mathsf{I}}}_{\mathbf{E}} {}^{\wedge}_{\wedge} \mathbf{e}_\tau \mathbf{d}\tau, . \tag{4.69}$$

Inserted in (4.63), we can expand

$$\begin{aligned}
\overline{\overline{\mathsf{M}}}_a &= A^{-2}\overline{\overline{\mathsf{I}}}_{\mathbf{E}}^{(2)T}|\overline{\overline{\epsilon}} \wedge \mathbf{e}_\tau|AB\overline{\overline{\mathsf{I}}}_{\mathbf{E}}^T {}^{\wedge}_{\wedge} \mathbf{d}\tau\mathbf{e}_\tau + (AB)^{-1}\overline{\overline{\mathsf{I}}}_{\mathbf{E}}^T {}^{\wedge}_{\wedge} \mathbf{d}\tau\mathbf{e}_\tau|\mathbf{d}\tau \wedge \overline{\overline{\mu}}^{-1}|A^2\overline{\overline{\mathsf{I}}}_{\mathbf{E}}^{(2)T} \\
&= \frac{B}{A}(\overline{\overline{\epsilon}} \wedge \mathbf{e}_\tau)|(\overline{\overline{\mathsf{I}}}_{\mathbf{E}}^T {}^{\wedge}_{\wedge} \mathbf{d}\tau\mathbf{e}_\tau) + \frac{A}{B}(\overline{\overline{\mathsf{I}}}_{\mathbf{E}}^T {}^{\wedge}_{\wedge} \mathbf{d}\tau\mathbf{e}_\tau)|(\mathbf{d}\tau \wedge \overline{\overline{\mu}}^{-1}) \\
&= \frac{B}{A}\overline{\overline{\epsilon}} \wedge \mathbf{e}_\tau + \frac{A}{B}\mathbf{d}\tau \wedge \overline{\overline{\mu}}^{-1}.
\end{aligned} \tag{4.70}$$

From this is is seen that the transformed medium has a medium dyadic of a similar special form as (4.65), and we can identify the transformed permittivity and permeability dyadics $\overline{\overline{\epsilon}}_a = (B/A)\overline{\overline{\epsilon}}$ and $\overline{\overline{\mu}}_a = (B/A)\overline{\overline{\mu}}$.

Problems

4.4.1 Show that the potential $\mathbf{A}(\mathbf{r}) = f(x)\mathbf{d}\rho$ gives the same two-form $\mathbf{d} \wedge \mathbf{A}(\mathbf{r})$ when written in Cartesian coordinates or cylindrical coordinates.

4.4.2 Check the equality of (4.55) and (4.53) through a change of coordinates.

4.4.3 Show that a medium with medium equations of the form

$$\mathbf{D} = \overline{\overline{\epsilon}}|\mathbf{E}, \quad \mathbf{H} = \overline{\overline{\mu}}^{-1}|\mathbf{B}, \quad \overline{\overline{\epsilon}} \in \mathbb{F}_2\mathbb{E}_1, \ \overline{\overline{\mu}}^{-1} \in \mathbb{F}_1\mathbb{E}_2$$

corresponds to the special four-dimensional medium dyadic

$$\overline{\overline{\mathsf{M}}} = \overline{\overline{\epsilon}} \wedge \mathbf{e}_\tau + \mathbf{d}\tau \wedge \overline{\overline{\mu}}^{-1}.$$

4.4.4 Derive (4.68) and (4.69) in detail.

4.5 SUPER FORMS

In the differential-form notation, $\mathbf{E}$ and $\mathbf{B}$ are combined to the force-field two-form $\boldsymbol{\Phi}$ while $\mathbf{H}$ and $\mathbf{D}$ are combined to the source-field two-form $\boldsymbol{\Psi}$ and the medium equation connects these two-forms to one another. In engineering electromagnetics one sometimes combines the Gibbsian field vectors $\mathbf{D}$ and $\mathbf{B}$ to a six-vector and $\mathbf{E}$ and $\mathbf{H}$ to another six-vector, and these two six-vectors are coupled in terms of a six-dyadic involving four conventional medium dyadics [40]. This arrangement emphasizes the symmetry of the Maxwell equations when both electric and magnetic sources are present. It can also be introduced to the differential-form formalism and the resulting source and field quantities are called super forms.

4.5.1 Maxwell equations

Let us combine the two Maxwell equations (4.43) in matrix form to a single super equation

$$\mathbf{d}\wedge\begin{pmatrix}\boldsymbol{\Psi}\\ \boldsymbol{\Phi}\end{pmatrix}=\begin{pmatrix}\gamma_e\\ \gamma_m\end{pmatrix}. \tag{4.71}$$

Defining the antisymmetric matrix (calligraphic J)

$$\mathcal{J}=\begin{pmatrix}0 & 1\\ -1 & 0\end{pmatrix}, \tag{4.72}$$

Equation (4.71) can be decomposed as

$$(\mathbf{d}_\mathbf{E}+\mathbf{d}\tau\partial_\tau)\wedge\begin{pmatrix}\mathbf{D}\\ \mathbf{B}\end{pmatrix}-\mathcal{J}\mathbf{d}_\mathbf{E}\wedge\begin{pmatrix}\mathbf{E}\\ \mathbf{H}\end{pmatrix}\wedge\mathbf{d}\tau$$

$$=\begin{pmatrix}\varrho_e\\ \varrho_m\end{pmatrix}-\begin{pmatrix}\mathbf{J}_e\\ \mathbf{J}_m\end{pmatrix}\wedge\mathbf{d}\tau. \tag{4.73}$$

Separating the spatial and temporal parts, we arrive at the three-dimensional matrix equations

$$\mathbf{d}_\mathbf{E}\wedge\begin{pmatrix}\mathbf{D}\\ \mathbf{B}\end{pmatrix}=\begin{pmatrix}\varrho_e\\ \varrho_m\end{pmatrix}, \tag{4.74}$$

$$\partial_\tau\begin{pmatrix}\mathbf{D}\\ \mathbf{B}\end{pmatrix}-\mathcal{J}\mathbf{d}_\mathbf{E}\wedge\begin{pmatrix}\mathbf{E}\\ \mathbf{H}\end{pmatrix}=-\begin{pmatrix}\mathbf{J}_e\\ \mathbf{J}_m\end{pmatrix}. \tag{4.75}$$

It is convenient to introduce the shorthand notation[1] for the three-dimensional super forms

$$\mathsf{e}=\begin{pmatrix}\mathbf{E}\\ \mathbf{H}\end{pmatrix},\quad \mathsf{d}=\begin{pmatrix}\mathbf{D}\\ \mathbf{B}\end{pmatrix}, \tag{4.76}$$

[1] One must be careful not to mistake d for $\mathbf{d}$.

$$\boxed{\begin{array}{c} \mathbf{D} = \overline{\overline{\epsilon}}|\mathbf{E} + \overline{\overline{\xi}}|\mathbf{H} \\[4pt] \mathbf{B} = \overline{\overline{\zeta}}|\mathbf{E} + \overline{\overline{\mu}}|\mathbf{H} \end{array}}$$

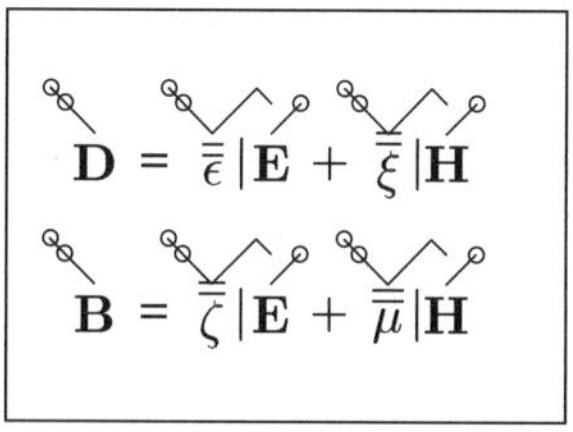

Fig. 4.6 *Visualizing the medium equations (4.83), (4.84) in three-dimensional representation. The medium dyadics are (generalized) Hodge dyadics.*

$$\rho = \begin{pmatrix} \varrho_e \\ \varrho_m \end{pmatrix}, \quad \mathsf{j} = \begin{pmatrix} \mathbf{J}_e \\ \mathbf{J}_m \end{pmatrix}, \tag{4.77}$$

where the quantity e is a super one-form, d and j are super two-forms, and ρ is a super three-form. The Maxwell equations (4.74), (4.75) can now be written as

$$\mathbf{d_E} \wedge \mathsf{d} = \rho, \tag{4.78}$$

$$\mathcal{J}\mathbf{d_E} \wedge \mathsf{e} = \partial_\tau \mathsf{d} + \mathsf{j}. \tag{4.79}$$

Differentiating the latter gives the continuity equation

$$\partial_\tau \rho + \mathbf{d_E} \wedge \mathsf{j} = 0. \tag{4.80}$$

4.5.2 Medium equations

Linear medium equations corresponding to the three-dimensional super forms can be expressed in terms of a dyadic matrix $\overline{\overline{\mathcal{M}}}$ as

$$\mathsf{d} = \overline{\overline{\mathcal{M}}}|\mathsf{e}. \tag{4.81}$$

The four elements of the medium matrix $\overline{\overline{\mathcal{M}}}$ are dyadics similar to those in Gibbsian electromagnetics [40, 41]

$$\overline{\overline{\mathcal{M}}} = \begin{pmatrix} \overline{\overline{\epsilon}} & \overline{\overline{\xi}} \\ \overline{\overline{\zeta}} & \overline{\overline{\mu}} \end{pmatrix}. \tag{4.82}$$

However, here the dyadics $\overline{\overline{\epsilon}} \cdots \overline{\overline{\mu}}$ are all elements of the same dyadic space $\mathbb{F}_2 \mathbb{E}_1$, which means that they are (generalized) Hodge dyadics, as seen in Figure 4.6. Written separately, the medium equations are

$$\begin{array}{rcl} \mathbf{D} & = & \overline{\overline{\epsilon}}|\mathbf{E} + \overline{\overline{\xi}}|\mathbf{H}, \end{array} \tag{4.83}$$

$$\begin{array}{rcl} \mathbf{B} & = & \overline{\overline{\zeta}}|\mathbf{E} + \overline{\overline{\mu}}|\mathbf{H}. \end{array} \tag{4.84}$$

In this form the medium equations become close to those familiar in the Gibbsian vector representation [38, 40, 41]. In Section 5.1, equivalence to Gibbsian medium dyadics will be defined by mapping the set of Hodge dyadics to a set of metric dyadics.

4.5.3 Time-harmonic sources

For time-harmonic sources with the time dependence $e^{jk_o\tau}$ the fields have the same dependence. In this case, in the previous expressions we can replace ∂_τ and $\mathbf{d}$ by jk_o and $\mathbf{d}_\mathbf{E} + jk_o\mathbf{d}\tau$, respectively. The Maxwell equations (4.79) in this case reduce to

$$\mathcal{J}\mathbf{d}_\mathbf{E} \wedge \mathsf{e} = jk_o\mathsf{d} + \mathsf{j} = jk_o\overline{\overline{\mathcal{M}}}|\mathsf{e} + \mathsf{j}. \tag{4.85}$$

The second set of equations (4.78) is not needed at all because it can be derived from (4.85) through the operation $\mathbf{d}_\mathbf{E}\wedge$ and using the continuity equation (4.80).

Problems

4.5.1 Find the inverse of the medium matrix $\overline{\overline{\mathcal{M}}}$ in a form which is valid when inverses of $\overline{\overline{\epsilon}}$ and $\overline{\overline{\mu}}$ exist but not necessarily those of $\overline{\overline{\xi}}$ and $\overline{\overline{\zeta}}$. Check the expression by setting $\overline{\overline{\xi}} = \overline{\overline{\zeta}} = 0$.

4.5.2 Derive the connection between the four dyadics $\overline{\overline{\epsilon}} \cdots \overline{\overline{\mu}}$ in (4.83) and (4.84) and the four dyadics $\overline{\overline{\alpha}} \cdots \overline{\overline{\beta}}$ defined by conditions of the form

$$\begin{aligned}\mathbf{D} &= \overline{\overline{\alpha}}|\mathbf{B} + \overline{\overline{\epsilon}}'|\mathbf{E}\\ \mathbf{H} &= \overline{\overline{\mu}}'^{-1}|\mathbf{B} + \overline{\overline{\beta}}|\mathbf{E}.\end{aligned}$$

4.5.3 Show that the most general antisymmetric six-dyadic consisting of four (generalized) metric dyadics in the space $\mathbb{E}_1\,\mathbb{E}_1$ can be expressed in the form

$$\overline{\overline{\mathcal{A}}} = \begin{pmatrix} \mathbf{A}\lfloor\overline{\overline{\mathsf{I}}}^T & \overline{\overline{\mathsf{C}}} \\ -\overline{\overline{\mathsf{C}}}^T & \mathbf{B}\lfloor\overline{\overline{\mathsf{I}}}^T \end{pmatrix},$$

where $\mathbf{A}$ and $\mathbf{B} \in \mathbb{E}_2$ are two bivectors and $\overline{\overline{\mathsf{C}}} \in \mathbb{E}_1\,\mathbb{E}_1$ is a dyadic.

4.5.4 Show that the inverse of the six-dyadic

$$\overline{\overline{\mathcal{D}}} = \begin{pmatrix} \overline{\overline{\mathsf{D}}}_{11} & \overline{\overline{\mathsf{D}}}_{12} \\ \overline{\overline{\mathsf{D}}}_{21} & \overline{\overline{\mathsf{D}}}_{22} \end{pmatrix}$$

where the dyadics $\overline{\overline{\mathsf{D}}}_{ij}$ are elements of the space $\mathbb{E}_1\,\mathbb{F}_1$ or $\mathbb{E}_1\,\mathbb{E}_1$, can be expressed as

$$\overline{\overline{\mathcal{D}}}^{-1} = \begin{pmatrix} (\overline{\overline{\mathsf{D}}}_{11} - \overline{\overline{\mathsf{D}}}_{12}|\overline{\overline{\mathsf{D}}}_{22}^{-1}|\overline{\overline{\mathsf{D}}}_{21})^{-1} & (\overline{\overline{\mathsf{D}}}_{21} - \overline{\overline{\mathsf{D}}}_{22}|\overline{\overline{\mathsf{D}}}_{12}^{-1}|\overline{\overline{\mathsf{D}}}_{11})^{-1} \\ (\overline{\overline{\mathsf{D}}}_{12} - \overline{\overline{\mathsf{D}}}_{11}|\overline{\overline{\mathsf{D}}}_{21}^{-1}|\overline{\overline{\mathsf{D}}}_{22})^{-1} & (\overline{\overline{\mathsf{D}}}_{22} - \overline{\overline{\mathsf{D}}}_{21}|\overline{\overline{\mathsf{D}}}_{11}^{-1}|\overline{\overline{\mathsf{D}}}_{12})^{-1} \end{pmatrix},$$

provided all the inverse dyadics exist.

4.5.5 Show that if only the dyadics $\overline{\overline{\mathsf{D}}}_{12}$ and $\overline{\overline{\mathsf{D}}}_{21}$ have inverses, the inverse six-dyadic of the previous problem must be expressed as

$$\overline{\overline{\mathcal{D}}}^{-1} = \begin{pmatrix} \overline{\overline{\mathsf{D}}}'_{11} & \overline{\overline{\mathsf{D}}}'_{12} \\ \overline{\overline{\mathsf{D}}}'_{21} & \overline{\overline{\mathsf{D}}}'_{22} \end{pmatrix},$$

with

$$\overline{\overline{\mathsf{D}}}'_{11} = -\overline{\overline{\mathsf{D}}}_{21}^{-1}|\overline{\overline{\mathsf{D}}}_{22}|(\overline{\overline{\mathsf{D}}}_{21} - \overline{\overline{\mathsf{D}}}_{22}|\overline{\overline{\mathsf{D}}}_{12}^{-1}|\overline{\overline{\mathsf{D}}}_{11})^{-1},$$

$$\overline{\overline{\mathsf{D}}}'_{12} = (\overline{\overline{\mathsf{D}}}_{21} - \overline{\overline{\mathsf{D}}}_{22}|\overline{\overline{\mathsf{D}}}_{12}^{-1}|\overline{\overline{\mathsf{D}}}_{11})^{-1},$$

$$\overline{\overline{\mathsf{D}}}'_{21} = (\overline{\overline{\mathsf{D}}}_{12} - \overline{\overline{\mathsf{D}}}_{11}|\overline{\overline{\mathsf{D}}}_{21}^{-1}|\overline{\overline{\mathsf{D}}}_{22})^{-1},$$

$$\overline{\overline{\mathsf{D}}}'_{22} = -\overline{\overline{\mathsf{D}}}_{12}^{-1}|\overline{\overline{\mathsf{D}}}_{11}|(\overline{\overline{\mathsf{D}}}_{12} - \overline{\overline{\mathsf{D}}}_{11}|\overline{\overline{\mathsf{D}}}_{21}^{-1}|\overline{\overline{\mathsf{D}}}_{22})^{-1}.$$

Using this formula, find the inverse of the antisymmetric six-dyadic $\overline{\overline{\mathcal{A}}}$ of Problem 4.5.3 when the inverse of $\overline{\overline{\mathsf{C}}}$ is assumed to exist.

5

Medium, Boundary, and Power Conditions

In this Chapter we consider medium and boundary conditions and concepts related to power and energy in electromagnetic fields.

5.1 MEDIUM CONDITIONS

The Maxwell equations $\mathbf{d}\wedge\boldsymbol{\Psi} = \gamma_m$ and $\mathbf{d}\wedge\boldsymbol{\Phi} = \gamma_e$ are independent of the medium. In fact, they remain the same for any medium: homogeneous or inhomogeneous, isotropic or anisotropic, linear or nonlinear. To solve for the fields, these equations must be augmented by an additional equation defining the connection between the two-forms $\boldsymbol{\Psi}$ and $\boldsymbol{\Phi}$. The simplest connection is the linear mapping (4.39) reproduced here as

$$\boldsymbol{\Psi} = \overline{\overline{\mathsf{M}}}|\boldsymbol{\Phi}. \tag{5.1}$$

$\overline{\overline{\mathsf{M}}} \in \mathbb{F}_2\mathbb{E}_2$ is the medium dyadic which maps a two-form onto a two-form. It can be interpreted as a Hodge dyadic. Because for $n = 4$ the space of two-forms has the dimension 6, the medium dyadic can be described in the most general case by 36 parameters M_{IJ}, as

$$\overline{\overline{\mathsf{M}}} = \sum_{I,J} M_{IJ}\boldsymbol{\varepsilon}_I\mathbf{e}_J, \tag{5.2}$$

where $I = \{i_1, i_2\}$ and $J = \{j_1, j_2\}$ are two ordered 2-indices and $\mathbf{e}_J, \boldsymbol{\varepsilon}_I$ are reciprocal bivector and dual bivector bases. Another form of the medium equation was given in (4.81) in terms of the dyadic matrix $\overline{\overline{\mathcal{M}}}$.

5.1.1 Modified medium dyadics

In Section 2.6 it was seen that a Hodge dyadic can be associated to a metric dyadic. In many cases it may be more convenient to work with the metric medium dyadic than with the Hodge medium dyadic $\overline{\overline{\mathsf{M}}}$. Change between the two representations is done through the the identity (1.121), which for the general p-vector $\mathbf{a}^p$ or dual p-vector $\boldsymbol{\alpha}^p$ can be written as

$$\begin{aligned} \mathbf{e}_N\lfloor(\boldsymbol{\varepsilon}_N\lfloor\mathbf{a}^p) &= (-1)^{p(n-p)}\mathbf{a}^p, \\ \boldsymbol{\varepsilon}_N\lfloor(\mathbf{e}_N\lfloor\boldsymbol{\alpha}^p) &= (-1)^{p(n-p)}\boldsymbol{\alpha}^p. \end{aligned} \tag{5.3}$$

In the interesting case $n = 4$ with $N = 1234$ the sign becomes $+1$ for $p = 2$ (two-forms and bivectors), and -1 for $p = 1, 3$ (one-forms, vectors, three-forms, and trivectors). For $n = 3$ with $N = 123$ the sign is $+1$ for all p.

Let us now define for $n = 4$ a modified medium dyadic $\overline{\overline{\mathsf{M}}}_g \in \mathbb{E}_2\mathbb{E}_2$ as

$$\overline{\overline{\mathsf{M}}}_g = \mathbf{e}_N\lfloor\overline{\overline{\mathsf{M}}}, \tag{5.4}$$

which is of the form of a metric dyadic because it maps two-forms to bivectors for $n = 4$. The converse relation is

$$\overline{\overline{\mathsf{M}}} = \boldsymbol{\varepsilon}_N\lfloor\overline{\overline{\mathsf{M}}}_g. \tag{5.5}$$

In case $\overline{\overline{\mathsf{M}}}_g$ is symmetric, it can be identified with a metric dyadic, and then, $\overline{\overline{\mathsf{M}}}$ is a Hodge dyadic. There is no need to define reciprocal bases here, because the transformation rules work for any quadrivector $\mathbf{e}_N$ and dual quadrivector $\boldsymbol{\varepsilon}_N$ satisfying $\mathbf{e}_N|\boldsymbol{\varepsilon}_N = 1$.

It also appears useful to define a modified form for the medium matrix $\overline{\overline{\mathcal{M}}}$ of the three-dimensional super-form representation (4.82) as

$$\overline{\overline{\mathcal{M}}}_g = \begin{pmatrix} \overline{\overline{\epsilon}}_g & \overline{\overline{\xi}}_g \\ \overline{\overline{\zeta}}_g & \overline{\overline{\mu}}_g \end{pmatrix} = \mathbf{e}_{123}\lfloor\overline{\overline{\mathcal{M}}} = \mathbf{e}_{123}\lfloor\begin{pmatrix} \overline{\overline{\epsilon}} & \overline{\overline{\xi}} \\ \overline{\overline{\zeta}} & \overline{\overline{\mu}} \end{pmatrix}. \tag{5.6}$$

In this case the dyadic elements $\overline{\overline{\epsilon}}_g, \overline{\overline{\xi}}_g, \overline{\overline{\zeta}}_g, \overline{\overline{\mu}}_g \in \mathbb{E}_1\mathbb{E}_1$ of the matrix $\overline{\overline{\mathcal{M}}}_g$ can be identified with those encountered in the Gibbsian electromagnetic analysis of bi-anisotropic media [40]. This is the reason why these dyadics will be occasionally called Gibbsian medium dyadics. The relation converse to (5.6) can be written as

$$\overline{\overline{\mathcal{M}}} = \boldsymbol{\varepsilon}_{123}\lfloor\overline{\overline{\mathcal{M}}}_g. \tag{5.7}$$

5.1.2 Bi-anisotropic medium

The general linear medium is called bi-anisotropic [38]. Like any Minkowskian $\mathbb{F}_2\mathbb{E}_2$ dyadic, $\overline{\overline{\mathsf{M}}}$ can be split in four terms, labeled as space–space, space–time, time–space and time–time components. The general expression can be written as (Figure 5.1)

$$\overline{\overline{\mathsf{M}}} = \overline{\overline{\alpha}} + \overline{\overline{\epsilon}}' \wedge \mathbf{e}_\tau + \mathbf{d}\tau \wedge \overline{\overline{\mu}}^{-1} + \mathbf{d}\tau \wedge \overline{\overline{\beta}} \wedge \mathbf{e}_\tau. \tag{5.8}$$

$$\overline{\overline{M}} = \overline{\overline{\alpha}} + \overline{\overline{\epsilon}}' \wedge \mathbf{e}_\tau + \mathbf{d}\tau \wedge \overline{\overline{\mu}}^{-1} + \mathbf{d}\tau \wedge \overline{\overline{\beta}} \wedge \mathbf{e}_\tau$$

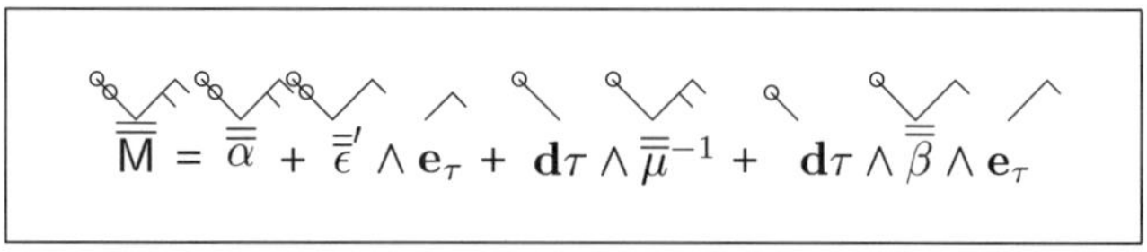

Fig. 5.1 *Representation of the four-dimensional medium dyadic in terms of four three-dimensional medium dyadics.*

The inverse of the medium dyadic is often needed in representing the relation as

$$\mathbf{\Phi} = \overline{\overline{M}}^{-1} | \mathbf{\Psi}. \tag{5.9}$$

To find the expansion of the inverse to the medium dyadic, it is expanded in a form similar to that in (5.8) as

$$\overline{\overline{M}}^{-1} = \overline{\overline{\alpha}}_1 + \overline{\overline{\epsilon}}'_1 \wedge \mathbf{e}_\tau + \mathbf{d}\tau \wedge \overline{\overline{\mu}}_1^{-1} + \mathbf{d}\tau \wedge \overline{\overline{\beta}}_1 \wedge \mathbf{e}_\tau. \tag{5.10}$$

Relations of the four dyadics $\overline{\overline{\alpha}}_1 \cdots \overline{\overline{\beta}}_1$ to $\overline{\overline{\alpha}} \cdots \overline{\overline{\beta}}$ can be expressed as

$$\overline{\overline{\epsilon}}'_1 = -(\overline{\overline{\mu}}^{-1} - \overline{\overline{\beta}} | \overline{\overline{\epsilon}}'^{-1} | \overline{\overline{\alpha}})^{-1}, \tag{5.11}$$

$$\overline{\overline{\mu}}_1^{-1} = -(\overline{\overline{\epsilon}}' - \overline{\overline{\alpha}} | \overline{\overline{\mu}} | \overline{\overline{\beta}})^{-1}, \tag{5.12}$$

$$\overline{\overline{\alpha}}_1 = \overline{\overline{\mu}} | \overline{\overline{\beta}} | \overline{\overline{\mu}}_1^{-1} = (\overline{\overline{\alpha}} - \overline{\overline{\epsilon}}' | \overline{\overline{\beta}}^{-1} | \overline{\overline{\mu}}^{-1})^{-1}, \tag{5.13}$$

$$\overline{\overline{\beta}}_1 = \overline{\overline{\epsilon}}'^{-1} | \overline{\overline{\alpha}} | \overline{\overline{\epsilon}}'_1 = (\overline{\overline{\beta}} - \overline{\overline{\mu}}^{-1} | \overline{\overline{\alpha}}^{-1} | \overline{\overline{\epsilon}}')^{-1}. \tag{5.14}$$

Derivation of these relations is straightforward and is left as an exercise. The latter expressions in the last two equations are valid only if the inverses of the dyadics $\overline{\overline{\alpha}}$ and $\overline{\overline{\beta}}$ exist. In the special case $\overline{\overline{\alpha}} = 0, \overline{\overline{\beta}} = 0$ we simply have

$$\overline{\overline{\epsilon}}'_1 = -\overline{\overline{\mu}}, \quad \overline{\overline{\mu}}_1 = -\overline{\overline{\epsilon}}', \tag{5.15}$$

and the inverse has the form

$$\overline{\overline{M}}^{-1} = (\overline{\overline{\epsilon}}' \wedge \mathbf{e}_\tau + \mathbf{d}\tau \wedge \overline{\overline{\mu}}^{-1})^{-1} = -\overline{\overline{\mu}} \wedge \mathbf{e}_\tau - \mathbf{d}\tau \wedge \overline{\overline{\epsilon}}'^{-1}. \tag{5.16}$$

5.1.3 Different representations

The special case $\overline{\overline{\alpha}} = 0, \overline{\overline{\beta}} = 0$ was already encountered in Section 4.4. The four three-dimensional medium dyadics mapping three-dimensional one-forms and two-forms belong to the respective dyadic spaces $\overline{\overline{\alpha}} \in \mathbb{F}_2 \mathbb{E}_2, \overline{\overline{\epsilon}}' \in \mathbb{F}_2 \mathbb{E}_1, \overline{\overline{\mu}}^{-1} \in \mathbb{F}_1 \mathbb{E}_2$ and $\overline{\overline{\beta}} \in \mathbb{F}_1 \mathbb{E}_1$.

Separating terms in (5.1) as

$$(\mathbf{D} + \mathbf{d}\tau \wedge \mathbf{H}) = \overline{\overline{M}} | (\mathbf{B} + \mathbf{E} \wedge \mathbf{d}\tau)$$

$$= (\overline{\overline{\alpha}} + \overline{\overline{\epsilon}}' \wedge \mathbf{e}_\tau + \mathbf{d}\tau \wedge \overline{\overline{\mu}}^{-1} + \mathbf{d}\tau \wedge \overline{\overline{\beta}} \wedge \mathbf{e}_\tau) | (\mathbf{B} + \mathbf{E} \wedge \mathbf{d}\tau)$$

$$= \overline{\overline{\alpha}} | \mathbf{B} + (\overline{\overline{\epsilon}}' \wedge \mathbf{e}_\tau) | (\mathbf{E} \wedge \mathbf{d}\tau) + (\mathbf{d}\tau \wedge \overline{\overline{\mu}}^{-1}) | \mathbf{B} + (\mathbf{d}\tau \wedge \overline{\overline{\beta}} \wedge \mathbf{e}_\tau) | (\mathbf{E} \wedge \mathbf{d}\tau)$$

$$= \overline{\overline{\alpha}} | \mathbf{B} + \overline{\overline{\epsilon}}' | \mathbf{E} + \mathbf{d}\tau \wedge (\overline{\overline{\mu}}^{-1} | \mathbf{B} + \overline{\overline{\beta}} | \mathbf{E}), \tag{5.17}$$

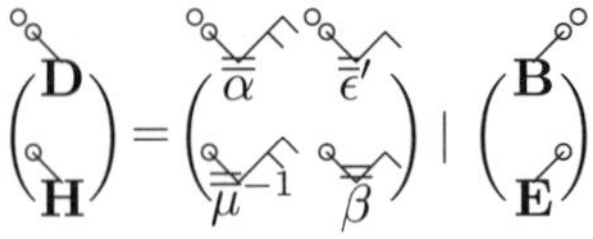

Fig. 5.2 *Visualization of the mapping between fields in terms of the three-dimensional medium dyadics.*

the space-like and time-like components of the two-form equation are split in two three-dimensional equations as

$$\mathbf{D} = \overline{\overline{\alpha}}|\mathbf{B} + \overline{\overline{\epsilon}}'|\mathbf{E}, \tag{5.18}$$

$$\mathbf{H} = \overline{\overline{\mu}}^{-1}|\mathbf{B} + \overline{\overline{\beta}}|\mathbf{E}. \tag{5.19}$$

The dyadic $\overline{\overline{\epsilon}}'$ differs somewhat from the permittivity dyadic $\overline{\overline{\epsilon}}$ defined through the equations (4.83) and (4.84) reproduced here as

$$\mathbf{D} = \overline{\overline{\epsilon}}|\mathbf{E} + \overline{\overline{\xi}}|\mathbf{H}, \tag{5.20}$$

$$\mathbf{B} = \overline{\overline{\zeta}}|\mathbf{E} + \overline{\overline{\mu}}|\mathbf{H}, \tag{5.21}$$

in which all four medium dyadics belong to the same space $\mathbb{F}_2\mathbb{E}_1$. However, the dyadic $\overline{\overline{\mu}}$ is the same in both representations. Also, the dyadics $\overline{\overline{\epsilon}}'$ and $\overline{\overline{\epsilon}}$ are the same in the special case $\overline{\overline{\alpha}} = \overline{\overline{\beta}} = 0$. The mappings defined by various medium dyadics are visualized in Figure 5.3. Relations between the two sets of medium dyadics can be written as

$$\overline{\overline{\epsilon}} = \overline{\overline{\epsilon}}' - \overline{\overline{\alpha}}|\overline{\overline{\mu}}|\overline{\overline{\beta}}, \quad \overline{\overline{\xi}} = \overline{\overline{\alpha}}|\overline{\overline{\mu}}, \quad \overline{\overline{\zeta}} = -\overline{\overline{\mu}}|\overline{\overline{\beta}} \tag{5.22}$$

and, conversely, as

$$\overline{\overline{\epsilon}}' = \overline{\overline{\epsilon}} - \overline{\overline{\xi}}|\overline{\overline{\mu}}^{-1}|\overline{\overline{\zeta}}, \quad \overline{\overline{\alpha}} = \overline{\overline{\xi}}|\overline{\overline{\mu}}^{-1}, \quad \overline{\overline{\beta}} = -\overline{\overline{\mu}}^{-1}|\overline{\overline{\zeta}}. \tag{5.23}$$

Relations to modified (Gibbsian) medium dyadics can be derived after some algebraic steps (left as an exercise) in the form

$$\begin{aligned} \overline{\overline{\epsilon}}' &= \boldsymbol{\varepsilon}_{123}\lfloor(\overline{\overline{\epsilon}}_g - \overline{\overline{\xi}}_g|\overline{\overline{\mu}}_g^{-1}|\overline{\overline{\zeta}}_g), \quad \overline{\overline{\mu}} = \boldsymbol{\varepsilon}_{123}\lfloor\overline{\overline{\mu}}_g, \\ \overline{\overline{\alpha}} &= \boldsymbol{\varepsilon}_{123}\lfloor(\overline{\overline{\xi}}_g|\overline{\overline{\mu}}_g^{-1})\rfloor\mathbf{e}_{123}, \quad \overline{\overline{\beta}} = -\overline{\overline{\mu}}_g^{-1}|\overline{\overline{\zeta}}_g. \end{aligned} \tag{5.24}$$

Combined, they give the medium dyadic $\overline{\overline{\mathsf{M}}}$ as

$$\overline{\overline{\mathsf{M}}} = \boldsymbol{\varepsilon}_{123}\lfloor\overline{\overline{\epsilon}}_g \wedge \mathbf{e}_\tau + (\boldsymbol{\varepsilon}_{123}\lfloor\overline{\overline{\xi}}_g + \mathbf{d}\tau \wedge \overline{\overline{\mathsf{I}}}^T)|\overline{\overline{\mu}}_g^{-1}|(\overline{\overline{\mathsf{I}}}\rfloor\mathbf{e}_{123} - \overline{\overline{\zeta}}_g \wedge \mathbf{e}_\tau). \tag{5.25}$$

It is left as an exercise to check the grades of the terms on the right-hand side. Applying the identity

$$\mathbf{e}_N\lfloor(\boldsymbol{\varepsilon}_{123}\lfloor\mathbf{a}) = -\mathbf{e}_\tau \wedge \mathbf{a}, \tag{5.26}$$

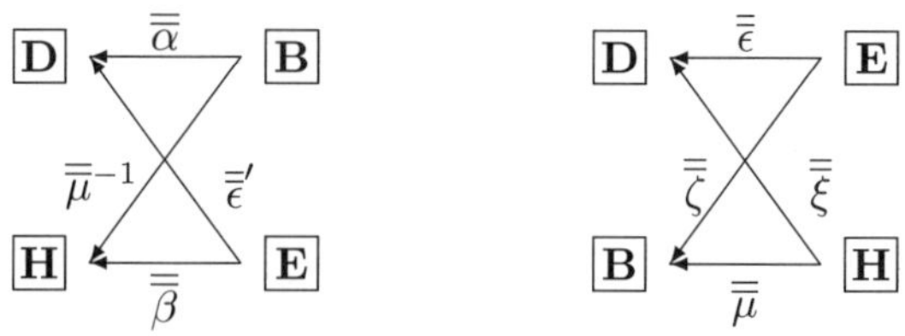

Fig. 5.3 *Comparison of two systems of medium relations between the electromagnetic field one and two-forms.*

valid for any vector $\mathbf{a}$ with $\mathbf{e}_N = \mathbf{e}_{1234}$, we can derive from (5.25) the modified medium dyadic $\overline{\overline{\mathsf{M}}}_g \in \mathbb{E}_2\mathbb{E}_2$ in terms of the Gibbsian medium dyadics as

$$\begin{aligned} \overline{\overline{\mathsf{M}}}_g &= \mathbf{e}_N\lfloor\overline{\overline{\mathsf{M}}} \\ &= \overline{\overline{\epsilon}}_g{}^{\wedge}_{\wedge}\mathbf{e}_\tau\mathbf{e}_\tau - (\mathbf{e}_{123}\lfloor\overline{\overline{\mathsf{I}}}^T + \mathbf{e}_\tau\wedge\overline{\overline{\xi}}_g)|\overline{\overline{\mu}}_g^{-1}|(\overline{\overline{\mathsf{I}}}\rfloor\mathbf{e}_{123} - \overline{\overline{\zeta}}_g\wedge\mathbf{e}_\tau). \end{aligned} \quad (5.27)$$

$\overline{\overline{\alpha}}, \overline{\overline{\beta}}$ and $\overline{\overline{\xi}}, \overline{\overline{\zeta}}$ are called magnetoelectric medium dyadics. In many practical cases they can be considered negligeable and omitted. In such cases the two permittivity dyadics are the same, $\overline{\overline{\epsilon}}' = \overline{\overline{\epsilon}}$. In the following we call a bi-anisotropic medium "anisotropic" or "simply anisotropic" if $\overline{\overline{\alpha}} = \overline{\overline{\beta}} = 0$ or $\overline{\overline{\xi}} = \overline{\overline{\zeta}} = 0$. In this case the Minkowskian medium dyadic can be expressed as

$$\overline{\overline{\mathsf{M}}} = \overline{\overline{\epsilon}}\wedge\mathbf{e}_\tau + \mathbf{d}\tau\wedge\overline{\overline{\mu}}^{-1} = \boldsymbol{\varepsilon}_{123}\lfloor\overline{\overline{\epsilon}}_g\wedge\mathbf{e}_\tau + \mathbf{d}\tau\wedge\overline{\overline{\mu}}_g^{-1}\rfloor\mathbf{e}_{123}, \quad (5.28)$$

where $\overline{\overline{\epsilon}}, \overline{\overline{\mu}} \in \mathbb{F}_2\mathbb{E}_1$ are Euclidean Hodge dyadics and $\overline{\overline{\epsilon}}_g, \overline{\overline{\mu}}_g \in \mathbb{E}_1\mathbb{E}_1$ are metric dyadics. One must note that the property $\overline{\overline{\alpha}} = 0$ and $\overline{\overline{\beta}} = 0$ is not preserved by the Lorentz transformation. Actually, the concept of bi-anisotropic media was originally launched as a means to describe "simple" media in uniform motion [38]. However, assuming only spatial affine transformations (5.35), the conditions of simple anisotropy remain valid.

5.1.4 Isotropic medium

The concept of "isotropic medium" in terms of Gibbsian field vectors implies a medium in which the electric and magnetic polarizations are parallel to the respective exciting electric and magnetic fields. For example, in an isotropic dielectric medium we have $\mathbf{D} = \overline{\overline{\epsilon}}\cdot\mathbf{E} = \epsilon\mathbf{E}$, whence $\mathbf{D}$ and $\mathbf{E}$ are scalar multiples of each other, and thus, parallel vectors. This happens when $\overline{\overline{\epsilon}}$ and $\overline{\overline{\mu}}$ are multiples of the Gibbsian unit dyadic. If all four medium dyadics $\overline{\overline{\epsilon}}, \overline{\overline{\xi}}, \overline{\overline{\zeta}}, \overline{\overline{\mu}}$ are multiples of the Gibbsian unit dyadic, the medium is called bi-isotropic.

General isotropy In terms of differential forms, the previous concept of isotropy is meaningless because, for example, $\mathbf{D}$ is a two-form and $\mathbf{E}$ isa one-form and they cannot be scalar multiples of one another. More general isotropy can, however, in principle be defined by requiring that the two electromagnetic two-forms $\mathbf{\Psi}$ and $\mathbf{\Phi}$ be scalar multiples of one another. This requirement leads to a medium condition of the form

$$\mathbf{\Psi} = \overline{\overline{\mathsf{M}}}|\mathbf{\Phi} = M\mathbf{\Phi}, \tag{5.29}$$

where M is some scalar. In this definition the medium dyadic $\overline{\overline{\mathsf{M}}}$ becomes a scalar multiple of the Minkowskian unit dyadic $\bar{\bar{\mathsf{I}}}_{\mathbf{M}}^{(2)T}$, which can be represented in terms of the Euclidean unit dyadic $\bar{\bar{\mathsf{I}}}_{\mathbf{E}}$ as

$$\begin{aligned} \overline{\overline{\mathsf{M}}} &= M\bar{\bar{\mathsf{I}}}_{\mathbf{M}}^{(2)T} = (\bar{\bar{\mathsf{I}}}_{\mathbf{E}} + \mathbf{d}\tau\mathbf{e}_\tau)^{(2)T} \\ &= M(\bar{\bar{\mathsf{I}}}_{\mathbf{E}}^{(2)T} + \bar{\bar{\mathsf{I}}}_{\mathbf{E}}^{T}{}^{\wedge}_{\wedge}\mathbf{d}\tau\mathbf{e}_\tau). \end{aligned} \tag{5.30}$$

Comparison with (5.8) leads to the medium dyadics

$$\bar{\bar{\alpha}} = M\bar{\bar{\mathsf{I}}}_{\mathbf{E}}^{(2)T}, \quad \bar{\bar{\epsilon}}' = 0, \quad \bar{\bar{\mu}}^{-1} = 0, \quad \bar{\bar{\beta}} = -M\bar{\bar{\mathsf{I}}}_{\mathbf{E}}^{T}. \tag{5.31}$$

These represent a strange medium because the medium equations now reduce to

$$\mathbf{D} = M\mathbf{B}, \quad \mathbf{H} = -M\mathbf{E}. \tag{5.32}$$

This kind of isotropy is invariant in the Lorentz transformation: the medium appears isotropic for any observer moving with constant velocity. This property is valid even more generally for any affine transformation, as is seen from the rule (4.63) by inserting (5.30), whence the transformed medium dyadic becomes

$$\overline{\overline{\mathsf{M}}}_a = \overline{\overline{\mathsf{A}}}^{(-2)T}|M\bar{\bar{\mathsf{I}}}_{\mathbf{M}}^{(2)T}|\overline{\overline{\mathsf{A}}}^{(2)T} = M\bar{\bar{\mathsf{I}}}_{\mathbf{M}}^{(2)T}. \tag{5.33}$$

It can also be noted that the condition $\mathbf{\Psi} = M\mathbf{\Phi}$ also imposes a condition between the electric and magnetic sources, $\gamma_e = M\gamma_m$, because otherwise a contradiction would arise in the Maxwell equations.

Spatial isotropy A slightly more general case would be the spatially isotropic medium, defined by taking $\mathbf{e}_4 = \mathbf{e}_\tau$ as a special axis. In this case we have two scalar constants and the medium dyadics are of the form

$$\bar{\bar{\alpha}} = \alpha\bar{\bar{\mathsf{I}}}_{\mathbf{E}}^{(2)T}, \quad \bar{\bar{\epsilon}}' = 0, \quad \bar{\bar{\mu}}^{-1} = 0, \quad \bar{\bar{\beta}} = \beta\bar{\bar{\mathsf{I}}}_{\mathbf{E}}^{T}. \tag{5.34}$$

This kind of medium does not remain invariant in the Lorentz transformation because time and space become mixed. However, for spatial affine transformations,

$$\overline{\overline{\mathsf{A}}}_{\mathbf{M}} = \overline{\overline{\mathsf{A}}}_{\mathbf{E}} + \mathbf{e}_\tau\mathbf{d}\tau, \tag{5.35}$$

the medium remains spatially isotropic. This is seen by applying the rules (details left as a problem)

$$\overline{\overline{\mathsf{A}}}_{\mathbf{M}}^{(2)} = \overline{\overline{\mathsf{A}}}_{\mathbf{E}}^{(2)} + \overline{\overline{\mathsf{A}}}_{\mathbf{E}}{}^{\wedge}_{\wedge}\mathbf{e}_\tau\mathbf{d}\tau, \tag{5.36}$$

$$\overline{\overline{\mathsf{A}}}_{\mathbf{M}}^{-1} = \overline{\overline{\mathsf{A}}}_{\mathbf{E}}^{-1} + \mathbf{e}_\tau\mathbf{d}\tau, \tag{5.37}$$

$$\overline{\overline{\mathsf{A}}}_{\mathbf{M}}^{(-2)} = \overline{\overline{\mathsf{A}}}_{\mathbf{E}}^{(-2)} + \overline{\overline{\mathsf{A}}}_{\mathbf{E}}^{-1}{}^{\wedge}_{\wedge}\mathbf{e}_\tau\mathbf{d}\tau. \tag{5.38}$$

where the latter inverses are understood in the three-dimensional Euclidean sense. In fact, by inserting these in the expression of the transformed medium dyadic,

$$\begin{aligned}\overline{\overline{\mathsf{M}}}_a &= (\overline{\overline{\mathsf{A}}}_{\mathbf{E}}^{(-2T)} + \overline{\overline{\mathsf{A}}}_{\mathbf{E}}^{-1} {}^\wedge_\wedge \mathbf{d}\tau\mathbf{e}_\tau)|(\alpha\overline{\overline{\mathsf{I}}}_{\mathbf{E}}^{(2)T} + \beta\overline{\overline{\mathsf{I}}}_{\mathbf{E}}^{T} {}^\wedge_\wedge \mathbf{d}\tau\mathbf{e}_\tau)|(\overline{\overline{\mathsf{A}}}^{(2)T} + \overline{\overline{\mathsf{A}}}_{\mathbf{E}} {}^\wedge_\wedge \mathbf{d}\tau\mathbf{e}_\tau) \\ &= \alpha\overline{\overline{\mathsf{I}}}_{\mathbf{E}}^{(2)T} + \beta\overline{\overline{\mathsf{I}}}_{\mathbf{E}}^{T} {}^\wedge_\wedge \mathbf{d}\tau\mathbf{e}_\tau, \end{aligned} \tag{5.39}$$

invariance of the spatially isotropic medium can be verified.

Gibbsian isotropy Let us finally check what is the counterpart of a medium which is considered isotropic in Gibbsian electromagnetics. The Gibbsian unit dyadic $\sum \mathbf{u}_i\mathbf{u}_i \in \mathbb{E}_1\mathbb{E}_1$ can actually be replaced by any Euclidean metric dyadic $\overline{\overline{\mathsf{G}}}_{\mathbf{E}}$ (symmetric by definition), whence the medium is defined by the modified medium dyadics

$$\overline{\overline{\epsilon}}_g = \epsilon\overline{\overline{\mathsf{G}}}_{\mathbf{E}}, \quad \overline{\overline{\mu}}_g = \mu\overline{\overline{\mathsf{G}}}_{\mathbf{E}}, \quad \overline{\overline{\xi}}_g = \overline{\overline{\zeta}}_g = 0. \tag{5.40}$$

Substituting in (5.24), we obtain

$$\overline{\overline{\epsilon}}' = \epsilon\varepsilon_{123}\lfloor\overline{\overline{\mathsf{G}}}_{\mathbf{E}}, \quad \overline{\overline{\mu}} = \mu\varepsilon_{123}\lfloor\overline{\overline{\mathsf{G}}}_{\mathbf{E}}, \quad \overline{\overline{\alpha}} = 0, \quad \overline{\overline{\beta}} = 0, \tag{5.41}$$

which are Hodge dyadics. The Minkowskian medium dyadic has now the form

$$\overline{\overline{\mathsf{M}}} = \epsilon\varepsilon_{123}\lfloor\overline{\overline{\mathsf{G}}}_{\mathbf{E}} \wedge \mathbf{e}_\tau + \frac{1}{\mu}\mathbf{d}\tau \wedge \overline{\overline{\Gamma}}_{\mathbf{E}}\rfloor\mathbf{e}_{123}, \tag{5.42}$$

where $\overline{\overline{\Gamma}}_{\mathbf{E}}$ is the inverse of the metric dyadic $\overline{\overline{\mathsf{G}}}_{\mathbf{E}}$.

5.1.5 Bi-isotropic medium

As a generalization of both Gibbsian isotropic and spatially isotropic media defined above, let us consider a medium defined by the three-dimensional medium dyadics:

$$\overline{\overline{\alpha}} = \alpha\overline{\overline{\mathsf{I}}}_{\mathbf{E}}^{(2)T}, \quad \overline{\overline{\epsilon}}' = \epsilon'\overline{\overline{\mathsf{B}}}, \quad \overline{\overline{\mu}} = \mu\overline{\overline{\mathsf{B}}}, \quad \overline{\overline{\beta}} = \beta\overline{\overline{\mathsf{I}}}_{\mathbf{E}}^{T}. \tag{5.43}$$

where $\overline{\overline{\mathsf{B}}} \in \mathbb{F}_2\mathbb{E}_1$ is a three-dimensional Euclidean Hodge dyadic. The corresponding four-dimensional medium dyadic is, then,

$$\overline{\overline{\mathsf{M}}} = \alpha\overline{\overline{\mathsf{I}}}_{\mathbf{E}}^{(2)T} + \epsilon'\overline{\overline{\mathsf{B}}} \wedge \mathbf{e}_\tau + \frac{1}{\mu}\mathbf{d}\tau \wedge \overline{\overline{\mathsf{B}}}^{-1} + \beta\mathbf{d}\tau \wedge \overline{\overline{\mathsf{I}}}_{\mathbf{E}}^{T} \wedge \mathbf{e}_\tau. \tag{5.44}$$

Inserting these in (5.22), the second system of parameter dyadics becomes

$$\overline{\overline{\epsilon}} = \overline{\overline{\epsilon}}' - \overline{\overline{\alpha}}|\overline{\overline{\mu}}|\overline{\overline{\beta}} = (\epsilon' - \alpha\mu\beta)\overline{\overline{\mathsf{B}}} = \epsilon\overline{\overline{\mathsf{B}}}, \tag{5.45}$$

$$\overline{\overline{\xi}} = \overline{\overline{\alpha}}|\overline{\overline{\mu}} = \alpha\mu\overline{\overline{\mathsf{B}}} = \xi\overline{\overline{\mathsf{B}}}, \tag{5.46}$$

$$\overline{\overline{\zeta}} = -\overline{\overline{\mu}}|\overline{\overline{\beta}} = -\mu\beta\overline{\overline{\mathsf{B}}} = \zeta\overline{\overline{\mathsf{B}}}, \tag{5.47}$$

$$\overline{\overline{\mu}} = \mu\overline{\overline{\mathsf{B}}}. \tag{5.48}$$

Thus, in the presentation (4.83), (4.84), all medium dyadics become multiples of the same Euclidean dyadic $\overline{\overline{\mathsf{B}}}$. Now it is known that, if all four Gibbsian medium dyadics are multiples of the same metric dyadic $\mathbf{e}_{123}\lfloor\overline{\overline{\mathsf{B}}} \in \mathbb{E}_1\,\mathbb{E}_1$ (symmetric), the medium can be transformed through a suitable three-dimensional affine transformation to a bi-isotropic medium whose medium dyadics $\overline{\overline{\epsilon}}_g, \overline{\overline{\xi}}_g, \overline{\overline{\zeta}}_g, \overline{\overline{\mu}}_g$ are multiples of the Gibbsian unit dyadic $\sum \mathbf{u}_i\mathbf{u}_i$. Such media have been called affine bi-isotropic in the past [40, 41]. We may call a medium defined by (5.43) or (5.44) spatially bi-isotropic because it is invariant in any spatial affine transformation.

To see this, let us again apply a spatial affine transformation to the medium dyadic (5.44) with the transformation dyadic $\overline{\overline{\mathsf{A}}}_{\mathbf{M}}$ defined by (5.35). Applying (5.38) to the medium dyadic (4.63) gives

$$
\begin{aligned}
\overline{\overline{\mathsf{M}}}_a &= \overline{\overline{\mathsf{A}}}_{\mathbf{M}}^{(-2)T}|\overline{\overline{\mathsf{M}}}|\overline{\overline{\mathsf{A}}}_{\mathbf{M}}^{(2)T} \\
&= \alpha\overline{\overline{\mathsf{A}}}_{\mathbf{E}}^{(-2)T}|\overline{\overline{\mathsf{I}}}^{(2)T}|\overline{\overline{\mathsf{A}}}_{\mathbf{E}}^{(2)T} + \epsilon'\overline{\overline{\mathsf{A}}}_{\mathbf{E}}^{(-2)T}|(\overline{\overline{\mathsf{B}}}\wedge\mathbf{e}_\tau)|\overline{\overline{\mathsf{A}}}_{\mathbf{E}}^{T}{}^{\wedge}_{\wedge}\mathbf{d}\tau\mathbf{e}_\tau \\
&\quad +\frac{1}{\mu}\overline{\overline{\mathsf{A}}}_{\mathbf{E}}^{-1T}{}^{\wedge}_{\wedge}\mathbf{d}\tau\mathbf{e}_\tau|(\mathbf{d}\tau\wedge\overline{\overline{\mathsf{B}}}^{-1})|\overline{\overline{\mathsf{A}}}_{\mathbf{E}}^{(2)T} \\
&\quad +\beta\overline{\overline{\mathsf{A}}}_{\mathbf{E}}^{-1T}{}^{\wedge}_{\wedge}\mathbf{d}\tau\mathbf{e}_\tau|(\mathbf{d}\tau\wedge\overline{\overline{\mathsf{I}}}_{\mathbf{E}}^{T}\wedge\mathbf{d}\tau)|\overline{\overline{\mathsf{A}}}_{\mathbf{E}}^{T}{}^{\wedge}_{\wedge}\mathbf{d}\tau\mathbf{e}_\tau \\
&= \alpha\overline{\overline{\mathsf{I}}}_{\mathbf{E}}^{(2)T} + \epsilon'\overline{\overline{\mathsf{A}}}_{\mathbf{E}}^{(-2)T}|\overline{\overline{\mathsf{B}}}|\overline{\overline{\mathsf{A}}}_{\mathbf{E}}^{T}\wedge\mathbf{e}_\tau \\
&\quad +\frac{1}{\mu}\mathbf{d}\tau\wedge\overline{\overline{\mathsf{A}}}_{\mathbf{E}}^{-1T}|\overline{\overline{\mathsf{B}}}^{-1}|\overline{\overline{\mathsf{A}}}_{\mathbf{E}}^{(2)T} + \beta\mathbf{d}\tau\wedge\overline{\overline{\mathsf{I}}}_{\mathbf{E}}^{T}\wedge\mathbf{e}_\tau.
\end{aligned}
\tag{5.49}
$$

Defining the transformed dyadic $\overline{\overline{\mathsf{B}}}_a$ as

$$\overline{\overline{\mathsf{B}}}_a = \overline{\overline{\mathsf{A}}}_{\mathbf{E}}^{(-2)T}|\overline{\overline{\mathsf{B}}}|\overline{\overline{\mathsf{A}}}_{\mathbf{E}}^{T}, \tag{5.50}$$

the affine-transformed medium dyadic can be written as

$$\overline{\overline{\mathsf{M}}}_a = \alpha\overline{\overline{\mathsf{I}}}_{\mathbf{E}}^{(2)T} + \epsilon'\overline{\overline{\mathsf{B}}}_a\wedge\mathbf{e}_\tau + \frac{1}{\mu}\mathbf{d}\tau\wedge\overline{\overline{\mathsf{B}}}_a{}^{-1} + \beta\mathbf{d}\tau\wedge\overline{\overline{\mathsf{I}}}_{\mathbf{E}}^{T}\wedge\mathbf{e}_\tau, \tag{5.51}$$

which is of the same form as (5.44). Expressing the Hodge dyadic $\overline{\overline{\mathsf{B}}} = \varepsilon_{123}\lfloor\overline{\overline{\mathsf{G}}}_{\mathbf{E}}$ in terms of a Euclidean metric dyadic $\overline{\overline{\mathsf{G}}}_{\mathbf{E}}$, a basis $\{\mathbf{e}_i\}$ can be defined as $\overline{\overline{\mathsf{G}}}_{\mathbf{E}} = \sum \mathbf{e}_i\mathbf{e}_i$, which corresponds to the unit dyadic in Gibbsian formalism. Thus, through the metric dyadic the spatially bi-isotropic medium can be conceived as bi-isotropic in the Gibbsian sense.

5.1.6 Uniaxial medium

The medium dyadics $\overline{\overline{\epsilon}}_g \cdots \overline{\overline{\mu}}_g$ are called Gibbsian because they are elements of the $\mathbb{E}_1\,\mathbb{E}_1$ space like in the Gibbsian vector electromagnetics. So we can take any set of Gibbsian medium dyadics and transform them to other systems through the preceding formulas. As an example, let us consider a uniaxial bi-anisotropic medium defined through the set of relative medium dyadics [41]

$$\overline{\overline{\epsilon}}_g = \epsilon_\perp(\mathbf{e}_1\mathbf{e}_1 + \mathbf{u}_2\mathbf{e}_2) + \epsilon_3\mathbf{e}_3\mathbf{e}_3, \tag{5.52}$$

$$\overline{\overline{\mu}}_g = \mu_\perp(\mathbf{e}_1\mathbf{e}_1 + \mathbf{e}_2\mathbf{e}_2) + \mu_3\mathbf{e}_3\mathbf{e}_3, \tag{5.53}$$

$$\overline{\overline{\xi}}_g = \xi_3\mathbf{e}_3\mathbf{e}_3, \tag{5.54}$$

$$\overline{\overline{\zeta}}_g = \zeta_3\mathbf{e}_3\mathbf{e}_3, \tag{5.55}$$

where $\mathbf{e}_1, \mathbf{e}_2, \mathbf{e}_3$ is a basis of Gibbsian unit vectors. This is a set of metric dyadics which defines the four-dimensional Hodge dyadic $\overline{\overline{\mathsf{M}}}$ through (5.25). However, let us form the three-dimensional Hodge medium dyadics. Forming the reciprocal dual basis $\{\varepsilon_i\}$, from (5.24) we can first express the permittivity dyadic as

$$\overline{\overline{\mu}} = \varepsilon_{123}\lfloor\overline{\overline{\mu}}_g = \mu_\perp(\varepsilon_{23}\mathbf{e}_1 + \varepsilon_{31}\mathbf{e}_2) + \mu_3\varepsilon_{12}\mathbf{e}_3. \tag{5.56}$$

To find the other dyadics the inverse of $\overline{\overline{\mu}}_g$ must be inserted, which now is an element of the $\mathbb{F}_1\mathbb{F}_1$ space,

$$\overline{\overline{\mu}}_g^{-1} = \mu_\perp^{-1}(\varepsilon_1\varepsilon_1 + \varepsilon_2\varepsilon_2) + \mu_3^{-1}\varepsilon_3\varepsilon_3. \tag{5.57}$$

Thus, from (5.24) we have

$$\overline{\overline{\beta}} = -\overline{\overline{\mu}}_g^{-1}|\overline{\overline{\zeta}}_g = -\frac{\zeta_3}{\mu_3}\varepsilon_3\mathbf{e}_3, \tag{5.58}$$

$$\overline{\overline{\alpha}} = \varepsilon_{123}\lfloor(\overline{\overline{\xi}}_g|\overline{\overline{\mu}}_g^{-1})\rfloor\mathbf{e}_{123} = \frac{\xi_3}{\mu_3}\varepsilon_{12}\mathbf{e}_{12}, \tag{5.59}$$

$$\begin{aligned}\overline{\overline{\epsilon}}' &= \varepsilon_{123}\lfloor(\overline{\overline{\epsilon}}_g - \overline{\overline{\xi}}_g|\overline{\overline{\mu}}_g^{-1}|\overline{\overline{\zeta}}_g) \\ &= \epsilon_\perp(\varepsilon_{23}\mathbf{e}_1 + \varepsilon_{31}\mathbf{e}_2) + (\epsilon_3 - \frac{\xi_3\zeta_3}{\mu_3})\varepsilon_{12}\mathbf{e}_3.\end{aligned} \tag{5.60}$$

The four-dimensional medium dyadic $\overline{\overline{\mathsf{M}}}$ can also be found through the expression (5.8) using the just derived Hodge dyadics. Here we need the inverse of the permittivity dyadic:

$$\begin{aligned}\overline{\overline{\mu}}^{-1} &= (\mu_\perp\varepsilon_{23}\mathbf{e}_1 + \varepsilon_{31}\mathbf{e}_2) + \mu_3\varepsilon_{12}\mathbf{e}_3)^{-1} \\ &= \mu_\perp^{-1}(\varepsilon_1\mathbf{e}_{23} + \varepsilon_2\mathbf{e}_{31}) + \mu_3^{-1}\varepsilon_3\mathbf{e}_{12}.\end{aligned} \tag{5.61}$$

It is left as an exercise to check that the four-dimensional Hodge medium dyadic expression obtained from (5.8) and (5.25) gives the same result.

5.1.7 Q-medium

Let us finally introduce an important class of bi-anisotropic media called for simplicity as that of Q-media [50]. By definition, a medium belongs to this class if there exists a dyadic $\overline{\overline{\mathsf{Q}}} \in \mathbb{E}_1\mathbb{E}_1$ such that the modified medium dyadic $\overline{\overline{\mathsf{M}}}_g \in \mathbb{E}_2\mathbb{E}_2$ can be expressed as

$$\overline{\overline{\mathsf{M}}}_g = \overline{\overline{\mathsf{Q}}}^{(2)}. \tag{5.62}$$

In the three-dimensional space ($n = 3$), any bivector dyadic $\overline{\overline{\mathsf{A}}} \in \mathbb{E}_2\mathbb{E}_2$ possessing an inverse can be expressed in the form $\overline{\overline{\mathsf{A}}} = \overline{\overline{\mathsf{Q}}}^{(2)}$ (see Problem 2.4.12). However, in the four-dimensional space ($n = 4$) this is no more the case and there is a special class of such dyadics. Let us now study what are the corresponding conditions for the Gibbsian medium dyadics $\overline{\overline{\epsilon}}_g \cdots \overline{\overline{\mu}}_g$ by equating the three-dimensional components on each side of (5.62) [51]. First we write

$$\overline{\overline{\mathsf{Q}}} = \overline{\overline{\mathsf{A}}} + \mathbf{e}_\tau\mathbf{a} + \mathbf{b}\mathbf{e}_\tau + c\mathbf{e}_\tau\mathbf{e}_\tau, \tag{5.63}$$

where $\overline{\overline{\mathsf{A}}}$, $\mathbf{a}$, and $\mathbf{b}$ denote a Euclidean dyadic and two Euclidean vectors, respectively. The double-wedge square can be expanded as

$$\overline{\overline{\mathsf{Q}}}^{(2)} = \overline{\overline{\mathsf{A}}}^{(2)} - \mathbf{e}_\tau \wedge \overline{\overline{\mathsf{A}}} \wedge \mathbf{a} - \mathbf{b} \wedge \overline{\overline{\mathsf{A}}} \wedge \mathbf{e}_\tau + (c\overline{\overline{\mathsf{A}}} - \mathbf{ba})^{\wedge}_{\wedge}\mathbf{e}_\tau\mathbf{e}_\tau. \tag{5.64}$$

The expression (5.27) for the modified medium dyadic can be expanded as

$$\begin{aligned} \overline{\overline{\mathsf{M}}}_g &= -\mathbf{e}_{123}\mathbf{e}_{123}\lfloor\lfloor\overline{\overline{\mu}}_g^{-1} - \mathbf{e}_\tau \wedge \overline{\overline{\xi}}_g|\overline{\overline{\mu}}_g^{-1}\rfloor\mathbf{e}_{123} \\ &\quad +\mathbf{e}_{123}\lfloor\overline{\overline{\mu}}_g^{-1}|\overline{\overline{\zeta}}_g \wedge \mathbf{e}_\tau + (\overline{\overline{\epsilon}}_g - \overline{\overline{\xi}}_g|\overline{\overline{\mu}}_g^{-1}|\overline{\overline{\zeta}}_g)^{\wedge}_{\wedge}\mathbf{e}_\tau\mathbf{e}_\tau. \end{aligned} \tag{5.65}$$

Requiring that the expressions (5.65) and (5.64) be equal leads to the following set of conditions for the three-dimensional dyadic components:

$$-\overline{\overline{\mu}}_g^{-1}\rfloor\rfloor\mathbf{e}_{123}\mathbf{e}_{123} = \overline{\overline{\mathsf{A}}}^{(2)}, \tag{5.66}$$

$$\overline{\overline{\xi}}_g|\overline{\overline{\mu}}_g^{-1}\rfloor\mathbf{e}_{123} = \overline{\overline{\mathsf{A}}} \wedge \mathbf{a}, \tag{5.67}$$

$$\mathbf{e}_{123}\lfloor\overline{\overline{\mu}}_g^{-1}|\overline{\overline{\zeta}}_g = -\mathbf{b} \wedge \overline{\overline{\mathsf{A}}}, \tag{5.68}$$

$$\overline{\overline{\epsilon}}_g - \overline{\overline{\xi}}_g|\overline{\overline{\mu}}_g^{-1}|\overline{\overline{\zeta}}_g = c\overline{\overline{\mathsf{A}}} - \mathbf{ba}. \tag{5.69}$$

Let us now solve these equations for the four Gibbsian medium dyadics in terms of the dyadic $\overline{\overline{\mathsf{A}}}$, vectors $\mathbf{a}$, $\mathbf{b}$ and the scalar c.

Solving for $\overline{\overline{\mu}}_g$ Applying the expression (2.169) for the inverse of a metric dyadic, we can write

$$\overline{\overline{\mathsf{A}}}^{-1} = \frac{\overline{\overline{\mathsf{A}}}^{(2)T}\rfloor\rfloor\varepsilon_{123}\varepsilon_{123}}{\overline{\overline{\mathsf{A}}}^{(3)}||\varepsilon_{123}\varepsilon_{123}}, \tag{5.70}$$

the condition (5.66) can be rewritten as

$$\overline{\overline{\mu}}_g^{-1} = -\overline{\overline{\mathsf{A}}}^{(2)}\rfloor\rfloor\varepsilon_{123}\varepsilon_{123} = -(\overline{\overline{\mathsf{A}}}^{(3)}||\varepsilon_{123}\varepsilon_{123})\overline{\overline{\mathsf{A}}}^{-1T}. \tag{5.71}$$

From this, the permeability dyadic is found in the form

$$\overline{\overline{\mu}}_g = -\frac{\overline{\overline{\mathsf{A}}}^T}{\overline{\overline{\mathsf{A}}}^{(3)}||\varepsilon_{123}\varepsilon_{123}}. \tag{5.72}$$

We assume that $\overline{\overline{\mathsf{A}}}^{(3)}||\varepsilon_{123}\varepsilon_{123} \neq 0$, because the converse case would lead to a rather pathological medium. The dyadic $\overline{\overline{\mathsf{A}}}$ can thus be expressed in terms of $\overline{\overline{\mu}}_g$ as

$$\overline{\overline{\mathsf{A}}} = \pm j \frac{\overline{\overline{\mu}}_g^T}{\sqrt{\overline{\overline{\mu}}_g^{(3)}||\varepsilon_{123}\varepsilon_{123}}}. \tag{5.73}$$

Solving for $\overline{\overline{\xi}}_g$ and $\overline{\overline{\zeta}}_g$ From the two conditions (5.67), (5.68), we obtain

$$\overline{\overline{\xi}}_g = ((\overline{\overline{\mathsf{A}}} \wedge \mathbf{a})\rfloor\varepsilon_{123})|\overline{\overline{\mu}}_g = -\frac{((\overline{\overline{\mathsf{A}}} \wedge \mathbf{a})\rfloor\varepsilon_{123})|\overline{\overline{\mathsf{A}}}^T}{\overline{\overline{\mathsf{A}}}^{(3)}||\varepsilon_{123}\varepsilon_{123}}, \tag{5.74}$$

$$\overline{\overline{\zeta}}_g = -\overline{\overline{\mu}}_g|(\varepsilon_{123}\lfloor(\mathbf{b} \wedge \overline{\overline{\mathsf{A}}})) = \frac{\overline{\overline{\mathsf{A}}}^T|(\varepsilon_{123}\lfloor(\mathbf{b} \wedge \overline{\overline{\mathsf{A}}}))}{\overline{\overline{\mathsf{A}}}^{(3)}||\varepsilon_{123}\varepsilon_{123}}. \tag{5.75}$$

The form of the right-hand sides in these expressions indicates that both $\overline{\overline{\xi}}_g$ and $\overline{\overline{\zeta}}_g$ must be antisymmetric dyadics. In fact, applying the identity

$$\begin{aligned}(\overline{\overline{\mathsf{A}}}\rfloor(\boldsymbol{\alpha} \wedge \boldsymbol{\beta}))|\overline{\overline{\mathsf{A}}}^T &= \overline{\overline{\mathsf{A}}}|(\boldsymbol{\beta}\boldsymbol{\alpha} - \boldsymbol{\alpha}\boldsymbol{\beta})|\overline{\overline{\mathsf{A}}}^T = (\overline{\overline{\mathsf{A}}}|\boldsymbol{\beta})(\overline{\overline{\mathsf{A}}}|\boldsymbol{\alpha}) - (\overline{\overline{\mathsf{A}}}|\boldsymbol{\alpha})(\overline{\overline{\mathsf{A}}}|\boldsymbol{\beta}) \\ &= \overline{\overline{\mathsf{I}}}\rfloor((\overline{\overline{\mathsf{A}}}|\boldsymbol{\alpha}) \wedge (\overline{\overline{\mathsf{A}}}|\boldsymbol{\beta})) = \overline{\overline{\mathsf{I}}}\rfloor(\overline{\overline{\mathsf{A}}}^{(2)}|(\boldsymbol{\alpha} \wedge \boldsymbol{\beta})),\end{aligned} \tag{5.76}$$

and replacing the dual bivector $\boldsymbol{\alpha} \wedge \boldsymbol{\beta}$ by $\mathbf{a}\rfloor\varepsilon_{123}$, we can write

$$\overline{\overline{\xi}}_g = -\frac{(\overline{\overline{\mathsf{A}}}\rfloor(\mathbf{a}\rfloor\varepsilon_{123}))|\overline{\overline{\mathsf{A}}}^T}{\overline{\overline{\mathsf{A}}}^{(3)}||\varepsilon_{123}\varepsilon_{123}} = -\frac{\overline{\overline{\mathsf{I}}}\rfloor(\overline{\overline{\mathsf{A}}}^{(2)}|(\mathbf{a}\rfloor\varepsilon_{123}))}{\overline{\overline{\mathsf{A}}}^{(3)}||\varepsilon_{123}\varepsilon_{123}}. \tag{5.77}$$

This means that the antisymmetric dyadic $\overline{\overline{\xi}}_g$ can be expressed in terms of a bivector $\mathbf{X}_g$ as

$$\overline{\overline{\xi}}_g = \overline{\overline{\mathsf{I}}}\rfloor\mathbf{X}_g = \mathbf{X}_g\lfloor\overline{\overline{\mathsf{I}}}^T, \tag{5.78}$$

with

$$\mathbf{X}_g = -\frac{\overline{\overline{\mathsf{A}}}^{(2)}|(\mathbf{a}\rfloor\varepsilon_{123})}{\overline{\overline{\mathsf{A}}}^{(3)}||\varepsilon_{123}\varepsilon_{123}} = -(\overline{\overline{\mathsf{A}}}^{-1T}\rfloor\rfloor\mathbf{e}_{123}\mathbf{e}_{123})|(\mathbf{a}\rfloor\varepsilon_{123}) = -\mathbf{e}_{123}\lfloor\overline{\overline{\mathsf{A}}}^{-1T}|\mathbf{a}. \tag{5.79}$$

Similarly, we can write the dyadic $\overline{\overline{\zeta}}_g$ in terms of another bivector $\mathbf{Z}_g$ as

$$\overline{\overline{\zeta}}_g = \overline{\overline{\mathsf{I}}}\rfloor\mathbf{Z}_g = \mathbf{Z}_g\lfloor\overline{\overline{\mathsf{I}}}^T, \tag{5.80}$$

with

$$\mathbf{Z}_g = \frac{\overline{\overline{\mathsf{A}}}^{(2)T}|(\varepsilon_{123}\lfloor\mathbf{b})}{\overline{\overline{\mathsf{A}}}^{(3)}||\varepsilon_{123}\varepsilon_{123}} = \mathbf{e}_{123}\lfloor\overline{\overline{\mathsf{A}}}^{-1}|\mathbf{b}. \tag{5.81}$$

Conversely, the vectors $\mathbf{a}$ and $\mathbf{b}$ can be expressed in terms of $\mathbf{X}_g$ and $\mathbf{Z}_g$ as

$$\mathbf{a} = -\overline{\overline{\mathsf{A}}}^T|(\varepsilon_{123}\lfloor\mathbf{X}_g), \quad \mathbf{b} = \overline{\overline{\mathsf{A}}}|(\varepsilon_{123}\lfloor\mathbf{Z}_g), \tag{5.82}$$

where we can substitute $\overline{\overline{\mathsf{A}}}$ in terms of $\overline{\overline{\mu}}_g$ from (5.73).

Solving for $\overline{\overline{\epsilon}}_g$ To find a representation for the Gibbsian permittivity dyadic $\overline{\overline{\epsilon}}_g$, we apply the following bac cab rule valid for any bivector $\mathbf{A}$, vector $\mathbf{c}$ and dual bivector $\mathbf{\Psi}$:

$$\mathbf{A}\lfloor(\mathbf{\Psi}\lfloor\mathbf{c}) = \mathbf{\Psi}\rfloor(\mathbf{A}\wedge\mathbf{c}) - (\mathbf{A}|\mathbf{\Psi})\mathbf{c}. \tag{5.83}$$

Because the vector $\mathbf{c}$ appears in every term as the last symbol, it can be replaced by a dyadic $\overline{\overline{\mathsf{A}}} \in \mathbb{E}_1\mathbb{E}_1$,

$$\mathbf{A}\lfloor(\mathbf{\Psi}\lfloor\overline{\overline{\mathsf{A}}}) = \mathbf{\Psi}\rfloor(\mathbf{A}\wedge\overline{\overline{\mathsf{A}}}) - (\mathbf{\Psi}|\mathbf{A})\overline{\overline{\mathsf{A}}}. \tag{5.84}$$

Now, let us apply this to the condition (5.69):

$$\begin{aligned}\overline{\overline{\epsilon}}_g &= c\overline{\overline{\mathsf{A}}} - \mathbf{ba} + \overline{\overline{\xi}}_g|\overline{\overline{\mu}}_g^{-1}|\overline{\overline{\zeta}}_g \\ &= c\overline{\overline{\mathsf{A}}} - \mathbf{ba} - (\mathbf{X}_g\lfloor\overline{\overline{\mathsf{I}}}^T)|(\varepsilon_{123}\lfloor(\mathbf{b}\wedge\overline{\overline{\mathsf{A}}})) \\ &= c\overline{\overline{\mathsf{A}}} - \mathbf{ba} - \mathbf{X}_g\lfloor((\varepsilon_{123}\lfloor\mathbf{b})\lfloor\overline{\overline{\mathsf{A}}}) \\ &= c\overline{\overline{\mathsf{A}}} - \mathbf{ba} - (\varepsilon_{123}\lfloor\mathbf{b})\rfloor(\mathbf{X}_g\wedge\overline{\overline{\mathsf{A}}}) + ((\varepsilon_{123}\lfloor\mathbf{b})|\mathbf{X}_g)\overline{\overline{\mathsf{A}}}.\end{aligned} \tag{5.85}$$

Applying the identity

$$(\mathbf{e}_{123}\lfloor\boldsymbol{\alpha})\wedge\mathbf{a} = \mathbf{e}_{123}(\boldsymbol{\alpha}|\mathbf{a}), \tag{5.86}$$

valid for any three-dimensional vector $\mathbf{a}$ and dual vector $\boldsymbol{\alpha}$, we can expand

$$\mathbf{X}_g\wedge\overline{\overline{\mathsf{A}}} = -(\mathbf{e}_{123}\lfloor(\overline{\overline{\mathsf{A}}}^{-1T}|\mathbf{a}))\wedge\overline{\overline{\mathsf{A}}} = -\mathbf{e}_{123}(\mathbf{a}|\overline{\overline{\mathsf{A}}}^{-1}|\overline{\overline{\mathsf{A}}}) = -\mathbf{e}_{123}\mathbf{a}, \tag{5.87}$$

which inserted in the third term of the last expression in (5.85) gives

$$-(\varepsilon_{123}\lfloor\mathbf{b})\rfloor(\mathbf{X}_g\wedge\overline{\overline{\mathsf{A}}}) = (\varepsilon_{123}\lfloor\mathbf{b})\rfloor\mathbf{e}_{123}\mathbf{a} = \mathbf{ba}. \tag{5.88}$$

This cancels the term $-\mathbf{ba}$ in the expression (5.85) for $\overline{\overline{\epsilon}}_g$. Expanding finally

$$(\varepsilon_{123}\lfloor\mathbf{b})|\mathbf{X}_g = (\varepsilon_{123}\lfloor\mathbf{X}_g)|\mathbf{b} = -(\varepsilon_{123}\lfloor(\mathbf{e}_{123}\lfloor\overline{\overline{\mathsf{A}}}^{-1T}|\mathbf{a}))|\mathbf{b} = -\mathbf{a}|\overline{\overline{\mathsf{A}}}^{-1}|\mathbf{b}, \tag{5.89}$$

(5.85) is simplified to

$$\overline{\overline{\epsilon}}_g = (c - \mathbf{a}|\overline{\overline{\mathsf{A}}}^{-1}|\mathbf{b})\overline{\overline{\mathsf{A}}}. \tag{5.90}$$

Because the permittivity dyadic is a multiple of the dyadic $\overline{\overline{\mathsf{A}}}$, a relation between the dyadics $\overline{\overline{\epsilon}}_g$ and $\overline{\overline{\mu}}_g$ of the form

$$\overline{\overline{\epsilon}}_g = a\overline{\overline{\mu}}_g^T, \tag{5.91}$$

with some scalar a, must exist in a Q-medium.

Medium conditions From the above analysis we see that the Q-medium conditions for the Gibbsian medium dyadics can be expressed as

$$\overline{\overline{\epsilon}}_g = a\overline{\overline{\mu}}_g^T, \quad \overline{\overline{\xi}}_g^T = -\overline{\overline{\xi}}_g, \quad \overline{\overline{\zeta}}_g^T = -\overline{\overline{\zeta}}_g, \tag{5.92}$$

for some scalar a. Expressing the antisymmetric dyadics $\overline{\overline{\xi}}_g$ and $\overline{\overline{\zeta}}_g$ in terms of the respective bivectors $\mathbf{X}_g$ and $\mathbf{Z}_g$ as in (5.78) and (5.80), and using (5.63), (5.73), (5.82), and (5.90) the $\overline{\overline{\mathsf{Q}}}$-dyadic can be constructed in the form

$$\overline{\overline{\mathsf{Q}}} = \frac{a}{A}\mathbf{e}_\tau\mathbf{e}_\tau + A(\overline{\overline{\mathsf{I}}} - \mathbf{e}_\tau(\varepsilon_{123}\lfloor\mathbf{X}_g))|\overline{\overline{\mu}}_g^T|(\overline{\overline{\mathsf{I}}}^T + (\varepsilon_{123}\lfloor\mathbf{Z}_g)\mathbf{e}_\tau), \tag{5.93}$$

where the scalar A is defined by

$$A = \pm\frac{j}{\sqrt{\overline{\overline{\mu}}_g^{(3)}||\varepsilon_{123}\varepsilon_{123}}} \tag{5.94}$$

with either sign. It is left as a problem to show that (5.93) substituted in (5.62) reproduces the expression (5.27) for the modified Q-medium dyadic $\overline{\overline{\mathsf{M}}}_g$.

One can show that the modified medium dyadic of a Q-medium satisfies a conditon of the form

$$A\overline{\overline{\mathsf{M}}}_g + B\overline{\overline{\mathsf{M}}}_g^{-1T}\rfloor\rfloor\mathbf{e}_N\mathbf{e}_N = 0 \tag{5.95}$$

for some scalar coefficients A and B and, conversely, if the medium dyadic satisfies a condition of the form (5.95), it belongs to the class of Q-media (see problems 5.1.13 and 5.1.14).

Media corresponding to the class of Q-media have been considered previously in terms of three-dimensional Gibbsian vector analysis when trying to define media in which the time-harmonic Green dyadic can be expressed in analytic form. In the special case of vanishing magnetoelectric parameters $\overline{\overline{\xi}}_g = \overline{\overline{\zeta}}_g = 0$, the class of media satisfying $\overline{\overline{\epsilon}}_g = a\overline{\overline{\mu}}_g^T$ was introduced in [43]. The more general class of bi-anisotropic media with added antisymmetric magnetoelectric dyadics $\overline{\overline{\xi}}_g$ and $\overline{\overline{\zeta}}_g$ was later introduced in [45]. In both cases the Green dyadic could actually be expressed in terms of analytic functions. In Chapter 7 it will be shown that the wave equation is greatly simplified when the bi-anisotropic medium belongs to the class of Q-media.

5.1.8 Generalized Q-medium

As a straighforward generalization of the Q-medium, let us instead of (5.62) define

$$\overline{\overline{\mathsf{M}}}_g = \overline{\overline{\mathsf{Q}}}^{(2)} + \mathbf{AB}, \tag{5.96}$$

where $\mathbf{A}$ and $\mathbf{B}$ are two Minkowskian bivectors. Let us just briefly study the Gibbsian interpretation for the special case

$$\mathbf{A} = \mathbf{p}\wedge\mathbf{e}_\tau, \quad \mathbf{B} = \mathbf{q}\wedge\mathbf{e}_\tau, \tag{5.97}$$

where $\mathbf{p}$ and $\mathbf{q}$ are two Euclidean vectors, by extending the results of the previous section. In this case, (5.96) becomes

$$\overline{\overline{\mathsf{M}}}_g = \overline{\overline{\mathsf{Q}}}^{(2)} + \mathbf{pq}{}^{\wedge}_{\wedge}\mathbf{e}_\tau\mathbf{e}_\tau. \tag{5.98}$$

It can be seen that the first three conditions (5.66)–(5.68) for the Gibbsian medium dyadics are not changed at all in the generalization, while (5.69) takes the form

$$\overline{\overline{\epsilon}}_g - \overline{\overline{\xi}}_g|\overline{\overline{\mu}}_g^{-1}|\overline{\overline{\zeta}}_g = c\overline{\overline{\mathsf{A}}} - \mathbf{ba} + \mathbf{pq}. \tag{5.99}$$

Thus, the previous solutions for $\overline{\overline{\mu}}_g$, $\overline{\overline{\xi}}_g$, and $\overline{\overline{\zeta}}_g$ remain the same and (5.90) is changed to

$$\overline{\overline{\epsilon}}_g = (c - \mathbf{a}|\overline{\overline{\mathsf{A}}}^{-1}|\mathbf{b})\overline{\overline{\mathsf{A}}} + \mathbf{pq}. \tag{5.100}$$

As a conclusion, the generalized Q-medium defined by (5.98) corresponds to a medium whose Gibbsian dyadics satisfy the following conditions:

$$\overline{\overline{\epsilon}}_g = a\overline{\overline{\mu}}_g^T + \mathbf{pq}, \quad \overline{\overline{\xi}}_g^T = -\overline{\overline{\xi}}_g, \quad \overline{\overline{\zeta}}_g^T = -\overline{\overline{\zeta}}_g, \tag{5.101}$$

for some scalar a and Euclidean vectors $\mathbf{p}, \mathbf{q}$. Obviously, the conditions (5.101) are more general than those of (5.92). Another special case of (5.101) is considered in Problem 5.1.12.

The conditions for the Gibbsian medium dyadics corresponding to the general case (5.96) have been derived in reference 52 and were shown to correspond to those of the class of decomposable media defined in reference 44, see Problem 7.4.5. Because the partial differential operator defining the fourth-order differential equation of the electromagnetic field in a decomposable medium can be factorized to a product of two second-order operators, solution processes in such media can be essentially simplified [60, 61].

Problems

5.1.1 Defining the medium dyadics of a simple anisotropic medium as

$$\overline{\overline{\epsilon}} = \epsilon\overline{\overline{\mathsf{H}}}_2^T, \quad \overline{\overline{\mu}} = \mu\overline{\overline{\mathsf{H}}}_2^T,$$

with the Euclidean Hodge dyadic

$$\overline{\overline{\mathsf{H}}}_2 = \mathbf{e}_x(\mathbf{d}y \wedge \mathbf{d}z) + \mathbf{e}_y(\mathbf{d}z \wedge \mathbf{d}x) + \mathbf{e}_z(\mathbf{d}x \wedge \mathbf{d}y),$$

express the four-dimensional medium equation $\mathbf{\Psi} = \overline{\overline{\mathsf{M}}}|\mathbf{\Phi}$ in explicit form and derive the three-dimensional medium equations $\mathbf{D} = \overline{\overline{\epsilon}}|\mathbf{E}$, $\mathbf{B} = \overline{\overline{\mu}}|\mathbf{H}$ in detail.

5.1.2 Show that if a Minkowskian affine transformation dyadic is spatial, that is, of the form

$$\overline{\overline{\mathsf{A}}}_{\mathrm{M}} = \overline{\overline{\mathsf{A}}}_{\mathrm{E}} + \mathbf{e}_\tau\mathbf{d}\tau,$$

we can write

$$\overline{\overline{\mathsf{A}}}_{\mathrm{M}}^{(-2)} = (\overline{\overline{\mathsf{A}}}_{\mathrm{M}}^{-1})^{(2)} = (\overline{\overline{\mathsf{A}}}_{\mathrm{M}}^{(2)})^{-1} = (\overline{\overline{\mathsf{A}}}_{\mathrm{E}}^{-1} + \mathbf{e}_\tau\mathbf{d}\tau)^{(2)},$$

where $\overline{\overline{\mathsf{A}}}_\epsilon^{-1}$ is the inverse in the three-dimensional Euclidean space.

5.1.3 Find the conditions for the dyadic medium parameters $\overline{\overline{\alpha}} \cdots \overline{\overline{\beta}}$ corresponding to the conditions of reciprocity $\overline{\overline{\epsilon}}_g^T = \overline{\overline{\epsilon}}_g$, $\overline{\overline{\mu}}_g^T = \overline{\overline{\mu}}_g$ and $\overline{\overline{\xi}}_g^T = -\overline{\overline{\zeta}}_g$.

5.1.4 Derive the transformation formulas (5.24) in detail. Prove first the relation $(\varepsilon_{123}\lfloor\overline{\overline{\mu}}_g)^{-1} = \overline{\overline{\mu}}_g^{-1}\rfloor\mathbf{e}_{123}$.

5.1.5 Derive the following representation for the medium dyadic $\overline{\overline{\mathsf{M}}}$ in terms of the dyadic parameters of the medium dyadic matrix $\overline{\overline{\mathcal{M}}}$:

$$\overline{\overline{\mathsf{M}}} = \overline{\overline{\epsilon}} \wedge \mathbf{e}_\tau + (\overline{\overline{\xi}} + \mathbf{d}\tau \wedge \overline{\overline{\mathsf{I}}}^T)|\overline{\overline{\mu}}^{-1}|(\overline{\overline{\mathsf{I}}}^{(2)T} - \overline{\overline{\zeta}} \wedge \mathbf{e}_\tau)$$

5.1.6 Derive the representation (5.25) for the medium dyadic $\overline{\overline{\mathsf{M}}}$ in terms of the dyadic parameters of the modified medium dyadic matrix $\overline{\overline{\mathcal{M}}}_g$.

5.1.7 Find the inverse medium dyadic $\overline{\overline{\mathsf{M}}}^{-1}$ by writing its expansion in a form similar to that of (5.8) and studying the terms from $\overline{\overline{\mathsf{M}}}|\overline{\overline{\mathsf{M}}}^{-1} = \overline{\overline{\mathsf{I}}}_{\mathbf{E}}^{(2)T} + \overline{\overline{\mathsf{I}}}_{\mathbf{E}}^T{}^\wedge_\wedge \mathbf{d}\tau\mathbf{e}_\tau$.

5.1.8 Find the dyadic $\overline{\overline{\mathsf{M}}}^{(2)}$ corresponding to the expansion of the bi-anisotropic medium (5.8).

5.1.9 Find the four-dimensional medium dyadic $\overline{\overline{\mathsf{M}}}$ for the medium discussed in the example above by using the Gibbsian dyadics (5.52) – (5.55) and using the expression (5.25) and the Hodge dyadics (5.58) – (5.60) and the expression (5.8). Check that the expressions are the same.

5.1.10 Check the compatibility of (5.92) and (5.62) by counting the number of free parameters on each side of (5.62).

5.1.11 Show that (5.93) substituted in (5.62) reproduces the expression (5.27) for the modified Q-medium dyadic $\overline{\overline{\mathsf{M}}}_g$.

5.1.12 Study a special case of the generalized Q-medium with condition (5.96) specified by

$$\mathbf{AB} = (\mathbf{a}_1 \wedge \mathbf{a}_2)(\mathbf{b}_1 \wedge \mathbf{b}_2)$$

and

$$\overline{\overline{\mathsf{Q}}} = \overline{\overline{\mathsf{A}}} + c\mathbf{e}_\tau\mathbf{e}_\tau,$$

where $\overline{\overline{\mathsf{A}}}$ is a three-dimensional dyadic and $\mathbf{a}_1 \cdots \mathbf{b}_2$ are three-dimensional vectors. Show that the corresponding Gibbsian dyadics satisfy conditions of the form

$$\overline{\overline{\mu}}_g = a\overline{\overline{\epsilon}}_g^T + \mathbf{fg}, \quad \overline{\overline{\xi}}_g = 0, \quad \overline{\overline{\zeta}}_g = 0,$$

with some scalar a and three-dimensional vectors $\mathbf{f}, \mathbf{g}$.

5.1.13 Show that for a Q-medium scalars A and B can be found such that a condition of the form

$$A\overline{\overline{\mathsf{M}}}_g + B\overline{\overline{\mathsf{M}}}_g^{-1T}\rfloor\rfloor\mathbf{e}_N\mathbf{e}_N = 0$$

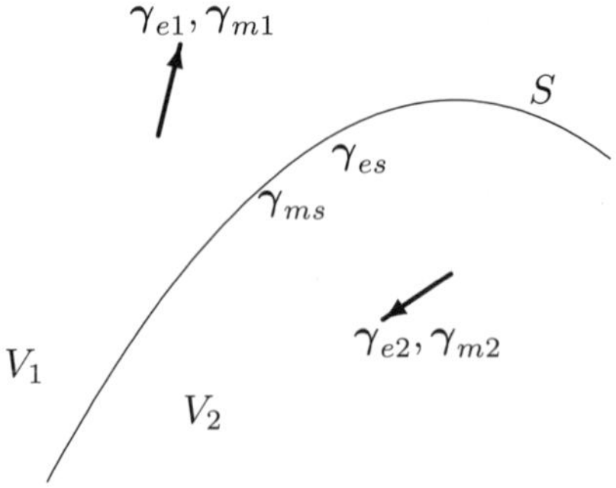

Fig. 5.4 *Two given electromagnetic source-field systems in regions V_1 and V_2 can be combined to one system by adding suitable surface sources γ_{es}, γ_{ms} on the boundary surface S.*

is valid for the modified medium dyadic $\overline{\overline{\mathsf{M}}}_g$. *Hint*: Prove first the inverse formula (valid for $n = 4$)

$$(\overline{\overline{\mathsf{Q}}}^{(2)})^{-1} = (\overline{\overline{\mathsf{Q}}}^{-1})^{(2)} = \frac{\overline{\overline{\mathsf{Q}}}^{(2)T}\rfloor\rfloor\boldsymbol{\kappa}_N\boldsymbol{\kappa}_N}{\overline{\overline{\mathsf{Q}}}^{(4)}||\boldsymbol{\kappa}_N\boldsymbol{\kappa}_N}.$$

5.1.14 Prove the converse of the previous problem. Assuming that there exist two scalars A, B such that the modified medium dyadic satisfies the condition

$$A\overline{\overline{\mathsf{M}}}_g + B\overline{\overline{\mathsf{M}}}_g^{-1T}\rfloor\rfloor\mathbf{e}_N\mathbf{e}_N = 0,$$

show that there exists a dyadic $\overline{\overline{\mathsf{Q}}} \in \mathbb{E}_1\mathbb{E}_1$ such the we can write $\overline{\overline{\mathsf{M}}}_g = \overline{\overline{\mathsf{Q}}}^{(2)}$. *Hint:* Derive first the expansion

$$\mathbf{e}_N\mathbf{e}_N\lfloor\lfloor\overline{\overline{\mathsf{M}}}_g^{-1} = -\overline{\overline{\mu}}_g{}^{\wedge}_{\wedge}\mathbf{e}_4\mathbf{e}_4 + (\mathbf{e}_{123}\lfloor\overline{\mathsf{I}}^T - \mathbf{e}_4 \wedge \overline{\overline{\zeta}}_g)|\overline{\overline{\epsilon}}_g^{-1}|(\overline{\mathsf{I}}\rfloor\mathbf{e}_{123} + \overline{\overline{\xi}}_g \wedge \mathbf{e}_4).$$

5.2 CONDITIONS ON BOUNDARIES AND INTERFACES

Behavior of fields at surfaces of medium discontinuity can be extracted from the Maxwell equations. Because Huygens' principle of equivalent surface sources can be formulated through the same procedure, we start by a general consideration of two electromagnetic source-field systems and their combination into one system.

5.2.1 Combining source-field systems

Let us assume that there exist two electromagnetic source-field systems denoted by susbscripts $i = 1$ and 2. They consist of electric and magnetic source three-forms γ_{ei}, γ_{mi} which create electromagnetic two-forms $\mathbf{\Phi}_i$, $\mathbf{\Psi}_i$ in the respective media

1 and 2 filling the whole space. The two media do not have any restrictions, they may be isotropic, anisotropic or bi-anisotropic and homogeneous or inhomogeneous. However, we assume a linear medium everywhere. Now let us consider the possibility of making a new combined electromagnetic source-field system by joining together part of the system 1 in region V_1 and part of the system 2 in the region V_2. The two regions are separated by a nonmoving surface S (Figure 5.4). Now we will show that, to make a single electromagnetic source-field system, a surface source γ_{es}, γ_{ms} on S is required as kind of glue to join the two systems together.

The regions 1 and 2 are defined by two complementary pulse functions denoted by $P_1(\mathbf{r})$ and $P_2(\mathbf{r})$:

$$P_1(\mathbf{r}) = \begin{cases} 1, & \mathbf{r} \in V_1 \\ 0, & \mathbf{r} \in V_2 \\ 1/2, & \mathbf{r} \in S \end{cases} \tag{5.102}$$

$$P_2(\mathbf{r}) = \begin{cases} 0, & \mathbf{r} \in V_1 \\ 1, & \mathbf{r} \in V_2 \\ 1/2, & \mathbf{r} \in S \end{cases} \tag{5.103}$$

The pulse functions satisfy

$$P_1(\mathbf{r}) + P_2(\mathbf{r}) = 1. \tag{5.104}$$

The combined source-field system is defined by

$$\mathbf{\Phi} = \mathbf{\Phi}_1 P_1 + \mathbf{\Phi}_2 P_2, \tag{5.105}$$

$$\mathbf{\Psi} = \mathbf{\Psi}_1 P_1 + \mathbf{\Psi}_2 P_2, \tag{5.106}$$

$$\gamma_e = \gamma_{e1} P_1 + \gamma_{e2} P_2 + \gamma_{es}, \tag{5.107}$$

$$\gamma_m = \gamma_{m1} P_1 + \gamma_{m2} P_2 + \gamma_{ms}. \tag{5.108}$$

Here we have added the anticipated surface-source terms γ_{es}, γ_{ms} assumed to vanish outside the surface S, in case they are needed to make the combination system satisfy the Maxwell equations. Also, we assume that the original sources 1 and 2 have no delta singularities on S.

Let us now require that the combined system $\{\mathbf{\Phi}, \mathbf{\Psi}, \gamma_e, \gamma_m\}$ satisfy the Maxwell equations:

$$\begin{aligned} \mathbf{d} \wedge \mathbf{\Phi} - \gamma_m &= (\mathbf{d} \wedge \mathbf{\Phi}_1) P_1 + (\mathbf{d} \wedge \mathbf{\Phi}_2) P_2 \\ &\quad + (\mathbf{d} P_1) \wedge \mathbf{\Phi}_1 + (\mathbf{d} P_2) \wedge \mathbf{\Phi}_2 \\ &\quad - \gamma_{m1} P_1 - \gamma_{m2} P_2 - \gamma_{ms} \\ &= (\mathbf{d} P_1) \wedge \mathbf{\Phi}_1 + (\mathbf{d} P_2) \wedge \mathbf{\Phi}_2 - \gamma_{ms} = 0, \end{aligned} \tag{5.109}$$

$$\begin{aligned} \mathbf{d} \wedge \mathbf{\Psi} - \gamma_e &= (\mathbf{d} \wedge \mathbf{\Psi}_1) P_1 + (\mathbf{d} \wedge \mathbf{\Psi}_2) P_2 \\ &\quad + (\mathbf{d} P_1) \wedge \mathbf{\Psi}_1 + (\mathbf{d} P_2) \wedge \mathbf{\Psi}_2 \\ &\quad - \gamma_{e1} P_1 - \gamma_{e2} P_2 - \gamma_{es} \\ &= (\mathbf{d} P_1) \wedge \mathbf{\Psi}_1 + (\mathbf{d} P_2) \wedge \mathbf{\Psi}_2 - \gamma_{es} = 0. \end{aligned} \tag{5.110}$$

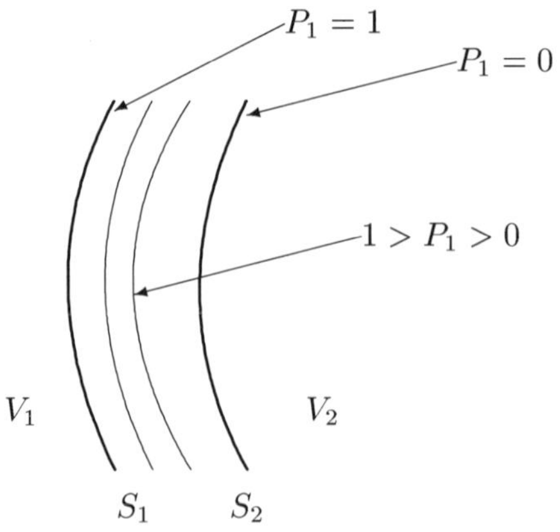

Fig. 5.5 *Surface S between regions V_1 and V_2 expanded to a layer. The one-form $\mathbf{d}P_1(\mathbf{r})$ is a set of surfaces between S_1 and S_2 which are squeezed to a single surface S.*

Here we have applied the Maxwell equations for the systems 1 and 2 to cancel four terms in each equation.

Because the pulse functions are constant outside S, the two one-forms $\mathbf{d}P_1(\mathbf{r})$ and $\mathbf{d}P_2(\mathbf{r})$ must have a singularity at the surface S and vanish outside. They are expressed compactly as

$$\mathbf{d}P_1(\mathbf{r}) = \mathbf{n}_1(\mathbf{r}), \quad \mathbf{d}P_2(\mathbf{r}) = \mathbf{n}_2(\mathbf{r}) = -\mathbf{n}_1(\mathbf{r}), \tag{5.111}$$

where $\mathbf{n}_1, \mathbf{n}_2$ are certain one-forms whose counterparts are Gibbsian unit normal vectors. Writing (5.109), (5.110) as

$$\mathbf{n}_1 \wedge \mathbf{\Phi}_1 + \mathbf{n}_2 \wedge \mathbf{\Phi}_2 = \gamma_{ms}, \tag{5.112}$$

$$\mathbf{n}_1 \wedge \mathbf{\Psi}_1 + \mathbf{n}_2 \wedge \mathbf{\Psi}_2 = \gamma_{es}, \tag{5.113}$$

the required additional electric and magnetic sources γ_{es}, γ_{ms} now become defined as surface sources. Equations (5.112) and (5.113) give conditions between the fields 1 and 2 at the surface S in symmetric form which helps in memorizing the formula. They also have a close similarity to the corresponding Gibbsian vector conditions.

A visual picture of the one-forms $\mathbf{n}_i(\mathbf{r}) = \mathbf{d}P_i(\mathbf{r})$ is obtained by broadening the sharp surface S to a layer of finite thickness between surfaces S_1 and S_2 separating V_1 and V_2 (Figure 5.5). Assuming $P_1(\mathbf{r})$ as a function changing continuously from 0 on S_2 to 1 on S_1, $\mathbf{n}_1(\mathbf{r})$ becomes a one-form defined in the region between S_1 and S_2. Recalling that a dual vector $\boldsymbol{\alpha} = \mathbf{d}(\boldsymbol{\alpha}|\mathbf{r})$ can be visualized by a set of parallel planes on which $\boldsymbol{\alpha}|\mathbf{r}$ has a constant value, $\mathbf{n}_1 = \mathbf{d}P_1(\mathbf{r})$ similarly consists of a set of surfaces $P_1(\mathbf{r})$ = constant. The surfaces of the one-form $\mathbf{n}_2(\mathbf{r})$ are the same with opposite orientation. When S_1 and S_2 approach to become a single surface S, $\mathbf{n}_1(\mathbf{r})$ can be visualized by an infinitely dense set of constant P_1 surfaces in S, a representation of a surface-delta function. In Gibbsian vector formalism, one introduces unit vectors normal to the surface; that has the drawback that, at any corner points of the surface with a discontinuous tangent, the normal direction is not unique. The one-form counterpart $\mathbf{n}_1(\mathbf{r})$ does not have this defect, since it is defined on the surface S without defining any normal direction.

5.2.2 Interface conditions

Equations (5.112), (5.113) can be interpreted in many ways. Let us first assume that the surface sources γ_{es}, γ_{ms} are given, whence the equations give conditions for the discontinuity of fields at the inteface of media 1 and 2. As a special case, if the surface sources vanish, they become the continuity conditions

$$\mathbf{n}_1 \wedge \mathbf{\Phi}_1 + \mathbf{n}_2 \wedge \mathbf{\Phi}_2 = 0, \tag{5.114}$$

$$\mathbf{n}_1 \wedge \mathbf{\Psi}_1 + \mathbf{n}_2 \wedge \mathbf{\Psi}_2 = 0. \tag{5.115}$$

Expanding (5.112) and (5.113) in terms of three-dimensional fields and sources as

$$\mathbf{n}_1 \wedge (\mathbf{B}_1 + \mathbf{E}_1 \wedge \mathbf{d}\tau) + \mathbf{n}_2 \wedge (\mathbf{B}_2 + \mathbf{E}_2 \wedge \mathbf{d}\tau) = \varrho_{ms} - \mathbf{J}_{ms} \wedge \mathbf{d}\tau, \tag{5.116}$$

$$\mathbf{n}_1 \wedge (\mathbf{D}_1 - \mathbf{H}_1 \wedge \mathbf{d}\tau) + \mathbf{n}_2 \wedge (\mathbf{D}_2 - \mathbf{H}_2 \wedge \mathbf{d}\tau) = \varrho_{es} - \mathbf{J}_{es} \wedge \mathbf{d}\tau \tag{5.117}$$

and separating the spatial and temporal components, we arrive at the three-dimensional conditions

$$\mathbf{n}_1 \wedge \mathbf{B}_1 + \mathbf{n}_2 \wedge \mathbf{B}_2 = \varrho_{ms}, \tag{5.118}$$

$$\mathbf{n}_1 \wedge \mathbf{E}_1 + \mathbf{n}_2 \wedge \mathbf{E}_2 = -\mathbf{J}_{ms}, \tag{5.119}$$

$$\mathbf{n}_1 \wedge \mathbf{D}_1 + \mathbf{n}_2 \wedge \mathbf{D}_2 = \varrho_{es}, \tag{5.120}$$

$$\mathbf{n}_1 \wedge \mathbf{H}_1 + \mathbf{n}_2 \wedge \mathbf{H}_2 = \mathbf{J}_{es}. \tag{5.121}$$

These correspond to the following classical interface conditions written for Gibbsian vector fields and scalar or vector surface sources with $\mathbf{n}_1, \mathbf{n}_2$ denoting normal unit vectors pointing toward media 1 and 2, respectively:

$$\mathbf{n}_1 \cdot \mathbf{B}_1 + \mathbf{n}_2 \cdot \mathbf{B}_2 = \varrho_{ms}, \tag{5.122}$$

$$\mathbf{n}_1 \times \mathbf{E}_1 + \mathbf{n}_2 \times \mathbf{E}_2 = -\mathbf{J}_{ms}, \tag{5.123}$$

$$\mathbf{n}_1 \cdot \mathbf{D}_1 + \mathbf{n}_2 \cdot \mathbf{D}_2 = \varrho_{es}, \tag{5.124}$$

$$\mathbf{n}_1 \times \mathbf{H}_1 + \mathbf{n}_2 \times \mathbf{H}_2 = \mathbf{J}_{es}. \tag{5.125}$$

The symmetric form helps in memorizing the formulas.

Example Let us study the geometrical content of the surface sources by considering one of the equations. Taking (5.120) as an example, the electric surface charge three-form is related to the field $\mathbf{D}$ two-forms by

$$\varrho_{es} = \mathbf{n}_1 \wedge \mathbf{D}_\Delta, \qquad \mathbf{D}_\Delta = \mathbf{D}_1 - \mathbf{D}_2. \tag{5.126}$$

The two-form $\mathbf{D}_\Delta$ fills all space and $\mathbf{n}_1 = \mathbf{d}P_1$ picks its value at the surface S. Considering a point $\mathbf{r}$ on S, the one-form $\mathbf{n}_1$ defines a local tangent plane, say the xy plane. Hence, it is a multiple of the one-form $\mathbf{d}z$ and satisfies $\mathbf{d}z \wedge \mathbf{n}_1 = 0$ at the point $\mathbf{r}$. Expanding the two-form $\mathbf{D}_\Delta$ at $\mathbf{r}$ as

$$\mathbf{D}_\Delta = D_x \mathbf{d}y \wedge \mathbf{d}z + D_y \mathbf{d}z \wedge \mathbf{d}x + D_z \mathbf{d}x \wedge \mathbf{d}y, \tag{5.127}$$

the charge three-form at the point $\mathbf{r}$ becomes

$$\varrho_{es} = \mathbf{n}_1 \wedge \mathbf{D}_\Delta = D_z \mathbf{n}_1 \wedge \mathbf{d}x \wedge \mathbf{d}y, \tag{5.128}$$

whence its numerical value equals the D_z component of the difference two-form $\mathbf{D}_\Delta$. This corresponds to the scalar value $\varrho_{es} = \mathbf{n}_1 \cdot (\mathbf{D}_1 - \mathbf{D}_2)$ of the Gibbsian formalism. Other source components can be handled in a similar way.

5.2.3 Boundary conditions

Boundary can be understood as a special case of the interface when the fields and sources are identically zero on the other side of the surface$S$, say V_2. In this case S acts as a boundary to the fields of region V_1. Setting $\mathbf{\Phi}_2 = \mathbf{\Psi}_2 = 0$, the conditions (5.112) – (5.113) become

$$\mathbf{n}_1 \wedge \mathbf{\Phi}_1 = \gamma_{ms}, \tag{5.129}$$

$$\mathbf{n}_1 \wedge \mathbf{\Psi}_1 = \gamma_{es}, \tag{5.130}$$

or, in their equivalent three-dimensional form,

$$\mathbf{n}_1 \wedge \mathbf{B}_1 = \varrho_{ms}, \tag{5.131}$$

$$\mathbf{n}_1 \wedge \mathbf{E}_1 = -\mathbf{J}_{ms}, \tag{5.132}$$

$$\mathbf{n}_1 \wedge \mathbf{D}_1 = \varrho_{es}, \tag{5.133}$$

$$\mathbf{n}_1 \wedge \mathbf{H}_1 = \mathbf{J}_{es}. \tag{5.134}$$

This tells us that if we want to terminate an electromagnetic field at a surface S, the sources above must be placed on S to cancel the fields penetrating into V_2.

PEC and PMC conditions The surface sources on the right-hand sides of (5.129) and (5.130) may be active sources canceling the fields for any medium 2 or they may appear because of medium 2 whose physical processes make the fields vanish. For example, if the medium 2 is made of perfect electrical conductor (PEC), it cannot support any magnetic charges and magnetic currents. In this case, $\gamma_{ms} = 0$ and the conditions (5.129) and (5.131), (5.132) become simpler:

$$\mathbf{n}_1 \wedge \mathbf{\Phi}_1 = 0, \tag{5.135}$$

or, in three-dimensional representation,

$$\mathbf{n}_1 \wedge \mathbf{B}_1 = 0, \quad \mathbf{n}_1 \wedge \mathbf{E}_1 = 0. \tag{5.136}$$

These are the PEC boundary conditions in differential-form language. In this case, there must exist electric charge and current at the surface as shown by (5.130) or (5.133) and (5.134). Correspondingly, medium 2 made of perfect magnetic conductor (PMC) requires vanishing of electric charges and currents, whence the corresponding boundary conditions are

$$\mathbf{n}_1 \wedge \mathbf{\Psi}_1 = 0, \tag{5.137}$$

$$\mathbf{n}_1 \wedge \mathbf{D}_1 = 0, \quad \mathbf{n}_1 \wedge \mathbf{H}_1 = 0 \tag{5.138}$$

and there must exist magnetic charge and current at the boundary as given by (5.129) or (5.131) and (5.132). More general boundary conditions will be discussed in the subsequent sections.

5.2.4 Huygens' principle

Another way to interpret equations (5.112) and (5.113) leads to the famous Huygens' principle. This assigns equivalent sources on a surface to replace true sources behind that surface. To formulate the principle, let us assume that the sources are confined in the region V_1 bounded by the surface S. Let us try to find surface sources on S which give rise to the same fields in V_2 as the original sources in V_1. Actually, we already know that the surface sources γ_{es}, γ_{ms} defined by (5.129), (5.130) create a field in V_2 which cancels that from the sources γ_{e1}, γ_{m1} in V_1 since the combined field vanishes. Thus, we simply change the signs of the surface sources and define the Huygens sources [2,40] compactly as (Figure 5.6)

$$\gamma_{mH} = -\mathbf{n}_1 \wedge \mathbf{\Phi}_1 = -\mathbf{d}P_1 \wedge \mathbf{\Phi}_1, \tag{5.139}$$

$$\gamma_{eH} = -\mathbf{n}_1 \wedge \mathbf{\Psi}_1 = -\mathbf{d}P_1 \wedge \mathbf{\Psi}_1. \tag{5.140}$$

To check this result, let us consider the field radiated by the difference of the original sources and the Huygens sources. In this case the difference field $\mathbf{\Phi}_o, \mathbf{\Psi}_o$ satisfies the symmetric equations

$$\mathbf{d} \wedge \mathbf{\Phi}_o = \gamma_{m1} - \gamma_{mH} = \gamma_{m1} + \mathbf{n}_1 \wedge \mathbf{\Phi}_1, \tag{5.141}$$

$$\mathbf{d} \wedge \mathbf{\Psi}_o = \gamma_{e1} - \gamma_{eH} = \gamma_{e1} + \mathbf{n}_1 \wedge \mathbf{\Psi}_1. \tag{5.142}$$

Replacing the original sources by

$$\gamma_{m1} = \mathbf{d} \wedge \mathbf{\Phi}_1 = P_1 \mathbf{d} \wedge \mathbf{\Phi}_1, \tag{5.143}$$

$$\gamma_{e1} = \mathbf{d} \wedge \mathbf{\Psi}_1 = P_1 \mathbf{d} \wedge \mathbf{\Psi}_1, \tag{5.144}$$

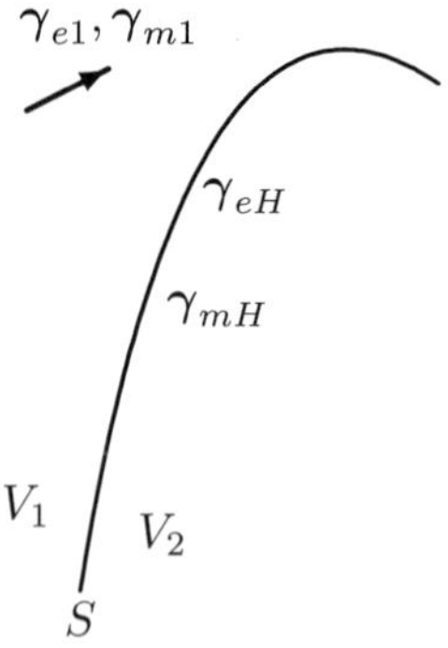

Fig. 5.6 *Huygens' principle replaces the sources γ_{e1}, γ_{m1} in region V_1 by equivalent sources γ_{eH}, γ_{mH} on surface S, producing the same fields $\mathbf{\Phi}_2, \mathbf{\Psi}_2$ in region V_2.*

where the P_1 functions can be inserted because the sources vanish in V_2, (5.141) and (5.142) can be expressed as

$$\mathbf{d} \wedge \mathbf{\Phi}_o = P_1 \mathbf{d} \wedge \mathbf{\Phi}_1 + \mathbf{n}_1 \wedge \mathbf{\Phi}_1 = \mathbf{d} \wedge (P_1 \mathbf{\Phi}_1), \quad (5.145)$$

$$\mathbf{d} \wedge \mathbf{\Psi}_o = P_1 \mathbf{d} \wedge \mathbf{\Psi}_1 + \mathbf{n}_1 \wedge \mathbf{\Psi} = \mathbf{d} \wedge (P_1 \mathbf{\Psi}_1). \quad (5.146)$$

If we write these as

$$\mathbf{d} \wedge (\mathbf{\Phi}_o - P_1 \mathbf{\Phi}_1) = 0, \quad (5.147)$$

$$\mathbf{d} \wedge (\mathbf{\Psi}_o - P_1 \mathbf{\Psi}_1) = 0 \quad (5.148)$$

and assume uniqueness of solution for the fields (no fields for no sources), the fields in square brackets must vanish because they have no sources on the right-hand side. Thus, the original sources minus the Huygens sources create the fields $\mathbf{\Phi}_o = P_1 \mathbf{\Phi}_1$, $\mathbf{\Psi}_o = P_1 \mathbf{\Psi}_1$ which vanish in V_2 because of the function P_1. Another consequence of this is that the Huygens sources γ_{eH}, γ_{mH} alone, as defined by (5.139) and (5.140), produce no field in V_1.

As can be noted, the electromagnetic medium plays no role in deriving Huygens' principle. Actually, the principle is independent of the medium and equally valid for all kinds of media, be they homogeneous or inhomogeneous, isotropic or anisotropic, linear or nonlinear.

Problems

5.2.1 Assume space regions V_1 and V_2 separated by a layer region V_o between two surfaces S_1, S_2 like in Figure 5.5 and defining $P_1(\mathbf{r}) = 1 - P_2(\mathbf{r})$ so that it changes continuously from 1 at S_1 to 0 at S_2. Define the boundary conditions so that the additional sources are volume sources in the layer region V_o.

5.2.2 Same as above but assume P_1 discontinuous at S_1 and S_2 so that $P_1(\mathbf{r}) = 1/2$ for $\mathbf{r} \in V_o$. In this case boundary sources lie on the two surfaces S_1 and S_2. Formulate Huygens' principle for fields in V_2.

5.3 POWER CONDITIONS

5.3.1 Three-dimensional formalism

Density of power flow (watts/m^2) in an electromagnetic field is represented in Gibbsian notation by the Poynting vector $\mathbf{S} = \mathbf{E} \times \mathbf{H}$. Because it is a quantity integrable over surfaces, its counterpart in differential forms is the Poynting two-form

$$\mathbf{S} = \mathbf{E} \wedge \mathbf{H}. \tag{5.149}$$

The Poynting vector is known to be associated with energy density W and power loss (Joule heat) $\mathbf{E} \cdot \mathbf{J}$ through the power balance equation

$$\nabla \cdot \mathbf{S} = -\partial_t W - \mathbf{E} \cdot \mathbf{J}. \tag{5.150}$$

Because the Gibbsian energy density is a scalar integrable over volumes, its counterpart is a three-form $\mathbf{w}$.

The power balance equation for differential forms in three-dimensional representation can be derived by starting from the spatial differential $\mathbf{d}_{\mathbf{E}}$ operating on the Poynting two-form and substituting from the Maxwell equations:

$$\begin{aligned} \mathbf{d}_{\mathbf{E}} \wedge \mathbf{S} &= \mathbf{d}_{\mathbf{E}} \wedge (\mathbf{E} \wedge \mathbf{H}) \\ &= (\mathbf{d}_{\mathbf{E}} \wedge \mathbf{E}) \wedge \mathbf{H} - \mathbf{E} \wedge (\mathbf{d}_{\mathbf{E}} \wedge \mathbf{H}) & (5.151) \\ &= -(\partial_\tau \mathbf{B}) \wedge \mathbf{H} - \mathbf{E} \wedge \partial_\tau \mathbf{D} - \mathbf{E} \wedge \mathbf{J}_e - \mathbf{J}_m \wedge \mathbf{H}. & (5.152) \end{aligned}$$

Here we have taken into account both electric and magnetic current sources. In analogy to (5.150), we would expect an equation of the form

$$\mathbf{d}_{\mathbf{E}} \wedge \mathbf{S} = -\partial_\tau \mathbf{w} - \mathbf{E} \wedge \mathbf{J}_e - \mathbf{H} \wedge \mathbf{J}_m. \tag{5.153}$$

However, there is a restriction concerning the medium. The energy-density three-form $\mathbf{w}$ cannot be defined unless the medium is linear and satisfies a certain symmetry condition.

Medium condition Let us consider the general linear medium using the three-dimensional super-form matrix formalism (4.81),

$$\begin{pmatrix} \mathbf{D} \\ \mathbf{B} \end{pmatrix} = \begin{pmatrix} \overline{\overline{\epsilon}} & \overline{\overline{\xi}} \\ \overline{\overline{\zeta}} & \overline{\overline{\mu}} \end{pmatrix} | \begin{pmatrix} \mathbf{E} \\ \mathbf{H} \end{pmatrix}, \quad \text{or} \quad \mathsf{d} = \overline{\overline{\mathcal{M}}}|\mathsf{e}. \tag{5.154}$$

We now express the four medium dyadics which are elements of the space $\mathbb{F}_2\mathbb{E}_1$ in terms of "Gibbsian" dyadics (5.6), elements of the space $\mathbb{E}_1\mathbb{E}_1$, in terms of the dual trivector ε_{123} as

$$\overline{\overline{\mathcal{M}}} = \varepsilon_{123}\lfloor\overline{\overline{\mathcal{M}}}_g = \begin{pmatrix} \varepsilon_{123}\lfloor\overline{\overline{\epsilon}}_g & \varepsilon_{123}\lfloor\overline{\overline{\xi}}_g \\ \varepsilon_{123}\lfloor\overline{\overline{\zeta}}_g & \varepsilon_{123}\lfloor\overline{\overline{\mu}}_g \end{pmatrix}. \tag{5.155}$$

Applying the three-dimensional identity

$$\boldsymbol{\alpha} \wedge (\varepsilon_{123}\lfloor\mathbf{a}) = \varepsilon_{123}(\boldsymbol{\alpha}|\mathbf{a}), \tag{5.156}$$

where $\mathbf{a}$ is a vector and $\boldsymbol{\alpha}$ a dual vector, we can write the three-form expression (5.152) in matrix form as

$$\begin{aligned} \mathbf{E} \wedge \partial_\tau \mathbf{D} + (\partial_\tau \mathbf{B}) \wedge \mathbf{H} &= \mathbf{E} \wedge \partial_\tau \mathbf{D} + \mathbf{H} \wedge \partial_\tau \mathbf{B} \\ &= (\mathbf{E}\;\mathbf{H}) \wedge \partial_\tau \begin{pmatrix} \mathbf{D} \\ \mathbf{B} \end{pmatrix} \\ &= (\mathbf{E}\;\mathbf{H}) \wedge (\varepsilon_{123}\lfloor\overline{\overline{\mathcal{M}}}_g)|\partial_\tau \begin{pmatrix} \mathbf{E} \\ \mathbf{H} \end{pmatrix} \\ &= \varepsilon_{123}\left[(\mathbf{E}\;\mathbf{H})|\overline{\overline{\mathcal{M}}}_g|\partial_\tau \begin{pmatrix} \mathbf{E} \\ \mathbf{H} \end{pmatrix}\right]. \end{aligned} \tag{5.157}$$

In shorthand the term in square brackets equals

$$\mathsf{e}|\overline{\overline{\mathcal{M}}}_g|\partial_\tau \mathsf{e} = \partial_\tau(\mathsf{e}|\overline{\overline{\mathcal{M}}}_g|\mathsf{e}) - \mathsf{e}|\overline{\overline{\mathcal{M}}}_g^T|\partial_\tau \mathsf{e}, \tag{5.158}$$

which can be rewritten as

$$\partial_\tau(\mathsf{e}|\overline{\overline{\mathcal{M}}}_g|\mathsf{e}) = \mathsf{e}|(\overline{\overline{\mathcal{M}}}_g + \overline{\overline{\mathcal{M}}}_g^T)|\partial_\tau \mathsf{e}. \tag{5.159}$$

From this we conclude that if the Gibbsian medium dyadic matrix $\overline{\overline{\mathcal{M}}}_g$ is symmetric,

$$\overline{\overline{\mathcal{M}}}_g^T = \overline{\overline{\mathcal{M}}}_g, \tag{5.160}$$

that is, if the four $\mathbb{E}_1\,\mathbb{E}_1$ Gibbsian medium dyadics satisfy

$$\overline{\overline{\epsilon}}_g^T = \overline{\overline{\epsilon}}_g, \quad \overline{\overline{\xi}}_g^T = \overline{\overline{\zeta}}_g, \quad \overline{\overline{\zeta}}_g^T = \overline{\overline{\xi}}_g, \quad \overline{\overline{\mu}}_g^T = \overline{\overline{\mu}}_g, \tag{5.161}$$

we can write

$$\mathsf{e}|\overline{\overline{\mathcal{M}}}_g|\partial_\tau \mathsf{e} = \partial_\tau \frac{1}{2}(\mathsf{e}|\overline{\overline{\mathcal{M}}}_g|\mathsf{e}). \tag{5.162}$$

Substituting in the right-hand side of (5.152), we have

$$\begin{aligned} \mathbf{E} \wedge \partial_\tau \mathbf{D} + (\partial_\tau \mathbf{B}) \wedge \mathbf{H} &= \varepsilon_{123}\mathsf{e}|\overline{\overline{\mathcal{M}}}_g|\partial_\tau \mathsf{e} \\ &= \frac{\varepsilon_{123}}{2}\partial_\tau(\mathsf{e}|\overline{\overline{\mathcal{M}}}_g|\mathsf{e}) = \partial_\tau \mathbf{w}, \end{aligned} \tag{5.163}$$

from which the energy-density three-form $\mathbf{w}$ of (5.153) can be identified as

$$\begin{aligned} \mathbf{w} &= \frac{\varepsilon_{123}}{2}(\mathsf{e}|\overline{\overline{\mathcal{M}}}_g|\mathsf{e}) = \frac{1}{2}(\mathsf{e} \wedge \overline{\overline{\mathcal{M}}}|\mathsf{e}) \\ &= \frac{1}{2}(\mathsf{e} \wedge \mathsf{d}) = \frac{1}{2}(\mathbf{E} \wedge \mathbf{D} + \mathbf{H} \wedge \mathbf{B}). \end{aligned} \tag{5.164}$$

Interpretation Integrating the divergence of the Poynting two-form over a volume $\mathcal{V}$ as

$$\mathbf{d_E}\wedge\mathbf{S}|\mathcal{V} = \mathbf{S}|\partial\mathcal{V} = -\partial_\tau\mathbf{w}|\mathcal{V} - (\mathbf{E}\wedge\mathbf{J}_e)|\mathcal{V} - (\mathbf{H}\wedge\mathbf{J}_m)|\mathcal{V}, \tag{5.165}$$

its terms can be given the following interpretation. $\mathbf{S}|\partial\mathcal{V}$ represents power flow out of the volume through its boundary surface $\partial\mathcal{V}$. The term $-\partial_\tau\mathbf{w}|\mathcal{V}$ gives the decrease of electromagnetic energy in the volume. Finally, $-(\mathbf{E}\wedge\mathbf{J}_e)|\mathcal{V} - (\mathbf{H}\wedge\mathbf{J}_m)|\mathcal{V}$ represents the loss of electric and magnetic energy in the volume as Joule heat. Thus, (5.153) is the equation for the conservation of electromagnetic energy. For other media not satisfying (5.160), equation (5.152) is still valid but it does not allow the same interpretation as (5.153).

Symmetry condition An interesting conclusion for the medium satisfying the symmetry condition (5.160) is obtained by considering the four-form $\boldsymbol{\pi}\in\mathbb{F}_4$, a function of field two-forms whose subscripts a and b refer to different sources,

$$\begin{aligned}\boldsymbol{\pi} &= \boldsymbol{\Phi}_a\wedge\boldsymbol{\Psi}_b - \boldsymbol{\Phi}_b\wedge\boldsymbol{\Psi}_a \\ &= (\mathbf{E}_a\wedge\mathbf{D}_b - \mathbf{E}_b\wedge\mathbf{D}_a + \mathbf{H}_a\wedge\mathbf{B}_b - \mathbf{H}_b\wedge\mathbf{B}_a)\wedge\mathbf{d}\tau.\end{aligned} \tag{5.166}$$

The corresponding scalar can be expanded as

$$\begin{aligned}\mathbf{e}_N|\boldsymbol{\pi} &= \mathbf{e}_{123}|(\mathbf{E}_a\wedge\mathbf{D}_b - \mathbf{E}_b\wedge\mathbf{D}_a + \mathbf{H}_a\wedge\mathbf{B}_b - \mathbf{H}_b\wedge\mathbf{B}_a) \\ &= \mathbf{E}_a|\mathbf{e}_{123}\lfloor\mathbf{D}_b - \mathbf{E}_b|\mathbf{e}_{123}\lfloor\mathbf{D}_a + \mathbf{H}_a|\mathbf{e}_{123}\lfloor\mathbf{B}_b - \mathbf{H}_b|\mathbf{e}_{123}\lfloor\mathbf{B}_a \\ &= \mathbf{E}_a|(\overline{\overline{\epsilon}}_g - \overline{\overline{\epsilon}}_g^T)|\mathbf{E}_b + \mathbf{H}_a|(\overline{\overline{\mu}}_g - \overline{\overline{\mu}}_g^T)|\mathbf{H}_b \\ &\quad + \mathbf{E}_a|(\overline{\overline{\xi}}_g - \overline{\overline{\zeta}}_g^T)|\mathbf{H}_2 + \mathbf{H}_a|(\overline{\overline{\zeta}}_g - \overline{\overline{\xi}}_g^T)|\mathbf{E}_2.\end{aligned} \tag{5.167}$$

From this it is seen that if (5.161) is valid, the four-form $\boldsymbol{\pi}$ vanishes and the following symmetry relation between the field two-forms arises:

$$\boldsymbol{\Phi}_a\wedge\boldsymbol{\Psi}_b = \boldsymbol{\Phi}_b\wedge\boldsymbol{\Psi}_a. \tag{5.168}$$

5.3.2 Four-dimensional formalism

In the four-dimensional representation the energy density three-form and the Poynting two-form are combined to the energy-power density three-form

$$\mathbf{W} = \mathbf{w} - \mathbf{S}\wedge\mathbf{d}\tau, \tag{5.169}$$

whose existence again presumes the medium condition (5.160) to be valid. Noting that $\mathbf{d_E}\wedge\mathbf{w} = 0$ because three-dimensional four-forms vanish, we can expand (5.153) as

$$\begin{aligned}(\mathbf{E}\wedge\mathbf{J}_e + \mathbf{H}\wedge\mathbf{J}_m)\wedge\mathbf{d}\tau &= -(\mathbf{d_E}\wedge\mathbf{S} + \partial_\tau\mathbf{w})\wedge\mathbf{d}\tau \\ &= \partial_\tau\mathbf{d}\tau\wedge\mathbf{w} - \mathbf{d_E}\wedge\mathbf{S}\wedge\mathbf{d}\tau \\ &= (\mathbf{d_E} + \partial_\tau\mathbf{d}\tau)\wedge(\mathbf{w} - \mathbf{S}\wedge\mathbf{d}\tau) \\ &= \mathbf{d}\wedge\mathbf{W}.\end{aligned} \tag{5.170}$$

Expanding further as

$$\begin{aligned} & \mathbf{E} \wedge (\mathbf{J}_e \wedge \mathbf{d}\tau) + \mathbf{H} \wedge (\mathbf{J}_m \wedge \mathbf{d}\tau) \\ = & \ -\mathbf{E} \wedge \gamma_e - \mathbf{H} \wedge \gamma_m \\ = & \ (\mathbf{\Phi} \lfloor \mathbf{e}_\tau) \wedge \gamma_e - (\mathbf{\Psi} \lfloor \mathbf{e}_\tau) \wedge \gamma_m, \end{aligned} \tag{5.171}$$

the power balance equation in four-dimensional quantities becomes

$$\mathbf{d} \wedge \mathbf{W} = (\mathbf{\Phi} \lfloor \mathbf{e}_\tau) \wedge \gamma_e - (\mathbf{\Psi} \lfloor \mathbf{e}_\tau) \wedge \gamma_m. \tag{5.172}$$

This is a four-form equation. Multiplying by $\mathbf{e}_N \rfloor$ it can be transformed to an equivalent scalar equation containing exactly the same information as the three-dimensional power-balance equation (5.153). When there are no sources, the energy-power density three-form $\mathbf{W}$ obeys the simple conservation law

$$\mathbf{d} \wedge \mathbf{W} = 0. \tag{5.173}$$

The form of this simple law suggests that there must exist a power-potential two-form Ξ such that $\mathbf{W} = \mathbf{d} \wedge \Xi$.

5.3.3 Complex power relations

In time-harmonic field analysis, complex time variable $e^{j\omega t} = e^{jk_o\tau}$ is suppressed in the linear expressions. Quadratic expressions like the Poynting vector contain a constant (time-independent) part and a time-harmonic part varying with double frequency:

$$\Re\{fe^{jk_o\tau}\}\Re\{ge^{jk_o\tau}\} = \frac{1}{2}\Re\{fg^*\} + \frac{1}{2}\Re\{fge^{j2k_o\tau}\}. \tag{5.174}$$

Here f and g may be any complex multivectors or multiforms and there may be any multiplication sign in between. Normally, one is interested only in the time-average quantity and the last time-harmonic term is omitted. It is customary to do the analysis for the full complex quantity $fg^*/2$ even if only its real part is of final interest.

For complex quantities with suppressed time dependence the field equations are best handled in three-dimensional differential forms. Since the operator $\mathbf{d}$ does not operate on time, $\mathbf{d}_\mathrm{E}$ of previous expressions can be replaced by $\mathbf{d}$. The complex Poynting vector defined by $\mathbf{S} = (1/2)\mathbf{E} \times \mathbf{H}^*$ now becomes the complex Poynting two-form

$$\mathbf{S} = \frac{1}{2}\mathbf{E} \wedge \mathbf{H}^* \tag{5.175}$$

whose real part represents the average power flow. Allowing also magnetic currents $\mathbf{J}_m$ to enter the analysis, the complex power balance equation now becomes

$$\begin{aligned} \mathbf{d} \wedge \mathbf{S} &= \frac{1}{2}\mathbf{d} \wedge (\mathbf{E} \wedge \mathbf{H}^*) = \frac{1}{2}(\mathbf{d} \wedge \mathbf{E}) \wedge \mathbf{H}^* - \frac{1}{2}\mathbf{E} \wedge (\mathbf{d} \wedge \mathbf{H}^*) \\ &= -jk_o\frac{1}{2}(\mathbf{B} \wedge \mathbf{H}^* - \mathbf{E} \wedge \mathbf{D}^*) - \frac{1}{2}(\mathbf{E} \wedge \mathbf{J}_e^* + \mathbf{H}^* \wedge \mathbf{J}_m). \end{aligned} \tag{5.176}$$

Taking the real part leaves us with

$$\begin{aligned}\Re\{\mathbf{d}\wedge\mathbf{S}\} &= \frac{jk_o}{4}(\mathbf{B}^*\wedge\mathbf{H}-\mathbf{B}\wedge\mathbf{H}^*+\mathbf{E}^*\wedge\mathbf{D}-\mathbf{E}\wedge\mathbf{D}^*)\\ &\quad -\frac{1}{4}(\mathbf{E}\wedge\mathbf{J}_e^*+\mathbf{J}_e\wedge\mathbf{E}^*+\mathbf{H}\wedge\mathbf{J}_m^*+\mathbf{J}_m\wedge\mathbf{H}^*). \quad (5.177)\end{aligned}$$

In terms of three-dimensional super forms and defining the $\wedge$ product as $\mathsf{e}\wedge\mathsf{d} = \mathbf{E}\wedge\mathbf{D}+\mathbf{H}\wedge\mathbf{B}$, this equals

$$\Re\{\mathbf{d}\wedge\mathbf{S}\} = \frac{jk_o}{4}(\mathsf{e}\wedge\mathsf{d}^* - \mathsf{e}^*\wedge\mathsf{d}) - \frac{1}{4}(\mathsf{e}\wedge\mathsf{j}^* + \mathsf{e}^*\wedge\mathsf{j}). \quad (5.178)$$

Inserting the medium dyadic matrix $\overline{\overline{\mathcal{M}}}$ in the modified Gibbsian form

$$\mathsf{d} = \overline{\overline{\mathcal{M}}}|\mathsf{e} = \varepsilon_{123}\lfloor\overline{\overline{\mathcal{M}}}_g|\mathsf{e}, \quad (5.179)$$

where the dyadic elements of $\overline{\overline{\mathcal{M}}}_g$ are in the space $\mathbb{E}_1\,\mathbb{E}_1$, with (5.156) this leads to

$$\Re\{\mathbf{d}\wedge\mathbf{S}\} = \varepsilon_{123}\frac{jk_o}{4}\mathsf{e}|(\overline{\overline{\mathcal{M}}}_g^* - \overline{\overline{\mathcal{M}}}_g^T)|\mathsf{e}^* - \frac{1}{4}(\mathsf{e}\wedge\mathsf{j}^* + \mathsf{e}^*\wedge\mathsf{j}). \quad (5.180)$$

The term after the $=$ sign corresponds to exchange of average power between the medium and the electromagnetic field when the sources are not present, $\mathsf{j}=0$. The medium is lossless (does not produce or consume energy) if the term vanishes for any possible fields e, in which case the medium must satisfy the condition

$$\overline{\overline{\mathcal{M}}}_g^T = \overline{\overline{\mathcal{M}}}_g^*, \quad (5.181)$$

that is, the Gibbsian medium dyadic matrix is Hermitian. For the four Gibbsian medium dyadics this corresponds to the relations

$$\overline{\overline{\epsilon}}_g^T = \overline{\overline{\epsilon}}_g^*, \quad \overline{\overline{\mu}}_g^T = \overline{\overline{\mu}}_g^*, \quad \overline{\overline{\xi}}_g^T = \overline{\overline{\zeta}}_g^*. \quad (5.182)$$

5.3.4 Ideal boundary conditions

A surface $\mathcal{S}$ can be called an ideal boundary if it does not let any electromagetic power pass trough [47]. In terms of the time-dependent Poynting vector $\mathbf{S}$ the ideal boundary conditions can be expressed as

$$\mathbf{S}|\Delta\mathcal{S} = (\mathbf{E}\wedge\mathbf{H})|\Delta\mathcal{S} = 0, \quad (5.183)$$

which must be satisfied for every surface element $\Delta\mathcal{S}$ on $\mathcal{S}$. The definition is extended to complex fields as

$$\mathbf{S}|\Delta\mathcal{S} = \frac{1}{2}(\mathbf{E}\wedge\mathbf{H}^*)|\Delta\mathcal{S} = 0. \quad (5.184)$$

Here the real part corresponds to the condition of suppressing the average power flow through the surface.

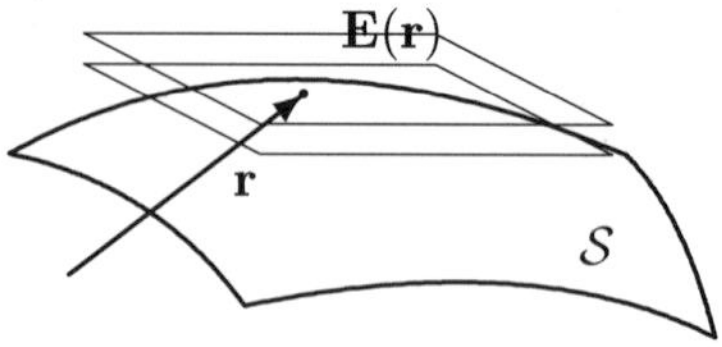

Fig. 5.7 *Field one-form* $\mathbf{E}$ *satisfying* $\mathbf{E}|\Delta\mathcal{S} = 0$ *at a point* $\mathbf{r}$ *is locally visualized as a set of planes parallel to the tangent plane of surface* $\mathcal{S}$ *at* $\mathbf{r}$.

PEC and PMC boundaries A perfect electric conductor (PEC) boundary is one for which (5.183) is satisfied with no restriction to the magnetic field $\mathbf{H}$. The electric field must then satisfy

$$\mathbf{H}|(\mathbf{E}\rfloor\Delta\mathcal{S}) = 0, \ \forall\mathbf{H}, \ \Rightarrow \ \mathbf{E}\rfloor\Delta\mathcal{S} = 0. \tag{5.185}$$

Taking the parallelogram $\Delta\mathcal{S} = \Delta\mathbf{r}_1 \wedge \Delta\mathbf{r}_2 \neq 0$ on the boundary, we have from the bac cab formula

$$\mathbf{E}\rfloor(\Delta\mathbf{r}_1 \wedge \Delta\mathbf{r}_2) = (\mathbf{E}|\Delta\mathbf{r}_1)\Delta\mathbf{r}_2 - (\mathbf{E}|\Delta\mathbf{r}_2)\Delta\mathbf{r}_1 = 0, \tag{5.186}$$

which is satisfied for all $\Delta\mathcal{S}$ only when $\mathbf{E}|\Delta\mathbf{r}_1 = 0$ and $\mathbf{E}|\Delta\mathbf{r}_2 = 0$. This implies that $\mathbf{E}|\Delta\mathbf{r} = 0$ for any $\Delta\mathbf{r}$ on the surface. Thus, the surface making the electric one-form $\mathbf{E}$ must be locally orthogonal to the surface $\mathcal{S}$ when orthogonality of a vector $\mathbf{a}$ and a dual vector $\boldsymbol{\alpha}$ is defined by $\mathbf{a}|\boldsymbol{\alpha} = 0$. Visualizing a one-form as a set of local parallel planes at each point, they must be parallel to the tangent plane of $\mathcal{S}$ at that point (Figure 5.7). Similarly, a perfect magnetic conductor (PMC) boundary requires another condition for the magnetic one-form:

$$\mathbf{H}\rfloor\Delta\mathcal{S} = 0. \tag{5.187}$$

Isotropic ideal boundary A generalized form of the previous ideal boundary conditions for real fields can be written as

$$(\alpha\mathbf{E} + \beta\mathbf{H})\rfloor\Delta\mathcal{S} = 0, \tag{5.188}$$

with scalars α, β depending on the nature of the surface. In fact, operating as

$$(\gamma\mathbf{E} + \delta\mathbf{H})|\{(\alpha\mathbf{E} + \beta\mathbf{H})\rfloor\Delta\mathcal{S}\} = (\gamma\beta - \delta\alpha)(\mathbf{E} \wedge \mathbf{H})|\mathcal{S} = 0, \tag{5.189}$$

with scalars γ, δ chosen so that $\gamma\beta - \delta\alpha \neq 0$, gives (5.183), the condition of no energy transmission through the surface $\mathcal{S}$.

Similar conditions are obtained for the complex fields as is seen if $\mathbf{H}$ is replaced by $\mathbf{H}^*$ in the above expressions. However, in the case of (5.188) it must be noted that the condition for the fields is not linear because it involves a condition between $\mathbf{E}$ and $\mathbf{H}^*$ [48].

Anisotropic ideal boundary An ideal boundary can also be anisotropic. Let us assume that, on the surface it locally has a special direction denoted by $\Delta\mathbf{r}_1$. The ideal boundary condition written as

$$\mathbf{S}|(\Delta\mathbf{r}_1 \wedge \Delta\mathbf{r}_2) = (\mathbf{S}\lfloor\Delta\mathbf{r}_1)|\Delta\mathbf{r}_2 = 0 \tag{5.190}$$

is now required to be valid for any $\Delta\mathbf{r}_2$ on the surface. This requires that on the surface we must have

$$\mathbf{S}\lfloor\Delta\mathbf{r}_1 = (\mathbf{E} \wedge \mathbf{H})\lfloor\Delta\mathbf{r}_1 = \mathbf{H}(\mathbf{E}|\Delta\mathbf{r}_1) - \mathbf{E}(\mathbf{H}|\Delta\mathbf{r}_1) = 0. \tag{5.191}$$

Assuming that $\mathbf{S} = \mathbf{E} \wedge \mathbf{H} \neq 0$, this splits in two conditions for the fields $\mathbf{E}$ and $\mathbf{H}$:

$$\mathbf{E}|\Delta\mathbf{r}_1 = 0, \quad \mathbf{H}|\Delta\mathbf{r}_1 = 0. \tag{5.192}$$

The same conditions are obtained for the time-harmonic case as well when $\Delta\mathbf{r}_1$ is assumed to be a real vector. In Gibbsian notation $\mathbf{E} \cdot \Delta\mathbf{r}_1 = \mathbf{H} \cdot \Delta\mathbf{r}_1 = 0$ implies $\Delta\mathbf{r}_1 \times (\mathbf{E} \times \mathbf{H}) = 0$, or the Poynting vector $\mathbf{S}$ is parallel to the local direction $\Delta\mathbf{r}_1$ on the surface. Similar result is valid for the complex Poynting vector. This kind of surface has been called the soft and hard (SHS) surface in the literature [47]. Its realization by a tuned corrugated surface is valid, however, only to a time-harmonic field for a certain frequency.

Problems

5.3.1 Find conditions for the parameters of the medium dyadic $\overline{\overline{\mathsf{M}}}$ which correspond to the symmetry condition (5.160) for the medium dyadic matrix $\overline{\overline{\mathcal{M}}}$.

5.3.2 Derive (5.172) by working through all the detailed steps.

5.3.3 Show that the anisotropic ideal surface condition (5.192) for time-harmonic fields is

$$\mathbf{E}|\Delta\mathbf{r}_c = 0, \quad \mathbf{H}|\Delta\mathbf{r}_c^* = 0,$$

where $\Delta\mathbf{r}_c = \Delta\mathbf{r}_1 + j\Delta\mathbf{r}_2$ is a complex vector on the surface $\mathcal{S}$ and $\Delta\mathcal{S} = \Delta\mathbf{r}_1 \wedge \Delta\mathbf{r}_2$.

5.4 THE LORENTZ FORCE LAW

The mechanical force between distributed electromagnetic sources obeys a law which in differential-form formalism can be best expressed in dyadic form.

5.4.1 Three-dimensional representation

Electric force In Gibbsian vector representation, the electrostatic force of the electric field $\mathbf{E}$[1] on a point charge Q is the vector

$$\mathbf{F}_e = Q\mathbf{E}. \tag{5.193}$$

The same expression appears to be valid also when both $\mathbf{F}_e$ and $\mathbf{E}$ are one-forms and Q is a scalar. However, things are not so simple when we consider the force on a distributed charge. The Gibbsian force density vector $\mathbf{f}_e$ created on the electric charge density ϱ_e given by

$$\mathbf{f}_e = \varrho_e \mathbf{E} \tag{5.194}$$

cannot be transformed so simply to differential-form representation because the charge density is now a three-form $\boldsymbol{\varrho}_e$. Because the point charge can be conceived as the integral of charge density three-form $\boldsymbol{\varrho}_e$ over a small volume element trivector $\Delta\mathbf{V}$, the resulting force one-form is

$$\Delta\mathbf{F}_e = \mathbf{E}(\boldsymbol{\varrho}_e|\Delta\mathbf{V}) = (\Delta\mathbf{V}|\boldsymbol{\varrho}_e)\mathbf{E}. \tag{5.195}$$

To obtain a formula resembling (5.194) as closely as possible, we define the electric force density as a $\mathbb{F}_3\mathbb{F}_1$ dyadic through the dyadic product of $\boldsymbol{\varrho}_e$ and $\mathbf{E}$,

$$\overline{\overline{\mathsf{f}}}_e = \boldsymbol{\varrho}_e \mathbf{E}. \tag{5.196}$$

The net electric force acting on a volume $\mathcal{V}$ is then obtained by integrating from the left as

$$\mathbf{F}_e = \mathcal{V}|\overline{\overline{\mathsf{f}}}_e = (\mathcal{V}|\boldsymbol{\varrho}_e)\mathbf{E}. \tag{5.197}$$

Here we must note that, in spite of the brackets, also $\mathbf{E}(\mathbf{r})$ is integrated and the result is constant in $\mathbf{r}$.

Magnetic force The magnetic force acting on a line current segment with the moment vector $\mathbf{p} = I\mathbf{L}$ (current I and directed length vector $\mathbf{L}$) in Gibbsian vector notation reads

$$\mathbf{F}_m = \mathbf{p} \times \mathbf{B} = I\mathbf{L} \times \mathbf{B}, \tag{5.198}$$

where $\mathbf{B}$ is the magnetic flux-density vector. For a distributed electric current with Gibbsian density vector $\mathbf{J}_e$, let us consider a small cylindrical volume ΔV of cross section ΔS and length ΔL, parallel to the current (Figure 5.8), which creates the magnetic force ΔF_m. The moment of the small current element is obviously

$$\Delta\mathbf{p} = \mathbf{J}_e \Delta S \Delta L = \mathbf{J}_e \Delta V, \tag{5.199}$$

which means that the vector $\mathbf{J}_e$ can be interpreted as the volume density of current moment. Thus, the Gibbsian magnetic force density is obtained as the limit

$$\mathbf{f}_m = \lim_{\Delta V \to 0} \frac{\Delta \mathbf{F}_m}{\Delta V} = \mathbf{J}_e \times \mathbf{B}. \tag{5.200}$$

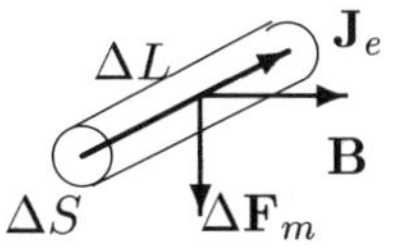

Fig. 5.8 *Gibbsian vector representation for magnetic force law.*

In differential-form language the electric current density is a two-form whence the magnetic moment corresponding to a small volume trivector $\Delta\mathbf{V}$ is a vector,

$$\Delta\mathbf{p} = \Delta\mathbf{V}\lfloor\mathbf{J}_e = \Delta\mathbf{V}|\bar{\bar{\mathsf{I}}}^{(3)T}\lfloor\mathbf{J}_e. \tag{5.201}$$

From the last expression we can identify the magnetic moment density as the dyadic quantity $\bar{\bar{\mathsf{I}}}^{(3)T}\lfloor\mathbf{J}_e \in \mathbb{F}_3\mathbb{E}_1$. Since the magnetic force $\Delta\mathbf{F}_m$ is a one-form (integrable over a line) which is linearly dependent on the moment vector $\Delta\mathbf{p}$ and the magnetic field two-form $\mathbf{B}$, the relation is a contraction and can be represented as

$$\Delta\mathbf{F}_m = \Delta\mathbf{p}\rfloor\mathbf{B} = (\Delta\mathbf{V}\lfloor\mathbf{J}_e)\rfloor\mathbf{B}. \tag{5.202}$$

Inserting a unit dyadic in between, from

$$= (\Delta\mathbf{V}\lfloor\mathbf{J}_e)|\bar{\bar{\mathsf{I}}}_{\mathbf{E}}^T\rfloor\mathbf{B} = \Delta\mathbf{V}|(\mathbf{J}_e\wedge\bar{\bar{\mathsf{I}}}_{\mathbf{E}}^T\rfloor\mathbf{B}), \tag{5.203}$$

the magnetic force density can finally be identified as the $\mathbb{F}_3\mathbb{F}_1$ dyadic

$$\bar{\bar{\mathsf{f}}}_m = \mathbf{J}_e\wedge\bar{\bar{\mathsf{I}}}_{\mathbf{E}}^T\rfloor\mathbf{B}. \tag{5.204}$$

This result is not very obvious from from the Gibbsian vector form of the magnetic force law (5.200). To check (5.204), let us see what corresponds to the fact that if the current and magnetic field are parallel Gibbsian vectors, the force (5.200) vanishes. Assuming that $\mathbf{J}_e$ and $\mathbf{B}$ are parallel two-forms (scalar multiples of the same two-form say Ξ) and applying the identity (2.107) in the form

$$\mathbf{J}_e\wedge\bar{\bar{\mathsf{I}}}_{\mathbf{E}}^T\rfloor\mathbf{B} = -\mathbf{B}\wedge\bar{\bar{\mathsf{I}}}_{\mathbf{E}}^T\rfloor\mathbf{J}_e \tag{5.205}$$

shows us that the magnetic force is zero because $\Xi\wedge\bar{\bar{\mathsf{I}}}_{\mathbf{E}}^T\rfloor\Xi$ vanishes for any Ξ. Thus, (5.204) is the dyadic counterpart of the classical vector formula (5.200).

Total force The Lorentz electromagnetic force-density, which in the vector language has the familiar expression

$$\mathbf{f} = \varrho_e\mathbf{E} + \mathbf{J}_e\times\mathbf{B}, \tag{5.206}$$

[1]Gibbsian vectors and corresponding differential forms for the same physical quantities are here expressed by the same symbols without fear of confusion.

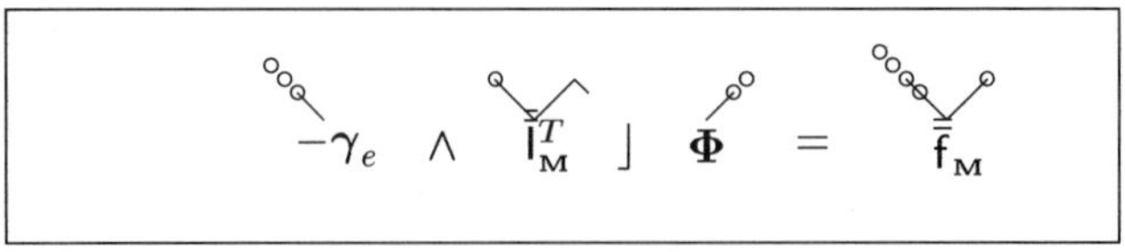

Fig. 5.9 *Visualization of the four-dimensional force-energy density dyadic.*

becomes in the differential form notation the dyadic [42],

$$\overline{\overline{\mathsf{f}}}_{\mathbf{E}} = \varrho_e \mathbf{E} + \mathbf{J}_e \wedge \overline{\overline{\mathsf{I}}}{}^T_{\mathbf{E}} \rfloor \mathbf{B}. \tag{5.207}$$

This expression could be added by terms corresponding to the magnetic charge and current, omitted here for simplicity.

5.4.2 Four-dimensional representation

The Euclidean expression (5.207) can now be extended to Minkowskian four-dimensional notation. In this case, the force density is understood as a quantity integrable over a four-volume in space-time denoted by the quadrivector element $\mathbf{e}_\tau \Delta\tau \wedge \Delta\mathbf{V}$. It turns out that the corresponding expression obeys the form of the following $\mathbb{F}_4\mathbb{F}_1$ dyadic (Figure 5.9):

$$\overline{\overline{\mathsf{f}}}_{\mathbf{M}} = -\gamma_e \wedge \overline{\overline{\mathsf{I}}}{}^T_{\mathbf{M}} \rfloor \boldsymbol{\Phi}. \tag{5.208}$$

Here $\overline{\overline{\mathsf{I}}}_{\mathbf{M}}$ denotes the four-dimensional unit dyadic. To check this compact expression, we insert the expansions for the four-current γ_e and the force-field two-form $\boldsymbol{\Phi}$ as

$$\overline{\overline{\mathsf{f}}}_{\mathbf{M}} = -(\varrho_e - \mathbf{J}_e \wedge \mathbf{d}\tau) \wedge (\overline{\overline{\mathsf{I}}}{}^T_{\mathbf{E}} + \mathbf{d}\tau \mathbf{e}_\tau)\rfloor(\mathbf{B} + \mathbf{E} \wedge \mathbf{d}\tau). \tag{5.209}$$

Noting that $\varrho_e \wedge \overline{\overline{\mathsf{I}}}{}^T_{\mathbf{E}} = 0$ and $\mathbf{e}_\tau \rfloor \mathbf{B} = 0$, this can be reduced to

$$\begin{aligned} \overline{\overline{\mathsf{f}}}_{\mathbf{M}} &= \mathbf{d}\tau \wedge (\varrho\mathbf{E} + \mathbf{J}_e \wedge \overline{\overline{\mathsf{I}}}{}^T_{\mathbf{E}} \rfloor \mathbf{B}) - \mathbf{d}\tau \wedge (\mathbf{J}_e \wedge \mathbf{E})\mathbf{d}\tau \\ &= \mathbf{d}\tau \wedge (\overline{\overline{\mathsf{f}}}_{\mathbf{E}} - (\mathbf{J}_e \wedge \mathbf{E})\mathbf{d}\tau). \end{aligned} \tag{5.210}$$

Integration over the quadrivector volume element $\mathbf{e}_\tau \Delta\tau \wedge \Delta\mathbf{V}$ gives the one-form

$$\mathbf{e}_\tau \Delta\tau \wedge \Delta\mathbf{V} | \overline{\overline{\mathsf{f}}}_{\mathbf{M}} = \Delta\mathbf{V} | \overline{\overline{\mathsf{f}}}_{\mathbf{E}} - \Delta\mathbf{V} | (\mathbf{J}_e \wedge \mathbf{E})\mathbf{d}\tau, \tag{5.211}$$

of which the first term (spatial part) is the three-dimensional force one-form. The second term (temporal part) corresponds to the integral of the Gibbsian $\mathbf{E} \cdot \mathbf{J}_e$ term, which gives the Joule power loss in watts. Thus, $\overline{\overline{\mathsf{f}}}_{\mathbf{M}}$ can be called the Lorentz force-energy density dyadic.

Problems

5.4.1 Derive (5.210) in detail.

5.4.2 Repeat the analysis leading to (5.207) by taking also the magnetic sources into account.

5.5 STRESS DYADIC

In four-dimensional formalism, the Poynting two-form and Lorentz force-energy can be combined to a compact quantity, the stress dyadic. Since the publication of Maxwell's *Treatise on Electricity and Magnetism* [56], the corresponding physical quantity has been called variously the stress tensor, the stress-energy tensor, the energy-momentum tensor or the stress-energy-momentum tensor, and it has been expressed in the literature in terms of Gibbsian dyadics or tensor (index) notation [34,38,55,56,62,68,70,72,79]. Because of our dyadic approach, we choose to call the quantity the stress dyadic. It is a "machine that contains a knowledge of the energy density, momentum density and stress, as measured by any observer" [58].

In differential-form notation, under certain restrictions on the electromagnetic medium, the Minkowskian force-energy dyadic $\overline{\overline{\mathsf{f}}}_{\mathrm{M}} \in \mathbb{F}_4\mathbb{F}_1$ can be expressed in terms of the respective stress dyadic $\overline{\overline{\overline{\mathsf{T}}}}_{\mathrm{M}} \in \mathbb{F}_3\mathbb{F}_1$ as

$$\overline{\overline{\mathsf{f}}}_{\mathrm{M}} = -\mathbf{d}\wedge\overline{\overline{\overline{\mathsf{T}}}}_{\mathrm{M}}. \tag{5.212}$$

In the literature the stress tensor in three dimensions is usually given in Gibbsian representation [70], while in four dimensions the expression is generally expressed in tensor form [34,38,79]. While in references 6 and 58 differential forms are widely used, the stress dyadic is, however, given in tensor notation. In reference 33, the stress tensor is expressed in terms of the Clifford algebra. A differential-form approach was introduced in reference 42 in a form reproduced here.

5.5.1 Stress dyadic in four dimensions

The stress dyadic can be expressed in a very compact form in the four-dimensional representation. To derive this we start from the Lorentz force-energy density four-dyadic (5.208) and replace the source three-form through the Maxwell equation as $\gamma_e = \mathbf{d}\wedge\mathbf{\Psi}$. Denoting quantities kept constant in differentiation by subscript c, we can expand

$$\begin{aligned}\overline{\overline{\mathsf{f}}}_{\mathrm{M}} &= -\gamma_e\wedge\overline{\overline{\mathsf{I}}}{}^T_{\mathrm{M}}\rfloor\mathbf{\Phi} = -(\mathbf{d}\wedge\mathbf{\Psi})\wedge\overline{\overline{\mathsf{I}}}{}^T_{\mathrm{M}}\rfloor\mathbf{\Phi}_c = -\mathbf{d}\wedge(\mathbf{\Psi}\wedge\overline{\overline{\mathsf{I}}}{}^T_{\mathrm{M}}\rfloor\mathbf{\Phi}_c)\\ &= -\mathbf{d}\wedge(\mathbf{\Psi}\wedge\overline{\overline{\mathsf{I}}}{}^T_{\mathrm{M}}\rfloor\mathbf{\Phi}) + \mathbf{d}\wedge(\mathbf{\Psi}_c\wedge\overline{\overline{\mathsf{I}}}{}^T_{\mathrm{M}}\rfloor\mathbf{\Phi}).\end{aligned} \tag{5.213}$$

Now we apply the identity (2.102) reproduced here as

$$(\mathbf{\Phi}\wedge\overline{\overline{\mathsf{I}}}{}^T_{\mathrm{M}})\rfloor\mathbf{\Psi} + (\mathbf{\Psi}\wedge\overline{\overline{\mathsf{I}}}{}^T_{\mathrm{M}})\rfloor\mathbf{\Phi} = (\mathbf{\Phi}\wedge\mathbf{\Psi})\lfloor\overline{\overline{\mathsf{I}}}_{\mathrm{M}} = (\mathbf{\Psi}\wedge\mathbf{\Phi})\lfloor\overline{\overline{\mathsf{I}}}_{\mathrm{M}} \tag{5.214}$$

to the last term of (5.213):

$$\mathbf{d}\wedge(\mathbf{\Psi}_c\wedge\overline{\overline{\mathsf{I}}}{}^T_{\mathrm{M}}\rfloor\mathbf{\Phi}) = \mathbf{d}\wedge((\mathbf{\Psi}_c\wedge\mathbf{\Phi})\lfloor\overline{\overline{\mathsf{I}}}_{\mathrm{M}}) - (\mathbf{d}\wedge\mathbf{\Phi})\wedge\overline{\overline{\mathsf{I}}}{}^T_{\mathrm{M}}\rfloor\mathbf{\Psi}_c. \tag{5.215}$$

The last term here vanishes because $\mathbf{d}\wedge\mathbf{\Phi} = 0$ is one of the Maxwell equations in the case of no magnetic sources. The force-energy density dyadic has thus been

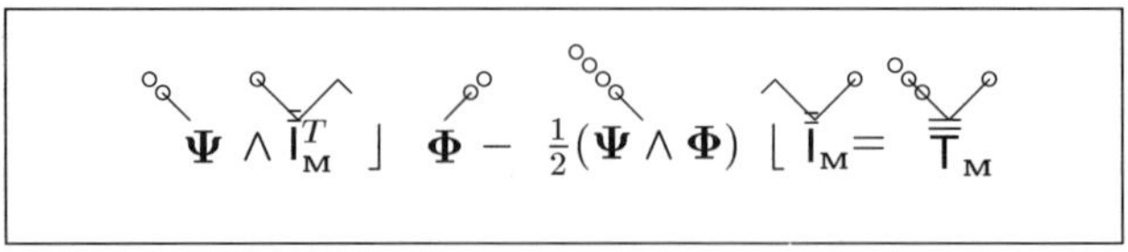

Fig. 5.10 *Visualization of the four-dimensional stress dyadic.*

transformed to a form depending on the field two-forms as

$$\overline{\overline{\mathsf{f}}}_{\mathrm{M}}=-\mathbf{d}\wedge(\mathbf{\Psi}\wedge\overline{\overline{\mathsf{I}}}{}^T_{\mathrm{M}}\rfloor\mathbf{\Phi})+\mathbf{d}\wedge((\mathbf{\Psi}_c\wedge\mathbf{\Phi})\lfloor\overline{\overline{\mathsf{I}}}_{\mathrm{M}}). \tag{5.216}$$

From (5.216) we cannot yet identify the stress dyadic as $-\mathbf{d}\wedge\overline{\overline{\mathsf{T}}}_{\mathrm{M}}$ because of the subindex c in the last term. To proceed, it is necessary to make an assumption which restricts the nature of the electromagnetic medium. In particular, after substituting $\mathbf{\Psi}=\overline{\overline{\mathsf{M}}}|\mathbf{\Phi}$, where $\overline{\overline{\mathsf{M}}}$ is a constant medium dyadic, we require that the following be valid for all two-forms $\mathbf{\Phi}$:

$$\mathbf{d}\wedge((\mathbf{\Psi}_c\wedge\mathbf{\Phi})\lfloor\overline{\overline{\mathsf{I}}}_{\mathrm{M}})=\mathbf{d}\wedge\frac{1}{2}((\mathbf{\Psi}\wedge\mathbf{\Phi})\lfloor\overline{\overline{\mathsf{I}}}_{\mathrm{M}}). \tag{5.217}$$

Since from $\mathbf{\Phi}\wedge\mathbf{\Psi}=\mathbf{\Psi}\wedge\mathbf{\Phi}$ the right-hand side is symmetric in $\mathbf{\Phi}$ and $\mathbf{\Psi}$, so must be the left-hand side, whence the condition

$$\mathbf{d}\wedge((\mathbf{\Psi}_c\wedge\mathbf{\Phi})\lfloor\overline{\overline{\mathsf{I}}}_{\mathrm{M}})=\mathbf{d}\wedge((\mathbf{\Phi}_c\wedge\mathbf{\Psi})\lfloor\overline{\overline{\mathsf{I}}}_{\mathrm{M}}) \tag{5.218}$$

must be valid as well. The assumption is valid for a certain class of media defined by a restriction to the medium dyadic $\overline{\overline{\mathsf{M}}}$ as will be shown in Section 5.5.3.

Substituting now (5.217) in (5.216) gives us

$$\overline{\overline{\mathsf{f}}}_{\mathrm{M}}=-\mathbf{d}\wedge(\mathbf{\Psi}\wedge\overline{\overline{\mathsf{I}}}{}^T_{\mathrm{M}}\rfloor\mathbf{\Phi}-\frac{1}{2}(\mathbf{\Psi}\wedge\mathbf{\Phi})\lfloor\overline{\overline{\mathsf{I}}}_{\mathrm{M}})=-\mathbf{d}\wedge\overline{\overline{\mathsf{T}}}_{\mathrm{M}}, \tag{5.219}$$

from which the stress dyadic $\overline{\overline{\mathsf{T}}}_{\mathrm{M}}$ can be identified as (Figure 5.10)

$$\overline{\overline{\mathsf{T}}}_{\mathrm{M}}=\mathbf{\Psi}\wedge\overline{\overline{\mathsf{I}}}{}^T_{\mathrm{M}}\rfloor\mathbf{\Phi}-\frac{1}{2}(\mathbf{\Psi}\wedge\mathbf{\Phi})\lfloor\overline{\overline{\mathsf{I}}}_{\mathrm{M}}. \tag{5.220}$$

This is the final result. The stress dyadic is not unique since a dyadic of the form $\mathbf{d}\wedge\overline{\overline{\mathsf{A}}}$ with any dyadic $\overline{\overline{\mathsf{A}}}\in\mathbb{F}_2\mathbb{F}_1$ can be added without changing the force-energy dyadic. Also, applying (2.102), the expression (5.220) can be changed to other equivalent forms under the special condition for the medium:

$$\begin{aligned}\overline{\overline{\mathsf{T}}}_{\mathrm{M}}&=-\mathbf{\Phi}\wedge\overline{\overline{\mathsf{I}}}{}^T_{\mathrm{M}}\rfloor\mathbf{\Psi}+\frac{1}{2}(\mathbf{\Phi}\wedge\mathbf{\Psi})\lfloor\overline{\overline{\mathsf{I}}}_{\mathrm{M}}\\&=\frac{1}{2}\mathbf{\Psi}\wedge\overline{\overline{\mathsf{I}}}{}^T_{\mathrm{M}}\rfloor\mathbf{\Phi}-\frac{1}{2}\mathbf{\Phi}\wedge\overline{\overline{\mathsf{I}}}{}^T_{\mathrm{M}}\rfloor\mathbf{\Psi}.\end{aligned} \tag{5.221}$$

Comparing this to (5.220) it is seen that the stress dyadic $\overline{\overline{\mathsf{T}}}_{\mathrm{M}}(\mathbf{\Phi}, \mathbf{\Psi})$ is an antisymmetric bilinear function of $\mathbf{\Phi}$ and $\mathbf{\Psi}$:

$$\overline{\overline{\mathsf{T}}}_{\mathrm{M}}(\mathbf{\Phi}, \mathbf{\Psi}) = -\overline{\overline{\mathsf{T}}}_{\mathrm{M}}(\mathbf{\Psi}, \mathbf{\Phi}), \tag{5.222}$$

which implies $\overline{\overline{\mathsf{T}}}_{\mathrm{M}}(\mathbf{\Phi}, \mathbf{\Phi}) = 0$. Thus, in the strange isotropic medium defined by $\overline{\overline{\mathsf{M}}} = M\overline{\overline{\mathsf{I}}}_{\mathrm{M}}$, the stress dyadic vanishes for any fields.

5.5.2 Expansion in three dimensions

To gain some physical insight, the expression for the stress dyadic (5.220) will now be expanded in terms of three-dimensional source and field quantities. Decomposing the different terms in their spatial and temporal parts in (5.220) can be written as

$$\begin{aligned} \overline{\overline{\mathsf{T}}}_{\mathrm{M}} &= (\mathbf{D} - \mathbf{H} \wedge \mathbf{d}\tau) \wedge \overline{\overline{\mathsf{I}}}^T_{\mathrm{M}} \rfloor (\mathbf{B} + \mathbf{E} \wedge \mathbf{d}\tau) \\ &\quad -\frac{1}{2}\{(\mathbf{D} - \mathbf{H} \wedge \mathbf{d}\tau) \wedge (\mathbf{B} + \mathbf{E} \wedge \mathbf{d}\tau)\} \lfloor \overline{\overline{\mathsf{I}}}_{\mathrm{M}} \\ &= \mathbf{D} \wedge \overline{\overline{\mathsf{I}}}^T_{\mathrm{M}} \rfloor \mathbf{B} + \mathbf{D} \wedge \overline{\overline{\mathsf{I}}}^T_{\mathrm{M}} \rfloor (\mathbf{E} \wedge \mathbf{d}\tau) \\ &\quad -(\mathbf{H} \wedge \mathbf{d}\tau) \wedge \overline{\overline{\mathsf{I}}}^T_{\mathrm{M}} \rfloor \mathbf{B} - (\mathbf{H} \wedge \mathbf{d}\tau) \wedge \overline{\overline{\mathsf{I}}}^T_{\mathrm{M}} \rfloor (\mathbf{E} \wedge \mathbf{d}\tau) \\ &\quad -\frac{1}{2}(\mathbf{D} \wedge \mathbf{B} + \mathbf{D} \wedge \mathbf{E} \wedge \mathbf{d}\tau \\ &\quad -\mathbf{H} \wedge \mathbf{d}\tau \wedge \mathbf{B} - \mathbf{H} \wedge \mathbf{d}\tau \wedge \mathbf{E} \wedge \mathbf{d}\tau) \lfloor \overline{\overline{\mathsf{I}}}_{\mathrm{M}}. \end{aligned} \tag{5.223}$$

Inserting $\overline{\overline{\mathsf{I}}}_{\mathrm{M}} = \overline{\overline{\mathsf{I}}}_{\mathrm{E}} + \mathbf{e}_\tau \mathbf{d}\tau$, the number of terms will be doubled. However, some of the terms are directly seen to vanish, like the last one containing $\mathbf{d}\tau \wedge \mathbf{d}\tau$, those containing the spacelike four-form $\mathbf{D} \wedge \mathbf{B}$, and certain contractions like $\mathbf{e}_\tau \rfloor \mathbf{B}$. After applying the bac-cab rule, we can write

$$\begin{aligned} \overline{\overline{\mathsf{T}}}_{\mathrm{M}} &= \mathbf{D} \wedge \overline{\overline{\mathsf{I}}}^T_{\mathrm{E}} \rfloor \mathbf{B} + \mathbf{D} \wedge (\mathbf{d}\tau \mathbf{E} - \mathbf{E} \mathbf{d}\tau) \\ &\quad +\mathbf{d}\tau \wedge (\mathbf{H} \wedge \overline{\overline{\mathsf{I}}}^T_{\mathrm{E}} \rfloor \mathbf{B}) + \mathbf{d}\tau \wedge (\mathbf{E} \wedge \mathbf{H})\mathbf{d}\tau \\ &\quad -\frac{1}{2}((\mathbf{D} \wedge \mathbf{E} \wedge \mathbf{d}\tau) \lfloor \overline{\overline{\mathsf{I}}}_{\mathrm{M}} - (\mathbf{B} \wedge \mathbf{H} \wedge \mathbf{d}\tau) \lfloor \overline{\overline{\mathsf{I}}}_{\mathrm{M}}). \end{aligned} \tag{5.224}$$

By applying (2.102), the last two terms can be further processed as

$$(\mathbf{D} \wedge \mathbf{E} \wedge \mathbf{d}\tau) \lfloor \overline{\overline{\mathsf{I}}}_{\mathrm{M}} = \mathbf{d}\tau \wedge \mathbf{D}\mathbf{E} - (\mathbf{D} \wedge \mathbf{E})\mathbf{d}\tau - \mathbf{d}\tau \wedge (\mathbf{E} \wedge \overline{\overline{\mathsf{I}}}^T_{\mathrm{E}} \rfloor \mathbf{D}), \tag{5.225}$$

$$(\mathbf{B} \wedge \mathbf{H} \wedge \mathbf{d}\tau) \lfloor \overline{\overline{\mathsf{I}}}_{\mathrm{M}} = \mathbf{d}\tau \wedge \mathbf{B}\mathbf{H} - (\mathbf{H} \wedge \mathbf{B})\mathbf{d}\tau - \mathbf{d}\tau \wedge (\mathbf{H} \wedge \overline{\overline{\mathsf{I}}}^T_{\mathrm{E}} \rfloor \mathbf{B}). \tag{5.226}$$

After all this, the stress dyadic can finally be written in terms of three-dimensional quantities in the form

$$\begin{aligned} \overline{\overline{\mathsf{T}}}_{\mathrm{M}} &= \mathbf{D} \wedge \overline{\overline{\mathsf{I}}}^T_{\mathrm{E}} \rfloor \mathbf{B} - \frac{1}{2}(\mathbf{E} \wedge \mathbf{D} + \mathbf{H} \wedge \mathbf{B})\mathbf{d}\tau + \mathbf{d}\tau \wedge (\mathbf{E} \wedge \mathbf{H})\mathbf{d}\tau \\ &\quad +\mathbf{d}\tau \wedge \frac{1}{2}(\mathbf{D}\mathbf{E} + \mathbf{B}\mathbf{H} + \mathbf{H} \wedge \overline{\overline{\mathsf{I}}}^T_{\mathrm{E}} \rfloor \mathbf{B} + \mathbf{E} \wedge \overline{\overline{\mathsf{I}}}^T_{\mathrm{E}} \rfloor \mathbf{D}). \end{aligned} \tag{5.227}$$

The economy of the four-dimensional expression (5.220) over (5.227) is obvious.

From (5.227) we can identify familiar three-dimensional physical quantities. The first term represents the electromagnetic momentum density, and the second term is associated with the energy density. Here one must note that existence of the three-form $\mathbf{w}$ as in (5.164) requires again a condition for the medium. The third term involves the Poynting vector and thus represents the density of power flow. The fourth term on the second line describes the electromagnetic stress. Every term belongs to the class of $\mathbb{F}_3\mathbb{F}_1$ dyadics. To compare with the same terms in the index notation one can consult, for example, reference 38. Examples of the use of the different terms in various physical situations can be found, for example, in 58.

5.5.3 Medium condition

Four-dimensional condition The nature of the condition to the electromagnetic medium dyadic $\overline{\overline{\mathsf{M}}}$ as required by (5.217) or (5.218) will now be studied. Applying on (5.218) the identity (2.101) with $\boldsymbol{\alpha}$ replaced by the operator one-form $\mathbf{d}$,

$$\mathbf{d} \wedge ((\boldsymbol{\Phi} \wedge \boldsymbol{\Psi})\lfloor\bar{\bar{\mathsf{I}}}_{\mathbf{M}}) = -(\mathbf{d}(\boldsymbol{\Phi} \wedge \boldsymbol{\Psi}))^T, \tag{5.228}$$

the condition in transposed form becomes

$$\mathbf{d}(\boldsymbol{\Psi}_c \wedge \boldsymbol{\Phi} - \boldsymbol{\Phi}_c \wedge \boldsymbol{\Psi}) = 0. \tag{5.229}$$

Assuming a linear medium with medium dyadic $\overline{\overline{\mathsf{M}}}$ and applying the identity (1.125), we can write

$$\begin{aligned} \boldsymbol{\Phi}_1 \wedge \overline{\overline{\mathsf{M}}}|\boldsymbol{\Phi}_2 &= \varepsilon_N \mathbf{e}_N|(\boldsymbol{\Phi}_1 \wedge \overline{\overline{\mathsf{M}}}|\boldsymbol{\Phi}_2) \\ &= \varepsilon_N(\boldsymbol{\Phi}_1|\mathbf{e}_N\lfloor\overline{\overline{\mathsf{M}}}|\boldsymbol{\Phi}_2) = \varepsilon_N(\boldsymbol{\Phi}_1|\overline{\overline{\mathsf{M}}}_g|\boldsymbol{\Phi}_2), \end{aligned} \tag{5.230}$$

where $N = 1234$ and $\overline{\overline{\mathsf{M}}}_g \in \mathbb{E}_2\mathbb{E}_2$ is the modified medium dyadic. Now (5.229) requires that the condition

$$\mathbf{d}\boldsymbol{\Phi}|(\overline{\overline{\mathsf{M}}}_g - \overline{\overline{\mathsf{M}}}_g^T)|\boldsymbol{\Phi}_c = 0 \tag{5.231}$$

be valid for any two-form function $\boldsymbol{\Phi}$. Multiplying this by some vector $\mathbf{a}$ and denoting $(\mathbf{a}|\mathbf{d})\boldsymbol{\Phi} = \boldsymbol{\Phi}'$ it is obvious that $\boldsymbol{\Phi}'$ can be made independent of $\boldsymbol{\Phi}$ whence this leads to the scalar condition of the form

$$\boldsymbol{\Phi}_1|\overline{\overline{\mathsf{M}}}_g|\boldsymbol{\Phi}_2 - \boldsymbol{\Phi}_2|\overline{\overline{\mathsf{M}}}_g|\boldsymbol{\Phi}_1 = 0 \tag{5.232}$$

for any two-forms $\boldsymbol{\Phi}_1, \boldsymbol{\Phi}_2$. This again requires that the modified medium dyadic be symmetric,

$$\overline{\overline{\mathsf{M}}}_g^T = \overline{\overline{\mathsf{M}}}_g. \tag{5.233}$$

In Chapter 6 we will see that this equals the condition for the reciprocal medium. Because of symmetry, $\overline{\overline{\mathsf{M}}}_g$ is a regular metric dyadic and can be used to define a dot product for two-forms as $\boldsymbol{\Phi} \cdot \boldsymbol{\Phi} = \boldsymbol{\Phi}|\overline{\overline{\mathsf{M}}}_g|\boldsymbol{\Phi}$ [36].

The symmetry condition (5.233) required by the medium for the existence of the stress dyadic is not the same as (5.160) except for the special case $\overline{\overline{\xi}}_g = \overline{\overline{\zeta}}_g = 0$. For the original medium dyadic $\overline{\overline{\mathsf{M}}}$ the condition (5.233) can be expressed as

$$(\mathbf{e}_N\lfloor\overline{\overline{\mathsf{M}}})^T = \overline{\overline{\mathsf{M}}}^T\rfloor\mathbf{e}_N = \mathbf{e}_N\lfloor\overline{\overline{\mathsf{M}}}, \tag{5.234}$$

where $\mathbf{e}_N$ can be any quadrivector. Finally, one can show that the condition (5.232) is equivalent to the condition

$$\mathbf{\Phi}_1 \wedge \mathbf{\Psi}_2 - \mathbf{\Phi}_2 \wedge \mathbf{\Psi}_1 = 0, \tag{5.235}$$

valid for any two electromagnetic fields 1 and 2 in the same medium.

Three-dimensional conditions If the dyadic $\overline{\overline{\mathsf{M}}}$, expanded in terms of the three-dimensional medium parameter dyadics as defined by (5.25), is inserted in the symmetry condition (5.234), corresponding symmetry conditions for the four Gibbsian medium dyadics $\overline{\overline{\epsilon}}_g$, $\overline{\overline{\mu}}_g$, $\overline{\overline{\xi}}_g$, and $\overline{\overline{\zeta}}_g$ can be obtained. In fact, from (5.27) we can evaluate for $\mathbf{e}_N = \mathbf{e}_{1234}$ (details left as a problem)

$$\mathbf{e}_N\lfloor\overline{\overline{\mathsf{M}}} = \mathbf{e}_\tau\mathbf{e}_\tau{}^\wedge_\wedge(\overline{\overline{\epsilon}}_g - \overline{\overline{\xi}}_g|\overline{\overline{\mu}}_g^{-1}|\overline{\overline{\zeta}}_g) - \mathbf{e}_\tau \wedge \overline{\overline{\xi}}_g|\overline{\overline{\mu}}_g^{-1}\rfloor\mathbf{e}_{123}$$

$$+\mathbf{e}_{123}\lfloor\overline{\overline{\mu}}_g^{-1}|\overline{\overline{\zeta}}_g \wedge \mathbf{e}_\tau - \mathbf{e}_{123}\mathbf{e}_{123}\lfloor\lfloor\overline{\overline{\mu}}_g^{-1}, \tag{5.236}$$

which from (5.234) must equal its transpose. This gives rise to four relations for the gibbsian medium dyadics:

$$\mathbf{e}_\tau\mathbf{e}_\tau{}^\wedge_\wedge(\overline{\overline{\epsilon}}_g - \overline{\overline{\xi}}_g|\overline{\overline{\mu}}_g^{-1}|\overline{\overline{\zeta}}_g) = \mathbf{e}_\tau\mathbf{e}_\tau{}^\wedge_\wedge(\overline{\overline{\epsilon}}_g^T - \overline{\overline{\zeta}}_g^T|\overline{\overline{\mu}}_g^{-1T}|\overline{\overline{\xi}}_g^T), \tag{5.237}$$

$$\mathbf{e}_\tau \wedge \overline{\overline{\xi}}_g|\overline{\overline{\mu}}_g^{-1}\rfloor\mathbf{e}_{123} = \mathbf{e}_\tau \wedge \overline{\overline{\zeta}}_g^T|\overline{\overline{\mu}}_g^{-1T}\rfloor\mathbf{e}_{123}, \tag{5.238}$$

$$\mathbf{e}_{123}\lfloor\overline{\overline{\mu}}_g^{-1}|\overline{\overline{\zeta}}_g \wedge \mathbf{e}_\tau = \mathbf{e}_{123}\lfloor\overline{\overline{\mu}}_g^{-1T}|\overline{\overline{\xi}}_g^T \wedge \mathbf{e}_\tau, \tag{5.239}$$

$$\mathbf{e}_{123}\mathbf{e}_{123}\lfloor\lfloor\overline{\overline{\mu}}_g^{-1} = \mathbf{e}_{123}\mathbf{e}_{123}\lfloor\lfloor\overline{\overline{\mu}}_g^{-1T}. \tag{5.240}$$

From equation (5.240) we first have

$$\overline{\overline{\mu}}_g = \overline{\overline{\mu}}_g^T. \tag{5.241}$$

The two conditions (5.238), and (5.239) are the same and they give

$$\overline{\overline{\xi}}_g = \overline{\overline{\zeta}}_g^T, \tag{5.242}$$

while (5.237) finally gives

$$\overline{\overline{\epsilon}}_g = \overline{\overline{\epsilon}}_g^T. \tag{5.243}$$

The symmetry condition (5.233) or (5.234) thus corresponds to the conditions (5.241)–(5.243) for the Gibbsian medium dyadics. It coincides with the medium conditions (5.161) assumed for the existence of the energy density three-form $\mathbf{w}$.

5.5.4 Complex force and stress

For time-harmonic sources and fields we can again use complex formalism. The Lorentz force dyadic can be rewritten in complex form as

$$\overline{\overline{\mathsf{f}}}_{\mathbf{E}} = \frac{1}{2}(\varrho_e \mathbf{E}^* + \mathbf{J}_e \wedge \overline{\overline{\mathsf{I}}}^T \rfloor \mathbf{B}^*), \tag{5.244}$$

and the force-energy dyadic can be rewritten as

$$\overline{\overline{\mathsf{f}}}_{\mathbf{M}} = -\frac{1}{2}\gamma_e \wedge \overline{\overline{\mathsf{I}}}_{\mathbf{M}}^T \rfloor \mathbf{\Phi}^*, \tag{5.245}$$

meaning that the real part of each gives the corresponding time-average quantity. In four-dimensional representation the $\mathbf{d}$ operator can be replaced by

$$\mathbf{d} \rightarrow \mathbf{d}_{\mathbf{E}} + jk_o \mathbf{d}\tau, \tag{5.246}$$

when the factor $e^{jk_o\tau}$ is suppressed. Thus, the complex form of the four-dimensional force-energy dyadic can be expressed by

$$\overline{\overline{\mathsf{f}}}_{\mathbf{M}} = -\mathbf{d} \wedge \overline{\overline{\mathsf{T}}}_M = -(\mathbf{d}_{\mathbf{E}} + jk_o \mathbf{d}\tau) \wedge \overline{\overline{\mathsf{T}}}_{\mathbf{M}}, \tag{5.247}$$

when the complex form of the stress dyadic $\overline{\overline{\mathsf{T}}}_{\mathbf{M}}$ is defined as

$$\overline{\overline{\mathsf{T}}}_{\mathbf{M}} = \frac{1}{2}\mathbf{\Psi} \wedge \overline{\overline{\mathsf{I}}}_{\mathbf{M}}^T \rfloor \mathbf{\Phi}^* - \frac{1}{4}(\mathbf{\Psi} \wedge \mathbf{\Phi}^*) \lfloor \overline{\overline{\mathsf{I}}}_{\mathbf{M}}. \tag{5.248}$$

The condition required to be satisfied by the medium dyadic for the stress dyadic to exist becomes now

$$\mathbf{\Phi}^* | \overline{\overline{\mathsf{M}}}_g | \mathbf{\Phi} = \mathbf{\Phi} | \overline{\overline{\mathsf{M}}}_g^* | \mathbf{\Phi}^*, \tag{5.249}$$

which should be valid for all complex two-forms $\mathbf{\Phi}$. This leads to the condition for the modified medium dyadic,

$$\overline{\overline{\mathsf{M}}}_g^T = \overline{\overline{\mathsf{M}}}_g^*, \tag{5.250}$$

which coincides with that of lossless media.

Problems

5.5.1 Repeat the analysis leading to (5.220) by taking also the magnetic sources into account.

5.5.2 Show that (5.235) follows from (5.232).

5.5.3 Derive (5.236).

5.5.4 Derive (5.249).

5.5.5 Show that the condition (5.250) corresponds to that of a lossless medium. Start by considering the complex Poynting two-form $\mathbf{S}$ and require that $\Re\{\mathbf{d} \wedge \mathbf{S}\}$ vanishes in the medium for all possible fields.

5.5.6 Check that the stress dyadic for a time-harmonic plane-wave field satisfies $\mathbf{d} \wedge \overline{\overline{\mathsf{T}}}_{\mathrm{M}} = 0$. The field two-forms are of the form

$$\boldsymbol{\Phi}(\mathbf{r}, \tau) = \boldsymbol{\Phi}_o \exp(\boldsymbol{\nu}|\mathbf{r} + jk_o\tau), \quad \boldsymbol{\Psi}(\mathbf{r}, \tau) = \boldsymbol{\Psi}_o \exp(\boldsymbol{\nu}|\mathbf{r} + jk_o\tau),$$

where $\boldsymbol{\nu}$ is a dual vector and $\boldsymbol{\Phi}_o, \boldsymbol{\Psi}_o$ are dual bivectors.

6

Theorems and Transformations

In this chapter we consider the duality transformation and the concepts of reciprocity and source equivalence which can be applied in solving electromagnetic problems.

6.1 DUALITY TRANSFORMATION

The term "duality" has two meanings. Here it refers to certain transformations of electromagnetic fields and sources and not to duality of multivector spaces. Because the former is related to electromagnetic field quantities while the latter is related to vector spaces, the danger of confusion should be minimal (changing an algebraic identity valid for multivectors to a form valid for dual multivectors could also be called a duality transformation in the latter sense). Duality transformations are useful in giving a possibility to derive solutions to new problems from solutions of old problems without having to repeat the whole solution procedure (Figure 6.1). The classical concept of duality in electromagnetic theory refers to the apparent symmetry in the Maxwell equations, which, written with magnetic sources included in Gibbsian vector form, are

$$-\nabla \times \mathbf{E} = \partial_t \mathbf{B} + \mathbf{J}_m, \quad \nabla \cdot \mathbf{B} = \varrho_m, \tag{6.1}$$

$$\nabla \times \mathbf{H} = \partial_t \mathbf{D} + \mathbf{J}_e, \quad \nabla \cdot \mathbf{D} = \varrho_e. \tag{6.2}$$

Historically, it may be of interest that the concept of duality was not known to James Clark Maxwell because he presented his equations in a very nonsymmetric form, in a mixed set of equations for scalar field and potential quantities as shown in Figure 1.1. Duality was first introduced by Oliver Heaviside in 1886 when he reformulated the Maxwell equations in a form which he called "the duplex method" [6, 32]. In fact,

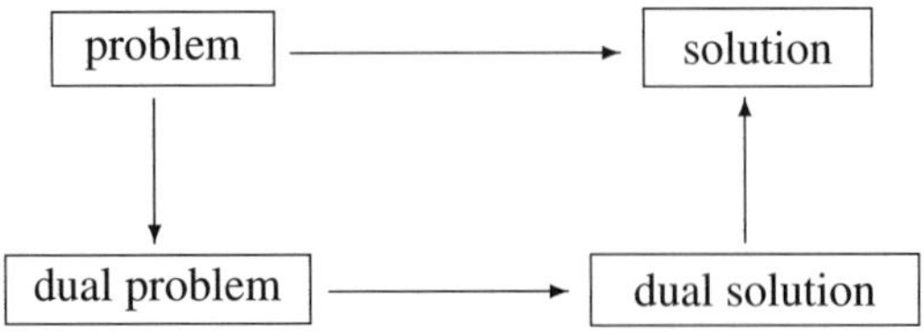

Fig. 6.1 *Solving a problem can be done without performing the solution process if it has already been done for the dual problem. The problem is first transformed to the dual problem, and the known dual solution is transformed back.*

by introducing the vector notation and eliminating the auxiliary potential quantities, Heaviside could reduce the original set of equations to a form equivalent to (6.1) and (6.2). The cross and dot products were introduced by Gibbs, which is why we refer to Gibbsian instead of Heavisidean representation. Heaviside listed four "great advantages" in the use of vector equations as follows.[1]

1. First, the abolition of the two potentials.

2. Next, we are brought into immediate contact with $\mathbf{E}$ and $\mathbf{H}$, which have physical significance in really defining the state of the medium anywhere (ϵ, μ and σ of course to be known), which $\mathbf{A}$ and ϕ do not, and cannot, even if given over all space.

3. Third, by reason of the close parallelism between (6.1) and (6.2), electric force being related to magnetic current, as magnetic force to electric current, we are enabled to perceive easily many important relations which are not at all obvious when the potentials $\mathbf{A}$ and ϕ are used and (6.2) is ignored.

4. Fourth, we are enabled with considerable ease, if we have obtained solutions relating to variable states in which the lines of $\mathbf{E}$ and $\mathbf{H}$ are related in one way, to at once get the solutions of problems quite different physical meaning, in which $\mathbf{E}$ and $\mathbf{H}$, or quantities directly related to them, change places. For example, the variation of magnetic force in a core placed in a coil, and of electric current in a round wire; and many others.

6.1.1 Dual substitution

In his fourth "advantage," Heaviside describes the use of duality. To transform electromagnetic field equations, a simple dual substitution method can be used. In fact, substituting symbols in (6.1) and (6.2) as

$$\mathbf{E} \rightarrow -\mathbf{H}, \quad \mathbf{H} \rightarrow -\mathbf{E}, \quad \mathbf{B} \rightarrow \mathbf{D}, \quad \mathbf{D} \rightarrow \mathbf{B},$$

[1]Quoted from reference 32, except that symbols are written in modern notation.

$$\mathbf{J}_e \to \mathbf{J}_m, \quad \mathbf{J}_m \to \mathbf{J}_e, \quad \varrho_e \to \varrho_m, \quad \varrho_m \to \varrho_e \tag{6.3}$$

does not change the whole system of equations. Thus, for the same change of symbols, solutions for these equations remain to be valid. Of course, boundary values and medium parameters must be changed accordingly.

In the three-dimensional differential-form formalism the simple substitution rules (6.3) can be applied as such for the multiform field and source quantities. In the four-dimensional representation the substitutions become more elegant as the disturbing minus sign of (6.3) lies hidden inside the field two-forms:

$$\mathbf{\Phi} \to \mathbf{\Psi}, \quad \mathbf{\Psi} \to \mathbf{\Phi}, \quad \gamma_e \to \gamma_m, \quad \gamma_m \to \gamma_e. \tag{6.4}$$

It then follows that the medium dyadic must be substituted by its inverse,

$$\overline{\overline{\mathsf{M}}} \to \overline{\overline{\mathsf{M}}}^{-1}. \tag{6.5}$$

This simple substitution method is not a transformation since the nature physical quantities is not preserved. For example, an electric field is replaced by a magnetic field, $\mathbf{E} \to -\mathbf{H}$. This defect could be remedied by inserting certain constant factors like in $\mathbf{E} \to Z\mathbf{H}$, where Z is some impedance factor. Let us, however, consider a still more general form of duality transformation.

Duality substitution can be applied to transform physical relations to another form. As a simple example, the electric field from an electric point charge Q_e obeys the familiar formula

$$\mathbf{E}(\mathbf{r}) = -\mathbf{d}\frac{Q}{4\pi\epsilon_o r}. \tag{6.6}$$

Applying the duality substitutions we obtain

$$\mathbf{H}(\mathbf{r}) = \mathbf{d}\frac{Q_m}{4\pi\mu_o r}, \tag{6.7}$$

the corresponding formula for the magnetostatic field from a magnetic point charge (monopole).

6.1.2 General duality

Four-dimensional formulation In the most general form, duality transformation can be understood as a linear mapping between field quantities and source quantities,

$$\begin{pmatrix} \mathbf{\Psi}_d \\ \mathbf{\Phi}_d \end{pmatrix} = \mathcal{D}\begin{pmatrix} \mathbf{\Psi} \\ \mathbf{\Phi} \end{pmatrix}, \quad \begin{pmatrix} \gamma_{ed} \\ \gamma_{md} \end{pmatrix} = \mathcal{D}\begin{pmatrix} \gamma_e \\ \gamma_m \end{pmatrix}, \tag{6.8}$$

where the transformation matrix

$$\mathcal{D} = \begin{pmatrix} A & B \\ C & D \end{pmatrix} \tag{6.9}$$

is defined by the four scalars $A \cdots D$. However, in many cases, change of the magnitude level is not important since it can be changed by a multiplying factor and we can restrict the transformation by some condition. Requiring that the determinant of the matrix be $\det\mathcal{D} = AD - BC = 1$, we can reduce the number of parameters to three. It is convenient to define the transformation in terms of three parameters Z, Z_d and θ as

$$\begin{pmatrix} Z_d\boldsymbol{\Psi}_d \\ \boldsymbol{\Phi}_d \end{pmatrix} = \mathcal{R}(\theta)\begin{pmatrix} Z\boldsymbol{\Psi} \\ \boldsymbol{\Phi} \end{pmatrix}, \tag{6.10}$$

where $\mathcal{R}(\theta)$ is the rotation matrix

$$\mathcal{R}(\theta) = \begin{pmatrix} \cos\theta & \sin\theta \\ -\sin\theta & \cos\theta \end{pmatrix}, \tag{6.11}$$

satisfying

$$\mathcal{R}^{-1}(\theta) = \mathcal{R}^T(\theta) = \mathcal{R}(-\theta). \tag{6.12}$$

In this form, the transformation consists of three steps: first, an impedance transformation by Z which makes the physical dimension of $\boldsymbol{\Phi}$ and $\boldsymbol{\Psi}$ the same, second, a duality rotation by angle θ which replaces $\boldsymbol{\Phi}$ and $\boldsymbol{\Psi}$ by their linear combination; and, third, another impedance transformation by $1/Z_d$ which returns the correct physical dimensions.

Expressing the transformation as (6.8), the duality-transformation matrix $\mathcal{D}$ is defined as

$$\begin{aligned} \mathcal{D} &= \begin{pmatrix} 1/Z_d & 0 \\ 0 & 1 \end{pmatrix}\mathcal{R}(\theta)\begin{pmatrix} Z & 0 \\ 0 & 1 \end{pmatrix} \\ &= \begin{pmatrix} (Z/Z_d)\cos\theta & (1/Z_d)\sin\theta \\ -Z\sin\theta & \cos\theta \end{pmatrix}. \end{aligned} \tag{6.13}$$

More explicitly, the transformed electromagnetic two-forms are

$$\boldsymbol{\Psi}_d = (Z/Z_d)\cos\theta\boldsymbol{\Psi} + (1/Z_d)\sin\theta\boldsymbol{\Phi}, \tag{6.14}$$

$$\boldsymbol{\Phi}_d = -Z\sin\theta\boldsymbol{\Psi} + \cos\theta\boldsymbol{\Phi}. \tag{6.15}$$

The inverse transformation is defined by the inverse matrix

$$\begin{aligned} \mathcal{D}^{-1} &= \begin{pmatrix} 1/Z & 0 \\ 0 & 1 \end{pmatrix}\mathcal{R}(-\theta)\begin{pmatrix} Z_d & 0 \\ 0 & 1 \end{pmatrix} \\ &= \begin{pmatrix} (Z_d/Z)\cos\theta & -(1/Z)\sin\theta \\ Z_d\sin\theta & \cos\theta \end{pmatrix}. \end{aligned} \tag{6.16}$$

The transformation law for the medium dyadic $\overline{\overline{\mathsf{M}}}$ can be found by replacing $\boldsymbol{\Psi}$ by $\overline{\overline{\mathsf{M}}}|\boldsymbol{\Phi}$, whence (6.14) and (6.15) become

$$\boldsymbol{\Phi}_d = (\cos\theta\, \bar{\mathsf{I}}^{(2)T} - \sin\theta Z\overline{\overline{\mathsf{M}}})|\boldsymbol{\Phi}, \tag{6.17}$$

$$Z_d \boldsymbol{\Psi}_d = (\sin\theta \bar{\bar{\mathsf{I}}}^{(2)T} + \cos\theta Z \overline{\overline{\mathsf{M}}})|\boldsymbol{\Phi}. \tag{6.18}$$

Eliminating $\boldsymbol{\Phi}$, we obtain

$$Z_d \overline{\overline{\mathsf{M}}}_d = (\sin\theta\, \bar{\bar{\mathsf{I}}}^{(2)T} + \cos\theta\, Z\overline{\overline{\mathsf{M}}})|(\cos\theta\, \bar{\bar{\mathsf{I}}}^{(2)T} - \sin\theta\, Z\overline{\overline{\mathsf{M}}})^{-1}. \tag{6.19}$$

Finding the corresponding rules for the medium dyadics $\overline{\overline{\alpha}}, \overline{\overline{\epsilon}}', \overline{\overline{\mu}}, \overline{\overline{\beta}}$ is left as a problem.

Three-dimensional formulation Breaking (6.8) in its three-dimensional components, the duality transformation rule becomes

$$\begin{aligned} \begin{pmatrix} \mathbf{D} \\ \mathbf{B} \end{pmatrix}_d - \mathcal{J}\begin{pmatrix} \mathbf{E} \\ \mathbf{H} \end{pmatrix}_d \wedge \mathbf{d}\tau &= \mathcal{D}\begin{pmatrix} \mathbf{D} \\ \mathbf{B} \end{pmatrix} - \mathcal{D}\mathcal{J}\begin{pmatrix} \mathbf{E} \\ \mathbf{H} \end{pmatrix} \wedge \mathbf{d}\tau \\ &= \mathcal{D}\begin{pmatrix} \mathbf{D} \\ \mathbf{B} \end{pmatrix} - \mathcal{J}\mathcal{D}'\begin{pmatrix} \mathbf{E} \\ \mathbf{H} \end{pmatrix} \wedge \mathbf{d}\tau. \end{aligned} \tag{6.20}$$

From this we obtain the rules

$$\begin{pmatrix} \mathbf{D} \\ \mathbf{B} \end{pmatrix}_d = \mathcal{D}\begin{pmatrix} \mathbf{D} \\ \mathbf{B} \end{pmatrix}, \quad \begin{pmatrix} \mathbf{E} \\ \mathbf{H} \end{pmatrix}_d = \mathcal{D}'\begin{pmatrix} \mathbf{E} \\ \mathbf{H} \end{pmatrix}. \tag{6.21}$$

For the fields $\mathbf{E}$ and $\mathbf{H}$ the transformation matrix is

$$\begin{aligned} \mathcal{D}' &= -\mathcal{J}\mathcal{D}\mathcal{J} \\ &= -\begin{pmatrix} 0 & 1 \\ -1 & 0 \end{pmatrix}\begin{pmatrix} (Z/Z_d)\cos\theta & (1/Z_d)\sin\theta \\ -Z\sin\theta & \cos\theta \end{pmatrix}\begin{pmatrix} 0 & 1 \\ -1 & 0 \end{pmatrix} \\ &= \begin{pmatrix} \cos\theta & Z\sin\theta \\ -(1/Z_d)\sin\theta & (Z/Z_d)\cos\theta \end{pmatrix}, \end{aligned} \tag{6.22}$$

which in explicit form gives

$$\mathbf{E}_d = \cos\theta \mathbf{E} + Z\sin\theta \mathbf{H}, \tag{6.23}$$

$$\mathbf{H}_d = -(1/Z_d)\sin\theta \mathbf{E} + (Z/Z_d)\cos\theta \mathbf{H}, \tag{6.24}$$

$$\mathbf{D}_d = (Z/Z_d)\cos\theta \mathbf{D} + (1/Z_d)\sin\theta \mathbf{B}, \tag{6.25}$$

$$\mathbf{B}_d = -Z\sin\theta \mathbf{D} + \cos\theta \mathbf{B}. \tag{6.26}$$

The Poynting two-form obeys the rule

$$\mathbf{S}_d = \mathbf{E}_d \wedge \mathbf{H}_d = \frac{Z}{Z_d}\mathbf{E}\wedge\mathbf{H} = (Z/Z_d)\mathbf{S}. \tag{6.27}$$

Ideal boundaries remain ideal boundaries in the duality transformation because $\mathbf{S}|\mathcal{S} = 0$ implies $\mathbf{S}_d|\mathcal{S} = 0$ for any boundary surface $\mathcal{S}$. However, the nature of the ideal boundary is changed. For example, the PEC condition $\mathbf{E}\rfloor\mathcal{S} = 0$ is changed to the more general form $(\cos\theta\mathbf{E} - Z_d \sin\theta\mathbf{H})\rfloor\mathcal{S} = 0$. For $\theta = \pm\pi/2$ a PEC boundary is transformed to a PMC boundary, and conversely.

The electromagnetic sources are transformed through the transformation matrix $\mathcal{D}$, whence their formulas are similar to those of $\mathbf{D}$ and $\mathbf{B}$:

$$\begin{pmatrix} \mathbf{J}_e \\ \mathbf{J}_m \end{pmatrix}_d = \mathcal{D}\begin{pmatrix} \mathbf{J}_e \\ \mathbf{J}_m \end{pmatrix}, \quad \begin{pmatrix} \varrho_e \\ \varrho_m \end{pmatrix}_d = \mathcal{D}\begin{pmatrix} \varrho_e \\ \varrho_m \end{pmatrix}. \tag{6.28}$$

The transformed medium dyadic matrix $\overline{\overline{\mathcal{M}}}_d$ can be found by writing

$$\begin{aligned} \begin{pmatrix} \mathbf{D} \\ \mathbf{B} \end{pmatrix}_d &= \mathcal{D}\begin{pmatrix} \mathbf{D} \\ \mathbf{B} \end{pmatrix} = \mathcal{D}\overline{\overline{\mathcal{M}}}|\begin{pmatrix} \mathbf{E} \\ \mathbf{H} \end{pmatrix} \\ &= \mathcal{D}\overline{\overline{\mathcal{M}}}\mathcal{D}'^{-1}|\begin{pmatrix} \mathbf{E} \\ \mathbf{H} \end{pmatrix}_d, \end{aligned} \tag{6.29}$$

which leads to the rule

$$\overline{\overline{\mathcal{M}}}_d = \mathcal{D}\overline{\overline{\mathcal{M}}}\mathcal{D}'^{-1}. \tag{6.30}$$

This can be also expressed as

$$\begin{pmatrix} \overline{\overline{\epsilon}}_d Z_d & \overline{\overline{\xi}}_d \\ \overline{\overline{\zeta}}_d & \overline{\overline{\mu}}_d/Z_d \end{pmatrix} = \mathcal{R}(\theta)\begin{pmatrix} \overline{\overline{\epsilon}} Z & \overline{\overline{\xi}} \\ \overline{\overline{\zeta}} & \overline{\overline{\mu}}/Z \end{pmatrix}\mathcal{R}(-\theta) \tag{6.31}$$

from which the duality transforms of the four medium dyadics are easily obtained in the form

$$\overline{\overline{\epsilon}}_d Z_d = \overline{\overline{\epsilon}} Z \cos^2\theta + (\overline{\overline{\mu}}/Z)\sin^2\theta + (\overline{\overline{\xi}} + \overline{\overline{\zeta}})\cos\theta\sin\theta, \tag{6.32}$$

$$\overline{\overline{\mu}}_d/Z_d = \overline{\overline{\epsilon}} Z \sin^2\theta + (\overline{\overline{\mu}}/Z)\cos^2\theta - (\overline{\overline{\xi}} + \overline{\overline{\zeta}})\cos\theta\sin\theta, \tag{6.33}$$

$$\overline{\overline{\xi}}_d = -(\overline{\overline{\epsilon}} Z - \overline{\overline{\mu}}/Z)\cos\theta\sin\theta + \overline{\overline{\xi}}\cos^2\theta - \overline{\overline{\zeta}}\sin^2\theta, \tag{6.34}$$

$$\overline{\overline{\zeta}}_d = -(\overline{\overline{\epsilon}} Z - \overline{\overline{\mu}}/Z)\cos\theta\sin\theta - \overline{\overline{\xi}}\sin^2\theta + \overline{\overline{\zeta}}\cos^2\theta. \tag{6.35}$$

Combining these, we can write the following more suggestive relations:

$$\overline{\overline{\epsilon}}_d Z_d + \overline{\overline{\mu}}_d/Z_d = \overline{\overline{\epsilon}} Z + \overline{\overline{\mu}}/Z, \tag{6.36}$$

$$\overline{\overline{\xi}}_d - \overline{\overline{\zeta}}_d = \overline{\overline{\xi}} - \overline{\overline{\zeta}}, \tag{6.37}$$

$$\overline{\overline{\epsilon}}_d Z_d - \overline{\overline{\mu}}_d/Z_d = (\overline{\overline{\epsilon}} Z - \overline{\overline{\mu}}/Z)\cos 2\theta + (\overline{\overline{\xi}} + \overline{\overline{\zeta}})\sin 2\theta, \tag{6.38}$$

$$\overline{\overline{\xi}}_d + \overline{\overline{\zeta}}_d = (\overline{\overline{\xi}} + \overline{\overline{\zeta}})\cos 2\theta - (\overline{\overline{\epsilon}} Z - \overline{\overline{\mu}}/Z)\sin 2\theta. \tag{6.39}$$

It is interesting to note that the dyadic $\overline{\overline{\xi}} - \overline{\overline{\zeta}}$ is invariant in any duality transformation while the dyadic $\overline{\overline{\epsilon}} Z + \overline{\overline{\mu}}/Z$ is transformed to $\overline{\overline{\epsilon}}_d Z_d + \overline{\overline{\mu}}_d/Z_d$. The dyadics $\overline{\overline{\xi}} + \overline{\overline{\zeta}}$ and $\overline{\overline{\epsilon}} Z - \overline{\overline{\mu}}/Z$ are transformed through a two-dimensional rotation by the angle 2θ.

6.1.3 Simple duality

There are some important limiting cases of the duality transformation in terms of the θ parameter. The case $\theta = 0$ gives a pure impedance transformation

$$\mathbf{\Phi}_d = \mathbf{\Phi}, \quad \mathbf{\Psi}_d = \frac{Z}{Z_d}\mathbf{\Psi}, \quad \overline{\overline{\mathsf{M}}}_d = \frac{Z}{Z_d}\overline{\overline{\mathsf{M}}} \tag{6.40}$$

which for $Z_d = Z$ reduces to the identity transformation.

Two more interesting limiting cases are obtained for $\theta = \pm\pi/2$, which augmented by the choice $Z_d = -Z$, lead to the transformation rules

$$\mathbf{\Phi}_d = \pm Z\mathbf{\Psi}, \quad \mathbf{\Psi}_d = \pm\frac{1}{Z}\mathbf{\Phi}, \quad \overline{\overline{\mathsf{M}}}_d = \frac{1}{Z^2}\overline{\overline{\mathsf{M}}}^{-1}. \tag{6.41}$$

It is seen that the two transformations defined by (6.41) coincide with their inverses because $(\mathbf{\Phi}_d)_d = \mathbf{\Phi}$ for both signs. Since there is no reason to prefer one of the $\pm$ signs, we consider these as a pair of duality transformations. In both transformations the four medium dyadics obey the same rules obtained from (6.32)–(6.35):

$$\overline{\overline{\epsilon}}_d = -\overline{\overline{\mu}}/Z^2, \quad \overline{\overline{\mu}}_d = -\overline{\overline{\epsilon}}Z^2, \quad \overline{\overline{\xi}}_d = -\overline{\overline{\zeta}}, \quad \overline{\overline{\zeta}}_d = -\overline{\overline{\xi}}. \tag{6.42}$$

Since the rules (6.41), (6.42) are so simple, the transformations defined by $\theta = \pm\pi/2$ are called simple duality transformations and the corresponding relations are called simple duality. Basically they replace electrical field, source, and medium quantities by the corresponding magnetic quantities and conversely.

A quantity q is called self-dual if it satisfies $q_d = q$ and it is called anti-self-dual if it satisfies $q_d = -q$. Denoting self-dual quantities corresponding to the two simple transformations by their corresponding signs $\pm$ as $q_\pm$, they satisfy

$$\mathbf{\Psi}_\pm = \pm\frac{1}{Z}\mathbf{\Phi}_\pm. \tag{6.43}$$

Thus, $\mathbf{\Phi}_\pm$ are eigen-two-forms, and $\pm 1/Z$ the corresponding eigenvalues, of the medium dyadic:

$$\overline{\overline{\mathsf{M}}}|\mathbf{\Phi}_\pm = \pm\frac{1}{Z}\mathbf{\Phi}_\pm. \tag{6.44}$$

If the medium is self-dual in either of the two simple duality transformations, from (6.42) we can see that its medium dyadics must satisfy $\overline{\overline{\xi}} = -\overline{\overline{\zeta}}$ and $\overline{\overline{\epsilon}}, \overline{\overline{\mu}}$ must be multiples of one another.

Example As an example of a self-dual medium satisfying $\overline{\overline{\mathsf{M}}}_d = \overline{\overline{\mathsf{M}}}$, we consider

$$\overline{\overline{\epsilon}} = \epsilon\overline{\overline{\mathsf{A}}}, \quad \overline{\overline{\mu}} = \mu\overline{\overline{\mathsf{A}}}, \quad \overline{\overline{\xi}} = \overline{\overline{\zeta}} = 0, \tag{6.45}$$

where $\overline{\overline{\mathsf{A}}} \in \mathbb{F}_2\mathbb{E}_1$ is a dyadic and ϵ, μ are two scalars. The Z parameter of the transformation must now satisfy

$$Z^2 = -\mu/\epsilon. \tag{6.46}$$

Thus, there are two simple transformations with respect to which the medium is self-dual,

$$\mathbf{\Phi}_d = \pm j\eta\mathbf{\Psi}, \quad \mathbf{\Psi}_d = \pm\frac{1}{j\eta}\mathbf{\Phi}, \quad \eta = \sqrt{\mu/\epsilon}. \tag{6.47}$$

Self-dual media are of interest because the original problem and the dual problem are in the same medium. This means that if we can solve a problem in this kind of a medium, we immediately have another problem with a solution in the same medium.

6.1.4 Duality rotation

The previous duality transformation is independent of any metric. One could also describe it as $\mathbf{E} \leftrightarrow \mathbf{H}$ duality. Another type is what could be called as $\mathbf{B} \leftrightarrow \mathbf{E}$ duality. Because $\mathbf{B}$ is a two-form and $\mathbf{E}$ is a one-form, a transformation between them necessarily depends on some metric through some preferred basis system. Such a duality transformation is defined in terms of a Minkowskian Hodge dyadic $\overline{\overline{\mathsf{H}}}{}^T_{\mathrm{M}2} \in \mathbb{F}_2\mathbb{E}_2$ defined in terms of a given basis as (2.272) reproduced here for convenience ($\mathbf{e}_4 = \mathbf{e}_\tau$, $\varepsilon_4 = \mathbf{d}\tau$):

$$\begin{aligned}\overline{\overline{\mathsf{H}}}{}^T_{\mathrm{M}2} &= (\varepsilon_{41}\mathbf{e}_{23} + \varepsilon_{42}\mathbf{e}_{31} + \varepsilon_{43}\mathbf{e}_{12}) + (\varepsilon_{23}\mathbf{e}_{14} + \varepsilon_{31}\mathbf{e}_{24} + \varepsilon_{12}\mathbf{e}_{34}) \\ &= \varepsilon_4 \wedge \overline{\overline{\mathsf{H}}}{}^T_{\mathrm{E}1} + \overline{\overline{\mathsf{H}}}{}^T_{\mathrm{E}2} \wedge \mathbf{e}_4. \end{aligned}\tag{6.48}$$

The last representation is based on the three-dimensional Euclidean Hodge dyadics (2.231), (2.232)

$$\overline{\overline{\mathsf{H}}}{}^T_{\mathrm{E}1} = \varepsilon_3\mathbf{e}_{12} + \varepsilon_1\mathbf{e}_{23} + \varepsilon_2\mathbf{e}_{31}, \tag{6.49}$$

$$\overline{\overline{\mathsf{H}}}{}^T_{\mathrm{E}2} = \varepsilon_{23}\mathbf{e}_1 + \varepsilon_{31}\mathbf{e}_2 + \varepsilon_{12}\mathbf{e}_3. \tag{6.50}$$

Because the Minkowskian Hodge dyadic satisfies $(\overline{\overline{\mathsf{H}}}{}^T_{\mathrm{M}2})^2 = \overline{\overline{\mathsf{I}}}{}^{(2)T}_{\mathrm{M}}$, it behaves like the imaginary unit. Let us use a more suggestive notation $\overline{\overline{\mathsf{J}}}{}^T_{\mathrm{M}}$ by defining

$$\overline{\overline{\mathsf{J}}}_{\mathrm{M}} = \overline{\overline{\mathsf{H}}}_{\mathrm{M}2}, \quad \overline{\overline{\mathsf{J}}}{}^2_{\mathrm{M}} = -\overline{\overline{\mathsf{I}}}{}^{(2)}_{\mathrm{M}}. \tag{6.51}$$

Applied to the Maxwell force two-form $\mathbf{\Phi}$, we obtain after substituting from (6.48)

$$\overline{\overline{\mathsf{J}}}{}^T_{\mathrm{M}}|\mathbf{\Phi} = \overline{\overline{\mathsf{H}}}{}^T_{\mathrm{M}2}|(\mathbf{B} + \mathbf{E}\wedge\mathbf{d}\tau) = \mathbf{d}\tau \wedge \overline{\overline{\mathsf{H}}}{}^T_{\mathrm{E}1}|\mathbf{B} + \overline{\overline{\mathsf{H}}}{}^T_{\mathrm{E}2}|\mathbf{E}, \tag{6.52}$$

which is of the form

$$\mathbf{\Phi}' = \mathbf{B}' + \mathbf{E}'\wedge\mathbf{d}\tau, \quad \mathbf{B}' = \overline{\overline{\mathsf{H}}}{}^T_{\mathrm{E}2}|\mathbf{E}, \quad \mathbf{E}' = -\overline{\overline{\mathsf{H}}}{}^T_{\mathrm{E}1}|\mathbf{B}. \tag{6.53}$$

Thus, $\overline{\overline{\mathsf{J}}}{}^T_{\mathrm{M}}$ has the property of transforming $\mathbf{E}$ to $\mathbf{B}'$ and $\mathbf{B}$ to $-\mathbf{E}'$.

More general transformation can be made by combining $\overline{\overline{\mathsf{I}}}{}^{(2)T}_{\mathrm{M}}$ and $\overline{\overline{\mathsf{J}}}{}^T_{\mathrm{M}}$ to a rotation dyadic as defined by (2.281)

$$\overline{\overline{\mathsf{R}}}_{\mathrm{M}}(\theta) = \overline{\overline{\mathsf{I}}}{}^{(2)T}_{\mathrm{M}}\cos\theta + \overline{\overline{\mathsf{J}}}{}^T_{\mathrm{M}}\sin\theta. \tag{6.54}$$

Expanded in terms of Euclidean dyadics, this becomes

$$\overline{\overline{\mathsf{R}}}_{\mathbf{M}}(\theta) = (\overline{\overline{\mathsf{I}}}_{\mathbf{E}}^{(2)T} + \overline{\overline{\mathsf{I}}}_{\mathbf{E}}^{T} {}^{\wedge}_{\wedge} \varepsilon_\tau \mathbf{e}_\tau)\cos\theta - (\mathbf{d}\tau \wedge \overline{\overline{\mathsf{H}}}_{\mathbf{E}1}^{T} + \overline{\overline{\mathsf{H}}}_{\mathbf{E}2}^{T} \wedge \mathbf{e}_\tau)\sin\theta$$

$$= \overline{\overline{\mathsf{I}}}_{\mathbf{E}}^{(2)T}\cos\theta - \mathbf{d}\tau \wedge \overline{\overline{\mathsf{H}}}_{\mathbf{E}1}^{T}\sin\theta - \overline{\overline{\mathsf{H}}}_{\mathbf{E}2}^{T} \wedge \mathbf{e}_\tau \sin\theta - \mathbf{e}_\tau \wedge \overline{\overline{\mathsf{I}}}_{\mathbf{E}}^{T} \wedge \varepsilon_\tau \cos\theta. \quad (6.55)$$

Applied to the Maxwell two-form $\boldsymbol{\Phi}$, the rotated two-form becomes

$$\boldsymbol{\Phi}_r = \overline{\overline{\mathsf{R}}}_{\mathbf{M}}(\theta)|\boldsymbol{\Phi} = \boldsymbol{\Phi}\cos\theta + \overline{\overline{\mathsf{J}}}_{\mathbf{M}}^{T}|\boldsymbol{\Phi}\sin\theta = \mathbf{B}_r + \mathbf{E}_r \wedge \mathbf{d}\tau, \quad (6.56)$$

with

$$\mathbf{B}_r = \mathbf{B}\cos\theta + \overline{\overline{\mathsf{H}}}_{\mathbf{E}2}^{T}|\mathbf{E}\sin\theta, \quad \mathbf{E}_r = \mathbf{E}\cos\theta - \overline{\overline{\mathsf{H}}}_{\mathbf{E}1}^{T}|\mathbf{B}\sin\theta. \quad (6.57)$$

The transformed field quantities depend on the preferred basis $\{\mathbf{e}_i\}$ and its reciprocal basis $\{\varepsilon_i\}$ as is evident from their explicit expressions

$$\mathbf{B}' = \overline{\overline{\mathsf{H}}}_{\mathbf{E}2}^{T}|\mathbf{E} = \varepsilon_{23}E_1 + \varepsilon_{31}E_2 + \varepsilon_{12}E_3, \quad (6.58)$$

$$\mathbf{E}' = \overline{\overline{\mathsf{H}}}_{\mathbf{E}1}^{T}|\mathbf{B} = \varepsilon_3 B_{12} + \varepsilon_1 B_{23} + \varepsilon_2 B_{31}, \quad (6.59)$$

when the components are defined by

$$E_i = \mathbf{e}_i|\mathbf{E}, \quad B_{ij} = \mathbf{e}_{ij}|\mathbf{B}. \quad (6.60)$$

The duality rotation is limited to Minkowskian two-forms and it cannot be applied to source three-forms or potential one-forms. Thus, it does not represent a transformation to the Maxwell equations.

Problems

6.1.1 Show from the medium transformation formula (6.19) that if the medium dyadic is of the form $\overline{\overline{\mathsf{M}}} = M\overline{\overline{\mathsf{I}}}^{(2)T}$, the transformed medium dyadic has the form $\overline{\overline{\mathsf{M}}}_d = M_d\overline{\overline{\mathsf{I}}}^{(2)T}$.

6.1.2 Find the duality transformation rules for the medium dyadics $\overline{\overline{\alpha}}, \overline{\overline{\epsilon}}', \overline{\overline{\mu}}, \overline{\overline{\beta}}$ corresponding to (6.19).

6.1.3 Under what conditions can a bianisotropic medium with dyadics $\overline{\overline{\epsilon}}, \overline{\overline{\mu}}, \overline{\overline{\xi}}, \overline{\overline{\zeta}}$ be transformed to a dual medium with $\overline{\overline{\xi}}_d = \overline{\overline{\zeta}}_d = 0$?

6.1.4 Reduce the eigenvalue problem (6.44) to three-dimensional Euclidean form for the fields $\mathbf{E}_\pm$.

6.1.5 Expanding the magnetoelectric dyadic parameters as

$$\overline{\overline{\xi}} = \overline{\overline{\chi}} - j\overline{\overline{\kappa}}\sqrt{\mu_o\epsilon_o}, \quad \overline{\overline{\zeta}} = \overline{\overline{\chi}} + j\overline{\overline{\kappa}}\sqrt{\mu_o\epsilon_o},$$

show that the chirality dyadic $\overline{\overline{\kappa}}$ is invariant in any duality transformation while the Tellegen parameter $\overline{\overline{\chi}}$ can be changed.

Fig. 6.2 *Reciprocity in Gibbsian electromagnetics relates source vectors and their field vectors. For example, if field $\mathbf{E}_1$ from source $\mathbf{J}_{e1}$ is orthogonal to source $\mathbf{J}_{e2}$, its field $\mathbf{E}_2$ is orthogonal to source $\mathbf{J}_{e1}$ provided medium satisfies reciprocity conditions.*

6.2 RECIPROCITY

6.2.1 Lorentz reciprocity

In its basic form, reciprocity (or Lorentz reciprocityıLorentz reciprocity [1]) in electromagnetics allows a change in the position of the source and the field-measuring instrument without any change in the reading on the instrument. For example, in Gibbsian electromagnetics, the electric field can be measured by a small test dipole with vector current moment $\Delta\mathbf{p} = \mathbf{J}_e \Delta V$. Applied to the electric field vector $\mathbf{E}$ it gives the result as the scalar $\Delta\mathbf{p}\cdot\mathbf{E} = (\mathbf{J}_e\cdot\mathbf{E})\Delta V$. In the differential-form electromagnetics $\Delta\mathbf{V}$ is a small trivector and the current moment is the vector $\Delta\mathbf{p} = \Delta\mathbf{V}\lfloor\mathbf{J}_e$. In this case the measurement gives the scalar

$$\Delta\mathbf{p}|\mathbf{E} = (\Delta\mathbf{V}\lfloor\mathbf{J}_e)|\mathbf{E} = \Delta\mathbf{V}|(\mathbf{J}_e\wedge\mathbf{E}). \tag{6.61}$$

The magnetic field can be measured with a small magnetic current (voltage source) as $\Delta\mathbf{V}|(\mathbf{J}_m\wedge\mathbf{H})$. Put together, the interaction of a source $\mathbf{J}_{e1}, \mathbf{J}_{m1}$ with the field $\mathbf{E}_2, \mathbf{H}_2$ due to a source $\mathbf{J}_{e2}, \mathbf{J}_{m2}$ is given by the quantity

$$<1,2>= \Delta\mathbf{V}|(\mathbf{J}_{e1}\wedge\mathbf{E}_2 - \mathbf{J}_{m1}\wedge\mathbf{H}_2), \tag{6.62}$$

called the reaction of field 2 to source 1 by Rumsey [67]. Reciprocity can then be expressed as symmetry of the reaction,

$$<1,2>=<2,1>. \tag{6.63}$$

Reciprocity can be effectively used in formulation and solution of electromagnetic problems [16]. It reduces the dimension of numerical procedures because certain system quantities like Green dyadics and scattering matrices can be made symmetric [78]. However, this requires that the medium satisfy certain conditions.

6.2.2 Medium conditions

Let us consider time-harmonic fields and their sources with suppressed time dependence $e^{jk_o\tau}$ in three-dimensional representation. Assuming two sets of electric and magnetic currents $\mathbf{J}_{e1}$, $\mathbf{J}_{m2}$ and $\mathbf{J}_{e2}$, $\mathbf{J}_{m2}$ creating corresponding fields in a linear medium, reciprocity requires that the reaction is symmetric [38,67]

$$(\mathbf{J}_{e1} \wedge \mathbf{E}_2 - \mathbf{J}_{m1} \wedge \mathbf{H}_2)|\mathcal{V} = (\mathbf{J}_{e2} \wedge \mathbf{E}_1 - \mathbf{J}_{m2} \wedge \mathbf{H}_1)|\mathcal{V}, \qquad (6.64)$$

integrated over all space. This gives a certain condition for the medium. To find it, let us proceed by inserting the Maxwell equations and expanding the following three-form expression as

$$\begin{aligned}\mathbf{\Gamma}_{12} &= \mathbf{J}_{e1} \wedge \mathbf{E}_2 - \mathbf{J}_{m1} \wedge \mathbf{H}_2 - \mathbf{J}_{e2} \wedge \mathbf{E}_1 + \mathbf{J}_{m2} \wedge \mathbf{H}_1 \\ &= (\mathbf{d} \wedge \mathbf{H}_1 - jk_o\mathbf{D}_1) \wedge \mathbf{E}_2 + (\mathbf{d} \wedge \mathbf{E}_1 + jk_o\mathbf{B}_1) \wedge \mathbf{H}_2 \\ &\quad -(\mathbf{d} \wedge \mathbf{H}_2 - jk_o\mathbf{D}_2) \wedge \mathbf{E}_1 - (\mathbf{d} \wedge \mathbf{E}_2 + jk_o\mathbf{B}_2) \wedge \mathbf{H}_1 \\ &= \mathbf{d} \wedge (\mathbf{E}_1 \wedge \mathbf{H}_2 - \mathbf{E}_2 \wedge \mathbf{H}_1) \\ &\quad +jk_o(-\mathbf{D}_1 \wedge \mathbf{E}_2 + \mathbf{B}_1 \wedge \mathbf{H}_2 + \mathbf{D}_2 \wedge \mathbf{E}_1 - \mathbf{B}_2 \wedge \mathbf{H}_1). \qquad (6.65)\end{aligned}$$

Picking a typical term from the last bracketed expression and applying the identity (5.156) rewritten for convenience as

$$\boldsymbol{\alpha} \wedge (\varepsilon_{123}\lfloor\mathbf{a}) = \varepsilon_{123}(\boldsymbol{\alpha}|\mathbf{a}), \qquad (6.66)$$

we can expand

$$\begin{aligned}\mathbf{D}_2 \wedge \mathbf{E}_1 &= \mathbf{E}_1 \wedge \mathbf{D}_2 = \mathbf{E}_1 \wedge (\overline{\overline{\epsilon}}|\mathbf{E}_2 + \overline{\overline{\xi}}|\mathbf{H}_2) \\ &= \mathbf{E}_1 \wedge \varepsilon_{123}\lfloor(\overline{\overline{\epsilon}}_g|\mathbf{E}_2 + \overline{\overline{\xi}}_g|\mathbf{H}_2) \\ &= \varepsilon_{123}\mathbf{E}_1|(\overline{\overline{\epsilon}}_g|\mathbf{E}_2 + \overline{\overline{\xi}}_g|\mathbf{H}_2), \qquad (6.67)\end{aligned}$$

where the modified (Gibbsian) medium dyadics have been introduced. Combining terms, we can finally write

$$\begin{aligned}\mathbf{\Gamma}_{12} &= \mathbf{d} \wedge (\mathbf{E}_1 \wedge \mathbf{H}_2 - \mathbf{E}_2 \wedge \mathbf{H}_1) \\ &\quad +jk_o\varepsilon_{123}\Big[\mathbf{E}_1|(\overline{\overline{\epsilon}}_g - \overline{\overline{\epsilon}}_g^T)|\mathbf{E}_2 - \mathbf{H}_1|(\overline{\overline{\mu}}_g - \overline{\overline{\mu}}_g^T)|\mathbf{H}_2 \\ &\quad +\mathbf{E}_1|(\overline{\overline{\xi}}_g + \overline{\overline{\zeta}}_g^T)|\mathbf{H}_2 - \mathbf{H}_1|(\overline{\overline{\zeta}}_g + \overline{\overline{\zeta}}_g^T)|\mathbf{E}_2\Big]. \qquad (6.68)\end{aligned}$$

For the medium to be reciprocal, the integral $\mathcal{V}|\mathbf{\Gamma}_{12}$ should vanish for any sources. Extending the integral over all space, from Stokes' theorem we can write

$$\mathcal{V}|\mathbf{d} \wedge (\mathbf{E}_1 \wedge \mathbf{H}_2 - \mathbf{E}_2 \wedge \mathbf{H}_1) = \partial\mathcal{V}|(\mathbf{E}_1 \wedge \mathbf{H}_2 - \mathbf{E}_2 \wedge \mathbf{H}_1), \qquad (6.69)$$

where $\partial\mathcal{V}$ denotes the surface in infinity. Assuming that the integral of the fields vanishes in the infinity, we are left with the equality of the volume integrals. Now the right-hand side vanishes totally when the Gibbsian medium dyadics satisfy a set of the conditions written as

$$\begin{pmatrix} \overline{\overline{\epsilon}}_g & \overline{\overline{\xi}}_g \\ \overline{\overline{\zeta}}_g & \overline{\overline{\mu}}_g \end{pmatrix}^T = \begin{pmatrix} \overline{\overline{\epsilon}}_g^T & \overline{\overline{\zeta}}_g^T \\ \overline{\overline{\xi}}_g^T & \overline{\overline{\mu}}_g^T \end{pmatrix} = \begin{pmatrix} \overline{\overline{\epsilon}}_g & -\overline{\overline{\xi}}_g \\ -\overline{\overline{\zeta}}_g & \overline{\overline{\mu}}_g \end{pmatrix}, \qquad (6.70)$$

In this case (6.64) is valid for any sources, which is the reciprocity condition for fields and sources: sources 1 as measured by the fields 2 equal sources 2 as measured by the fields 1. Equation (6.70) gives the reciprocity conditions for the bi-anisotropic medium [1, 38, 40]. It must be noted that it does not coincide with the symmetry conditions (5.241)–(5.243) associated with the existence of the Maxwell stress dyadic. Actually, the latter conditions correspond to those of (6.70) with missing minus signs, which is equivalent to the symmetry conditions

$$\overline{\overline{\mathsf{M}}}_g^T = \overline{\overline{\mathsf{M}}}_g$$

or

$$\overline{\overline{\mathcal{M}}}_g^T = \overline{\overline{\mathcal{M}}}_g. \tag{6.71}$$

Problems

6.2.1 Find conditions for the medium dyadics corresprondiing to a bi-anisotropic medium which is both reciprocal and lossless in the time-harmonic case.

6.2.2 Find the reciprocity conditions for the parameter dyadics $\overline{\overline{\alpha}}, \overline{\overline{\epsilon}}', \overline{\overline{\mu}}, \overline{\overline{\beta}}$ of the medium dyadic $\overline{\overline{\mathsf{M}}}$ corresponding to the general bi-anisotropic medium.

6.2.3 Check with the help of (5.27) that $\overline{\overline{\mathsf{M}}}_g \in \mathbb{E}_2\mathbb{E}_2$ is a symmetric dyadic under the reciprocity conditions (6.70) satisfied by the Gibbsian medium dyadics.

6.2.4 Show that the reciprocity condition is equivalent to requiring that the condition

$$\mathbf{E}_1 \wedge \mathbf{D}_2 - \mathbf{H}_1 \wedge \mathbf{B}_2 = \mathbf{E}_2 \wedge \mathbf{D}_1 - \mathbf{H}_2 \wedge \mathbf{B}_1$$

be valid for any two sets of fields 1 and 2.

6.2.5 Show that the reciprocity condition is equivalent to requiring that the condition

$$\mathbf{\Phi}_1 \wedge \mathbf{\Psi}_2^+ = \mathbf{\Phi}_2 \wedge \Psi_1^+$$

be valid for any two sets of fields 1 and 2, when we define the time-reversion operation as

$$\Psi^+ = (\mathbf{D} - \mathbf{H} \wedge \mathbf{d}\tau)^+ = \mathbf{D} + \mathbf{H} \wedge \mathbf{d}\tau.$$

6.3 EQUIVALENCE OF SOURCES

Two sources are called equivalent with respect to a region in space if they produce the same field in that region. Let us consider two systems of sources and their fields in terms of superforms,

$$\mathbf{d} \wedge \begin{pmatrix} \mathbf{\Psi}_1 \\ \mathbf{\Phi}_1 \end{pmatrix} = \begin{pmatrix} \gamma_{e1} \\ \gamma_{m1} \end{pmatrix}, \tag{6.72}$$

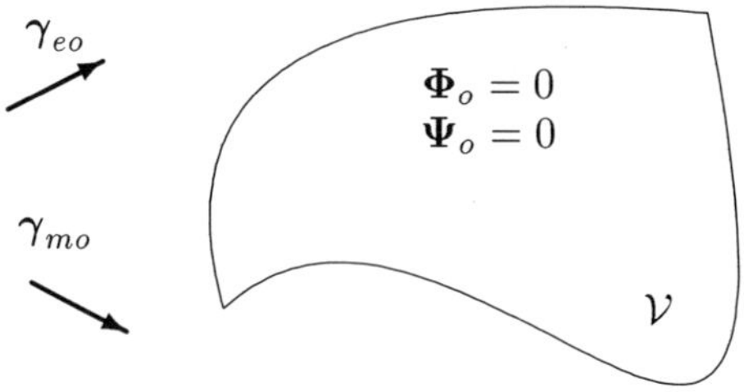

Fig. 6.3 *Nonradiating sources radiate zero field in the region $\mathcal{V}$.*

$$\mathbf{d} \wedge \begin{pmatrix} \mathbf{\Psi}_2 \\ \mathbf{\Phi}_2 \end{pmatrix} = \begin{pmatrix} \gamma_{e2} \\ \gamma_{m2} \end{pmatrix}. \tag{6.73}$$

Their difference is

$$\mathbf{d} \wedge \begin{pmatrix} \mathbf{\Psi}_o \\ \mathbf{\Phi}_o \end{pmatrix} = \begin{pmatrix} \gamma_{eo} \\ \gamma_{mo} \end{pmatrix}, \tag{6.74}$$

with

$$\mathbf{\Phi}_o = \mathbf{\Phi}_1 - \mathbf{\Phi}_2, \quad \mathbf{\Psi}_o = \mathbf{\Psi}_1 - \mathbf{\Psi}_2, \tag{6.75}$$

$$\gamma_{mo} = \gamma_{m1} - \gamma_{m2}, \quad \gamma_{eo} = \gamma_{e1} - \gamma_{e2}. \tag{6.76}$$

Now if the two fields 1 and 2 are the same in some region in space $\mathcal{V}$, the difference field vanishes, $\mathbf{\Phi}_o = \mathbf{\Psi}_o = 0$ in $\mathcal{V}$. The corresponding sources γ_{eo}, γ_{mo} are then called nonradiating in $\mathcal{V}$, and the source systems 1 and 2 are called equivalent with respect to $\mathcal{V}$ (Figure6.3). Obviously, in this case the sources γ_{eo}, γ_{mo} must vanish in $\mathcal{V}$, which means that the original sources must coincide within $\mathcal{V}$.

6.3.1 Nonradiating sources

Because of the conservation laws (4.35), (4.41), electromagnetic sources must be of the form

$$\begin{pmatrix} \gamma_{eo} \\ \gamma_{mo} \end{pmatrix} = \mathbf{d} \wedge \begin{pmatrix} \mathbf{\Gamma}_e \\ \mathbf{\Gamma}_m \end{pmatrix}, \tag{6.77}$$

for some two-forms $\mathbf{\Gamma}_e, \mathbf{\Gamma}_m$. Let us assume that both two-forms $\mathbf{\Gamma}_e, \mathbf{\Gamma}_m = 0$ vanish in the spatial region $\mathcal{V}$. Now it can be shown that sources of the form (6.77) are always nonradiating in $\mathcal{V}$. To see this, we write the Maxwell equations as

$$\mathbf{d} \wedge \begin{pmatrix} \mathbf{\Psi} - \mathbf{\Gamma}_e \\ \mathbf{\Phi} - \mathbf{\Gamma}_m \end{pmatrix} = \begin{pmatrix} 0 \\ 0 \end{pmatrix}, \tag{6.78}$$

from which we note that $\mathbf{\Psi} - \mathbf{\Gamma}_e$, $\mathbf{\Phi} - \mathbf{\Gamma}_m$ are field two-forms with no sources whatever. Assuming uniqueness conditions for the fields in infinity which prevent the existence of non-null fields, we conclude that the fields due to the sources (6.77) must actually be $\mathbf{\Psi} = \mathbf{\Gamma}_e$, $\mathbf{\Phi} = \mathbf{\Gamma}_m$ which vanish in $\mathcal{V}$ due to the assumption. Thus, sources of the form (6.77) are nonradiating in $\mathcal{V}$.

The converse can also be proven: any sources nonradiating in $\mathcal{V}$ can be presented in the form (6.77) through some two-forms $\mathbf{\Gamma}_e, \mathbf{\Gamma}_m$ which vanish in $\mathcal{V}$. The proof is trivial, because the sources of any nonradiating fields $\mathbf{\Psi}_{NR}, \mathbf{\Phi}_{NR}$ are

$$\begin{pmatrix} \gamma_e \\ \gamma_m \end{pmatrix} = \mathbf{d} \wedge \begin{pmatrix} \mathbf{\Psi}_{NR} \\ \mathbf{\Phi}_{NR} \end{pmatrix}. \tag{6.79}$$

These are of the form (6.77) because $\mathbf{\Psi}_{NR}, \mathbf{\Phi}_{NR}$ vanish in $\mathcal{V}$.

It is obvious that there exist an infinity of possible sources nonradiating in any given region. Certain symmetric current configurations do not radiate outside their support. For example, a radially symmetric current due to a nuclear detonation in a homogeneous space does not radiate [40]. The electromagnetic pulse (EMP) due to a nuclear blast in the atmosphere is due to the unsymmetry created by the atmospheric inhomogeneity.

6.3.2 Equivalent sources

Two sources are equivalent with respect to a region $\mathcal{V}$ if their difference is nonradiating in $\mathcal{V}$. Thus, with respect to a given region, there exists a class of infinitely many equivalent sources because nonradiating sources can be added to the original sources without changing the fields in $\mathcal{V}$. For example, a dipole source at a point outside $\mathcal{V}$ can be replaced by a multipole source at another point outside $\mathcal{V}$ or by Huygens' source on a chosen surface outside $\mathcal{V}$.

As an example, let us consider equivalence of electric and magnetic current non-static sources. For this we must borrow the wave equation (7.63) for the electric field one-form from Chapter 7:

$$\overline{\overline{\mathsf{W}}}_e(\mathbf{d_E}, \partial_\tau)|\mathbf{E} = -\partial_\tau \mathbf{J}_e - (\mathbf{d_E} \wedge \overline{\overline{\mathsf{I}}}^T - \overline{\overline{\xi}}\partial_\tau)|\overline{\overline{\mu}}^{-1}|\mathbf{J}_m, \tag{6.80}$$

where $\overline{\overline{\mathsf{W}}}_e(\mathbf{d_E}, \partial_\tau)$ is the wave-operator dyadic. Now let us assume that the magnetic source $\mathbf{J}_m(\mathbf{r}, \tau)$ is a two-form vanishing in the region $\mathcal{V}$ and the electric source $\mathbf{J}_e(\mathbf{r}, \tau)$ satisfies

$$\partial_\tau \mathbf{J}_e = -(\mathbf{d_E} \wedge \overline{\overline{\mathsf{I}}}^T - \overline{\overline{\xi}}\partial_\tau)|\overline{\overline{\mu}}^{-1}|\mathbf{J}_m, \tag{6.81}$$

whence it also vanishes in $\mathcal{V}$ if it vanishes for $\tau = -\infty$. Because in this case the source of the wave equation (6.80) vanishes everywhere, assuming uniqueness of solution, we conclude that the electric field must vanish everywhere. The magnetic field satisfies another equation similar to (6.80) but with right-hand side which does not necessarily vanish except in $\mathcal{V}$. Because sources and the electric field vanish in $\mathcal{V}$, it turns out that the magnetic field vanishes in $\mathcal{V}$. Because this all may appear quite confusing, let us summarize:

- $\mathbf{J}_m = 0$ in $\mathcal{V}$

- $\mathbf{J}_e$ satisfies (6.81), thus also $\mathbf{J}_e = 0$ in $\mathcal{V}$
- From (6.80) $\mathbf{E} = 0$ everywhere
- $\mathbf{H} = 0$ in $\mathcal{V}$

Because electric and magnetic fields vanish in $\mathcal{V}$, $\mathbf{J}_e$ satisfying (6.81) is equivalent to the source $\mathbf{J}_m$ with respect to $\mathcal{V}$. Thus, the electric source equivalent to the magnetic source $\mathbf{J}_m$ can be written as

$$\mathbf{J}_e^{eq} = -(\frac{1}{\partial_\tau}\mathbf{d}_\mathbf{E} \wedge \bar{\bar{\mathsf{I}}}^T - \bar{\bar{\xi}})|\bar{\bar{\mu}}^{-1}|\mathbf{J}_m, \tag{6.82}$$

where $1/\partial_\tau$ denotes integration with respect to τ. The dual form of this gives another equivalence relation

$$\mathbf{J}_m^{eq} = (\frac{1}{\partial_\tau}\mathbf{d}_\mathbf{E} \wedge \bar{\bar{\mathsf{I}}}^T + \bar{\bar{\zeta}})|\bar{\bar{\epsilon}}^{-1}|\mathbf{J}_e. \tag{6.83}$$

For time-harmonic sources with the time dependence $e^{jk_o\tau}$, these reduce to

$$\mathbf{J}_e^{eq} = -(\frac{1}{jk_o}\mathbf{d}_\mathbf{E} \wedge \bar{\bar{\mathsf{I}}}^T - \bar{\bar{\xi}})|\bar{\bar{\mu}}^{-1}|\mathbf{J}_m, \tag{6.84}$$

$$\mathbf{J}_m^{eq} = (\frac{1}{jk_o}\mathbf{d}_\mathbf{E} \wedge \bar{\bar{\mathsf{I}}}^T + \bar{\bar{\zeta}})|\bar{\bar{\epsilon}}^{-1}|\mathbf{J}_e \tag{6.85}$$

and, for simply anisotropic media, to

$$\mathbf{J}_e^{eq} = -\frac{1}{jk_o}\mathbf{d}_\mathbf{E} \wedge \bar{\bar{\mu}}^{-1}|\mathbf{J}_m, \tag{6.86}$$

$$\mathbf{J}_m^{eq} = \frac{1}{jk_o}\mathbf{d}_\mathbf{E} \wedge \bar{\bar{\epsilon}}^{-1}|\mathbf{J}_e. \tag{6.87}$$

Example As an example, let us find the equivalent of a time-harmonic magnetic current source of square cross section $2a \times 2a$ (Figure 6.4),

$$\mathbf{J}_m(\mathbf{r}) = I_m U(a^2 - x^2)U(a^2 - y^2)\mathbf{d}x \wedge \mathbf{d}y, \tag{6.88}$$

with respect to the region $\mathcal{V}$ outside the cylindrical region where $\mathbf{J}_m = 0$. Here $U(z)$ denotes the Heaviside unit step function. Assuming anisotropic medium, from (6.86) we have

$$\begin{aligned}\mathbf{J}_e^{eq} &= -\frac{I_m}{jk_o}\mathbf{d} \wedge \bar{\bar{\mu}}^{-1}|\mathbf{d}x \wedge \mathbf{d}y U(a^2 - x^2)U(a^2 - y^2) \\ &= -\frac{I_m}{jk_o}\mathbf{f}(x,y) \wedge \bar{\bar{\mu}}^{-1}|\mathbf{d}x \wedge \mathbf{d}y. \end{aligned} \tag{6.89}$$

Here we denote the one-form

$$\begin{aligned}\mathbf{f}(x,y) &= \mathbf{d}(U(a^2 - x^2)U(a^2 - y^2)) \\ &= \mathbf{d}x(\delta(x+a) - \delta(x-a))U(a^2 - y^2) \\ &\quad + \mathbf{d}y(\delta(y+a) - \delta(y-a))U(a^2 - x^2), \end{aligned} \tag{6.90}$$

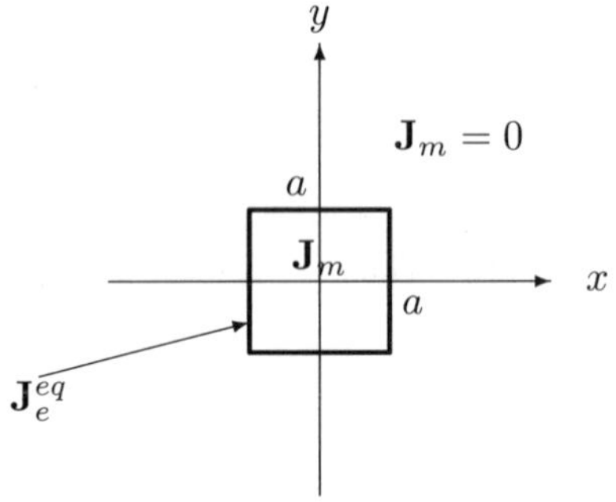

Fig. 6.4 *Square-cylindrical magnetic current $\mathbf{J}_m$ can be replaced by an equivalent electric current $\mathbf{J}_{\mathrm{e}}^{eq}$ which exists at the surface of the cylinder. The field from the two sources is the same outside the cylinder.*

which is a function vanishing everywhere except at the square cylindrical surface.

For the special permeability dyadic

$$\overline{\overline{\mu}} = \mu_{xx}(\mathbf{d}y \wedge \mathbf{d}z)\mathbf{e}_x + \mu_{yy}(\mathbf{d}z \wedge \mathbf{d}x)\mathbf{e}_y + \mu_{zz}(\mathbf{d}x \wedge \mathbf{d}y)\mathbf{e}_z, \tag{6.91}$$

with

$$\overline{\overline{\mu}}^{-1} = \mu_{xx}^{-1}\mathbf{d}x(\mathbf{e}_y \wedge \mathbf{e}_z) + \mu_{yy}^{-1}\mathbf{d}y(\mathbf{e}_z \wedge \mathbf{e}_x) + \mu_{zz}^{-1}\mathbf{d}z(\mathbf{e}_x \wedge \mathbf{e}_y), \tag{6.92}$$

the equivalent source becomes

$$\begin{aligned} \mathbf{J}_{\mathrm{e}}^{eq}(x,y) &= -\frac{I_m}{jk_o\mu_{zz}}\mathbf{f}(x,y) \wedge \mathbf{d}z(\mathbf{e}_x \wedge \mathbf{e}_y)\rfloor(\mathbf{d}x \wedge \mathbf{d}y) \\ &= -\frac{I_m}{jk_o\mu_{zz}}(\mathbf{d}y \wedge \mathbf{d}z)(\delta(y+a) - \delta(y-a))U(a^2 - x^2) \\ &\quad + \frac{I_m}{jk_o\mu_{zz}}(\mathbf{d}z \wedge \mathbf{d}x)(\delta(x+a) - \delta(x-a))U(a^2 - y^2). \end{aligned} \tag{6.93}$$

Outside the cylinder, both sources give rise to the same field while inside the cylinder the fields are different. A nonradiating source is the combination of $\mathbf{J}_m$ and $-\mathbf{J}_{\mathrm{e}}^{eq}$ which creates the null field outside the square cylinder.

Problems

6.3.1 Show that a current system depending on the space variable vector $\mathbf{r}$ through the scalar function $\boldsymbol{\nu}|\mathbf{r}$, where $\boldsymbol{\nu}$ is a constant dual vector, does not radiate outside the current region.

6.3.2 Show that a radially directed cylindrically symmetric current system does not radiate outside the current region.

6.3.3 Study the equivalence of electric charge density ϱ_e and magnetic current density $\mathbf{J}_m$ for the electrostatic field.

7

Electromagnetic Waves

In this chapter, basic electromagnetic problems are discussed based on multivector and dyadic operations. Wave equations for the potentials and field two-forms as well as basic solutions for plane waves and fields from point sources, are considered.

7.1 WAVE EQUATION FOR POTENTIALS

Wave equations for the potentials can be easily constructed in the four-dimensional formulation. In case there are both electric and magnetic sources present, the problem represented by the basic equations

$$\mathbf{d} \wedge \mathbf{\Phi} = \gamma_m, \quad \mathbf{d} \wedge \mathbf{\Psi} = \gamma_e, \quad \mathbf{\Psi} = \overline{\overline{\mathsf{M}}}|\mathbf{\Phi} \tag{7.1}$$

can be split in two:

$$\mathbf{d} \wedge \mathbf{\Phi}_e = 0, \quad \mathbf{d} \wedge \mathbf{\Psi}_e = \gamma_e, \quad \mathbf{\Psi}_e = \overline{\overline{\mathsf{M}}}|\mathbf{\Phi}_e \tag{7.2}$$

$$\mathbf{d} \wedge \mathbf{\Phi}_m = \gamma_m, \quad \mathbf{d} \wedge \mathbf{\Psi}_m = 0, \quad \mathbf{\Psi}_m = \overline{\overline{\mathsf{M}}}|\mathbf{\Phi}_m, \tag{7.3}$$

with

$$\mathbf{\Phi} = \mathbf{\Phi}_e + \Phi_m, \quad \mathbf{\Psi} = \mathbf{\Psi}_e + \Psi_m. \tag{7.4}$$

From (7.2) there exists an electric four-potential $\boldsymbol{\alpha}_e$, generated by the electric source γ_e, and from (7.3) there exists a magnetic four-potential $\boldsymbol{\alpha}_m$, generated by the magnetic source γ_m:

$$\mathbf{\Phi}_e = \mathbf{d} \wedge \boldsymbol{\alpha}_e, \quad \mathbf{\Psi}_m = \mathbf{d} \wedge \boldsymbol{\alpha}_m. \tag{7.5}$$

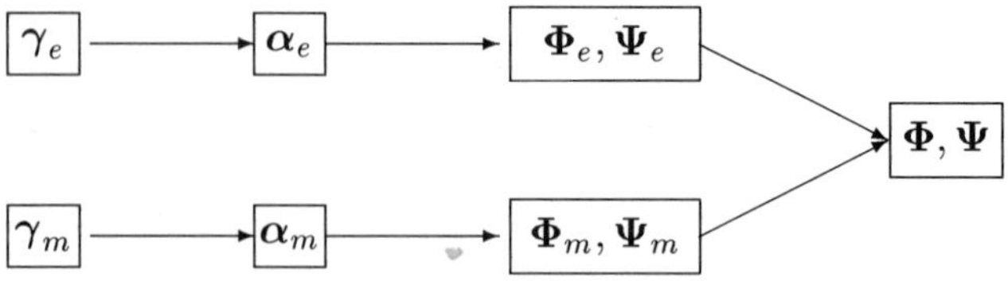

Fig. 7.1 *There is no single potential $\boldsymbol{\alpha}$ in the case when both electric and magnetic sources are present. However, the problem can be split into two parts, solving the electric potential $\boldsymbol{\alpha}_e$ from electric sources and solving the magnetic potential $\boldsymbol{\alpha}_m$ from magnetic sources.*

The total field two-form $\boldsymbol{\Phi}$ is obtained as the sum of both components (Figure 7.1).

The wave equation corresponding to the electric source can be formed by substituting $\boldsymbol{\Phi}_e = \mathbf{d}\wedge\boldsymbol{\alpha}_e$ and the medium equation $\boldsymbol{\Psi}_e = \overline{\overline{\mathsf{M}}}|\boldsymbol{\Phi}_e$ in the Maxwell equation $\mathbf{d}\wedge\boldsymbol{\Psi}_e = \gamma_e$, whence we obtain

$$\mathbf{d}\wedge\overline{\overline{\mathsf{M}}}|(\mathbf{d}\wedge\boldsymbol{\alpha}_e) = \gamma_e. \tag{7.6}$$

Decomposed into its three-dimensional parts for the four-potential $\boldsymbol{\alpha}_e = \mathbf{A}_e - \phi_e\mathbf{d}\tau$ this gives (details left as a problem)

$$\mathbf{d}\wedge\overline{\overline{\alpha}}|(\mathbf{d}\wedge\mathbf{A}_e) - \mathbf{d}\wedge\overline{\overline{\epsilon}}'|\mathbf{d}\wedge\phi_e = \varrho_e, \tag{7.7}$$

$$\mathbf{d}\wedge\overline{\overline{\mu}}^{-1}|(\mathbf{d}\wedge\mathbf{A}_e) - \mathbf{d}\wedge\overline{\overline{\beta}}|\mathbf{d}\phi_e = \mathbf{J}_e. \tag{7.8}$$

The corresponding wave equation for the magnetic potential one-form $\boldsymbol{\alpha}_m$ is

$$\mathbf{d}\wedge\overline{\overline{\mathsf{M}}}^{-1}|(\mathbf{d}\wedge\boldsymbol{\alpha}_m) = \gamma_m, \tag{7.9}$$

which can also be obtained through the dual substitution of Section 6.1. Both (7.6) and (7.9) are three-form equations of the second order for one-form potentials. Thus, the dimension of the equation matches that of the unknown in both cases.

7.1.1 Electric four-potential

Let us first consider the electric four-potential $\boldsymbol{\alpha}_e$ and try to simplify the wave equation (7.6). Here we can apply the following vector identity valid for any dual vector $\boldsymbol{\beta}$ and dual bivector $\boldsymbol{\Pi}$

$$\mathbf{e}_N\lfloor(\boldsymbol{\beta}\wedge\boldsymbol{\Pi}) = (\mathbf{e}_N\lfloor\boldsymbol{\Pi})\lfloor\boldsymbol{\beta}. \tag{7.10}$$

In fact, writing (7.6) as

$$\mathbf{e}_N\lfloor(\mathbf{d}\wedge\overline{\overline{\mathsf{M}}})|(\mathbf{d}\wedge\boldsymbol{\alpha}_e) = \mathbf{e}_N\lfloor\gamma_e, \tag{7.11}$$

we can define the second-order dyadic wave operator $\overline{\overline{\mathsf{V}}}_e(\mathbf{d}) \in \mathbb{E}_1\mathbb{E}_1$ by

$$\overline{\overline{\mathsf{V}}}_e(\mathbf{d}) = (\mathbf{e}_N\lfloor\overline{\overline{\mathsf{M}}})\lfloor\lfloor\mathbf{dd} = \overline{\overline{\mathsf{M}}}_g\lfloor\lfloor\mathbf{dd}. \tag{7.12}$$

$\overline{\overline{\mathsf{V}}}_e(\mathbf{d})$ is a metric dyadic operator mapping one-forms to vectors and depending only on the medium dyadic $\overline{\overline{\mathsf{M}}}$ or $\overline{\overline{\mathsf{M}}}_g$. The wave equation has now the following vector form for the potential one-form $\boldsymbol{\alpha}_e$:

$$\overline{\overline{\mathsf{V}}}_e(\mathbf{d})|\boldsymbol{\alpha}_e = \mathbf{e}_N\lfloor\boldsymbol{\gamma}_e, \tag{7.13}$$

whose operator appears simpler than that in (7.6). Solutions to this equation are not unique since we can add to $\boldsymbol{\alpha}_e$ the one-form $\mathbf{d}\phi$ with any scalar ϕ. Thus, we must impose an extra gauge condition for $\boldsymbol{\alpha}_e$ to make it unique. For example, we could choose any vector $\mathbf{a}$ and require $\boldsymbol{\alpha}_e$ to satisfy the condition $\mathbf{a}|\boldsymbol{\alpha}_e = 0$. However, more proper gauge conditions leading to simpler wave equations can be chosen for some electromagnetic media.

7.1.2 Magnetic four-potential

The magnetic four-potential obeys an equation somewhat similar to (7.13). Operating on (7.9) as

$$\mathbf{e}_N\lfloor(\mathbf{d}\wedge\overline{\overline{\mathsf{M}}}^{-1})|(\mathbf{d}\wedge\boldsymbol{\alpha}_m) = \mathbf{e}_N\lfloor\boldsymbol{\gamma}_m \tag{7.14}$$

and denoting

$$\overline{\overline{\mathsf{V}}}_m(\mathbf{d}) = \mathbf{e}_N\lfloor(\mathbf{d}\wedge\overline{\overline{\mathsf{M}}}^{-1})\lfloor\mathbf{d} = (\mathbf{e}_N\lfloor\overline{\overline{\mathsf{M}}}^{-1})\lfloor\lfloor\mathbf{dd}, \tag{7.15}$$

we obtain an equation dual to (7.13)

$$\overline{\overline{\mathsf{V}}}_m(\mathbf{d})|\boldsymbol{\alpha}_m = \mathbf{e}_N\lfloor\boldsymbol{\gamma}_m. \tag{7.16}$$

Applying the identity

$$\mathbf{e}_N\lfloor\overline{\overline{\mathsf{M}}}^{-1} = (\overline{\overline{\mathsf{M}}}\rfloor\boldsymbol{\varepsilon}_N)^{-1}, \tag{7.17}$$

the operator can be given another form connected to the modified medium dyadic $\overline{\overline{\mathsf{M}}}_g$,

$$\overline{\overline{\mathsf{V}}}_m(\mathbf{d}) = (\overline{\overline{\mathsf{M}}}\rfloor\boldsymbol{\varepsilon}_N)^{-1}\lfloor\lfloor\mathbf{dd} = (\boldsymbol{\varepsilon}_N\boldsymbol{\varepsilon}_N\lfloor\lfloor\overline{\overline{\mathsf{M}}}_g)^{-1}\lfloor\lfloor\mathbf{dd}. \tag{7.18}$$

7.1.3 Anisotropic medium

As an example of a wave operator $\overline{\overline{\mathsf{V}}}_e(\mathbf{d})$ in the wave equation for the electric four-potential (7.13), let us consider the simple anisotropic medium defined by $\overline{\overline{\xi}} = \overline{\overline{\zeta}} = 0$ or $\overline{\overline{\alpha}} = \overline{\overline{\beta}} = 0$. In this case we have $\overline{\overline{\epsilon}}' = \overline{\overline{\epsilon}}$ and the modified medium dyadic becomes

$$\overline{\overline{\mathsf{M}}}_g = \mathbf{e}_{1234}\lfloor(\overline{\overline{\epsilon}}\wedge\mathbf{e}_\tau + \mathbf{d}\tau\wedge\overline{\overline{\mu}}^{-1}). \tag{7.19}$$

Let us first expand these two terms separately. Applying the identities

$$\mathbf{e}_{1234}\lfloor\boldsymbol{\Pi}_{\mathbf{E}} = -\mathbf{e}_4\wedge(\mathbf{e}_{123}\lfloor\boldsymbol{\Pi}_{\mathbf{E}}), \tag{7.20}$$

$$\mathbf{e}_{1234}\lfloor(\varepsilon_4\wedge\boldsymbol{\alpha}_{\mathbf{E}}) = -\mathbf{e}_{4123}\lfloor(\varepsilon_4\wedge\boldsymbol{\alpha}_{\mathbf{E}}) = -\mathbf{e}_{123}\lfloor\boldsymbol{\alpha}_{\mathbf{E}}, \tag{7.21}$$

valid for any Euclidean two-form $\boldsymbol{\Pi}_{\mathbf{E}}$ and one-form $\boldsymbol{\alpha}_{\mathbf{E}}$, we obtain

$$\begin{aligned}\mathbf{e}_{1234}\lfloor\overline{\overline{\epsilon}}\wedge\mathbf{e}_\tau &= -\mathbf{e}_\tau\wedge(\mathbf{e}_{123}\lfloor\overline{\overline{\epsilon}})\wedge\mathbf{e}_\tau = -\mathbf{e}_\tau\wedge\overline{\overline{\epsilon}}_g\wedge\mathbf{e}_\tau\\ &= \overline{\overline{\epsilon}}_g{}^\wedge_\wedge\mathbf{e}_\tau\mathbf{e}_\tau,\end{aligned} \tag{7.22}$$

$$\begin{aligned}\mathbf{e}_{1234}\lfloor(\mathbf{d}\tau\wedge\overline{\overline{\mu}}^{-1}) &= -\mathbf{e}_{123}\lfloor(\varepsilon_{123}\lfloor\overline{\overline{\mu}}_g)^{-1} = -\mathbf{e}_{123}\lfloor\overline{\overline{\mu}}_g^{-1}\rfloor\mathbf{e}_{123}\\ &= -\mathbf{e}_{123}\mathbf{e}_{123}\lfloor\lfloor\overline{\overline{\mu}}_g^{-1}\\ &= -\mathbf{e}_{123}\mathbf{e}_{123}\lfloor\lfloor\frac{\overline{\overline{\mu}}_g^{(2)T}\rfloor\rfloor\varepsilon_{123}\varepsilon_{123}}{\overline{\overline{\mu}}_g^{(3)}||\varepsilon_{123}\varepsilon_{123}}\\ &= -\frac{\overline{\overline{\mu}}_g^{(2)T}}{\overline{\overline{\mu}}_g^{(3)}||\varepsilon_{123}\varepsilon_{123}},\end{aligned} \tag{7.23}$$

where, again, $\overline{\overline{\epsilon}}_g$ and $\overline{\overline{\mu}}_g$ are the Gibbsian permittivity and permeability dyadics. Here we have also applied the formula for the inverse of a dyadic, (2.169). The wave operator now becomes

$$\overline{\overline{\mathsf{V}}}_e(\mathbf{d}) = \left(\overline{\overline{\epsilon}}_g{}^\wedge_\wedge\mathbf{e}_\tau\mathbf{e}_\tau - \frac{\overline{\overline{\mu}}_g^{(2)T}}{\overline{\overline{\mu}}_g^{(3)}||\varepsilon_{123}\varepsilon_{123}}\right)\lfloor\lfloor\mathbf{dd}. \tag{7.24}$$

The nature of this operator can be seen more closely by applying the identity (2.91) reproduced here as

$$(\overline{\overline{\mathsf{A}}}{}^\wedge_\wedge\overline{\overline{\mathsf{B}}})\lfloor\lfloor\overline{\overline{\mathsf{C}}}^T = (\overline{\overline{\mathsf{A}}}||\overline{\overline{\mathsf{C}}}^T)\overline{\overline{\mathsf{B}}} + (\overline{\overline{\mathsf{B}}}||\overline{\overline{\mathsf{C}}}^T)\overline{\overline{\mathsf{A}}} - \overline{\overline{\mathsf{A}}}|\overline{\overline{\mathsf{C}}}|\overline{\overline{\mathsf{B}}} - \overline{\overline{\mathsf{B}}}|\overline{\overline{\mathsf{C}}}|\overline{\overline{\mathsf{A}}}, \tag{7.25}$$

which can be applied also in the case when $\overline{\overline{\mathsf{A}}}, \overline{\overline{\mathsf{B}}}$ are $\mathbb{E}_1\mathbb{E}_1$ dyadics and $\overline{\overline{\mathsf{C}}}$ is a $\mathbb{F}_1\mathbb{F}_1$ dyadic. The two terms of (7.24) can now be expanded as

$$\overline{\overline{\epsilon}}_g{}^\wedge_\wedge\mathbf{e}_\tau\mathbf{e}_\tau\lfloor\lfloor\mathbf{dd} = (\overline{\overline{\epsilon}}_g||\mathbf{dd})\mathbf{e}_\tau\mathbf{e}_\tau + \partial_\tau^2\overline{\overline{\epsilon}}_g - (\overline{\overline{\epsilon}}_g|\mathbf{d})\partial_\tau\mathbf{e}_\tau - \mathbf{e}_\tau\partial_\tau(\mathbf{d}|\overline{\overline{\epsilon}}_g), \tag{7.26}$$

$$\overline{\overline{\mu}}_g^{(2)T}\lfloor\lfloor\mathbf{dd} = (\overline{\overline{\mu}}_g||\mathbf{dd})\overline{\overline{\mu}}_g{}^T - (\mathbf{d}|\overline{\overline{\mu}}_g)(\overline{\overline{\mu}}_g|\mathbf{d}). \tag{7.27}$$

The wave operator corresponding to the magnetic potential can be obtained from (7.24) through dual substition. In fact, applying the substitution for the medium,

$$\overline{\overline{\mathsf{M}}} \to \overline{\overline{\mathsf{M}}}^{-1} = (\overline{\overline{\epsilon}}\wedge\mathbf{e}_\tau + \mathbf{d}\tau\wedge\overline{\overline{\mu}}^{-1})^{-1} = -\overline{\overline{\mu}}\wedge\mathbf{e}_\tau - \mathbf{d}\tau\wedge\overline{\overline{\epsilon}}^{-1}, \tag{7.28}$$

which corresponds to

$$\overline{\overline{\epsilon}} \to -\overline{\overline{\mu}}, \quad \overline{\overline{\mu}} \to -\overline{\overline{\epsilon}}, \tag{7.29}$$

the magnetic wave operator is obtained in the form

$$\overline{\overline{\mathsf{V}}}_m(\mathbf{d}) = -\left(\overline{\overline{\mu}}_g{}^\wedge_\wedge\mathbf{e}_\tau\mathbf{e}_\tau - \frac{\overline{\overline{\epsilon}}_g^{(2)T}}{\overline{\overline{\epsilon}}_g^{(3)}||\varepsilon_{123}\varepsilon_{123}}\right)\lfloor\lfloor\mathbf{dd}. \tag{7.30}$$

7.1.4 Special anisotropic medium

As a more specific example, let us consider a reciprocal anisotropic medium whose medium dyadics $\overline{\overline{\epsilon}}_g$ and $\overline{\overline{\mu}}_g$ are multiples of the same symmetric Euclidean metric dyadic $\overline{\overline{\mathsf{S}}} \in \mathbb{E}_1\,\mathbb{E}_1$ as defined by

$$\overline{\overline{\epsilon}}_g = \epsilon\overline{\overline{\mathsf{S}}}, \quad \overline{\overline{\mu}}_g = \mu\overline{\overline{\mathsf{S}}}, \quad \overline{\overline{\mathsf{S}}}^{(3)} = \mathbf{e}_{123}\mathbf{e}_{123}. \tag{7.31}$$

Here ϵ and μ are dimensionless numbers ("relative permittivity and permeability"). This choice of medium leads to the modified medium dyadic of the special form

$$\overline{\overline{\mathsf{M}}}_g = \frac{1}{\mu}(\mu\epsilon\overline{\overline{\mathsf{S}}}{}^{\wedge}_{\wedge}\mathbf{e}_\tau\mathbf{e}_\tau - \overline{\overline{\mathsf{S}}}^{(2)}) = -\frac{1}{\mu}\overline{\overline{\mathsf{P}}}^{(2)}, \tag{7.32}$$

where we have introduced for brevity the $\overline{\overline{\mathsf{P}}}$ dyadic defined by

$$\overline{\overline{\mathsf{P}}} = \overline{\overline{\mathsf{S}}} - \mu\epsilon\mathbf{e}_\tau\mathbf{e}_\tau. \tag{7.33}$$

Its inverse has the form

$$\overline{\overline{\mathsf{P}}}^{-1} = \overline{\overline{\mathsf{S}}}^{-1} - \frac{1}{\mu\epsilon}\mathbf{d}_T\mathbf{d}_T. \tag{7.34}$$

Actually, this medium is a special case of the Q-medium defined in Section 5.1 with $\overline{\overline{\xi}}_g = \overline{\overline{\zeta}}_g = 0$. Obviously, $\overline{\overline{\mathsf{P}}} \in \mathbb{E}_1\,\mathbb{E}_1$ is a Minkowskian metric dyadic, and it depends on the medium only. $\overline{\overline{\mathsf{S}}}^{-1}$ denotes the three-dimensional inverse.

Now we can again apply the identity (7.25) with the special case

$$\overline{\overline{\mathsf{P}}}^{(2)}\lfloor\lfloor\boldsymbol{\alpha}\boldsymbol{\alpha} = \overline{\overline{\mathsf{P}}}(\overline{\overline{\mathsf{P}}}||\boldsymbol{\alpha}\boldsymbol{\alpha}) - (\overline{\overline{\mathsf{P}}}|\boldsymbol{\alpha})(\boldsymbol{\alpha}|\overline{\overline{\mathsf{P}}}) \tag{7.35}$$

to the operator (7.24), which for this special medium becomes

$$\overline{\overline{\mathsf{V}}}_e(\mathbf{d}) = -\frac{1}{\mu}\overline{\overline{\mathsf{P}}}^{(2)}\lfloor\lfloor\mathbf{dd} = -\frac{1}{\mu}(\overline{\overline{\mathsf{P}}}(\mathbf{d}|\overline{\overline{\mathsf{P}}}|\mathbf{d}) - \overline{\overline{\mathsf{P}}}|\mathbf{dd}|\overline{\overline{\mathsf{P}}}). \tag{7.36}$$

The equation for the electric four-potential can thus be expressed as

$$(\mathbf{d}|\overline{\overline{\mathsf{P}}}|\mathbf{d})\overline{\overline{\mathsf{P}}}|\boldsymbol{\alpha}_e - (\overline{\overline{\mathsf{P}}}|\mathbf{d})(\mathbf{d}|\overline{\overline{\mathsf{P}}}|\boldsymbol{\alpha}_e) = -\mu\mathbf{e}_{1234}\lfloor\gamma_e. \tag{7.37}$$

Gauge condition The potential can be made unique by imposing an additional scalar-valued gauge condition. From (7.37) one can see that by choosing the condition

$$\mathbf{d}|\overline{\overline{\mathsf{P}}}|\boldsymbol{\alpha}_e = (\mathbf{d}_{\mathbf{E}}|\overline{\overline{\mathsf{S}}} - \mu\epsilon\partial_\tau\mathbf{e}_\tau)|\boldsymbol{\alpha}_e = 0, \tag{7.38}$$

the wave equation is reduced to

$$(\mathbf{d}|\overline{\overline{\mathsf{P}}}|\mathbf{d})\overline{\overline{\mathsf{P}}}|\boldsymbol{\alpha}_e = (\mathbf{d}_{\mathbf{E}}|\overline{\overline{\mathsf{S}}}|\mathbf{d}_{\mathbf{E}} - \mu\epsilon\partial_\tau^2)\overline{\overline{\mathsf{P}}}|\boldsymbol{\alpha}_e = -\mu\mathbf{e}_{1234}\lfloor\gamma_e \tag{7.39}$$

or, equivalently, to

$$(\mathbf{d}|\overline{\overline{\mathsf{P}}}|\mathbf{d})\boldsymbol{\alpha}_e = -\mu\overline{\overline{\mathsf{P}}}^{-1}|(\mathbf{e}_{1234}\lfloor\boldsymbol{\gamma}_e) = \mu\overline{\overline{\mathsf{P}}}^{-1} \wedge \boldsymbol{\gamma}_e|\mathbf{e}_{1234}. \tag{7.40}$$

Actually, the gauge condition (7.38) is satisfied by the solution of the wave equation (7.39). In fact, operating (7.39) by $\mathbf{d}|$ and applying the rule

$$\boldsymbol{\beta}|(\mathbf{e}_{1234}\lfloor\boldsymbol{\Gamma}) = \mathbf{e}_{1234}\lfloor(\boldsymbol{\Gamma} \wedge \boldsymbol{\beta}) = -\mathbf{e}_{1234}|(\boldsymbol{\beta} \wedge \boldsymbol{\Gamma}), \tag{7.41}$$

valid for any dual vector $\boldsymbol{\beta}$ and dual trivector $\boldsymbol{\Gamma}$, we have

$$(\mathbf{d}|\overline{\overline{\mathsf{P}}}|\mathbf{d})(\mathbf{d}|\overline{\overline{\mathsf{P}}}|\boldsymbol{\alpha}_e) = -\mu\mathbf{d}|(\mathbf{e}_{1234}\lfloor\boldsymbol{\gamma}_e) = \mu\mathbf{e}_{1234}|(\mathbf{d} \wedge \boldsymbol{\gamma}_e) = 0 \tag{7.42}$$

because of the conservation law $\mathbf{d} \wedge \boldsymbol{\gamma}_e = 0$. This means that there is no source for the scalar field $\mathbf{d}|\overline{\overline{\mathsf{P}}}|\boldsymbol{\alpha}_e$ and, assuming uniqueness for the wave equation, it must vanish. Thus, the gauge condition is already contained in the solution of (7.39).

Magnetic potential The corresponding equations for the magnetic four-potential are again obtained through the dual substitution $\mu \leftrightarrow -\epsilon$ which leaves the dyadic $\overline{\overline{\mathsf{P}}}$ invariant:

$$(\mathbf{d}|\overline{\overline{\mathsf{P}}}|\mathbf{d})\overline{\overline{\mathsf{P}}}|\boldsymbol{\alpha}_m = (\mathbf{d}|\overline{\overline{\mathsf{S}}}|\mathbf{d} - \mu\epsilon\partial_\tau^2)\overline{\overline{\mathsf{P}}}|\boldsymbol{\alpha}_m = \epsilon\mathbf{e}_{1234}\lfloor\boldsymbol{\gamma}_m, \tag{7.43}$$

$$(\mathbf{d}|\overline{\overline{\mathsf{P}}}|\mathbf{d})\boldsymbol{\alpha}_m = \epsilon\overline{\overline{\mathsf{P}}}^{-1}|(\mathbf{e}_{1234}\lfloor\boldsymbol{\gamma}_m) = \epsilon\overline{\overline{\mathsf{P}}}^{-1} \wedge \boldsymbol{\gamma}_m|\mathbf{e}_{1234}, \tag{7.44}$$

$$\mathbf{d}|\overline{\overline{\mathsf{P}}}|\boldsymbol{\alpha}_m = (\mathbf{d}_\mathbf{E}|\overline{\overline{\mathsf{S}}} - \mu\epsilon\partial_\tau\mathbf{e}_\tau)|\boldsymbol{\alpha}_m = 0. \tag{7.45}$$

The last one is the gauge condition which is satisfied by the solution of the equation (7.43).

7.1.5 Three-dimensional equations

By defining the vector and scalar operators in three dimensions as

$$\nabla = \overline{\overline{\mathsf{S}}}|\mathbf{d}_\mathbf{E} = \mathbf{d}_\mathbf{E}|\overline{\overline{\mathsf{S}}}, \quad \nabla^2 = \mathbf{d}_\mathbf{E}|\overline{\overline{\mathsf{S}}}|\mathbf{d}_\mathbf{E} \tag{7.46}$$

and in four dimensions as

$$\Box = \overline{\overline{\mathsf{P}}}|\mathbf{d} = \mathbf{d}|\overline{\overline{\mathsf{P}}}, \quad \Box^2 = \mathbf{d}|\overline{\overline{\mathsf{P}}}|\mathbf{d} = \mathbf{d}_\mathbf{E}|\overline{\overline{\mathsf{S}}}|\mathbf{d}_\mathbf{E} - \mu\epsilon\partial_\tau^2, \tag{7.47}$$

we can express (7.39) as

$$\Box^2\overline{\overline{\mathsf{P}}}|\boldsymbol{\alpha}_e = -\mu\mathbf{e}_{1234}\lfloor\boldsymbol{\gamma}_e. \tag{7.48}$$

Decomposition in its spatial and temporal parts gives the more familiar-looking wave equations (details left as a problem)

$$(\nabla^2 - \mu\epsilon\partial_\tau^2)\overline{\overline{\mathsf{S}}}|\boldsymbol{\alpha}_e = -\mu\mathbf{e}_{123}\lfloor\mathbf{J}_e, \tag{7.49}$$

$$(\nabla^2 - \mu\epsilon\partial_\tau^2)\phi_e = -\frac{1}{\epsilon}\mathbf{e}_{123}|\boldsymbol{\varrho}_e. \tag{7.50}$$

Here the vector $\mathbf{e}_{123}\lfloor\mathbf{J}_e$ and the scalar $\mathbf{e}_{123}\lfloor\varrho_e$ correspond to the Gibbsian current density vector and charge density scalar, and $\overline{\overline{\mathsf{S}}}|\mathbf{A}_e$ corresponds to the Gibbsian vector potential. Equations (7.49) and (7.50) look similar to the wave equations in the classical vector language,

$$(\nabla^2 - \mu\epsilon\partial_t^2)\mathbf{A} = -\mu\mathbf{J}, \tag{7.51}$$

$$(\nabla^2 - \mu\epsilon\partial_t^2)\phi = -\varrho/\epsilon. \tag{7.52}$$

Also, the gauge condition (7.38) has the form

$$(\nabla - \mu\epsilon\partial_\tau\mathbf{e}_\tau)|\boldsymbol{\alpha}_e = \nabla|\mathbf{A}_e + \mu\epsilon\partial_\tau\phi_e = 0, \tag{7.53}$$

which corresponds to the familiar Lorenz condition

$$\nabla\cdot\mathbf{A} + \mu\epsilon\partial_t\phi = 0. \tag{7.54}$$

The special medium considered here corresponds to what is known as an affine-isotropic medium in Gibbsian electromagnetics [40]. It can be generalized by relaxing the assumption of symmetry for the $\overline{\overline{\mathsf{S}}}$ dyadic, which in that case becomes a metric dyadic.

7.1.6 Equations for field two-forms

From the potential equations we can easily write equations for the corresponding field two-forms. Because the two-form $\mathbf{\Phi}$ is obtained from the electric four-potential as $\mathbf{d}\wedge\boldsymbol{\alpha}_e$, from (7.40) valid for the special anisotropic medium (7.31), we can form the wave equation for the field two-form as

$$(\mathbf{d}|\overline{\overline{\mathsf{P}}}|\mathbf{d})\mathbf{\Phi} = -\mu\mathbf{d}\wedge\overline{\overline{\mathsf{P}}}^{-1}|(\mathbf{e}_{1234}\lfloor\boldsymbol{\gamma}_e). \tag{7.55}$$

However, this equation assumes absence of magnetic sources. In the dual case with no electric sources present, the two-form $\mathbf{\Psi}$ is expressed as $\mathbf{d}\wedge\boldsymbol{\alpha}_m$ and it satisfies the dual wave equation

$$(\mathbf{d}|\overline{\overline{\mathsf{P}}}|\mathbf{d})\mathbf{\Psi} = \epsilon\mathbf{d}\wedge\overline{\overline{\mathsf{P}}}^{-1}|(\mathbf{e}_{1234}\lfloor\boldsymbol{\gamma}_m). \tag{7.56}$$

More general case will be considered in the next section.

Problems

7.1.1 Derive (7.7) and (7.8) from (7.6).

7.1.2 Express the operator $\overline{\overline{\mathsf{M}}}_g\lfloor\lfloor\mathbf{dd}$ in explicit form for the general bi-anisotropic medium.

7.1.3 Derive the dual substitution rule (7.29) for the anisotropic medium.

7.1.4 Do the details in the derivation of (7.49) and (7.50) from (7.40).

7.1.5 Find the wave equation (7.39) for the case that the symmetric dyadic $\overline{\overline{\mathsf{S}}}$ is replaced by a metric dyadic which is not necessarily symmetric.

7.2 WAVE EQUATION FOR FIELDS

7.2.1 Three-dimensional field equations

In Gibbsian vector notation the wave equation for one of the field vectors is derived from the Maxwell equations and medium equations through elimination of the other field vectors. In the three-dimensional differential-form notation the derivation follows quite similar steps. Applying the bi-anisotropic medium conditions in the form

$$\mathbf{D} = \overline{\overline{\epsilon}}|\mathbf{E} + \overline{\overline{\xi}}|\mathbf{H}, \tag{7.57}$$

$$\mathbf{B} = \overline{\overline{\zeta}}|\mathbf{E} + \overline{\overline{\mu}}|\mathbf{H}, \tag{7.58}$$

the Maxwell equations become

$$(\mathbf{d}_\mathbf{E} \wedge \overline{\overline{\mathsf{I}}}^T + \overline{\overline{\zeta}}\partial_\tau)|\mathbf{E} + \overline{\overline{\mu}}\partial_\tau|\mathbf{H} = -\mathbf{J}_m, \tag{7.59}$$

$$(\mathbf{d}_\mathbf{E} \wedge \overline{\overline{\mathsf{I}}}^T - \overline{\overline{\xi}}\partial_\tau)|\mathbf{H} - \overline{\overline{\epsilon}}\partial_\tau|\mathbf{E} = \mathbf{J}_e. \tag{7.60}$$

These equations show us that knowing one of the field one-forms the other one is known without an arbitrary static component. For example, knowing the electric one-form $\mathbf{E}$, from (7.59) we cab solve

$$\mathbf{H} = -\overline{\overline{\mu}}^{-1}|\frac{1}{\partial_\tau}(\mathbf{d}_\mathbf{E} \wedge \overline{\overline{\mathsf{I}}}^T + \overline{\overline{\zeta}}\partial_\tau)|\mathbf{E} - \overline{\overline{\mu}}^{-1}|\frac{1}{\partial_\tau}\mathbf{J}_m, \tag{7.61}$$

where $1/\partial_\tau$ stands for integration in τ. Knowing the initial value $\mathbf{H}(-\infty)$ solves the field for all times. Similarly, knowing the magnetic field one-form $\mathbf{H}$ gives

$$\mathbf{E} = \overline{\overline{\epsilon}}^{-1}|\frac{1}{\partial_\tau}(\mathbf{d}_\mathbf{E} \wedge \overline{\overline{\mathsf{I}}}^T - \overline{\overline{\xi}}\partial_\tau)|\mathbf{H} - \overline{\overline{\epsilon}}^{-1}|\frac{1}{\partial_\tau}\mathbf{J}_e. \tag{7.62}$$

In static problems, $\mathbf{E}$ and $\mathbf{H}$ do not depend on each other.

Eliminating successively $\mathbf{H}$ and $\mathbf{E}$ from the respective equations (7.59) and (7.60), the wave equation for the electric and magnetic field one-forms are obtained as

$$\overline{\overline{\mathsf{W}}}_e(\mathbf{d}_\mathbf{E}, \partial_\tau)|\mathbf{E} = -\partial_\tau\mathbf{J}_e - (\mathbf{d}_\mathbf{E} \wedge \overline{\overline{\mathsf{I}}}^T - \overline{\overline{\xi}}\partial_\tau)|\overline{\overline{\mu}}^{-1}|\mathbf{J}_m, \tag{7.63}$$

$$\overline{\overline{\mathsf{W}}}_m(\mathbf{d}_\mathbf{E}, \partial_\tau)|\mathbf{H} = -\partial_\tau\mathbf{J}_m + (\mathbf{d}_\mathbf{E} \wedge \overline{\overline{\mathsf{I}}}^T + \overline{\overline{\zeta}}\partial_\tau)|\overline{\overline{\epsilon}}^{-1}|\mathbf{J}_e. \tag{7.64}$$

The second-order wave-operator dyadics $\overline{\overline{\mathsf{W}}}_e(\mathbf{d}_\mathbf{E}, \partial_\tau), \overline{\overline{\mathsf{W}}}_m(\mathbf{d}_\mathbf{E}, \partial_\tau) \in \mathbb{F}_2\mathbb{E}_1$ have the form

$$\overline{\overline{\mathsf{W}}}_e(\mathbf{d}_\mathbf{E}, \partial_\tau) = (\mathbf{d}_\mathbf{E} \wedge \overline{\overline{\mathsf{I}}}^T - \overline{\overline{\xi}}\partial_\tau)|\overline{\overline{\mu}}^{-1}|(\mathbf{d}_\mathbf{E} \wedge \overline{\overline{\mathsf{I}}}^T + \overline{\overline{\zeta}}\partial_\tau) + \overline{\overline{\epsilon}}\partial_\tau^2, \tag{7.65}$$

$$\overline{\overline{\mathsf{W}}}_m(\mathbf{d}_\mathbf{E}, \partial_\tau) = (\mathbf{d}_\mathbf{E} \wedge \overline{\overline{\mathsf{I}}}^T + \overline{\overline{\zeta}}\partial_\tau)|\overline{\overline{\epsilon}}^{-1}|(\mathbf{d}_\mathbf{E} \wedge \overline{\overline{\mathsf{I}}}^T - \overline{\overline{\xi}}\partial_\tau) + \overline{\overline{\mu}}\partial_\tau^2, \tag{7.66}$$

which correspond the Gibbsian operators (3.32), (3.33) in reference 40. The other two Maxwell equations $\mathbf{d} \wedge \mathbf{D} = \varrho_e$ and $\mathbf{D} \wedge \mathbf{B} = \varrho_m$ serve as initial conditions for the field two-forms $\mathbf{E}, \mathbf{H}$ written as

$$\mathbf{d} \wedge \overline{\overline{\epsilon}}|\mathbf{E} + \mathbf{d} \wedge \overline{\overline{\xi}}|\mathbf{H} = \varrho_e, \tag{7.67}$$

$$\mathbf{d} \wedge \overline{\overline{\zeta}}|\mathbf{E} + \mathbf{d} \wedge \overline{\overline{\mu}}|\mathbf{H} = \varrho_e. \tag{7.68}$$

For the simple anisotropic medium with $\overline{\overline{\xi}} = \overline{\overline{\zeta}} = 0$, the wave operators are reduced to

$$\overline{\overline{\mathsf{W}}}_e(\mathbf{d}_\mathbf{E}, \partial_\tau) = -\mathbf{d}_\mathbf{E}\mathbf{d}_\mathbf{E}{}^{\wedge}_{\wedge}\overline{\overline{\mu}}^{-1} + \overline{\overline{\epsilon}}\partial_\tau^2, \quad (7.69)$$

$$\overline{\overline{\mathsf{W}}}_m(\mathbf{d}_\mathbf{E}, \partial_\tau) = -\mathbf{d}_\mathbf{E}\mathbf{d}_\mathbf{E}{}^{\wedge}_{\wedge}\overline{\overline{\epsilon}}^{-1} + \overline{\overline{\mu}}\partial_\tau^2. \quad (7.70)$$

7.2.2 Four-dimensional field equations

Differential-form formalism in the four-dimensional representation appears to be an ideal tool for electromagnetic analysis because the Maxwell equations for the two electromagnetic two-forms $\mathbf{\Phi}, \mathbf{\Psi}$ are represented by differential equations of the simplest possible form. However, constructing the wave equation for a single electromagnetic two-form $\mathbf{\Phi}$ or $\mathbf{\Psi}$ through elimination appears to be cumbersome. Because such a procedure could not be found in the literature, the following analysis based on various multivector identities has some novelty [50]. For convenience, the identities needed in the analysis are given in the course of the derivation.

To form the wave equation for the field two-form $\mathbf{\Phi}$, we start by operating the Maxwell–Faraday equation $\mathbf{d} \wedge \mathbf{\Phi} = \gamma_m$ as

$$\mathbf{d} \wedge (\overline{\overline{\mathsf{M}}}\rfloor(\mathbf{d} \wedge \mathbf{\Phi})) = \mathbf{d} \wedge (\overline{\overline{\mathsf{M}}}\rfloor\gamma_m). \quad (7.71)$$

Applying on the left-hand side the identity

$$\mathbf{A}\rfloor(\beta \wedge \mathbf{\Gamma}) = \beta(\mathbf{A}|\mathbf{\Gamma}) + \mathbf{\Gamma}\lfloor(\mathbf{A}\lfloor\beta) = \beta(\mathbf{A}|\mathbf{\Gamma}) - (\mathbf{A}\lfloor\beta)\rfloor\mathbf{\Gamma}, \quad (7.72)$$

valid for any bivector $\mathbf{A}$, dual vector β, and dual bivector $\mathbf{\Gamma}$, we obtain

$$\mathbf{d} \wedge (\overline{\overline{\mathsf{M}}}|\mathbf{\Phi})\mathbf{d} - \mathbf{d} \wedge (\overline{\overline{\mathsf{M}}}\lfloor\mathbf{d})\rfloor\mathbf{\Phi} = \mathbf{d} \wedge (\overline{\overline{\mathsf{M}}}\rfloor\gamma_m). \quad (7.73)$$

In the first term the last $\mathbf{d}$ operates on $\mathbf{\Phi}$ from the right and the term equals the dyadic $(\mathbf{d} \wedge \mathbf{\Psi})\mathbf{d} = \gamma_e\mathbf{d} \in \mathbb{F}_3\mathbb{F}_1$. This leads to the following second-order equation to the electromagnetic two-form $\mathbf{\Phi}$:

$$(\mathbf{d} \wedge \overline{\overline{\mathsf{M}}}\lfloor\mathbf{d})\rfloor\mathbf{\Phi} = \gamma_e\mathbf{d} - \mathbf{d} \wedge \overline{\overline{\mathsf{M}}}\rfloor\gamma_m, \quad (7.74)$$

which represents the wave equation. As a check we see that, operating by $\mathbf{d}\wedge$, the equation is identically satisfied because its every term vanishes.

Change of operator To obtain an equation with an operator similar to that in the equation for the electric potential, (7.12), let us represent the medium dyadic $\overline{\overline{\mathsf{M}}}$ in terms of the modified medium dyadic $\overline{\overline{\mathsf{M}}}_g$. Applying the following identity valid for $n = 4$,

$$\alpha \wedge (\kappa_N\lfloor\mathbf{A}) = \kappa_N\lfloor(\alpha\rfloor\mathbf{A}), \quad (7.75)$$

where α is a dual vector, $\mathbf{A}$ is a bivector and κ_N is a dual quadrivector, we can write

$$(\mathbf{d} \wedge \overline{\overline{\mathsf{M}}}\lfloor\mathbf{d})\rfloor\mathbf{\Phi} = \mathbf{d} \wedge (\varepsilon_N\lfloor\overline{\overline{\mathsf{M}}}_g\lfloor\mathbf{d})\rfloor\mathbf{\Phi} = \varepsilon_N\lfloor(\mathbf{d}\rfloor\overline{\overline{\mathsf{M}}}_g\lfloor\mathbf{d})\rfloor\mathbf{\Phi}, \quad (7.76)$$

whence (7.74) becomes

$$\varepsilon_N\lfloor(\mathbf{d}\rfloor\overline{\overline{\mathsf{M}}}_g\lfloor\mathbf{d})\rfloor\mathbf{\Phi} = \gamma_e\mathbf{d} - \varepsilon_N\lfloor(\mathbf{d}\rfloor\overline{\overline{\mathsf{M}}}_g))\rfloor\gamma_m. \tag{7.77}$$

Operating (7.77) by $\mathbf{e}_N\lfloor$ and applying

$$\mathbf{e}_N\lfloor(\varepsilon_N\lfloor\mathbf{a}^p) = (-1)^{p(n-p)}\mathbf{a}^p, \tag{7.78}$$

valid for any p-vector $\mathbf{a}^p$, we obtain

$$(\mathbf{d}\rfloor\overline{\overline{\mathsf{M}}}_g\lfloor\mathbf{d})\rfloor\mathbf{\Phi} = -(\mathbf{e}_N\lfloor\gamma_e)\mathbf{d} - (\mathbf{d}\rfloor\overline{\overline{\mathsf{M}}}_g))\rfloor\gamma_m. \tag{7.79}$$

This is the desired result. The second-order dyadic operator $\mathbf{d}\rfloor\overline{\overline{\mathsf{M}}}_g\lfloor\mathbf{d} \in \mathbb{E}_1\mathbb{E}_1$ is seen to equal that denoted by $\overline{\overline{\mathsf{V}}}_e(\mathbf{d})$ in the wave equation (7.12) for the potential. Note that the electric source on the right-hand side of (7.79) is being operated by $\mathbf{d}$ from the right.

Compatibility check Let us briefly check the compatibility of the wave equations for the field two-form, (7.79), and the potential one-form, (7.13), in the absense of magnetic sources. From $\gamma_m = 0$ we can introduce the electric potential as $\mathbf{\Phi} = \mathbf{d}\wedge\boldsymbol{\alpha}_e$. Inserting this in (7.79) and applying the bac cab rule (1.51), we can expand

$$\begin{aligned}(\mathbf{d}\rfloor\overline{\overline{\mathsf{M}}}_g\lfloor\mathbf{d})\rfloor(\mathbf{d}\wedge\boldsymbol{\alpha}_e) &= (\mathbf{d}\rfloor\overline{\overline{\mathsf{M}}}_g\lfloor\mathbf{d})|\boldsymbol{\alpha}_e\mathbf{d} - (\mathbf{d}\rfloor\overline{\overline{\mathsf{M}}}_g\lfloor\mathbf{d})|\mathbf{d}\boldsymbol{\alpha}_e \\ &= (\mathbf{d}\rfloor\overline{\overline{\mathsf{M}}}_g\lfloor\mathbf{d})|\boldsymbol{\alpha}_e\mathbf{d},\end{aligned} \tag{7.80}$$

whence (7.79) becomes

$$((\mathbf{d}\rfloor\overline{\overline{\mathsf{M}}}_g\lfloor\mathbf{d})|\boldsymbol{\alpha}_e)\mathbf{d} = -(\mathbf{e}_N\lfloor\gamma_e)\mathbf{d}. \tag{7.81}$$

Obviously, this coincides with the wave equation for the electric potential (7.13) when both sides of the latter are operated by $\mathbf{d}$ from the right. Similar test can be made for the case $\gamma_e = 0$.

Formal solution The unknown in (7.79) is the electromagnetic field two-form $\mathbf{\Phi}$, which is a six-dimensional quantity. Because (7.79) is a dyadic equation whose every term has the form of a dyadic $\mathbb{E}_1\mathbb{F}_1$, it corresponds to a set of $4\times 4 = 16$ scalar equations, 10 of which must be redundant. Actually, it is not easy to prove that the number of unknowns equals the number of independent equations, and this will only be shown for special media in the subsequent sections.

A formal solution for the wave equation can be expressed as follows. Let us first modify the left-hand side of (7.79) as

$$(\mathbf{d}\rfloor\overline{\overline{\mathsf{M}}}_g\lfloor\mathbf{d})\rfloor\mathbf{\Phi} = (\mathbf{d}\rfloor\overline{\overline{\mathsf{M}}}_g\lfloor\mathbf{d})|(\bar{\mathsf{I}}^T\rfloor\mathbf{\Phi}) = -(\overline{\overline{\mathsf{M}}}_g\lfloor\lfloor\mathbf{dd})|(\bar{\mathsf{I}}^T\rfloor\mathbf{\Phi}). \tag{7.82}$$

The unknown is here represented by the antisymmetric dyadic $\bar{\mathsf{I}}^T\rfloor\mathbf{\Phi} \in \mathbb{F}_1\mathbb{F}_1$, which also has six scalar components. The wave equation has now the form

$$(\overline{\overline{\mathsf{M}}}_g\lfloor\lfloor\mathbf{dd})|(\bar{\mathsf{I}}^T\rfloor\mathbf{\Phi}) = (\mathbf{e}_N\lfloor\gamma_e)\mathbf{d} + (\mathbf{d}\rfloor\overline{\overline{\mathsf{M}}}_g))\rfloor\gamma_m. \tag{7.83}$$

The formal solution can be expressed as

$$\overline{\overline{\mathsf{I}}}^T\rfloor\mathbf{\Phi} = (\overline{\overline{\mathsf{M}}}_g\lfloor\lfloor\mathbf{dd})^{-1}|((\mathbf{e}_N\lfloor\gamma_e)\mathbf{d} + (\mathbf{d}\rfloor\overline{\overline{\mathsf{M}}}_g))\rfloor\gamma_m), \tag{7.84}$$

where the inverse operator must be understood in generalized sense because a dyadic of the form $\overline{\overline{\mathsf{M}}}_g\lfloor\lfloor\boldsymbol{\alpha\alpha}$ maps the dual vector $\boldsymbol{\alpha}$ onto zero and, thus, does not have a regular inverse. Because the left-hand side of (7.84) is an antisymmetric dyadic, so must also be the right-hand side. This will be verified for special media in the following sections. According to the pattern given in Chapter 5 of reference 40, it appears useful for solving the wave equation if one can express the inverse operator in the form $\overline{\overline{\mathsf{V}}}_e{}^{-1} = \overline{\overline{\mathsf{L}}}(\mathbf{d})/L(\mathbf{d})$, where $\overline{\overline{\mathsf{L}}}(\mathbf{d})$ is a dyadic polynomial operator and $L(\mathbf{d})$ is a scalar polynomial operator. In this case the wave equation can be reduced to one with a scalar operator $L(\mathbf{d})$ which in general is of the fourth order.

7.2.3 Q-medium

Let us now assume that the modified medium dyadic $\overline{\overline{\mathsf{M}}}_g \in \mathbb{E}_2\mathbb{E}_2$ can be expressed in the form (5.63), that is,

$$\overline{\overline{\mathsf{M}}}_g = \overline{\overline{\mathsf{Q}}}^{(2)}, \tag{7.85}$$

for some dyadic $\overline{\overline{\mathsf{Q}}} \in \mathbb{E}_1\mathbb{E}_1$. Media satisfying (7.85) have been called Q-media for brevity [50, 51].

To simplify the operator in the wave equation (7.83) for Q-media, we start by applying the dyadic identity

$$\overline{\overline{\mathsf{Q}}}^{(2)}\lfloor\lfloor\overline{\overline{\mathsf{A}}} = (\overline{\overline{\mathsf{Q}}}||\overline{\overline{\mathsf{A}}})\overline{\overline{\mathsf{Q}}} - \overline{\overline{\mathsf{Q}}}|\overline{\overline{\mathsf{A}}}^T|\overline{\overline{\mathsf{Q}}}, \tag{7.86}$$

valid for any dyadic $\overline{\overline{\mathsf{A}}} \in \mathbb{F}_1\mathbb{F}_1$. Setting $\overline{\overline{\mathsf{A}}} = \mathbf{dd}$, (7.83) takes the form

$$\begin{aligned}(\overline{\overline{\mathsf{Q}}}^{(2)}\lfloor\lfloor\mathbf{dd})|(\overline{\overline{\mathsf{I}}}^T\rfloor\mathbf{\Phi}) &= (\mathbf{d}|\overline{\overline{\mathsf{Q}}}|\mathbf{d})\overline{\overline{\mathsf{Q}}}|(\overline{\overline{\mathsf{I}}}^T\rfloor\mathbf{\Phi}) - (\overline{\overline{\mathsf{Q}}}|\mathbf{d})(\mathbf{d}|\overline{\overline{\mathsf{Q}}})|(\overline{\overline{\mathsf{I}}}^T\rfloor\mathbf{\Phi}) \\ &= (\mathbf{e}_N\lfloor\gamma_e)\mathbf{d} + (\mathbf{d}\rfloor\overline{\overline{\mathsf{Q}}}^{(2)}))\rfloor\gamma_m. \end{aligned} \tag{7.87}$$

Multiplying by $\overline{\overline{\mathsf{Q}}}^{-1}|$ this can be written as

$$(\mathbf{d}|\overline{\overline{\mathsf{Q}}}|\mathbf{d})\overline{\overline{\mathsf{I}}}^T\rfloor\mathbf{\Phi} = \mathbf{d}(\mathbf{d}|\overline{\overline{\mathsf{Q}}}\rfloor\mathbf{\Phi}) + \overline{\overline{\mathsf{Q}}}^{-1}|(\mathbf{e}_N\lfloor\gamma_e)\mathbf{d} + \overline{\overline{\mathsf{Q}}}^{-1}|(\mathbf{d}\rfloor\overline{\overline{\mathsf{Q}}}^{(2)}))\rfloor\gamma_m. \tag{7.88}$$

Now the left-hand side of (7.88) is an antisymmetric dyadic because it has a scalar operator. To verify that the number of equations equals that of unknowns, we have to prove that the right-hand side of (7.88) is also antisymmetric.

Proof of antisymmetry Let us apply the identity

$$\boldsymbol{\alpha}\rfloor\overline{\overline{\mathsf{Q}}}^{(2)} = \overline{\overline{\mathsf{Q}}} \wedge (\boldsymbol{\alpha}|\overline{\overline{\mathsf{Q}}}) \tag{7.89}$$

to the last term of (7.88):

$$\overline{\overline{\mathsf{Q}}}^{-1}|(\mathbf{d}\rfloor\overline{\overline{\mathsf{Q}}}^{(2)})\rfloor\gamma_m = \overline{\overline{\mathsf{Q}}}^{-1}|(\overline{\overline{\mathsf{Q}}}\wedge(\mathbf{d}|\overline{\overline{\mathsf{Q}}}))\rfloor\gamma_m = \overline{\overline{\mathsf{I}}}^T\rfloor(\mathbf{d}|\overline{\overline{\mathsf{Q}}}\rfloor\gamma_m), \tag{7.90}$$

which obtains the form of an antisymmetric dyadic. Considering next the first term on the right-hand side of (7.88), we expand

$$\begin{aligned}\mathbf{d}|\overline{\overline{\mathsf{Q}}}\rfloor\boldsymbol{\Phi} &= (\overline{\overline{\mathsf{I}}}^T\wedge(\mathbf{d}|\overline{\overline{\mathsf{Q}}}))|\boldsymbol{\Phi}\\ &= \overline{\overline{\mathsf{Q}}}^{-1}|(\overline{\overline{\mathsf{Q}}}\wedge(\mathbf{d}|\overline{\overline{\mathsf{Q}}}))|\boldsymbol{\Phi} = \overline{\overline{\mathsf{Q}}}^{-1}|(\mathbf{d}\rfloor(\overline{\overline{\mathsf{Q}}}^{(2)}|\boldsymbol{\Phi})).\end{aligned} \tag{7.91}$$

Inserting $\overline{\overline{\mathsf{Q}}}^{(2)}|\boldsymbol{\Phi} = \overline{\overline{\mathsf{M}}}_g|\boldsymbol{\Phi} = \mathbf{e}_N\lfloor\boldsymbol{\Psi}$, we obtain

$$\begin{aligned}\mathbf{d}|\overline{\overline{\mathsf{Q}}}\rfloor\boldsymbol{\Phi} &= \overline{\overline{\mathsf{Q}}}^{-1}|(\mathbf{d}\rfloor(\mathbf{e}_N\lfloor\boldsymbol{\Psi})) = -\overline{\overline{\mathsf{Q}}}^{-1}|((\mathbf{e}_N\lfloor\boldsymbol{\Psi})\lfloor\mathbf{d})\\ &= -\overline{\overline{\mathsf{Q}}}^{-1}|(\mathbf{e}_N\lfloor(\mathbf{d}\wedge\boldsymbol{\Psi})) = -\overline{\overline{\mathsf{Q}}}^{-1}|(\mathbf{e}_N\lfloor\gamma_e).\end{aligned} \tag{7.92}$$

Adding this to the second term on the right-hand side of (7.88) is of the antisymmetric form $-\boldsymbol{\beta}\mathbf{d} + \mathbf{d}\boldsymbol{\beta}$ and the sum of the two terms can be written compactly as

$$\begin{aligned}\mathbf{d}(\mathbf{d}|\overline{\overline{\mathsf{Q}}}\rfloor\boldsymbol{\Phi}) + \overline{\overline{\mathsf{Q}}}^{-1}|(\mathbf{e}_N\lfloor\gamma_e)\mathbf{d} &= -\mathbf{d}\overline{\overline{\mathsf{Q}}}^{-1}|(\mathbf{e}_N\lfloor\gamma_e) + \overline{\overline{\mathsf{Q}}}^{-1}|(\mathbf{e}_N\lfloor\gamma_e)\mathbf{d}\\ &= \overline{\overline{\mathsf{I}}}^T\rfloor(\mathbf{d}\wedge\overline{\overline{\mathsf{Q}}}^{-1}|(\mathbf{e}_N\lfloor\gamma_e)).\end{aligned} \tag{7.93}$$

Two-form wave equation Because each term of the dyadic wave equation (7.88) are of the antisymmetric form $\overline{\overline{\mathsf{I}}}^T\rfloor\boldsymbol{\Gamma}$ where $\boldsymbol{\Gamma}$ is a two-form, dropping the algebraic operation $\overline{\overline{\mathsf{I}}}^T\rfloor$ the wave equation for the electromagnetic two-form can be presented as a two-form equation,

$$(\mathbf{d}|\overline{\overline{\mathsf{Q}}}|\mathbf{d})\boldsymbol{\Phi} = \mathbf{d}\wedge\overline{\overline{\mathsf{Q}}}^{-1}|(\mathbf{e}_N\lfloor\gamma_e) + \mathbf{d}|\overline{\overline{\mathsf{Q}}}\rfloor\gamma_m. \tag{7.94}$$

This is the final form for the wave equation for any Q-medium. It is equivalent to the dyadic equation (7.87) but, obviously, simpler in form because it contains a scalar second-order operator. As a simple check of (7.94), we operate it by $\mathbf{d}\wedge$ to arrive at the result

$$\begin{aligned}(\mathbf{d}|\overline{\overline{\mathsf{Q}}}|\mathbf{d})\mathbf{d}\wedge\boldsymbol{\Phi} &= \mathbf{d}\wedge(\mathbf{d}|\overline{\overline{\mathsf{Q}}}\rfloor\gamma_m)\\ &= (\mathbf{d}|\overline{\overline{\mathsf{Q}}})\rfloor(\mathbf{d}\wedge\gamma_m) + (\mathbf{d}|\overline{\overline{\mathsf{Q}}}|\mathbf{d})\gamma_m\\ &= (\mathbf{d}|\overline{\overline{\mathsf{Q}}}|\mathbf{d})\gamma_m,\end{aligned} \tag{7.95}$$

which reproduces one of the Maxwell equations operated by $(\mathbf{d}|\overline{\overline{\mathsf{Q}}}|\mathbf{d})$.

Example To be able to compare (7.94) with the corresponding wave equations known from the Gibbsian analysis, let us consider as an example the special (affine-isotropic) medium defined by (7.32) reproduced here as

$$\overline{\overline{\epsilon}}_g = \epsilon\overline{\overline{\mathsf{S}}}, \quad \overline{\overline{\mu}}_g = \mu\overline{\overline{\mathsf{S}}}, \quad \overline{\overline{\mathsf{S}}}^{(3)} = \mathbf{e}_{123}\mathbf{e}_{123}, \tag{7.96}$$

where $\overline{\overline{\mathsf{S}}} \in \mathbb{E}_1 \mathbb{E}_1$ is a three-dimensional symmetric dyadic. Because we can write

$$\begin{aligned} \overline{\overline{\mathsf{M}}}_g &= \epsilon \overline{\overline{\mathsf{S}}}{}^{\wedge}_{\wedge} \mathbf{e}_\tau \mathbf{e}_\tau - \frac{1}{\mu} \overline{\overline{\mathsf{S}}}{}^{-1} \rfloor\rfloor \mathbf{e}_{123} \mathbf{e}_{123} \\ &= \epsilon \overline{\overline{\mathsf{S}}}{}^{\wedge}_{\wedge} \mathbf{e}_\tau \mathbf{e}_\tau - \frac{1}{\mu} \overline{\overline{\mathsf{S}}}{}^{(2)} = -\frac{1}{\mu} (\overline{\overline{\mathsf{S}}} - \mu\epsilon \mathbf{e}_\tau \mathbf{e}_\tau)^{(2)}, \end{aligned} \tag{7.97}$$

such a medium is a special case of the Q-medium. Substituting now (note the slight difference in notation when compared to (7.32))

$$\overline{\overline{\mathsf{Q}}} = \frac{j}{\sqrt{\mu}} (\overline{\overline{\mathsf{S}}} - \mu\epsilon \mathbf{e}_\tau \mathbf{e}_\tau), \quad \overline{\overline{\mathsf{Q}}}{}^{(2)} = -\frac{1}{\mu} (\overline{\overline{\mathsf{S}}}{}^{(2)} - \mu\epsilon \overline{\overline{\mathsf{S}}}{}^{\wedge}_{\wedge} \mathbf{e}_\tau \mathbf{e}_\tau), \tag{7.98}$$

after some simple steps the wave equation (7.94) can be separated into space-like and time-like components as

$$\begin{aligned} &(\mathbf{d}_{\mathbf{E}} | \overline{\overline{\mathsf{S}}} | \mathbf{d}_{\mathbf{E}} - \mu\epsilon\partial_\tau^2) \mathbf{B} \\ &= \mu \mathbf{d}_{\mathbf{E}} \wedge (\overline{\overline{\mathsf{S}}}{}^{-1} | (\mathbf{e}_{123} \lfloor \mathbf{J}_e)) + \mu\epsilon\partial_\tau \mathbf{J}_m + (\mathbf{d}_{\mathbf{E}} | \overline{\overline{\mathsf{S}}}) \rfloor \varrho_m, \end{aligned} \tag{7.99}$$

$$\begin{aligned} &(\mathbf{d}_{\mathbf{E}} | \overline{\overline{\mathsf{S}}} | \mathbf{d}_{\mathbf{E}} - \mu\epsilon\partial_\tau^2) \mathbf{E} \\ &= (\mathbf{d}_{\mathbf{E}} | \overline{\overline{\mathsf{S}}}) \rfloor \mathbf{J}_m + \mu\partial_\tau (\overline{\overline{\mathsf{S}}}{}^{-1} | (\mathbf{e}_{123} \lfloor \mathbf{J}_e)) + \frac{1}{\epsilon} \mathbf{d}_{\mathbf{E}} (\mathbf{e}_{123} | \varrho_e). \end{aligned} \tag{7.100}$$

For example, for the isotropic Gibbsian medium with $\overline{\overline{\mathsf{S}}} = \mathbf{e}_1\mathbf{e}_1 + \mathbf{e}_2\mathbf{e}_2 + \mathbf{e}_3\mathbf{e}_3$ and $\mathbf{d}_{\mathbf{E}} | \overline{\overline{\mathsf{S}}} | \mathbf{d}_{\mathbf{E}} = \nabla^2$, these correspond to the familiar wave equations written in Gibbsian formalism as

$$(\nabla^2 - \mu\epsilon\partial_\tau^2)\mathbf{B} = -\mu\nabla \times \mathbf{J}_e + \mu\epsilon\partial_\tau \mathbf{J}_m + \nabla \varrho_m, \tag{7.101}$$

$$(\nabla^2 - \mu\epsilon\partial_\tau^2)\mathbf{E} = \nabla \times \mathbf{J}_m + \mu\partial_\tau \mathbf{J}_e + \frac{1}{\epsilon}\nabla \varrho_e. \tag{7.102}$$

7.2.4 Generalized Q-medium

The Q-medium case considered above is a simple example because the equation for the electromagnetic two-form reduces to one with a scalar second-order operator, (7.94). It is known from Gibbsian analysis that for the most general bi-anisotropic medium the corresponding time-harmonic equation can be reduced to a scalar equation of the fourth order [40]. Let us consider a generalization of the class of Q-media as defined by (5.96) reproduced here as

$$\overline{\overline{\mathsf{M}}}_g = \overline{\overline{\mathsf{Q}}}{}^{(2)} + \mathbf{AB}. \tag{7.103}$$

Here $\mathbf{A}$ and $\mathbf{B}$ are two bivectors. Applying again the identity (7.86), the wave equation (7.83) now becomes

$$(\overline{\overline{\mathsf{Q}}} || \mathbf{dd}) \overline{\overline{\mathsf{Q}}} | (\overline{\overline{\mathsf{I}}}{}^T \rfloor \mathbf{\Phi}) - (\overline{\overline{\mathsf{Q}}} | \mathbf{d})(\mathbf{d} | \overline{\overline{\mathsf{Q}}} \rfloor \mathbf{\Phi}) + (\mathbf{A} \lfloor \mathbf{d})(\mathbf{B} \lfloor \mathbf{d}) \rfloor \mathbf{\Phi}$$

$$= (\mathbf{e}_N\lfloor\gamma_e)\mathbf{d} + (\mathbf{d}\rfloor\overline{\overline{\mathsf{Q}}}^{(2)})\rfloor\gamma_m + (\mathbf{d}\rfloor\mathbf{A})(\mathbf{B}\rfloor\gamma_m). \tag{7.104}$$

Multiplying this by $\overline{\overline{\mathsf{Q}}}^{-1}|$ and applying previous results from the Q-medium case modified as

$$\mathbf{d}|\overline{\overline{\mathsf{Q}}}\rfloor\mathbf{\Phi} = -\overline{\overline{\mathsf{Q}}}^{-1}|(\mathbf{e}_N\lfloor\gamma_e) - \overline{\overline{\mathsf{Q}}}^{-1}|(\mathbf{d}\rfloor\mathbf{A})(\mathbf{B}|\mathbf{\Phi}) \tag{7.105}$$

$$(\mathbf{d}\rfloor\overline{\overline{\mathsf{M}}}_g)\rfloor\gamma_m = (\overline{\overline{\mathsf{Q}}}\wedge(\mathbf{d}|\overline{\overline{\mathsf{Q}}}))\rfloor\gamma_m + (\mathbf{d}\rfloor\mathbf{A})(\mathbf{B}\rfloor\gamma_m), \tag{7.106}$$

we can write the wave equation as

$$\begin{aligned}(\overline{\overline{\mathsf{Q}}}||\mathbf{dd})\bar{\bar{\mathsf{I}}}^T\rfloor\mathbf{\Phi} &= \bar{\bar{\mathsf{I}}}^T\rfloor(\mathbf{d}\wedge(\overline{\overline{\mathsf{Q}}}^{-1}|(\mathbf{e}_N\lfloor\gamma_e)) \\ &\quad +\bar{\bar{\mathsf{I}}}^T\rfloor(\mathbf{d}|\overline{\overline{\mathsf{Q}}}\rfloor\gamma_m) - \mathbf{d}\overline{\overline{\mathsf{Q}}}^{-1}|(\mathbf{d}\rfloor\mathbf{A})(\mathbf{B}|\mathbf{\Phi}) \\ &\quad -\overline{\overline{\mathsf{Q}}}^{-1}|(\mathbf{A}\lfloor\mathbf{d})[(\mathbf{B}\lfloor\mathbf{d})\rfloor\mathbf{\Phi} + \mathbf{B}\rfloor\gamma_m].\end{aligned} \tag{7.107}$$

Applying the identity (7.72), we can simplify the last square-bracketed term as

$$(\mathbf{B}\lfloor\mathbf{d})\rfloor\mathbf{\Phi} + \mathbf{B}\rfloor\gamma_m = (\mathbf{B}\lfloor\mathbf{d})\rfloor\mathbf{\Phi} + \mathbf{B}\rfloor(\mathbf{d}\wedge\mathbf{\Phi}) = \mathbf{d}(\mathbf{B}|\mathbf{\Phi}), \tag{7.108}$$

whence, again, the right-hand side of (7.107) has the form of an antisymmetric dyadic. Thus, we can replace the dyadic equation (7.104) by a two-form equation

$$L_1(\mathbf{d})\mathbf{\Phi} = \mathbf{d}\wedge(\overline{\overline{\mathsf{Q}}}^{-1}|(\mathbf{e}_N\lfloor\gamma_e)) + \mathbf{d}|\overline{\overline{\mathsf{Q}}}\rfloor\gamma_m + \mathbf{d}\wedge\overline{\overline{\mathsf{Q}}}^{-1}|(\mathbf{d}\rfloor\mathbf{A})(\mathbf{B}|\mathbf{\Phi}), \tag{7.109}$$

with a scalar second-order operator $L_1(\mathbf{d})$ defined by

$$L_1(\mathbf{d}) = \overline{\overline{\mathsf{Q}}}||\mathbf{dd}. \tag{7.110}$$

Fourth-order wave equation The right-hand side of (7.109) still contains the unknown in the form $\mathbf{B}|\mathbf{\Phi}$ which must be eliminated. This can be done by multiplying the whole equation by $\mathbf{B}|$, whence we obtain an equation for the scalar function $\mathbf{B}|\mathbf{\Phi}$

$$L_2(\mathbf{d})(\mathbf{B}|\mathbf{\Phi}) = \mathbf{B}|(\mathbf{d}\wedge(\overline{\overline{\mathsf{Q}}}^{-1}|(\mathbf{e}_N\lfloor\gamma_e)) + \mathbf{d}|\overline{\overline{\mathsf{Q}}}\rfloor\gamma_m). \tag{7.111}$$

The left-hand side contains a second-order scalar operator $L_2(\mathbf{d})$ defined by

$$\begin{aligned}L_2(\mathbf{d}) &= \overline{\overline{\mathsf{Q}}}||\mathbf{dd} - (\mathbf{B}\lfloor\mathbf{d})|\overline{\overline{\mathsf{Q}}}^{-1}|(\mathbf{d}\rfloor\mathbf{A}) \\ &= \overline{\overline{\mathsf{Q}}}||\mathbf{dd} + \overline{\overline{\mathsf{Q}}}^{-1}||(\mathbf{BA}\lfloor\lfloor\mathbf{dd}).\end{aligned} \tag{7.112}$$

Operating (7.109) by $L_2(\mathbf{d})$, we can finally write the wave equation for the two-form $\mathbf{\Phi}$ for the generalized Q medium in the form

$$L_1(\mathbf{d})L_2(\mathbf{d})\mathbf{\Phi}$$

$$= \big(L_2(\mathbf{d})\bar{\bar{\mathsf{I}}}^{(2)T} + \mathbf{d}\wedge\overline{\overline{\mathsf{Q}}}^{-1}|(\mathbf{d}\rfloor\mathbf{A})\mathbf{B}\big)|\big(\mathbf{d}\wedge\overline{\overline{\mathsf{Q}}}^{-1}|(\mathbf{e}_N\lfloor\gamma_e) + \mathbf{d}|\overline{\overline{\mathsf{Q}}}\rfloor\gamma_m\big). \tag{7.113}$$

The scalar operator on the left-hand side of (7.113) is now of the fourth order. It is not of the most general fourth-order form because the operator is factorized, as a product of two second-order operators. It is of interest to see if (7.113) reduces to the simpler wave equation (7.94) for vanishing dyad $\mathbf{AB}$, in which case the generalized Q-medium reduces to the Q-medium. In this case the two operators become the same, $L_2(\mathbf{d}) = L_1(\mathbf{d})$, and (7.113) reduces to (7.94) with the operator $L_1(\mathbf{d})$ operating both sides of the equation. Assuming suitable conditions of uniqueness for the solution we are allowed to cancel the operator. The present class of media bears resemblance to the class of uniaxial media encountered in the classical Gibbsian analysis.

Problems

7.2.1 Fill in the details in deriving (7.79) from (7.77).

7.2.2 Find the relations between the four Gibbsian medium dyadics $\overline{\overline{\epsilon}}_g, \overline{\overline{\mu}}_g, \overline{\overline{\xi}}_g, \overline{\overline{\zeta}}_g$ for the Q-medium.

7.2.3 Do the details in the derivation of (7.99) and (7.100).

7.2.4 Define the wave operator $L_2(\mathbf{d})$ in the special case of generalized Q-medium as defined by (5.98).

7.2.5 Show that if $\overline{\overline{\mathsf{M}}}_g \in \mathbb{E}_2\mathbb{E}_2$ satisfies the condition (7.103) of extended Q-medium, its inverse $\overline{\overline{\mathsf{M}}}_g^{-1} \in \mathbb{F}_2\mathbb{F}_2$ satisfies a condition of the form

$$\overline{\overline{\mathsf{M}}}_g^{-1} = (\overline{\overline{\mathsf{Q}}}^{-1})^{(2)} + \mathbf{\Gamma\Pi},$$

where $\mathbf{\Gamma}$ and $\mathbf{\Pi}$ are two dual bivectors. Find their relation to $\overline{\overline{\mathsf{Q}}}$, $\mathbf{A}$ and $\mathbf{B}$.

7.3 PLANE WAVES

7.3.1 Wave equations

Plane wave is a solution of the wave equation which depends only on a single Cartesian coordinate and time. Let us assume that there are no sources in the finite region. The coordinate is denoted by $w(\mathbf{r})$ and defined by a constant one-form $\boldsymbol{\nu}$ as

$$w(\mathbf{r}) = \boldsymbol{\nu}|\mathbf{r}. \tag{7.114}$$

Because in this case the spatial Euclidean operator can be replaced by

$$\mathbf{d}_{\mathbb{E}} = \partial_w \mathbf{d}w = \partial_w \boldsymbol{\nu}, \tag{7.115}$$

the wave equation for the field two-form (7.79) becomes

$$(\overline{\overline{\mathsf{M}}}_g \lfloor\lfloor(\partial_w \mathbf{d}w + \partial_\tau \mathbf{d}\tau)(\partial_w \mathbf{d}w + \partial_\tau \mathbf{d}\tau))\rfloor\mathbf{\Phi}$$
$$= (\overline{\overline{\mathsf{M}}}_g \lfloor\lfloor(\partial_w \boldsymbol{\nu} + \partial_\tau \mathbf{d}\tau)(\partial_w \boldsymbol{\nu} + \partial_\tau \mathbf{d}\tau))|(\overline{\overline{\mathsf{I}}}^T\rfloor\mathbf{\Phi}) = 0. \tag{7.116}$$

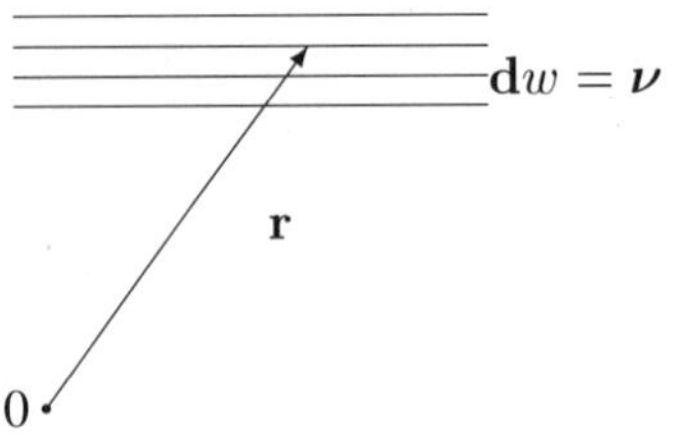

Fig. 7.2 *The one-form* $\mathbf{d}w = \boldsymbol{\nu}$ *corresponding to the coordinate* $w(\mathbf{r})$ *can be depicted by a set of planes of constant* $w(\mathbf{r}) = \boldsymbol{\nu}|\mathbf{r}$.

The corresponding three-dimensional representation for the electric one-form is obtained from (7.63) as

$$\overline{\overline{\mathsf{W}}}_e(\partial_w, \partial_\tau)|\mathbf{E} = 0, \tag{7.117}$$

with the wave-operator dyadic $\overline{\overline{\mathsf{W}}}_e(\partial_w, \partial_\tau) \in \mathbb{F}_2\,\mathbb{E}_1$ defined by

$$\overline{\overline{\mathsf{W}}}_e(\partial_w, \partial_\tau) = (\partial_w \boldsymbol{\nu} \wedge \overline{\overline{\mathsf{I}}}^T - \overline{\overline{\xi}}\partial_\tau)|\overline{\overline{\mu}}^{-1}|(\partial_w \boldsymbol{\nu} \wedge \overline{\overline{\mathsf{I}}}^T + \overline{\overline{\zeta}}\partial_\tau) + \overline{\overline{\epsilon}}\partial_\tau^2. \tag{7.118}$$

Both equations (7.116) and (7.117) are of the second order and valid for the general bi-anisotropic medium.

Three-dimensional representation In the Gibbsian analysis we know that the plane-wave equation can be reduced to one with a scalar operator, by multiplying with the adjoint of the dyadic operator [40]. The same can also be done here. In the latter case the operator $\overline{\overline{\mathsf{W}}}_e \in \mathbb{F}_2\,\mathbb{E}_1$ is of Hodge type, and it must first be transformed to an operator of the metric type $\overline{\overline{\mathsf{W}}}_{e1} \in \mathbb{E}_1\,\mathbb{E}_1$,

$$\overline{\overline{\mathsf{W}}}_{e1}(\partial_w, \partial_\tau) = \mathbf{e}_{123}\lfloor\overline{\overline{\mathsf{W}}}_e(\partial_w, \partial_\tau). \tag{7.119}$$

Applying (2.169), we can form its adjoint as

$$(\boldsymbol{\varepsilon}_{123}\boldsymbol{\varepsilon}_{123}||\overline{\overline{\mathsf{W}}}_{e1}^{(3)})\overline{\overline{\mathsf{W}}}_{e1}^{-1} = \boldsymbol{\varepsilon}_{123}\boldsymbol{\varepsilon}_{123}\lfloor\lfloor\overline{\overline{\mathsf{W}}}_{e1}^{(2)T}, \tag{7.120}$$

satisfying

$$(\boldsymbol{\varepsilon}_{123}\boldsymbol{\varepsilon}_{123}\lfloor\lfloor\overline{\overline{\mathsf{W}}}_{e1}^{(2)T})|\overline{\overline{\mathsf{W}}}_{e1} = (\boldsymbol{\varepsilon}_{123}\boldsymbol{\varepsilon}_{123}||\overline{\overline{\mathsf{W}}}_{e1}^{(3)})\overline{\overline{\mathsf{I}}}. \tag{7.121}$$

The plane-wave equation for the electric field now becomes

$$W_e(\partial_w, \partial_\tau)\mathbf{E}(w, \tau) = 0 \tag{7.122}$$

with a scalar operator $W_e(\partial_w, \partial_\tau)$ defined by

$$W_e(\partial_w, \partial_\tau) = \boldsymbol{\varepsilon}_{123}\boldsymbol{\varepsilon}_{123}||\overline{\overline{\mathsf{W}}}_{e1}^{(3)}(\partial_w, \partial_\tau). \tag{7.123}$$

At first sight this operator appears to be of the sixth order. A closer look, however, will reveal that it actually is of the fourth order. Even so, its analytical solution appears to be out of reach for the general bi-anisotropic medium. For certain classes of linear media, labeled as decomposable media, the fourth-order operator dyadics can be expressed as a product of two second-order operator dyadics, in which case the possible plane waves are split into two sets [49, 60]. The generalized Q-medium defined in the previous section is an example of such a medium as will be seen in an example below.

Four-dimensional representation The same trick can be done to the four-dimensional equation (7.116). The corresponding equation for the electric four-potential $\boldsymbol{\alpha}_e$ (7.6) with vanishing right-hand side is

$$\overline{\overline{\mathsf{V}}}_e(\partial_w \boldsymbol{\nu} + \partial_\tau \mathbf{d}\tau)|\boldsymbol{\alpha}_e(w,\tau) = 0. \tag{7.124}$$

The dyadic operator $\overline{\overline{\mathsf{V}}}_e(\partial_w \boldsymbol{\nu} + \partial_\tau \mathbf{d}\tau) \in \mathbb{E}_1\,\mathbb{E}_1$ is defined as

$$\overline{\overline{\mathsf{V}}}_e(\partial_w \boldsymbol{\nu} + \partial_\tau \mathbf{d}\tau) = \overline{\overline{\mathsf{M}}}_g \lfloor\lfloor(\partial_w \boldsymbol{\nu} + \mathbf{d}\tau\partial_\tau)(\partial_w \boldsymbol{\nu} + \mathbf{d}\tau\partial_\tau). \tag{7.125}$$

However, we cannot now use the inverse rule (2.169) for the four-dimensional operator $\overline{\overline{\mathsf{V}}}_e$ because $\overline{\overline{\mathsf{V}}}_e^{(4)} = 0$. This is because a dyadic of the form $\overline{\overline{\mathsf{A}}}\lfloor\lfloor\boldsymbol{\beta}\boldsymbol{\beta}$ satifies $(\overline{\overline{\mathsf{A}}}\lfloor\lfloor\boldsymbol{\beta}\boldsymbol{\beta})|\boldsymbol{\beta} = 0$ and does not have an inverse. Here we can use a modified form which is valid if $\overline{\overline{\mathsf{V}}}_e^{(3)} \neq 0$. However, because we cannot proceed with the most general bi-anisotropic medium, let us consider the four-dimensional representation in terms of special cases.

7.3.2 Q-medium

The class of bi-anisotropic media called Q-media was defined in (5.63) as $\overline{\overline{\mathsf{M}}}_g = \overline{\overline{\mathsf{Q}}}^{(2)}$. In this case, applying (7.94), the plane-wave equation can be expressed simply as

$$(\partial_w \mathbf{d}w + \partial_\tau \mathbf{d}\tau)|\overline{\overline{\mathsf{Q}}}|(\partial_w \mathbf{d}_w + \partial_\tau \mathbf{d}\tau)\boldsymbol{\Phi}(w,\tau) = 0, \tag{7.126}$$

where we now have a scalar second-order operator. This means that the magnetic and electric components $\mathbf{B}(w,\tau)$ and $\mathbf{E}(w,\tau)$, of the two-form $\boldsymbol{\Phi}$, satisfy the same equation. If the $\overline{\overline{\mathsf{Q}}}$ dyadic is expanded as

$$\overline{\overline{\mathsf{Q}}} = Q_{ww}\mathbf{e}_w\mathbf{e}_w + Q_{w\tau}\mathbf{e}_w\mathbf{e}_\tau + Q_{\tau w}\mathbf{e}_\tau\mathbf{e}_w + Q_{\tau\tau}\mathbf{e}_\tau\mathbf{e}_\tau, \tag{7.127}$$

the wave equation reads

$$(Q_{ww}\partial_w^2 + (Q_{w\tau} + Q_{\tau w})\partial_w\partial_\tau + Q_{\tau\tau}\partial_\tau^2)\boldsymbol{\Phi}(w,\tau) = 0. \tag{7.128}$$

To proceed we must assume properties of the parameters Q_{ij}.

Example Let us consider as an example of a Q medium the special anisotropic medium (7.32) in the form

$$\overline{\overline{\mathsf{Q}}} = \frac{j}{\sqrt{\mu}}(\overline{\overline{\mathsf{S}}} - \mu\epsilon \mathbf{e}_\tau \mathbf{e}_\tau), \tag{7.129}$$

with the symmetric dyadic defined by

$$\overline{\overline{\mathsf{S}}} = S_{xx}\mathbf{e}_x\mathbf{e}_x + S_{yy}\mathbf{e}_y\mathbf{e}_y + S_{zz}\mathbf{e}_z\mathbf{e}_z. \tag{7.130}$$

The plane-wave equation (7.126) now becomes

$$(\boldsymbol{\nu}|\overline{\overline{\mathsf{S}}}|\boldsymbol{\nu}\partial_w^2 - \mu\epsilon\partial_\tau^2)\Phi(w,\tau) = 0. \tag{7.131}$$

This is a wave equation, a hyperbolic partial differential equation, when the relative parameters μ, ϵ are real and positive numbers. Factorizing the second-order operator,

$$(\sqrt{\boldsymbol{\nu}|\overline{\overline{\mathsf{S}}}|\boldsymbol{\nu}}\partial_w - \sqrt{\mu\epsilon}\partial_\tau)(\sqrt{\boldsymbol{\nu}|\overline{\overline{\mathsf{S}}}|\boldsymbol{\nu}}\partial_w + \sqrt{\mu\epsilon}\partial_\tau)\boldsymbol{\Phi}(w,\tau) = 0, \tag{7.132}$$

we can express the general solution as a sum of solutions of the two first-order equations

$$(\sqrt{\boldsymbol{\nu}|\overline{\overline{\mathsf{S}}}|\boldsymbol{\nu}}\partial_w \pm \sqrt{\mu\epsilon}\partial_\tau)\boldsymbol{\Phi}_\pm(w,\tau) = 0, \tag{7.133}$$

$$\boldsymbol{\Phi}(w,t) = \boldsymbol{\Phi}_+(w,\tau) + \boldsymbol{\Phi}_-(w,\tau). \tag{7.134}$$

The classical method of solving the wave equation in terms of arbitrary scalar functions $\psi_+(\zeta)$ and $\psi_-(\zeta)$ can be applied as

$$\boldsymbol{\Phi}_\pm(w,t) = \psi_\pm(\sqrt{\boldsymbol{\nu}|\overline{\overline{\mathsf{S}}}|\boldsymbol{\nu}}\,\tau \mp \sqrt{\mu\epsilon}\,w)\boldsymbol{\Phi}_{o\pm}. \tag{7.135}$$

$\boldsymbol{\Phi}_{o\pm}$ are constant two-forms or dual-bivector amplitudes, which define the polarizations of the waves. Following a point of constant amplitude on the wave,

$$\sqrt{\boldsymbol{\nu}|\overline{\overline{\mathsf{S}}}|\boldsymbol{\nu}}\,\tau \mp \sqrt{\mu\epsilon}\,w = \text{const}, \tag{7.136}$$

the normalized velocity of the wave in both directions is the same,

$$v = \pm\frac{dw}{d\tau} = \sqrt{\frac{\boldsymbol{\nu}|\overline{\overline{\mathsf{S}}}|\boldsymbol{\nu}}{\mu\epsilon}}. \tag{7.137}$$

Its value depends on the metric dyadic $\overline{\overline{\mathsf{S}}}$ and on the one-form $\boldsymbol{\nu}$ which defines the two opposite directions of propagation of the plane wave.

The dual-bivector amplitudes $\boldsymbol{\Phi}_{o\pm}$ have arbitrary magnitudes, and they can be made unique through normalizing. The dual bivectors are restricted by algebraic equations obtained by substituting $\boldsymbol{\Phi}_\pm(w,\tau)$ in the source-free Maxwell equation:

$$\mathbf{d}\wedge\boldsymbol{\Phi}_\pm(w,\tau) = (\sqrt{\boldsymbol{\nu}|\overline{\overline{\mathsf{S}}}|\boldsymbol{\nu}}\mathbf{d}\tau \mp \sqrt{\mu\epsilon}\mathbf{d}w)\wedge\boldsymbol{\Phi}_{o\pm}\psi'_\pm = 0. \tag{7.138}$$

Here, prime denotes differentiation with respect to the scalar argument of the functions $\psi_\pm$. The polarization conditions for the dual-bivector amplitudes are thus of the form

$$\boldsymbol{\pi}_\pm \wedge \boldsymbol{\Phi}_{o\pm} = 0, \tag{7.139}$$

where the dual vectors $\boldsymbol{\pi}_\pm$ are defined by

$$\boldsymbol{\pi}_\pm = v\mathbf{d}\tau \mp \boldsymbol{\nu}, \tag{7.140}$$

and v is the wave velocity (7.137). The conditions (7.139) correspond to one denoted in Gibbsian notation by $\mathbf{k}\cdot\mathbf{E} = 0$ for a time-harmonic plane wave in a simple isotropic medium. It is known that this condition does not define the polarization of $\mathbf{E}$ uniquely but, rather, defines a plane orthogonal to the $\mathbf{k}$ vector and any $\mathbf{E}$ parallel to the plane is a possible field vector of the wave. A similar case is given here by the condition (7.139).

In the time-harmonic case the substitution $\partial_\tau \rightarrow jk_o$ leads to simplified plane-wave equations. For example, (7.132) becomes

$$(\sqrt{\boldsymbol{\nu}|\overline{\overline{\mathsf{S}}}|\boldsymbol{\nu}}\,\partial_w - jk_o\sqrt{\mu\epsilon})(\sqrt{\boldsymbol{\nu}|\overline{\overline{\mathsf{S}}}|\boldsymbol{\nu}}\,\partial_w + jk_o\sqrt{\mu\epsilon})\boldsymbol{\Phi}(w) = 0. \tag{7.141}$$

The general solution is a sum of two waves which are exponential functions

$$\boldsymbol{\Phi}(w) = \boldsymbol{\Phi}_+ e^{-jkw} + \boldsymbol{\Phi}_- e^{jkw}, \quad k = \frac{k_o\sqrt{\mu\epsilon}}{\sqrt{\boldsymbol{\nu}|\overline{\overline{\mathsf{S}}}|\boldsymbol{\nu}}}, \tag{7.142}$$

where the wave number $k = k(\boldsymbol{\nu})$ is a function of $\boldsymbol{\nu}$ as well as the metric dyadic $\overline{\overline{\mathsf{S}}}$.

7.3.3 Generalized Q-medium

A more general case is obtained by considering the generalization of the class of Q media as defined by (5.96) and (7.103) with

$$\overline{\overline{\mathsf{M}}}_g = \overline{\overline{\mathsf{Q}}}^{(2)} + \mathbf{AB}. \tag{7.143}$$

In this case the plane-wave equation is of the fourth order and it can be written from (7.113) as

$$L_1(\partial_w \mathbf{d}w + \partial_\tau \mathbf{d}\tau)L_2(\partial_w \mathbf{d}w + \partial_\tau \mathbf{d}\tau)\boldsymbol{\Phi}(w,\tau) = 0, \tag{7.144}$$

where the second-order operators L_1, L_2 are defined by (7.110) and (7.112). Because the fourth-order operator is factorized, the general solution can be written as a sum of two waves

$$\boldsymbol{\Phi}(w,\tau) = \boldsymbol{\Phi}_1(w,\tau) + \boldsymbol{\Phi}_2(w,\tau), \tag{7.145}$$

satisfying the second-order wave equations

$$L_1(\partial_w \boldsymbol{\nu} + \partial_\tau \mathbf{d}\tau)\boldsymbol{\Phi}_1(w,\tau) = 0, \tag{7.146}$$

$$L_2(\partial_w \boldsymbol{\nu} + \partial_\tau \mathbf{d}\tau)\boldsymbol{\Phi}_2(w,\tau) = 0. \tag{7.147}$$

Example As an example, let us consider the Q medium of the previous example generalized by adding the bivector dyad

$$\mathbf{AB} = jA(\mathbf{e}_x \wedge \mathbf{e}_y)(\mathbf{e}_x \wedge \mathbf{e}_y). \tag{7.148}$$

The second-order equation (7.146) corresponding to the operator L_1 equals that of the previous example, (7.132), and the solution can be expressed as

$$\mathbf{\Phi}_1(w,\tau) = \mathbf{\Phi}_{1+}(w,\tau) + \mathbf{\Phi}_{1-}(w,\tau), \tag{7.149}$$

$$\mathbf{\Phi}_{1\pm}(w,\tau) = \psi_\pm(\sqrt{\boldsymbol{\nu}|\overline{\overline{\mathsf{S}}}|\boldsymbol{\nu}}\tau \mp \sqrt{\mu\epsilon}w), \tag{7.150}$$

where $\psi_\pm(\zeta)$ are arbiirary functions. The second equation (7.147) is defined by the operator (7.112) which can be written as

$$L_2(\partial_w\boldsymbol{\nu} + \partial_\tau\mathbf{d}\tau) = L_1(\partial_w\boldsymbol{\nu} + \partial_\tau\mathbf{d}\tau) + \Delta L(\partial_w\boldsymbol{\nu} + \partial_\tau\mathbf{d}\tau), \tag{7.151}$$

$$\begin{aligned}\Delta L(\mathbf{d}) &= -(\mathbf{B}\lfloor\mathbf{d})|\overline{\overline{\mathsf{Q}}}^{-1}|(\mathbf{d}\rfloor\mathbf{A}) \\ &= -jA(\mathbf{e}_x \wedge \mathbf{e}_y\lfloor\mathbf{d})|\overline{\overline{\mathsf{Q}}}^{-1}|(\mathbf{d}\rfloor\mathbf{e}_x \wedge \mathbf{e}_y) \\ &= -jA\mathbf{d}|(\mathbf{e}_x \wedge \mathbf{e}_y)\lfloor\overline{\overline{\mathsf{Q}}}^{-1}\rfloor(\mathbf{e}_x \wedge \mathbf{e}_y)|\mathbf{d}.\end{aligned} \tag{7.152}$$

In this case the inverse dyadic reads

$$\overline{\overline{\mathsf{Q}}}^{-1} = -j\sqrt{\mu}(\overline{\overline{\mathsf{S}}} - \mu\epsilon\mathbf{e}_\tau\mathbf{e}_\tau)^{-1} = -j\sqrt{\mu}(\overline{\overline{\mathsf{S}}}^{-1} - \frac{1}{\mu\epsilon}\mathbf{e}_\tau\mathbf{e}_\tau), \tag{7.153}$$

$$\overline{\overline{\mathsf{S}}}^{-1} = \frac{1}{S_{xx}}\mathbf{d}x\mathbf{d}x + \frac{1}{S_{yy}}\mathbf{d}y\mathbf{d}y + \frac{1}{S_{zz}}\mathbf{d}z\mathbf{d}z, \tag{7.154}$$

which gives the operator

$$\Delta L(\mathbf{d}) = -A\sqrt{\mu}\mathbf{d}|(\frac{1}{S_{xx}}\mathbf{e}_y\mathbf{e}_y + \frac{1}{S_{yy}}\mathbf{e}_x\mathbf{e}_x)|\mathbf{d}. \tag{7.155}$$

Thus, the operator $\overline{\overline{\mathsf{L}}}_2$ becomes

$$\begin{aligned}&L_2(\partial_w\boldsymbol{\nu} + \partial_\tau\mathbf{d}\tau) \\ &= L_1(\partial_w\boldsymbol{\nu} + \partial_\tau\mathbf{d}\tau) - A\sqrt{\mu}(\frac{(\mathbf{e}_y|\boldsymbol{\nu})^2}{S_{xx}} + \frac{(\mathbf{e}_x|\boldsymbol{\nu})^2}{S_{yy}})\partial_w^2.\end{aligned} \tag{7.156}$$

This operator is similar to L_1 and the plane-wave equation (7.147) can be written as

$$(\boldsymbol{\nu}|\overline{\overline{\mathsf{S}}}'|\boldsymbol{\nu}\partial_w^2 - \mu\epsilon\partial_\tau^2)\Phi(w,\tau) = 0, \tag{7.157}$$

when we define the modified dyadic

$$\overline{\overline{\mathsf{S}}}' = \overline{\overline{\mathsf{S}}} - A\mu\frac{\mathbf{e}_y\mathbf{e}_y}{S_{xx}} - A\mu\frac{\mathbf{e}_x\mathbf{e}_x}{S_{yy}}. \tag{7.158}$$

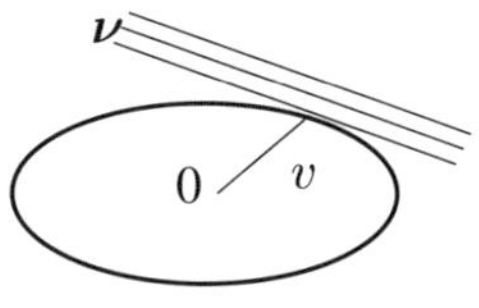

Fig. 7.3 *Wave velocity as a function of the dual vector $\boldsymbol{\nu} = \mathbf{d}w$ can be represented as a surface whose tangent plane is parallel to $\boldsymbol{\nu}$-planes, and the velocity $v(\boldsymbol{\nu})$ is shown by the radius vector. In the generalized Q medium there are two such surfaces corresponding to the two waves.*

The solution can be expressed as

$$\mathbf{\Phi}_2(w,\tau) = \mathbf{\Phi}_{2+}(w,\tau) + \mathbf{\Phi}_{2-}(w,\tau), \tag{7.159}$$

$$\mathbf{\Phi}_{2\pm}(w,\tau) = \xi_\pm(\sqrt{\boldsymbol{\nu}|\overline{\overline{\mathsf{S}}}'|\boldsymbol{\nu}}\,\tau \mp \sqrt{\mu\epsilon}\, w), \tag{7.160}$$

for arbitrary functions $\xi_\pm(\zeta)$. For these two waves the velocity is

$$v_2 = \sqrt{\frac{\boldsymbol{\nu}|\overline{\overline{\mathsf{S}}}'|\boldsymbol{\nu}}{\mu\epsilon}}, \tag{7.161}$$

which differs from that of the waves 1, in general. Thus, the difference in generalizing the Q medium is that there are two possible velocities for plane waves instead of only one in the Q medium. Obviously, in the limit $\mathbf{AB} \to 0$ we have $\overline{\overline{\mathsf{S}}}' \to \overline{\overline{\mathsf{S}}}$ and the two velocities coincide. One of the wave-velocity surfaces is depicted in Figure 7.3.

Problems

7.3.1 Expand $\overline{\overline{\mathsf{V}}}_e^{(4)}$ for the special anisotropic medium (7.31).

7.3.2 Show that the scalar operator (7.123) is at most of fourth order.

7.3.3 Find the equation for the surface of constant phase $k(\boldsymbol{\nu})w = a$ in (7.142), corresponding to the metric dyadic $\overline{\overline{\mathsf{S}}} = \mathbf{e}_x\mathbf{e}_x + \mathbf{e}_y\mathbf{e}_y + 2\mathbf{e}_z\mathbf{e}_z$. Express $\boldsymbol{\nu} = \mathbf{d}w$ in the basis $\mathbf{d}x, \mathbf{d}y, \mathbf{d}z$.

7.4 TE AND TM POLARIZED WAVES

In the previous section it was seen that the polarization of the plane wave depends on the medium parameters. Let us consider special polarization properties for the time-harmonic plane waves

$$\mathbf{\Phi}(\mathbf{r},\tau) = \mathbf{\Phi}_o \exp(\boldsymbol{\nu}|\mathbf{r} + jk_o\tau), \quad \mathbf{\Phi}_o = \mathbf{B}_o + \mathbf{E}_o \wedge \mathbf{d}\tau, \tag{7.162}$$

$$\mathbf{\Psi}(\mathbf{r},\tau) = \mathbf{\Psi}_o \exp(\boldsymbol{\nu}|\mathbf{r} + jk_o\tau), \quad \mathbf{\Psi}_o = \mathbf{D}_o - \mathbf{H}_o \wedge \mathbf{d}\tau, \tag{7.163}$$

and denote

$$\boldsymbol{\kappa} = \mathbf{d}(\boldsymbol{\nu}|\mathbf{r} + jk_o\tau) = \boldsymbol{\nu} + jk_o\mathbf{d}\tau. \tag{7.164}$$

7.4.1 Plane-wave equations

The Maxwell equations now reduce to algebraic equations for the amplitude dual bivectors $\mathbf{\Phi}_o, \mathbf{\Psi}_o$ as

$$\boldsymbol{\kappa} \wedge \mathbf{\Phi}_o = 0, \quad \boldsymbol{\kappa} \wedge \mathbf{\Psi}_o = 0. \tag{7.165}$$

From these we can show that $\mathbf{\Phi}_o$ and $\mathbf{\Psi}_o$ must be simple dual bivectors, which means that they have the form $\boldsymbol{\alpha} \wedge \boldsymbol{\beta}$. In fact, invoking the bac cab rule (1.91) reproduced here as

$$\mathbf{a}\rfloor(\boldsymbol{\beta} \wedge \mathbf{\Gamma}) = \boldsymbol{\beta} \wedge (\mathbf{c}\rfloor\mathbf{\Gamma}) + \mathbf{\Gamma}(\mathbf{a}|\boldsymbol{\beta}), \tag{7.166}$$

valid for any vector $\mathbf{a}$, dual vector $\boldsymbol{\beta}$, and dual bivector $\mathbf{\Gamma}$, we have

$$\mathbf{e}_\tau\rfloor(\boldsymbol{\kappa} \wedge \mathbf{\Phi}_o) = (\mathbf{e}_\tau|\boldsymbol{\kappa})\mathbf{\Phi}_o + \boldsymbol{\kappa} \wedge (\mathbf{e}_\tau\rfloor\mathbf{\Phi}_o) = 0. \tag{7.167}$$

From this and

$$\mathbf{e}_\tau\rfloor\mathbf{\Phi}_o = \mathbf{e}_\tau\rfloor\mathbf{B}_o + \mathbf{e}_\tau\rfloor(\mathbf{E}_o \wedge \mathbf{d}\tau) = \mathbf{E}_o \tag{7.168}$$

due to $\mathbf{e}_t\rfloor\mathbf{B}_o = 0$ and $\mathbf{e}_\tau\rfloor\mathbf{E}_o = 0$, we conclude that the amplitude $\mathbf{\Phi}_o$ of a plane wave can be expressed as

$$\mathbf{\Phi}_o = -\boldsymbol{\kappa} \wedge \frac{\mathbf{e}_\tau\rfloor\mathbf{\Phi}_o}{jk_o} = -\frac{1}{jk_o}\boldsymbol{\kappa} \wedge \mathbf{E}_o. \tag{7.169}$$

Similarly, we have

$$\mathbf{\Psi}_o = \frac{1}{jk_o}\boldsymbol{\kappa} \wedge \mathbf{H}_o. \tag{7.170}$$

From these representations it follows that the amplitude bivectors of any plane wave satisfy the simple conditions

$$\mathbf{\Phi}_o \wedge \mathbf{\Phi}_o = 0, \quad \mathbf{\Psi}_o \wedge \mathbf{\Psi}_o = 0, \quad \mathbf{\Phi}_o \wedge \mathbf{\Psi}_o = 0. \tag{7.171}$$

The corresponding three-dimensional conditions following from (7.171) are

$$\mathbf{E}_o \wedge \mathbf{B}_o = 0, \quad \mathbf{H}_o \wedge \mathbf{D}_o = 0, \quad \mathbf{E}_o \wedge \mathbf{D}_o - \mathbf{H}_o \wedge \mathbf{B}_o = 0, \tag{7.172}$$

and they have a form similar to that of the corresponding Gibbsian vectors:

$$\mathbf{E}_o \cdot \mathbf{B}_o = 0, \quad \mathbf{H}_o \cdot \mathbf{D}_o = 0, \quad \mathbf{E}_o \cdot \mathbf{D}_o - \mathbf{H}_o \cdot \mathbf{B}_o = 0. \tag{7.173}$$

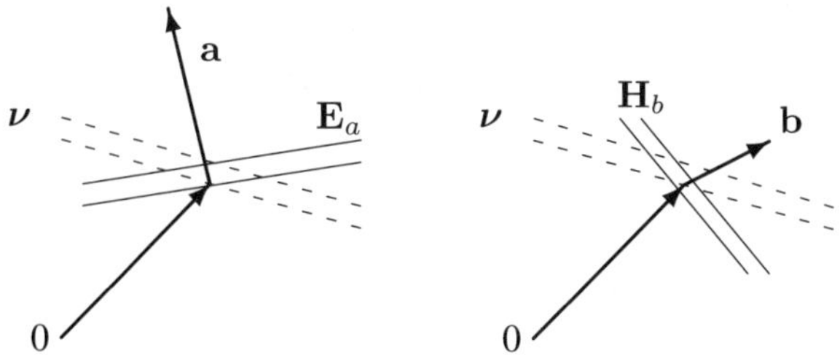

Fig. 7.4 *On the left: TE_a-polarized plane wave with electric one-form $\mathbf{E}_a$ denoted by parallel planes of constant $\mathbf{E}_a|\mathbf{r}$ orthogonal to the given vector $\mathbf{a}$. The planes of constant phase defined by the one-form $\boldsymbol{\nu}$ are not parallel to those of the electric field, in general. On the right: Similar figure for the TM_b wave.*

7.4.2 TE and TM polarizations

In electromagnetic field theory we often encounter plane waves with simple polarizations labeled as TE_a and TM_b. Denoting the corresponding amplitude one- and two-forms by the respective subscripts a and b, a TE_a plane wave is defined by the condition

$$\mathbf{a}|\mathbf{E}_a = 0, \tag{7.174}$$

and called transverse electric with respect to a given vector $\mathbf{a}$. Similarly, a TM_b plane wave satisfies

$$\mathbf{b}|\mathbf{H}_b = 0, \tag{7.175}$$

and it is called transverse magnetic with respect to another given vector $\mathbf{b}$. Here we assume that $\mathbf{a}$ and $\mathbf{b}$ are two space-like vectors satisfying

$$\mathbf{a}|\mathbf{d}\tau = \mathbf{b}|\mathbf{d}\tau = 0. \tag{7.176}$$

The dual vectors $\mathbf{E}_a$ and $\mathbf{H}_b$ ant the vectors $\mathbf{a}$ and $\mathbf{b}$ are depicted in Figure 7.4. From (7.169), (7.170) we can see that the correponding conditions for the four-dimensional two-forms are

$$(\mathbf{a} \wedge \mathbf{e}_\tau)|\mathbf{\Phi}_a = 0, \quad (\mathbf{b} \wedge \mathbf{e}_\tau)|\mathbf{\Psi}_b = 0. \tag{7.177}$$

A straightforward generalization to the TE/TM polarization conditions is obtained when $\mathbf{a} \wedge \mathbf{e}_\tau$ and $\mathbf{b} \wedge \mathbf{e}_\tau$ are replaced by more general bivectors $\mathbf{A}$ and $\mathbf{B}$ in (7.177).

7.4.3 Medium conditions

It is known from the corresponding Gibbsian analysis that TE and TM waves cannot exist in the general medium. Actually, requiring (7.174) and (7.175) or (7.177) creates some condition on the medium dyadics. To find the medium condition, let us consider only the anisotropic medium defined by $\overline{\overline{\alpha}} = 0$, $\overline{\overline{\beta}} = 0$, and

$$\overline{\overline{\mathsf{M}}} = \overline{\overline{\epsilon}} \wedge \mathbf{e}_\tau + \mathbf{d}\tau \wedge \overline{\overline{\mu}}^{-1}. \tag{7.178}$$

In this case it is easiest to use the three-dimensional representation starting from the three-form conditions (7.172). If we multiply them by $\mathbf{e}_{123}|$, the equivalent scalar conditions are

$$\mathbf{E}_o|(\mathbf{e}_{123}\lfloor\mathbf{B}_o) = 0, \quad \mathbf{H}_o|(\mathbf{e}_{123}\lfloor\mathbf{D}_o) = 0, \tag{7.179}$$

$$\mathbf{E}_o|(\mathbf{e}_{123}\lfloor\mathbf{D}_o) - \mathbf{H}_o|(\mathbf{e}_{123}\lfloor\mathbf{B}_o) = 0. \tag{7.180}$$

The last condition can be omitted here. It is needed when considering generalizations to the TE/TM polarization as was done in reference 60 in Gibbsian vector formalism. The condition for the anisotropic medium supporting TE and TM waves comes from (7.179) by combining the two conditions as one,

$$A\mathbf{E}_o|(\mathbf{e}_{123}\lfloor\overline{\overline{\mu}}|\mathbf{H}_o) + B\mathbf{H}_o|(\mathbf{e}_{123}\lfloor\overline{\overline{\epsilon}}|\mathbf{E}_o) = \mathbf{E}_o|(A\overline{\overline{\mu}}_g + B\overline{\overline{\epsilon}}_g^T)|\mathbf{H}_o = 0, \tag{7.181}$$

which is valid for any plane wave and for all values of the scalar coefficients A and B. If now vectors $\mathbf{a}$, $\mathbf{b}$ and scalars A, B can be found such that the modified permittivity and permeability dyadics satisfy the condition

$$A\overline{\overline{\mu}}_g + B\overline{\overline{\epsilon}}_g^T = \mathbf{ab}, \tag{7.182}$$

the plane-wave field one-forms satisfy

$$(\mathbf{E}_o|\mathbf{a})(\mathbf{b}|\mathbf{H}_o) = 0. \tag{7.183}$$

From this we see that any plane wave in such a medium must have either TE_a or TM_b polarization because it satisfies either (7.174) or (7.175). Comparing the condition (7.182) and that in (5.101), it is seen that the medium considered here belongs to the class of generalized Q-media.

Example To check the previous theory, let us consider an anisotropic medium satisfying the condition (7.182) and check whether the plane waves have TE_a and TM_b polarizations. The modified medium dyadics are defined by

$$\overline{\overline{\epsilon}}_g = \epsilon(\overline{\overline{\mathsf{S}}} + \mathbf{ba}), \quad \overline{\overline{\mu}}_g = \mu\overline{\overline{\mathsf{S}}}, \quad \overline{\overline{\mathsf{S}}}^{(3)}||\varepsilon_{123}\varepsilon_{123} = 1, \tag{7.184}$$

where $\overline{\overline{\mathsf{S}}} \in \mathbb{E}_1\,\mathbb{E}_1$ is a symmetric dyadic. Such a medium falls in the class of generalized Q-media, because we can write

$$\begin{aligned}\overline{\overline{\mathsf{M}}}_g &= \mathbf{e}_N\lfloor\overline{\overline{\mathsf{M}}} = \overline{\overline{\epsilon}}_g{}^{\wedge}_{\wedge}\mathbf{e}_\tau\mathbf{e}_\tau - \overline{\overline{\mu}}_g^{-1}\rfloor\rfloor\mathbf{e}_{123}\mathbf{e}_{123} \\ &= \epsilon\overline{\overline{\mathsf{S}}}{}^{\wedge}_{\wedge}\mathbf{e}_\tau\mathbf{e}_\tau + \epsilon\mathbf{ba}{}^{\wedge}_{\wedge}\mathbf{e}_\tau\mathbf{e}_\tau - \frac{1}{\mu}\overline{\overline{\mathsf{S}}}^{(2)} \\ &= -\frac{1}{\mu}(\overline{\overline{\mathsf{S}}} - \mu\epsilon\mathbf{e}_\tau\mathbf{e}_\tau)^{(2)} + \epsilon\mathbf{ba}{}^{\wedge}_{\wedge}\mathbf{e}_\tau\mathbf{e}_\tau.\end{aligned} \tag{7.185}$$

Here we have used the formula (2.169) for the inverse of the metric dyadic $\overline{\overline{\mu}}_g$. Obviously, $\overline{\overline{\mathsf{M}}}_g$ is of the form (7.143).

Now let us study the polarization of a plane wave in such a medium. The Maxwell equations for a plane wave (7.165) have the temporal parts

$$\boldsymbol{\nu} \wedge \mathbf{E}_o = -jk_o\overline{\overline{\mu}}|\mathbf{H}_o, \quad (7.186)$$

$$\boldsymbol{\nu} \wedge \mathbf{H}_o = jk_o\overline{\overline{\epsilon}}|\mathbf{E}_o. \quad (7.187)$$

Multiplying these by $\mathbf{e}_{123}\lfloor$ yields

$$(\mathbf{e}_{123}\lfloor\boldsymbol{\nu})\lfloor\mathbf{E}_o = -jk_o\overline{\overline{\mu}}_g|\mathbf{H}_o = -jk_o\mu\overline{\overline{\mathsf{S}}}|\mathbf{H}_o, \quad (7.188)$$

$$(\mathbf{e}_{123}\lfloor\boldsymbol{\nu})\lfloor\mathbf{H}_o = jk_o\overline{\overline{\epsilon}}_g|\mathbf{E}_o = jk_o\epsilon(\overline{\overline{\mathsf{S}}} + \mathbf{ba})|\mathbf{E}_o. \quad (7.189)$$

Multiplying (7.188) by $\mathbf{E}_o|$, its first term vanishes, whence the field one-forms must satisfy

$$\mathbf{E}_o|\overline{\overline{\mathsf{S}}}|\mathbf{H}_o = 0. \quad (7.190)$$

Multiplying (7.189) by $\mathbf{H}_o|$, its first term also vanishes. Applying the previous result, the last term gives

$$\mathbf{H}_o|(\overline{\overline{\mathsf{S}}} + \mathbf{ba})|\mathbf{E}_o = (\mathbf{H}_o|\mathbf{b})(\mathbf{a}|\mathbf{E}_o) = 0. \quad (7.191)$$

This shows that, indeed, any plane wave in a medium described above must satisfy either $\mathbf{a}|\mathbf{E}_o = 0$ or $\mathbf{b}|\mathbf{H}_o = 0$. Thus, all plane waves can be decomposed in two classes, those of TE$_a$ and TMb waves. A generalization of this decomposition is possible for the generalized Q-media, see Problem 7.4.5.

Problems

7.4.1 Show that the conditions (7.171) and (7.172) are equivalent.

7.4.2 Do the details in deriving (7.185).

7.4.3 Show that if two four-dimensional dual bivectors $\boldsymbol{\Phi}_o$ and $\boldsymbol{\Psi}_o$ satisfy the three conditions

$$\boldsymbol{\Phi}_o \wedge \boldsymbol{\Phi}_o = 0, \quad \boldsymbol{\Phi}_o \wedge \boldsymbol{\Psi}_o = 0, \quad \boldsymbol{\Psi}_o \wedge \boldsymbol{\Psi}_o = 0,$$

there exist dual vectors $\boldsymbol{\nu}$, $\boldsymbol{\alpha}$ and $\boldsymbol{\beta}$ such that we can write

$$\boldsymbol{\Phi}_o = \boldsymbol{\nu} \wedge \boldsymbol{\alpha}, \quad \boldsymbol{\Psi}_o = \boldsymbol{\nu} \wedge \boldsymbol{\beta}.$$

7.4.5 It can be shown that the Gibbsian medium dyadics corresponding to the generalized Q-medium condition (5.96) must be of the form [52]

$$\overline{\overline{\epsilon}}_g = c'(\overline{\overline{\mathsf{A}}} + \mathbf{p}_2\mathbf{q}_1),$$

$$\overline{\overline{\mu}}_g = \frac{1}{D}(-\overline{\overline{\mathsf{A}}}^T + \mathbf{p}_1\mathbf{q}_2),$$

$$\overline{\overline{\xi}}_g = \overline{\overline{\mathsf{I}}}\rfloor \mathbf{X}_g + c'\mathbf{p}_2\mathbf{q}_2,$$

$$\overline{\overline{\zeta}}_g = \overline{\overline{\mathsf{I}}}\rfloor \mathbf{Z}_g + \frac{1}{D}\mathbf{p}_1\mathbf{q}_1,$$

where $\overline{\overline{\mathsf{A}}} \in \mathbb{E}_1\mathbb{E}_1$ is a three-dimensional dyadic, $\mathbf{X}_g, \mathbf{Z}_g \in \mathbb{E}_2$ are three-dimensional bivectors and $\mathbf{p}_1, \mathbf{p}_2, \mathbf{q}_1, \mathbf{q}_2 \in \mathbb{E}_1$ are three-dimensional vectors. D and c' are two scalars. Show by expanding the scalar function $\mathbf{e}_{123}\lfloor(D\mathbf{E} \wedge \mathbf{B} + \mathbf{H} \wedge \mathbf{D}/c')$, which vanishes for any plane wave, that the plane-wave field one-forms $\mathbf{E}$ and $\mathbf{H}$ in such a medium must satisfy

$$(\mathbf{p}_1|\mathbf{E} + \mathbf{p}_2|\mathbf{H})(\mathbf{q}_1|\mathbf{E} + \mathbf{q}_2|\mathbf{H}) = 0.$$

This is a generalization to the TE/TM decomposition.

7.5 GREEN FUNCTIONS

Green function can be interpreted as the field corresponding to a point source of unit amplitude. It has significance only in a linear medium because solutions corresponding to extended sources can be expressed as an integral of the Green function. For example, in the classical time-harmonic electromagnetic analysis the relation between the Gibbsian current vector $\mathbf{J}(\mathbf{r})$ and the electric field vector $\mathbf{E}(\mathbf{r})$ is given by the linear differential equation

$$\overline{\overline{\mathsf{H}}}(\nabla) \cdot \mathbf{E}(\mathbf{r}) = j\omega\mu_o\mathbf{J}(\mathbf{r}), \tag{7.192}$$

where $\overline{\overline{\mathsf{H}}}(\nabla)$ is the dyadic Helmholtz operator [40]. The Green function mapping vectors to vectors is a dyadic function $\overline{\overline{\mathsf{G}}}(\mathbf{r})$ defined as the solution of the equation with dyadic delta-function source (minus sign added by convention)

$$\overline{\overline{\mathsf{H}}}(\nabla) \cdot \overline{\overline{\mathsf{G}}}(\mathbf{r} - \mathbf{r}') = -\overline{\overline{\mathsf{I}}}\delta(\mathbf{r} - \mathbf{r}'), \tag{7.193}$$

where $\overline{\overline{\mathsf{I}}}$ is the unit dyadic. The field corresponding to a given current source $\mathbf{J}$ can now be expressed as the integral

$$\mathbf{E}(\mathbf{r}) = -j\omega\mu_o \int \overline{\overline{\mathsf{G}}}(\mathbf{r} - \mathbf{r}') \cdot \mathbf{J}(\mathbf{r}')dV', \tag{7.194}$$

assuming that the field point is outside the source region. To check this, we can operate by $\overline{\overline{\mathsf{H}}}(\nabla)$, change the order of differentiation and integration and use the property of the delta function:

$$\begin{aligned}\overline{\overline{\mathsf{H}}}(\nabla) \cdot \mathbf{E}(\mathbf{r}) &= -j\omega\mu_o \int \overline{\overline{\mathsf{H}}}(\nabla) \cdot \overline{\overline{\mathsf{G}}}(\mathbf{r} - \mathbf{r}') \cdot \mathbf{J}(\mathbf{r}')dV' \\ &= j\omega\mu_o \int \delta(\mathbf{r} - \mathbf{r}')\mathbf{J}(\mathbf{r}')dV' = j\omega\mu_o\mathbf{J}(\mathbf{r}).\end{aligned} \tag{7.195}$$

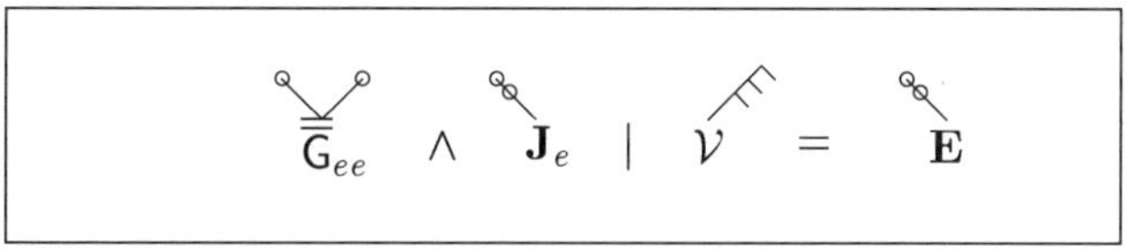

Fig. 7.5 *Checking the grades of Equation (7.197).*

7.5.1 Green function as a mapping

To define Green functions in differential-form formalism it is useful to consider the nature of the mapping. First, the spaces of the source and the field must have the same dimension, which is not always the case. For example, the Maxwell equation $\mathbf{d} \wedge \mathbf{\Psi} = \boldsymbol{\gamma}_e$ relates the source three-form $\boldsymbol{\gamma}_e$ with the dimension 4 to the field two-form $\mathbf{\Psi}$ with the dimension 6. Thus, there does not exist a simple Green dyadic mapping $\boldsymbol{\gamma}_e \to \mathbf{\Psi}$. On the other hand, the electric potential one-form $\boldsymbol{\alpha}_e$ has the same dimension 4 as the source three-form $\boldsymbol{\gamma}_e$. This makes it possible to define a Green function which gives the electric potential as an integral of the electric source over a four-domain $\mathcal{D}$. Thus, the Green function can be taken in the form of a dyadic $\overline{\overline{\mathsf{G}}}_e(\mathbf{x} - \mathbf{x}') \in \mathbb{F}_1\mathbb{F}_1$, which, combined with $\boldsymbol{\gamma}_e(\mathbf{x}')$ as $\overline{\overline{\mathsf{G}}}_e \wedge \boldsymbol{\gamma}_e \in \mathbb{F}_1\mathbb{F}_4$, can be integrated over $\mathcal{D}' \in \mathbb{E}_4$. In three-dimensional representation the dimensions of the electric one-form $\mathbf{E}$ and the current two-form $\mathbf{J}$ are again the same, which makes it possible to define a Green dyadic mapping between them. In this case the electric–electric Green dyadic must be taken as $\overline{\overline{\mathsf{G}}}_{ee}(\mathbf{r} - \mathbf{r}') \in \mathbb{F}_1\mathbb{F}_1$, which, combined with $\mathbf{J}_e(\mathbf{r}')$ as $\overline{\overline{\mathsf{G}}}_{ee} \wedge \mathbf{J}_e \in \mathbb{F}_1\mathbb{F}_3$, can be integrated over a volume $\mathcal{V}'$. The Green dyadics defined in this manner are metric dydics.

7.5.2 Three-dimensional representation

To define the Euclidean Green dyadics for electric and magnetic fields $\mathbf{E}$ and $\mathbf{H}$ we start from the wave equation for the electric field and assume no magnetic source. Let us assume time-harmonic sources and fields, leading to the substitution $\partial_\tau \to jk_o$ whence the wave equation (7.63) becomes the Helmholtz equation

$$\overline{\overline{\mathsf{W}}}_e(\mathbf{d}, jk_o)|\mathbf{E}(\mathbf{r}) = -jk_o\mathbf{J}_e(\mathbf{r}). \tag{7.196}$$

The operator is in the space of Hodge dyadics, $\overline{\overline{\mathsf{W}}}_e \in \mathbb{F}_2\mathbb{E}_1$. Expressing now the electric one-form as an integral of the form suggested above (Figure 7.5),

$$\mathbf{E}(\mathbf{r}) = \overline{\overline{\mathsf{G}}}_{ee}(\mathbf{r} - \mathbf{r}') \wedge \mathbf{J}_e(\mathbf{r}')|\mathcal{V}', \tag{7.197}$$

where $\mathcal{V}'$ means that integration is with respect to $\mathbf{r}'$, this substituted in (7.196) must satisfy

$$\overline{\overline{\mathsf{W}}}_e(\mathbf{d}, jk_o)|(\overline{\overline{\mathsf{G}}}_{ee}(\mathbf{r} - \mathbf{r}') \wedge \mathbf{J}_e(\mathbf{r}'))|\mathcal{V}' = -jk_o\mathbf{J}_e(\mathbf{r}). \tag{7.198}$$

Applying the identity

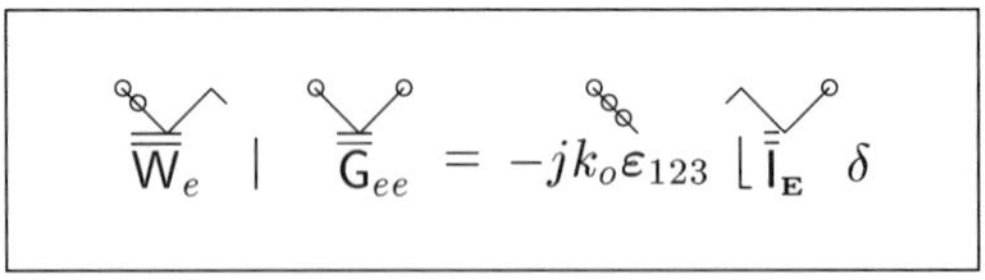
$\overline{\overline{\mathsf{W}}}_e \;|\; \overline{\overline{\mathsf{G}}}_{ee} = -jk_o\varepsilon_{123}\lfloor\overline{\overline{\mathsf{I}}}_{\mathbf{E}}\;\delta$

Fig. 7.6 *Checking the grades of Equation (7.200).*

$$\varepsilon_{123}\lfloor\overline{\overline{\mathsf{I}}}_{\mathbf{E}} \wedge \mathbf{J}_e = \mathbf{J}_e\varepsilon_{123}, \tag{7.199}$$

we can see that if the electric–electric Green dyadic $\overline{\overline{\mathsf{G}}}_{ee}(\mathbf{r}) \in \mathbb{F}_1\mathbb{F}_1$ satisfies the equation (Figure 7.6)

$$\overline{\overline{\mathsf{W}}}_e(\mathbf{d}, jk_o)|\overline{\overline{\mathsf{G}}}_{ee}(\mathbf{r}-\mathbf{r}') = -jk_o\varepsilon_{123}\lfloor\overline{\overline{\mathsf{I}}}_{\mathbf{E}}\delta(\mathbf{r}-\mathbf{r}'), \tag{7.200}$$

(7.198) is satisfied. Thus, (7.200) is the equation for the Green dyadic. Equation (7.200) is not of the standard form since the right-hand side is not a multiple of the unit dyadic. Written in standard form, the equation for the Green dyadic is

$$\overline{\overline{\mathsf{W}}}_{e1}(\mathbf{d}, jk_o)|\overline{\overline{\mathsf{G}}}_{ee}(\mathbf{r}-\mathbf{r}') = -\overline{\overline{\mathsf{I}}}_{\mathbf{E}}\delta(\mathbf{r}-\mathbf{r}'). \tag{7.201}$$

This is obtained by multiplying (7.200) with $(\mathbf{e}_{123}/jk_o)\lfloor$ and applying the identity

$$\mathbf{e}_{123}\lfloor(\varepsilon_{123}\lfloor\overline{\overline{\mathsf{I}}}_{\mathbf{E}}) = \overline{\overline{\mathsf{I}}}_{\mathbf{E}}. \tag{7.202}$$

The modified operator then has the form

$$\overline{\overline{\mathsf{W}}}_{e1}(\mathbf{d}, jk_o) = \frac{1}{jk_o}\mathbf{e}_{123}\lfloor\overline{\overline{\mathsf{W}}}_e(\mathbf{d}, jk_o), \tag{7.203}$$

which is a dyadic of the metric type, $\overline{\overline{\mathsf{W}}}_{e1} \in \mathbb{E}_1\mathbb{E}_1$. The corresponding magnetic–magnetic Green dyadic satisfies

$$\overline{\overline{\mathsf{W}}}_m(\mathbf{d}, jk_o)|\overline{\overline{\mathsf{G}}}_{mm}(\mathbf{r}) = \varepsilon_{123}\lfloor\overline{\overline{\mathsf{I}}}\delta(\mathbf{r}), \tag{7.204}$$

and the magnetic field from the magnetic current two-form is obtained as

$$\mathbf{H}(\mathbf{r}) = -k_o\overline{\overline{\mathsf{G}}}_{mm}(\mathbf{r}-\mathbf{r}') \wedge \mathbf{J}_m(\mathbf{r}')|\mathcal{V}'. \tag{7.205}$$

The general relations can be best written in the superform representation of the time-harmonic Maxwell equations

$$\begin{pmatrix} -jk_o\overline{\overline{\epsilon}} & \mathbf{d}\wedge\overline{\overline{\mathsf{I}}}^T_{\mathbf{E}} - jk_o\overline{\overline{\xi}} \\ -\mathbf{d}\wedge\overline{\overline{\mathsf{I}}}^T_{\mathbf{E}} - jk_o\overline{\overline{\zeta}} & -jk_o\overline{\overline{\mu}} \end{pmatrix} \Big| \begin{pmatrix} \mathbf{E}(\mathbf{r}) \\ \mathbf{H}(\mathbf{r}) \end{pmatrix} = \begin{pmatrix} \mathbf{J}_e(\mathbf{r}) \\ \mathbf{J}_m(\mathbf{r}) \end{pmatrix}, \tag{7.206}$$

in terms of a Green dyadic matrix which satisfies

$$\begin{pmatrix} -jk_o\overline{\overline{\epsilon}} & \mathbf{d}\wedge\overline{\overline{\mathsf{I}}}^T_{\mathbf{E}} - jk_o\overline{\overline{\xi}} \\ -\mathbf{d}\wedge\overline{\overline{\mathsf{I}}}^T_{\mathbf{E}} - jk_o\overline{\overline{\zeta}} & -jk_o\overline{\overline{\mu}} \end{pmatrix} \Big| \begin{pmatrix} \overline{\overline{\mathsf{G}}}_{ee}(\mathbf{r}) & \overline{\overline{\mathsf{G}}}_{em}(\mathbf{r}) \\ \overline{\overline{\mathsf{G}}}_{me}(\mathbf{r}) & \overline{\overline{\mathsf{G}}}_{mm}(\mathbf{r}) \end{pmatrix}$$

$$= \varepsilon_{123}\lfloor \begin{pmatrix} \overline{\overline{\mathsf{I}}}_{\mathbf{E}} & 0 \\ 0 & \overline{\overline{\mathsf{I}}}_{\mathbf{E}} \end{pmatrix} \delta(\mathbf{r}), \tag{7.207}$$

From this expression one can see that the off-diagonal Green dyadics $\overline{\overline{\mathsf{G}}}_{em}, \overline{\overline{\mathsf{G}}}_{me}$ can be directly obtained from the diagonal ones without solving any differential equations:

$$\overline{\overline{\mathsf{G}}}_{em}(\mathbf{r}) = (jk_o\overline{\overline{\epsilon}})^{-1}|(\mathbf{d}\wedge\overline{\overline{\mathsf{I}}}^T_{\mathbf{E}} - jk_o\overline{\overline{\xi}})|\overline{\overline{\mathsf{G}}}_{mm}(\mathbf{r}), \tag{7.208}$$

$$\overline{\overline{\mathsf{G}}}_{me}(\mathbf{r}) = -(jk_o\overline{\overline{\mu}})^{-1}|(\mathbf{d}\wedge\overline{\overline{\mathsf{I}}}^T_{\mathbf{E}} + jk_o\overline{\overline{\zeta}})|\overline{\overline{\mathsf{G}}}_{ee}(\mathbf{r}). \tag{7.209}$$

7.5.3 Four-dimensional representation

In the Minkowskian representation we consider the electric potential $\boldsymbol{\alpha}_e$ satisfying the wave equation (7.13)

$$\overline{\overline{\mathsf{V}}}_e(\mathbf{d})|\boldsymbol{\alpha}_e = \mathbf{e}_N\lfloor\boldsymbol{\gamma}_e, \tag{7.210}$$

with the dyadic operator defined by (7.12) as

$$\overline{\overline{\mathsf{V}}}_e(\mathbf{d}) = \overline{\overline{\mathsf{M}}}_g\lfloor\lfloor\mathbf{dd}. \tag{7.211}$$

As pointed out above, a Green dyadic $\overline{\overline{\mathsf{G}}}_e(\mathbf{x}) \in \mathbb{F}_1\mathbb{F}_1$ can now be defined to satisfy the equation

$$\overline{\overline{\mathsf{V}}}_e(\mathbf{d})|\overline{\overline{\mathsf{G}}}_e(\mathbf{x}) = -\overline{\overline{\mathsf{I}}}\delta(\mathbf{x}). \tag{7.212}$$

Because of the form of the operator $\overline{\overline{\mathsf{V}}}_e(\mathbf{d})$, the Green dyadic $\overline{\overline{\mathsf{G}}}_e$, like the four-potential, is not unique and it requires an extra gauge condition for uniqueness. Actually, if $\overline{\overline{\mathsf{G}}}_e(\mathbf{x})$ satisfies (7.212), so does $\overline{\overline{\mathsf{G}}}_e(\mathbf{x}) + \mathbf{d}\boldsymbol{\lambda}(\mathbf{x})$ for any one-form $\boldsymbol{\lambda}(\mathbf{x})$. The gauge condition depends on the medium.

The potential can be represented through integration over the four-domain $\mathcal{D}$ as

$$\boldsymbol{\alpha}_e(\mathbf{x}) = \overline{\overline{\mathsf{G}}}_e(\mathbf{x}-\mathbf{x}')\wedge\boldsymbol{\gamma}_e(\mathbf{x}')|\mathcal{D}'. \tag{7.213}$$

To check this, we assume field point $\mathbf{x}$ outside the source region $\mathcal{D}'$ and operate

$$\begin{aligned} \overline{\overline{\mathsf{V}}}_e(\mathbf{d})|\boldsymbol{\alpha}_e(\mathbf{x}) &= \overline{\overline{\mathsf{V}}}_e(\mathbf{d})|\overline{\overline{\mathsf{G}}}_e(\mathbf{x}-\mathbf{x}')\wedge\boldsymbol{\gamma}_e(\mathbf{x}')|\mathcal{D}' \\ &= -\overline{\overline{\mathsf{I}}}\delta(\mathbf{x}-\mathbf{x}')\wedge\boldsymbol{\gamma}_e(\mathbf{x}')|\mathcal{D}' \\ &= -\overline{\overline{\mathsf{I}}}\wedge\boldsymbol{\gamma}_e(\mathbf{x})|\mathbf{e}_{1234} = -\boldsymbol{\gamma}_e(\mathbf{x})\rfloor\mathbf{e}_{1234} \\ &= \mathbf{e}_{1234}\lfloor\boldsymbol{\gamma}_e(\mathbf{x}), \end{aligned} \tag{7.214}$$

which coincides with (7.210). Again, it is not possible to solve the Green dyadic for the general case and a simplified assumption about the medium must be made.

Example Let us consider as an example finding the electric four-potential Green dyadic corresponding to the special anisotropic medium defined by (7.32). In fact, this assumption allows us to solve the equation (7.212) for $\overline{\overline{\mathsf{G}}}_e(\mathbf{x})$. Denoting

$$\overline{\overline{\mathsf{P}}} = \overline{\overline{\mathsf{S}}} - \mu\epsilon\mathbf{e}_\tau\mathbf{e}_\tau, \quad \mathbf{d}|\overline{\overline{\mathsf{P}}}|\mathbf{d} = \mathbf{d}_\mathbf{E}|\overline{\overline{\mathsf{S}}}|\mathbf{d}_\mathbf{E} - \mu\epsilon\partial_\tau^2, \tag{7.215}$$

with

$$\overline{\overline{\mathsf{S}}} = \sum_{i=1}^{3} S_i\mathbf{e}_i\mathbf{e}_i, \quad \overline{\overline{\mathsf{S}}}^{-1} = \sum_{i=1}^{3} S_i^{-1}\boldsymbol{\varepsilon}_i\boldsymbol{\varepsilon}_i, \tag{7.216}$$

and applying the identity (7.35), the Green dyadic equation (7.212) can be expressed as

$$(\mathbf{d}|\overline{\overline{\mathsf{P}}}|\mathbf{d})\overline{\overline{\mathsf{P}}}|\overline{\overline{\mathsf{G}}}_e(\mathbf{x}) - (\overline{\overline{\mathsf{P}}}|\mathbf{d})(\mathbf{d}|\overline{\overline{\mathsf{P}}}|\overline{\overline{\mathsf{G}}}_e(\mathbf{x})) = -\mu\bar{\mathsf{I}}\delta(\mathbf{x}). \tag{7.217}$$

For uniqueness of solution it is convenient to choose the scalar gauge condition as

$$\mathbf{d}|\overline{\overline{\mathsf{P}}}|\overline{\overline{\mathsf{G}}}_e(\mathbf{x}) = (\mathbf{d}|\overline{\overline{\mathsf{S}}} - \mu\epsilon\partial_\tau\mathbf{e}_\tau)|\overline{\overline{\mathsf{G}}}_e(\mathbf{x}) = 0, \tag{7.218}$$

which is similar to that of the four-potential (7.38). Thus, (7.212) is simplified to

$$(\mathbf{d}|\overline{\overline{\mathsf{P}}}|\mathbf{d})\overline{\overline{\mathsf{P}}}|\overline{\overline{\mathsf{G}}}_e(\mathbf{x}) = (\mathbf{d}_\mathbf{E}|\overline{\overline{\mathsf{S}}}|\mathbf{d}_\mathbf{E} - \mu\epsilon\partial_\tau^2)\overline{\overline{\mathsf{P}}}|\overline{\overline{\mathsf{G}}}_e(\mathbf{r}) = -\mu\bar{\mathsf{I}}\delta(\mathbf{x}) \tag{7.219}$$

or, equivalently, to

$$\begin{aligned}(\mathbf{d}_\mathbf{E}|\overline{\overline{\mathsf{S}}}|\mathbf{d}_\mathbf{E} - \mu\epsilon\partial_\tau^2)\overline{\overline{\mathsf{G}}}_e(\mathbf{r}) &= -\mu\overline{\overline{\mathsf{P}}}^{-1}\delta(\mathbf{x}) \\ &= -\mu(\overline{\overline{\mathsf{S}}}^{-1} - \frac{1}{\mu\epsilon}\mathbf{d}\tau\mathbf{d}\tau)\delta(\mathbf{x}).\end{aligned} \tag{7.220}$$

As before, $\overline{\overline{\mathsf{S}}}$ is assumed to be a symmetric Euclidean metric dyadic and it defines the three-dimensional distance function

$$D_S(\mathbf{r}) = \sqrt{\mathbf{r}|\overline{\overline{\mathsf{S}}}^{-1}|\mathbf{r}}. \tag{7.221}$$

The retarded solution for the Green dyadic can now be written in the form

$$\overline{\overline{\mathsf{G}}}_e(\mathbf{x}) = -\mu(\overline{\overline{\mathsf{S}}}^{-1} - \frac{1}{\mu\epsilon}\mathbf{d}\tau\mathbf{d}\tau)\frac{\delta(\tau - D_S\sqrt{\mu\epsilon})}{4\pi D_S}, \tag{7.222}$$

as can be checked by inserting in (7.217).

The solution (7.222) describes a delta singularity on a moving surface defined by the vector $\mathbf{r}(\tau) = \mathbf{u}_r r(\tau)$ with

$$r(\tau) = \frac{\tau}{\sqrt{\mu\epsilon}\sqrt{\mathbf{u}_r|\overline{\overline{\mathsf{S}}}^{-1}|\mathbf{u}_r}}. \tag{7.223}$$

This has real and positive values for all directions $\mathbf{u}_r$ only when the components S_i in (7.216) are all real and positive, in which case the surface is an ellipsoid. However, when one or two of the components are negative, there is a cone of directions defined by

$$\mathbf{u}_r|\overline{\overline{\mathsf{S}}}^{-1}|\mathbf{u}_r < 0, \tag{7.224}$$

where $r(\tau)$ becomes imaginary and propagation occurs only in the directions where (7.224) is not valid. Such a case happens, for example, when a medium is in uniform motion [74].

Problems

7.5.1 Verify the steps leading to (7.200) in detail.

7.5.2 Check the solution (7.222) by inserting it in (7.219).

7.5.3 Generalize the solution (7.222) to that corresponding to the Q-medium defined by (7.85).

References

1. C. Altman and K. Suchy, *Reciprocity, Spatial Mapping and Time Reversal Electromagnetics*, Dordrecht: Kluwer, 1991.

2. B. B. Baker and E. T. Copson, *The Mathematical Theory of Huygens' Principle,* Cambridge: Cambridge University Press, 1939.

3. N. V. Balasubramanian, J. W. Lynn and D. P. Sen Gupta, *Differential Forms on Electromagnetic Networks*, London: Butterworths, 1970.

4. D. Baldomir, "Differential forms and electromagnetism in 3-dimensional Euclidean space R^3," *IEE Proc. A*, vol. 133, no. 3, pp. 139–143, May 1986.

5. D. Baldomir, "Global geometry of electromagnetic systems," *IEE Proc.*, vol. 140, Pt. A, no. 2, pp. 142–150, March 1993.

6. D. Baldomir and P. Hammond, *Geometry of Electromagnetic Systems*, Oxford: Clarendon Press, 1996.

7. G. N. Borzdov, "An intrinsic tensor technoque in Minkowski space with applications to boundary value problems," *J. Math. Phys.*, vol. 34, no. 7, pp. 3162–3196, July 1993.

8. A. Bossavit, "Whitney forms: a class of finite elements for three-dimensional computations in electromagnetism", *IEE Proc. A*, vol. 135, pp. 493–500, 1988.

9. Ph. Boulanger and M. Hayes, *Bivectors and Waves in Mechanics and Optics*, London: Chapman and Hall, 1993.

10. S. Brehmer and H. Haar, *Differentialformen und Vektoranalysis*, Berlin: VEB Deutscher Verlag der Wissenschaften, 1973.

11. I. Brevik, "Electromagnetic energy-momentum tensor within material media" (in two parts), *Det Kongelige Danske Videnskabernes Selskab, Matematisk-fysiske Meddelelser* vol. 37, no. 11 (part 1), no. 13 (part 2), Copenhagen 1970.

12. W. L. Burke, "Manifestly parity invariant electromagnetic theory and twisted tensors," *J. Math. Phys.*, vol. 24, no. 1, pp. 65–69, 1983.

13. H. Cartan, *Formes Différentielles*, Paris: Hermann, 1967.

14. D. K. Cheng, *Field and Wave Electromagnetics*, Reading MA: Addison-Wesley, 1983.

15. M. J. Crowe, *A History of Vector Analysis*, Notre Dame, IN: University of Notre Dame Press, 1967. (Reprint Dover, 1985.)

16. A. T. de Hoop, *Handbook of Radiation and Scattering of Waves*, London: Academic Press, 1995, Chapter 29.

17. G. A. Deschamps, "Exterior differential forms," in *Mathematics Applied to Sciences*, E. Roubine, ed, Berlin: Springer-Verlag, 1970, pp. 111–161.

18. G. A. Deschamps, "Electromagnetics and differential forms," *Proc. IEEE*, vol. 69, no. 6, pp. 676–696, June 1981.

19. J. Dieudonné, "The tragedy of Grassmann," *Linear and Multilinear Algebra*, vol. 8, pp. 1–14, 1979.

20. M. P. do Carmo, *Differential Forms and Applications*, Berlin: Springer-Verlag, 1994.

21. N. V. Efimov and E.R. Rozendorn, *Linear Algebra and Multidimensional Geometry*, Moscow: Mir Publishers, 1975. Translation of the original Russian 1974 edition.

22. D. Fearnley-Sander, "Hermann Grassmann and the creation of linear algebra," *Am. Math. Monthly*, vol. 86, pp. 809–817, 1979.

23. D. Fearnley-Sander, "Hermann Grassmann and the prehistory of universal algebra," *Am. Math. Monthly*, vol. 89, pp. 161–166, 1982.

24. H. Flanders, *Differential Forms with Applications to the Physical Sciences*, New York: Academic Press, 1963.

25. T. Frankel, *The Geometry of Physics*, Cambridge: Cambridge University Press, 1997.

26. J. W. Gibbs, *Element of Vector Analysis*, privately printed in two parts 1881, 1884; reprint in *The Scientific Papers of J. Willard Gibbs*, New York: Dover, 1961.

27. J. W. Gibbs, "On multiple algebra," *Proc. Am. Assoc. Adv. Sci.*, vol. 35, pp. 37–66, 1886; reprint in *The Scientific Papers of J. Willard Gibbs*, New York: Dover, 1961.

28. J. W. Gibbs and E.B. Wilson, *Vector Analysis*, New York: Scribner, 1909. Reprint, New York: Dover, 1960.

29. H. Grassmann, *Extension Theory*, Providence, R I: American Mathematical Society, 2000. Translation of *Die Ausdehnungslehre*, Berlin: Verlag von Th. Chr. Fr. Ensin, 1862.

30. P. Hammond and D. Baldomir, "Dual energy methods in electromagnetism using tubes and slices," *IEE Proc.*, vol. 135, Pt. A, no. 3, pp. 167–172, March 1988.

31. R. Harrington, *Time-Harmonic Electromagnetic Fields*, New York: McGraw-Hill, 1961.

32. O. Heaviside, *Electrical Papers*, New York: Chelsea, 1970, reprint of the first edition, London 1892; vol. 1 p.447, vol. 2, pp. 172–175. The original article was published in *Philos. Mag.*, August 1886, p.118.

33. D. Hestenes, *Space-Time Algebra*, New York: Gordon and Breach, 1966.

34. J. D. Jackson, *Classical Electrodynamics*, 3rd ed., New York: Wiley, 1998, pp. 258–262, 605–612.

35. B. Jancewicz, *Multivectors and Clifford Algebra in Electrodynamics*, Singapore: World Scientific, 1988.

36. B. Jancewicz, "A variable metric electrodynamics. The Coulomb and Biot-Savart laws in anisotropic media," *Ann. Phys.*, vol. 245, no. 2, February 1996, pp. 227–274.

37. M. Kline, *Mathematical Thought from Ancient to Modern Times*, New York: Oxford University Press, 1972, Chapter 32: Quaternions, Vectors and Linear Associative Algebras.

38. J. A. Kong, *Electromagnetic Wave Theory* 2nd ed., New York: Wiley 1990, Chapter 1.4.

39. *Lexikon bedeutender Mathematiker*, S. Gottwald, H-J. Ilgauds and K-H. Schlote, eds., Leipzig: Bibliographisches Institut, 1990, Grassmann.

40. I. V. Lindell, *Methods for Electromagnetic Field Analysis*, Oxford: Clarendon Press, 1992, Chapter 2.

41. I. V. Lindell, A.H. Sihvola, S.A. Tretyakov and A.J. Viitanen, *Electromagnetic Waves in Chiral and Bi-Isotropic Media*, Boston: Artech House, 1994.

42. I. V. Lindell and B. Jancewicz, "Maxwell stress dyadic in differential-form formalism," *IEE Proc. Sci. Meas. Technol.*, vol. 147, no. 1, pp. 19–26, January 2000.

43. I. V. Lindell and F. Olyslager, "Analytic Green dyadic for a class of nonreciprocal anisotropic media", *IEEE Trans. Antennas Propagat.*, vol. 45, no. 10, pp. 1563–1565, October 1997.

44. I. V. Lindell and F. Olyslager, "Generalized decomposition of electromagnetic fields in bi-anisotropic media", *IEEE Trans. Antennas Propagat.*, vol. 46, no. 10, pp. 1584–1585, October 1998.

45. I. V. Lindell and F. Olyslager, "Green dyadic for a class of nonreciprocal bianisotropic media", *Microwave Opt. Tech. Lett.*, vol. 19, no. 3, pp. 216–221, October 1998.

46. I.V. Lindell and L.H. Ruotanen, "Duality transformations and Green dyadics for bi-anisotropic media," *J. Electromagn. Waves Appl.*, vol. 12, no. 9, pp. 1131–1152, 1998.

47. I.V. Lindell, "Condition for the general ideal boundary," *Microwave Opt. Tech. Lett.*, vol. 26, no. 1, pp. 61–64, July 2000.

48. I.V. Lindell, "Image theory for the isotropic ideal boundary," *Microwave Opt. Tech. Lett.*, vol. 27, no. 1, pp. 68–72, October 2000.

49. I.V. Lindell and F. Olyslager, "Potentials in bi-anisotropic media," *J. Electromagn. Waves Appl.*, vol. 15, no. 1, pp. 3–18, 2001.

50. I. V. Lindell and H. Wallén, "Wave equations for bi-anisotropic media in differential forms," *J. Elecromagn. Waves Appl.*, vol. 16, no. 11, pp. 1615–1635, 2002.

51. I. V. Lindell and H. Wallén, "Differential-form electromagnetics and bi-anisotropic Q-media," *J. Elecromagn. Waves Appl.*, vol. 18, 2004.

52. I. V. Lindell and H. Wallén, "Generalized Q-media and field decomposition in differential-form approach," *Helsinki Univ. Tech. Elecetromagn. Lab. Rept.* 421, December 2003.

53. E. Lohr, *Vektor und Dyadenrechnung für Physiker und Techniker*, Berlin: de Gruyter, 1939.

54. P. Lounesto, *Clifford Algebras and Spinors*, Cambridge: Cambridge University Press, 1997.

55. F. E. Low, *Classical Field Theory*, New York: Wiley, 1997, pp. 273–277.

56. J. C. Maxwell, *A Treatise on Electricity and Magnetism*, vol. 2, 3rd ed., Oxford: Clarendon Press 1891, pp. 278–283.

57. K. Meetz and W.L. Engl, *Elektromagnetische Felder*, Berlin: Springer-Verlag, 1980.

58. C. W. Misner, K.S. Thorne and J.A. Wheeler, *Gravitation*, San Francisco: Freeman, 1973.

59. F. Olyslager and I.V. Lindell, "Field decomposition and factorization of the Helmholtz determinant operator for bianisotropic media," *IEEE Trans. Antennas Propag.*, vol. 49, no. 4, April 2001, pp. 660–665.

60. F. Olyslager and I.V. Lindell, "Green's dyadics and factorization of the Helmholtz determinant operator for a class of bianisotropic media," *Microwave Opt. Tech. Lett.*, vol. 21, no. 4, May 1999, pp. 304–309.

61. F. Olyslager and I.V. Lindell, "Electromagnetics and exotic media: a quest for the Holy Grail," *IEEE Ant. Propag. Mag.*, vol. 44, no. 2, April 2002, pp. 48–58.

62. W. K. H. Panofsky and M. Phillips, *Classical Electricity and Magnetism*, 2nd. ed., Reading, MA: Addison-Wesley, 1962, pp. 377–383.

63. E. J. Post, *Formal Structure of Electromagnetics*, Amsterdam: North Holland, 1962. Reprint, Mineola, NY: Dover, 1997.

64. E. J. Post, "The constitutive map and some of its ramifications," *Annals of Physics*, vol. 71, pp. 497–518, 1972.

65. H. Reichardt, *Vorlesungen über Vektor- und Tensorrechnung*, Berlin: VEB Deutscher Verlag der Wissenschaften, 1957.

66. F. N. H. Robinson, "Electromagnetic stress and momentum in matter," *Physics Reports*, vol. 16, no. 6, pp. 313–354, 1975.

67. V. H. Rumsey, "The reaction concept in electroimagnetic theory," *Phys. Rev.*, ser. 2, vol. 94, no. 6, pp. 1483–1491, June 1954.

68. G. Scharf, *From Electrostatics to Optics*, Berlin: Springer, 1994, pp. 139–143.

69. B. Schutz, *Geometrical Methods of Mathematical Physics,* Cambridge: Cambridge University Press, 1980.

70. J. Schwinger, L.L. DeRaad Jr., K.A. Milton and W. Tsai, *Classical Electrodynamics*, Reading, MA: Perseus Books, 1998, pp. 22–24, 75–76.

71. I. Stewart, "Hermann Grassmann was right," *Nature*, vol. 321, p.17, 1 May 1986.

72. J. A. Stratton, *Electromagnetic Theory*, New York: McGraw-Hill, 1941, pp. 97–104, 146–147.

73. B. W. Stuck, "On some mathematical aspects of deterministic classical electrodynamics," *J. Math. Phys*, vol. 15, no. 3, pp. 383–393, March 1974.

74. C-T. Tai, *Dyadic Green Functions in Electromagnetic Theory*, 2nd. ed., New York: IEEE Press, 1994.

75. P. G. Tait, *An Elementary Treatise on Quaternions,* 3rd ed., Cambridge: Cambridge University Press, 1890.

76. G. Temple, *100 Years of Mathematics*, London: Duckworth, 1981, Chapter 6: Vectors and Tensors.

77. W. Thirring, *A Course in Mathematical Physics 2: Classical Field Theory*, New York: Springer-Verlag, 1978.

78. J. Van Bladel, *Electromagnetic Fields*, New York: McGraw-Hill, 1964.

79. J. Vanderlinde, *Classical Electromagnetic Theory*, New York: Wiley, 1993, pp. 81–86, 205–207, 308–313.

80. K. F. Warnick and D. V. Arnold, "Electromagnetic Green functions using differential forms," *J. Electromagn. Waves Appl.*, vol. 10, no. 3, pp. 427–438, 1996.

81. J. A. Wheeler, *Geometrodynamics*, New York: Academic Press, 1962.

Appendix A Multivector and Dyadic Identities

In this Appendix some useful identities are collected for convenience. Where not otherwise stated, n is the dimension of the vector space and p, q are numbers in the range $0...n$.

Notation

$$\text{vectors } \mathbf{a}, \mathbf{b}, \mathbf{c}, ... \in \mathbb{E}_1, \quad \text{dual vectors } \boldsymbol{\alpha}, \boldsymbol{\beta}, \boldsymbol{\gamma}, ... \in \mathbb{F}_1$$

$$\text{bivectors } \mathbf{A}, \mathbf{B}, \mathbf{C}, ... \in \mathbb{E}_2, \quad \text{dual bivectors } \boldsymbol{\Phi}, \boldsymbol{\Psi}, \boldsymbol{\Gamma}, ... \in \mathbb{F}_2$$

$$p\text{–vectors } \mathbf{a}^p, \mathbf{b}^p, ... \in \mathbb{E}_p, \quad \text{dual } p\text{–vectors } \boldsymbol{\alpha}^p, \boldsymbol{\beta}^p, ... \in \mathbb{F}_p$$

$$p\text{–index } J = i_1 i_2 \cdots i_p, \quad N = 123 \cdots n, \quad \mathbf{a}_N = \mathbf{a}_1 \wedge \mathbf{a}_2 \cdots \wedge \mathbf{a}_n \in \mathbb{E}_n$$

$$\text{dimension of p} - \text{vectors} \quad C_p^n = \frac{n!}{p!(n-p)!} = C_{n-p}^n$$

$$\text{basis } \mathbf{e}_1, \mathbf{e}_2, ..., \mathbf{e}_n \in \mathbb{E}_1, \quad \text{reciprocal dual basis } \boldsymbol{\varepsilon}_1, \boldsymbol{\varepsilon}_2, ..., \boldsymbol{\varepsilon}_n \in \mathbb{F}_1$$

$$\text{dyadics } \overline{\overline{\mathsf{A}}}, \overline{\overline{\mathsf{B}}}, ... \in \mathbb{E}_1\mathbb{F}_1, \quad \overline{\overline{\mathsf{A}}}^T, \overline{\overline{\mathsf{B}}}^T, ... \in \mathbb{F}_1\mathbb{E}_1$$

$$\text{metric dyadics } \overline{\overline{\mathsf{A}}}, \overline{\overline{\mathsf{B}}}, ... \in \mathbb{E}_1\mathbb{E}_1, \quad \overline{\overline{\Gamma}}, \overline{\overline{\Pi}}, ... \in \mathbb{F}_1\mathbb{F}_1$$

Multivectors

Multiplication

$$\mathbf{a}^p|\boldsymbol{\alpha}^p = \boldsymbol{\alpha}^p|\mathbf{a}^p \in \mathbb{E}_0$$

$$\mathbf{a}^p \wedge \mathbf{b}^q = (-1)^{pq}\mathbf{b}^q \wedge \mathbf{a}^p \in \mathbb{E}_{p+q}, \quad = 0 \text{ for } p+q > n$$

$$p > q, \quad \mathbf{a}^p\lfloor\boldsymbol{\alpha}^q = (-1)^{q(p-q)}\boldsymbol{\alpha}^q\rfloor\mathbf{a}^p \in \mathbb{E}_{p-q}$$

$$p = q + r, \quad (\mathbf{a}^p\lfloor\boldsymbol{\alpha}^q)|\boldsymbol{\beta}^r = \mathbf{a}^p|(\boldsymbol{\alpha}^q \wedge \boldsymbol{\beta}^r)$$

$$p > q + r, \quad (\mathbf{a}^p\lfloor\boldsymbol{\alpha}^q)\lfloor\boldsymbol{\beta}^r = \mathbf{a}^p\lfloor(\boldsymbol{\alpha}^q \wedge \boldsymbol{\beta}^r)$$

Reciprocal bases

$$\mathbf{e}_i|\boldsymbol{\varepsilon}_j = \boldsymbol{\varepsilon}_j|\mathbf{e}_i = \delta_{ij}$$

$$\mathbf{e}_{K(i)} = \mathbf{e}_1 \wedge \mathbf{e}_2 \wedge \cdots \wedge \mathbf{e}_{i-1} \wedge \mathbf{e}_{i+1} \wedge \cdots \wedge \mathbf{e}_n \in \mathbb{E}_{n-1}$$

$$\mathbf{e}_N = \mathbf{e}_1 \wedge \cdots \wedge \mathbf{e}_n = (-1)^{i-1}\mathbf{e}_i \wedge \mathbf{e}_{K(i)} = (-1)^{n-i}\mathbf{e}_{K(i)} \wedge \mathbf{e}_i$$

$$\mathbf{e}_N\lfloor\boldsymbol{\varepsilon}_i = (-1)^{i-1}\mathbf{e}_{K(i)}, \quad \mathbf{e}_N\lfloor\boldsymbol{\varepsilon}_{K(i)} = (-1)^{n-i}\mathbf{e}_i$$

$$(\boldsymbol{\alpha}\rfloor\mathbf{e}_N) \wedge \mathbf{a} = (\boldsymbol{\alpha}|\mathbf{a})\mathbf{e}_N,$$

$$\mathbf{e}_N\lfloor(\boldsymbol{\varepsilon}_N\lfloor\mathbf{a}^p) = (-1)^{p(n-p)}\mathbf{a}^p$$

$$n = 4, \quad \mathbf{e}_{1234}\lfloor(\boldsymbol{\varepsilon}_{123}\lfloor\mathbf{a}^p) = \mathbf{a}^p \wedge \mathbf{e}_4$$

Bivector products

$$(\mathbf{a} \wedge \mathbf{b})|(\boldsymbol{\alpha} \wedge \boldsymbol{\beta}) = (\mathbf{a}|\boldsymbol{\alpha})(\mathbf{b}|\boldsymbol{\beta}) - (\mathbf{a}|\boldsymbol{\beta})(\mathbf{b}|\boldsymbol{\alpha})$$

$$(\mathbf{a} \wedge \mathbf{b})|(\boldsymbol{\alpha} \wedge \boldsymbol{\beta}) = \mathbf{a}|\{\mathbf{b}\rfloor(\boldsymbol{\alpha} \wedge \boldsymbol{\beta})\} = \{(\mathbf{a} \wedge \mathbf{b})\lfloor\boldsymbol{\alpha}\}|\boldsymbol{\beta}$$

$$(\mathbf{a} \wedge \mathbf{b})\lfloor\boldsymbol{\alpha} = \mathbf{b}(\mathbf{a}|\boldsymbol{\alpha}) - \mathbf{a}(\mathbf{b}|\boldsymbol{\alpha}) = (\mathbf{ba} - \mathbf{ab})|\boldsymbol{\alpha} = -\boldsymbol{\alpha}\rfloor(\mathbf{a} \wedge \mathbf{b})$$

$$\boldsymbol{\alpha} \wedge (\mathbf{a}\rfloor(\boldsymbol{\beta} \wedge \boldsymbol{\gamma})) = -(\mathbf{a}\rfloor(\boldsymbol{\beta} \wedge \boldsymbol{\gamma})) \wedge \boldsymbol{\alpha} = \mathbf{a}\rfloor(\boldsymbol{\alpha} \wedge \boldsymbol{\beta} \wedge \boldsymbol{\gamma}) - (\mathbf{a}|\boldsymbol{\alpha})(\boldsymbol{\beta} \wedge \boldsymbol{\gamma})$$

Bac cab rules

$$\boldsymbol{\alpha}\rfloor(\mathbf{b} \wedge \mathbf{c}) = \mathbf{b}(\boldsymbol{\alpha}|\mathbf{c}) - \mathbf{c}(\boldsymbol{\alpha}|\mathbf{b}) = (\mathbf{c} \wedge \mathbf{b})\lfloor\boldsymbol{\alpha}$$

$$\boldsymbol{\alpha}\rfloor(\mathbf{b} \wedge \mathbf{C}) = \mathbf{b} \wedge (\boldsymbol{\alpha}\rfloor\mathbf{C}) + \mathbf{C}(\boldsymbol{\alpha}|\mathbf{b}), \quad \mathbf{C} \in \mathbb{E}_2$$

$$\boldsymbol{\alpha}\rfloor(\mathbf{B} \wedge \mathbf{C}) = \mathbf{B} \wedge (\boldsymbol{\alpha}\rfloor\mathbf{C}) + \mathbf{C} \wedge (\boldsymbol{\alpha}\rfloor\mathbf{B}), \quad \mathbf{B}, \mathbf{C} \in \mathbb{E}_2$$

$$\mathbf{A}\rfloor(\boldsymbol{\beta} \wedge \boldsymbol{\Gamma}) = \boldsymbol{\beta}(\mathbf{A}|\boldsymbol{\Gamma}) + \boldsymbol{\Gamma}\lfloor(\mathbf{A}\lfloor\boldsymbol{\beta}), \quad \mathbf{A} \in \mathbb{E}_2, \boldsymbol{\Gamma} \in \mathbb{F}_2$$

$$\boldsymbol{\alpha}\rfloor(\mathbf{b}^p \wedge \mathbf{c}^q) = \mathbf{b}^p \wedge (\boldsymbol{\alpha}\rfloor\mathbf{c}^q) + (-1)^{pq}\mathbf{c}^q \wedge (\boldsymbol{\alpha}\rfloor\mathbf{b}^p)$$

$$(\boldsymbol{\alpha}\wedge\boldsymbol{\beta})\rfloor(\mathbf{B}\wedge\mathbf{C}) = \mathbf{B}(\boldsymbol{\alpha}\wedge\boldsymbol{\beta})|\mathbf{C} + \mathbf{C}(\boldsymbol{\alpha}\wedge\boldsymbol{\beta})|\mathbf{B} + (\boldsymbol{\beta}\rfloor\mathbf{B})\wedge(\boldsymbol{\alpha}\rfloor\mathbf{C}) + (\boldsymbol{\beta}\rfloor\mathbf{C})\wedge(\boldsymbol{\alpha}\rfloor\mathbf{B})$$

$$(\boldsymbol{\alpha} \wedge \boldsymbol{\beta})\rfloor(\mathbf{b}^p \wedge \mathbf{c}^q)$$

$$= (\mathbf{b}^p\lfloor(\boldsymbol{\alpha} \wedge \boldsymbol{\beta})) \wedge \mathbf{c}^q + \mathbf{b}^p \wedge ((\boldsymbol{\alpha} \wedge \boldsymbol{\beta})\rfloor\mathbf{c}^q) + (\boldsymbol{\alpha}\rfloor\mathbf{b}^p) \wedge (\mathbf{c}^q\lfloor\boldsymbol{\beta}) - (\boldsymbol{\beta}\rfloor\mathbf{b}^p) \wedge (\mathbf{c}^q\lfloor\boldsymbol{\alpha})$$

Trivector products

$$(\mathbf{a}\wedge\mathbf{b}\wedge\mathbf{c})|(\boldsymbol{\alpha}\wedge\boldsymbol{\beta}\wedge\boldsymbol{\gamma}) = ((\mathbf{a}\wedge\mathbf{b}\wedge\mathbf{c})\lfloor\boldsymbol{\alpha})|(\boldsymbol{\beta}\wedge\boldsymbol{\gamma}) = \boldsymbol{\alpha}|((\mathbf{a}\wedge\mathbf{b}\wedge\mathbf{c})\lfloor(\boldsymbol{\beta}\wedge\boldsymbol{\gamma}))$$

$$= \boldsymbol{\alpha}|(\mathbf{a}(\mathbf{b}\wedge\mathbf{c}) + \mathbf{b}(\mathbf{c}\wedge\mathbf{a}) + \mathbf{c}(\mathbf{a}\wedge\mathbf{b}))|(\boldsymbol{\beta}\wedge\boldsymbol{\gamma})$$

$$= (\mathbf{a}|\boldsymbol{\alpha})(\mathbf{b}\wedge\mathbf{c})|(\boldsymbol{\beta}\wedge\boldsymbol{\gamma}) - (\mathbf{a}\rfloor(\boldsymbol{\beta}\wedge\boldsymbol{\gamma}))|(\boldsymbol{\alpha}\rfloor(\mathbf{b}\wedge\mathbf{c}))$$

$$(\mathbf{a}\wedge\mathbf{C})|(\boldsymbol{\alpha}\wedge\boldsymbol{\Gamma}) = (\mathbf{a}|\boldsymbol{\alpha})(\mathbf{C}|\boldsymbol{\Gamma}) - (\mathbf{a}\rfloor\boldsymbol{\Gamma})|(\boldsymbol{\alpha}\rfloor\mathbf{C})$$

$$(\mathbf{a}\wedge\mathbf{b}\wedge\mathbf{c})\lfloor\boldsymbol{\alpha} = \{(\mathbf{a}\wedge\mathbf{b})\mathbf{c} + (\mathbf{b}\wedge\mathbf{c})\mathbf{a} + (\mathbf{c}\wedge\mathbf{a})\mathbf{b}\}|\boldsymbol{\alpha}$$

$$(\mathbf{a}\wedge\mathbf{b}\wedge\mathbf{c})\lfloor\boldsymbol{\Gamma} = \{\mathbf{a}(\mathbf{b}\wedge\mathbf{c}) + \mathbf{b}(\mathbf{c}\wedge\mathbf{a}) + \mathbf{c}(\mathbf{a}\wedge\mathbf{b})\}|\boldsymbol{\Gamma}$$

Quadrivector products

$$(\mathbf{a}\wedge\mathbf{b}\wedge\mathbf{c}\wedge\mathbf{d})|(\boldsymbol{\alpha}\wedge\boldsymbol{\beta}\wedge\boldsymbol{\gamma}\wedge\boldsymbol{\delta}) = ((\mathbf{a}\wedge\mathbf{b}\wedge\mathbf{c}\wedge\mathbf{d})\lfloor\boldsymbol{\alpha})|(\boldsymbol{\beta}\wedge\boldsymbol{\gamma}\wedge\boldsymbol{\delta})$$

$$= \boldsymbol{\alpha}|(\mathbf{a}(\mathbf{b}\wedge\mathbf{c}\wedge\mathbf{d}) - \mathbf{b}(\mathbf{c}\wedge\mathbf{d}\wedge\mathbf{a}) + \mathbf{c}(\mathbf{d}\wedge\mathbf{a}\wedge\mathbf{b}) - \mathbf{d}(\mathbf{a}\wedge\mathbf{b}\wedge\mathbf{c}))|(\boldsymbol{\beta}\wedge\boldsymbol{\gamma}\wedge\boldsymbol{\delta})$$

$$(\mathbf{a}\wedge\mathbf{b}\wedge\mathbf{c}\wedge\mathbf{d})\lfloor\boldsymbol{\alpha} = \mathbf{d}\wedge((\mathbf{b}\wedge\mathbf{c})\mathbf{a} + (\mathbf{c}\wedge\mathbf{a})\mathbf{b} + (\mathbf{a}\wedge\mathbf{b})\mathbf{c})|\boldsymbol{\alpha} - (\mathbf{a}\wedge\mathbf{b}\wedge\mathbf{c})(\mathbf{d}|\boldsymbol{\alpha})$$

$$(\mathbf{a}\wedge\mathbf{b}\wedge\mathbf{c}\wedge\mathbf{d})\lfloor(\boldsymbol{\beta}\wedge\boldsymbol{\gamma}\wedge\boldsymbol{\delta}) =$$

$$= (((\mathbf{a}(\mathbf{b}\wedge\mathbf{c} + \mathbf{b}(\mathbf{c}\wedge\mathbf{a}) + \mathbf{c}(\mathbf{a}\wedge\mathbf{b}))\wedge\mathbf{d} - \mathbf{d}(\mathbf{a}\wedge\mathbf{b}\wedge\mathbf{c}))|(\boldsymbol{\beta}\wedge\boldsymbol{\gamma}\wedge\boldsymbol{\delta})$$

$$(\mathbf{a}\wedge\mathbf{b}\wedge\mathbf{c}\wedge\mathbf{d})|(\boldsymbol{\alpha}\wedge\boldsymbol{\beta}\wedge\boldsymbol{\gamma}\wedge\boldsymbol{\delta}) = (\boldsymbol{\gamma}\wedge\boldsymbol{\delta})|((\mathbf{a}\wedge\mathbf{b}\wedge\mathbf{c}\wedge\mathbf{d})\lfloor(\boldsymbol{\alpha}\wedge\boldsymbol{\beta}))$$

$$= (\boldsymbol{\gamma}\wedge\boldsymbol{\delta})|((\mathbf{a}\wedge\mathbf{b})(\mathbf{c}\wedge\mathbf{d}) - (\mathbf{a}\wedge\mathbf{c})(\mathbf{b}\wedge\mathbf{d}) + (\mathbf{a}\wedge\mathbf{d})(\mathbf{b}\wedge\mathbf{c})$$

$$+(\mathbf{b}\wedge\mathbf{c})(\mathbf{a}\wedge\mathbf{d}) - (\mathbf{b}\wedge\mathbf{d})(\mathbf{a}\wedge\mathbf{c}) + (\mathbf{c}\wedge\mathbf{d})(\mathbf{a}\wedge\mathbf{b}))|(\boldsymbol{\alpha}\wedge\boldsymbol{\beta})$$

$$(\mathbf{a}\wedge\mathbf{b}\wedge\mathbf{c}\wedge\mathbf{d})\lfloor(\boldsymbol{\alpha}\wedge\boldsymbol{\beta}) = (((\mathbf{b}\wedge\mathbf{c})\mathbf{a} + (\mathbf{c}\wedge\mathbf{a})\mathbf{b} + (\mathbf{a}\wedge\mathbf{b})\mathbf{c})\wedge\mathbf{d})|(\boldsymbol{\alpha}\wedge\boldsymbol{\beta})$$

$$-(\mathbf{d}\wedge(\mathbf{a}(\mathbf{b}\wedge\mathbf{c}) + \mathbf{b}(\mathbf{c}\wedge\mathbf{a}) + \mathbf{c}(\mathbf{a}\wedge\mathbf{b})))|(\boldsymbol{\alpha}\wedge\boldsymbol{\beta})$$

Special cases for $n = 3$

$$\mathbf{a}(\mathbf{b}\wedge\mathbf{c}\wedge\mathbf{d}) - \mathbf{b}(\mathbf{c}\wedge\mathbf{d}\wedge\mathbf{a}) + \mathbf{c}(\mathbf{d}\wedge\mathbf{a}\wedge\mathbf{b}) - \mathbf{d}(\mathbf{a}\wedge\mathbf{b}\wedge\mathbf{c}) = 0$$

$$(\mathbf{a}(\mathbf{b}\wedge\mathbf{c}) + \mathbf{b}(\mathbf{c}\wedge\mathbf{a}) + \mathbf{c}(\mathbf{a}\wedge\mathbf{b}))\wedge\mathbf{d} = \mathbf{d}(\mathbf{a}\wedge\mathbf{b}\wedge\mathbf{c})$$

$$((\mathbf{a}\wedge\mathbf{b})\mathbf{c} + (\mathbf{c}\wedge\mathbf{a})\mathbf{b} + (\mathbf{b}\wedge\mathbf{c})\mathbf{a})\wedge\mathbf{d} = \mathbf{d}\wedge(\mathbf{a}(\mathbf{b}\wedge\mathbf{c}) + \mathbf{b}(\mathbf{c}\wedge\mathbf{a}) + \mathbf{c}(\mathbf{a}\wedge\mathbf{b}))$$

$$((\mathbf{a}\wedge\mathbf{b})\rfloor\boldsymbol{\kappa})\wedge((\mathbf{b}\wedge\mathbf{c})\rfloor\boldsymbol{\kappa}) = (\mathbf{b}\rfloor\boldsymbol{\kappa})((\mathbf{a}\wedge\mathbf{b}\wedge\mathbf{c})|\boldsymbol{\kappa}), \quad \boldsymbol{\kappa}\in\mathbb{F}_3$$

Quintivector products

$$\boldsymbol{\alpha}\rfloor(\mathbf{a}\wedge\mathbf{b}\wedge\mathbf{c}\wedge\mathbf{d}\wedge\mathbf{e}) = \boldsymbol{\alpha}|(\mathbf{a}(\mathbf{b}\wedge\mathbf{c}\wedge\mathbf{d}\wedge\mathbf{e}) - \mathbf{b}(\mathbf{a}\wedge\mathbf{c}\wedge\mathbf{d}\wedge\mathbf{e})$$

$$+\mathbf{c}(\mathbf{a}\wedge\mathbf{b}\wedge\mathbf{d}\wedge\mathbf{e}) - \mathbf{d}(\mathbf{a}\wedge\mathbf{b}\wedge\mathbf{c}\wedge\mathbf{e}) + \mathbf{e}(\mathbf{a}\wedge\mathbf{b}\wedge\mathbf{c}\wedge\mathbf{d}))$$

$$= (\boldsymbol{\alpha}\rfloor(\mathbf{a}\wedge\mathbf{b}\wedge\mathbf{c}))\wedge(\mathbf{d}\wedge\mathbf{e}) + (\mathbf{a}\wedge\mathbf{b}\wedge\mathbf{c})\wedge(\boldsymbol{\alpha}\rfloor(\mathbf{d}\wedge\mathbf{e}))$$

Special cases for $n = 4$

$$\mathbf{a}(\mathbf{b}\wedge\mathbf{c}\wedge\mathbf{d}\wedge\mathbf{e}) - \mathbf{b}(\mathbf{a}\wedge\mathbf{c}\wedge\mathbf{d}\wedge\mathbf{e}) + \mathbf{c}(\mathbf{a}\wedge\mathbf{b}\wedge\mathbf{d}\wedge\mathbf{e})$$

$$-\mathbf{d}(\mathbf{a}\wedge\mathbf{b}\wedge\mathbf{c}\wedge\mathbf{e}) + \mathbf{e}(\mathbf{a}\wedge\mathbf{b}\wedge\mathbf{c}\wedge\mathbf{d}) = 0$$

$$(((\mathbf{a}(\mathbf{b}\wedge\mathbf{c}) + \mathbf{b}(\mathbf{c}\wedge\mathbf{a}) + \mathbf{c}(\mathbf{a}\wedge\mathbf{b}))\wedge\mathbf{d}) - \mathbf{d}(\mathbf{a}\wedge\mathbf{b}\wedge\mathbf{c}))\wedge\mathbf{e} + \mathbf{e}(\mathbf{a}\wedge\mathbf{b}\wedge\mathbf{c}\wedge\mathbf{d}) = 0$$

$$(\boldsymbol{\alpha}\rfloor(\mathbf{a}\wedge\mathbf{b}\wedge\mathbf{c}))\wedge(\mathbf{d}\wedge\mathbf{e}) = -(\mathbf{a}\wedge\mathbf{b}\wedge\mathbf{c})\wedge(\boldsymbol{\alpha}\rfloor(\mathbf{d}\wedge\mathbf{e}))$$

p-vector rules

$$(\mathbf{a}_1\wedge\mathbf{a}_2\wedge\cdots\wedge\mathbf{a}_p)\lfloor\boldsymbol{\alpha} = (-1)^{p-1}\boldsymbol{\alpha}\rfloor(\mathbf{a}_1\wedge\mathbf{a}_2\wedge\cdots\wedge\mathbf{a}_p) = \boldsymbol{\alpha}|\sum_{i=1}^{p}(-1)^{i-1}\mathbf{a}_i\mathbf{a}_{K_p(i)}$$

$$= \boldsymbol{\alpha}|(\mathbf{a}_1\mathbf{a}_{K_p(1)} - \mathbf{a}_2\mathbf{a}_{K_p(2)} + \cdots + (-1)^{p-1}\mathbf{a}_p\mathbf{a}_{K_p(p)}), \quad 2 \le p \le n$$

$$\mathbf{a}_{K_p(i)} = \mathbf{a}_1\wedge\cdots\wedge\mathbf{a}_{i-1}\wedge\mathbf{a}_{i+1}\wedge\cdots\wedge\mathbf{a}_p$$

$$(\mathbf{a}^p\rfloor\varepsilon_N)|(\boldsymbol{\alpha}^p\rfloor\mathbf{e}_N) = \mathbf{a}^p|\boldsymbol{\alpha}^p$$

Dyadics

Basic rules

$$\mathbf{a}\boldsymbol{\alpha}\in\mathbb{E}_1\mathbb{F}_1, \quad \mathbf{a}\mathbf{b}\in\mathbb{E}_1\mathbb{E}_1, \quad (\mathbf{a}\boldsymbol{\alpha})^T = \boldsymbol{\alpha}\mathbf{a}\in\mathbb{F}_1\mathbb{E}_1, \quad \boldsymbol{\alpha}\boldsymbol{\beta}\in\mathbb{F}_1\mathbb{F}_1$$

$$\overline{\overline{\mathsf{A}}} = \sum\mathbf{a}_i\boldsymbol{\alpha}_i = \mathbf{a}_1\boldsymbol{\alpha}_1 + \mathbf{a}_2\boldsymbol{\alpha}_2 + \cdots + \mathbf{a}_n\boldsymbol{\alpha}_n$$

$$\overline{\overline{\mathsf{A}}}|\mathbf{a} = \mathbf{a}|\overline{\overline{\mathsf{A}}}^T, \quad \overline{\overline{\mathsf{A}}}|\overline{\overline{\mathsf{B}}} = \sum\mathbf{a}_i\boldsymbol{\alpha}_i|\sum\mathbf{b}_j\boldsymbol{\beta}_j = \sum_{i,j}(\boldsymbol{\alpha}_i|\mathbf{b}_j)\mathbf{a}_i\boldsymbol{\beta}_j$$

$$\overline{\overline{\mathsf{I}}} = \sum\mathbf{e}_i\varepsilon_i, \quad \overline{\overline{\mathsf{I}}}|\overline{\overline{\mathsf{A}}} = \overline{\overline{\mathsf{A}}}|\overline{\overline{\mathsf{I}}} = \overline{\overline{\mathsf{A}}}$$

$$(\overline{\overline{\mathsf{A}}}|\overline{\overline{\mathsf{B}}})|\overline{\overline{\mathsf{C}}} = \overline{\overline{\mathsf{A}}}|(\overline{\overline{\mathsf{B}}}|\overline{\overline{\mathsf{C}}}), \quad \overline{\overline{\mathsf{A}}}^p = \overline{\overline{\mathsf{A}}}|\overline{\overline{\mathsf{A}}}^{p-1}, \quad \overline{\overline{\mathsf{A}}}^0 = \overline{\overline{\mathsf{I}}}$$

Double-bar products

$$(\mathbf{a}\boldsymbol{\alpha})||(\mathbf{b}\boldsymbol{\beta})^T = (\mathbf{a}\boldsymbol{\alpha})||(\boldsymbol{\beta}\mathbf{b}) = (\mathbf{a}|\boldsymbol{\beta})(\mathbf{b}|\boldsymbol{\alpha})$$

$$\mathrm{tr}\,\overline{\overline{\mathsf{A}}} = \mathrm{tr}\sum\mathbf{a}_i\boldsymbol{\alpha}_i = \sum\mathbf{a}_i|\boldsymbol{\alpha}_i = \overline{\overline{\mathsf{A}}}||\overline{\overline{\mathsf{I}}}^T$$

$$\overline{\overline{\mathsf{A}}}||\overline{\overline{\mathsf{B}}}^T = \overline{\overline{\mathsf{B}}}||\overline{\overline{\mathsf{A}}}^T = \mathrm{tr}(\overline{\overline{\mathsf{A}}}|\overline{\overline{\mathsf{B}}}) = (\overline{\overline{\mathsf{A}}}|\overline{\overline{\mathsf{B}}})||\overline{\overline{\mathsf{I}}}^T$$

Double-wedge products

$$(\mathbf{a}\boldsymbol{\alpha}){}^{\wedge}_{\wedge}(\mathbf{b}\boldsymbol{\beta}) = (\mathbf{a}\wedge\mathbf{b})(\boldsymbol{\alpha}\wedge\boldsymbol{\beta})$$

$$\overline{\overline{\mathsf{A}}}{}^{\wedge}_{\wedge}\overline{\overline{\mathsf{B}}} = \overline{\overline{\mathsf{B}}}{}^{\wedge}_{\wedge}\overline{\overline{\mathsf{A}}}$$

$$(\overline{\overline{\mathsf{A}}}{}^{\wedge}_{\wedge}\overline{\overline{\mathsf{B}}}){}^{\wedge}_{\wedge}\overline{\overline{\mathsf{C}}} = \overline{\overline{\mathsf{A}}}{}^{\wedge}_{\wedge}(\overline{\overline{\mathsf{B}}}{}^{\wedge}_{\wedge}\overline{\overline{\mathsf{C}}}) = \overline{\overline{\mathsf{A}}}{}^{\wedge}_{\wedge}\overline{\overline{\mathsf{B}}}{}^{\wedge}_{\wedge}\overline{\overline{\mathsf{C}}} = \overline{\overline{\mathsf{B}}}{}^{\wedge}_{\wedge}\overline{\overline{\mathsf{A}}}{}^{\wedge}_{\wedge}\overline{\overline{\mathsf{C}}} = \cdots$$

$$(\overline{\overline{\mathsf{A}}}{}^{\wedge}_{\wedge}\overline{\overline{\mathsf{B}}})|(\mathbf{a}\wedge\mathbf{b}) = (\overline{\overline{\mathsf{A}}}|\mathbf{a})\wedge(\overline{\overline{\mathsf{B}}}|\mathbf{b}) + (\overline{\overline{\mathsf{B}}}|\mathbf{a})\wedge(\overline{\overline{\mathsf{A}}}|\mathbf{b})$$

$$(\overline{\overline{\mathsf{A}}}|\mathbf{a})\wedge(\overline{\overline{\mathsf{A}}}|\mathbf{b}) = \frac{1}{2}(\overline{\overline{\mathsf{A}}}{}^{\wedge}_{\wedge}\overline{\overline{\mathsf{A}}})|(\mathbf{a}\wedge\mathbf{b}) = \overline{\overline{\mathsf{A}}}{}^{(2)}|(\mathbf{a}\wedge\mathbf{b})$$

$$(\overline{\overline{\mathsf{A}}}{}^{\wedge}_{\wedge}\overline{\overline{\mathsf{B}}})\lfloor\mathbf{a} = (\overline{\overline{\mathsf{A}}}|\mathbf{a})\wedge\overline{\overline{\mathsf{B}}} + (\overline{\overline{\mathsf{B}}}|\mathbf{a})\wedge\overline{\overline{\mathsf{A}}}$$

Multiple-wedge powers

$$\overline{\overline{\mathsf{A}}}{}^{(2)} = \frac{1}{2}\overline{\overline{\mathsf{A}}}{}^{\wedge}_{\wedge}\overline{\overline{\mathsf{A}}} = \frac{1}{2}\sum \mathbf{a}_i\boldsymbol{\alpha}_i{}^{\wedge}_{\wedge}\sum \mathbf{a}_j\boldsymbol{\alpha}_j = \sum_{i<j}(\mathbf{a}_i\wedge\mathbf{a}_j)(\boldsymbol{\alpha}_i\wedge\boldsymbol{\alpha}_j) = \sum_{i<j}\mathbf{a}_{ij}\boldsymbol{\alpha}_{ij}$$

$$\overline{\overline{\mathsf{A}}}{}^{(p)} = \frac{1}{p!}\overline{\overline{\mathsf{A}}}{}^{\wedge}_{\wedge}\overline{\overline{\mathsf{A}}}{}^{\wedge}_{\wedge}\cdots{}^{\wedge}_{\wedge}\overline{\overline{\mathsf{A}}} = \frac{r!s!}{p!}\overline{\overline{\mathsf{A}}}{}^{(r)}{}^{\wedge}_{\wedge}\overline{\overline{\mathsf{A}}}{}^{(s)}, \quad r+s=p$$

$$\overline{\overline{\mathsf{A}}}{}^{(n)} = \mathbf{a}_N\boldsymbol{\alpha}_N = \mathbf{a}_{123\cdots n}\boldsymbol{\alpha}_{123\cdots n} = (\mathbf{a}_N|\boldsymbol{\alpha}_N)\mathbf{e}_N\boldsymbol{\varepsilon}_N = \det\overline{\overline{\mathsf{A}}}\,\overline{\overline{\mathsf{I}}}{}^{(n)}$$

$$(\overline{\overline{\mathsf{A}}}|\mathbf{a})\wedge(\overline{\overline{\mathsf{A}}}|\mathbf{b}) = \overline{\overline{\mathsf{A}}}{}^{(2)}|(\mathbf{a}\wedge\mathbf{b})$$

$$(\overline{\overline{\mathsf{A}}}|\mathbf{a}_1)\wedge(\overline{\overline{\mathsf{A}}}|\mathbf{a}_2)\cdots\wedge(\overline{\overline{\mathsf{A}}}|\mathbf{a}_p) = \overline{\overline{\mathsf{A}}}{}^{(p)}|(\mathbf{a}_1\wedge\mathbf{a}_2\wedge\cdots\wedge\mathbf{a}_p)$$

$$\overline{\overline{\mathsf{A}}}{}^{(2)}\lfloor\mathbf{a} = (\overline{\overline{\mathsf{A}}}|\mathbf{a})\wedge\overline{\overline{\mathsf{A}}}, \qquad \boldsymbol{\alpha}\rfloor\overline{\overline{\mathsf{A}}}{}^{(2)} = \overline{\overline{\mathsf{A}}}\wedge(\boldsymbol{\alpha}|\overline{\overline{\mathsf{A}}})$$

$$\overline{\overline{\mathsf{A}}}{}^{(3)}\lfloor(\mathbf{a}\wedge\mathbf{b}) = (\overline{\overline{\mathsf{A}}}|\mathbf{a})\wedge(\overline{\overline{\mathsf{A}}}|\mathbf{b})\wedge\overline{\overline{\mathsf{A}}} = (\overline{\overline{\mathsf{A}}}{}^{(2)}|(\mathbf{a}\wedge\mathbf{b}))\wedge\overline{\overline{\mathsf{A}}}$$

$$(\overline{\overline{\mathsf{A}}}|\overline{\overline{\mathsf{B}}})^{(p)} = \overline{\overline{\mathsf{A}}}{}^{(p)}|\overline{\overline{\mathsf{B}}}{}^{(p)}$$

$$(\frac{1}{2}\overline{\overline{\mathsf{A}}}{}^{\wedge}_{\wedge}\overline{\overline{\mathsf{B}}})^{(2)} = \frac{1}{2}\overline{\overline{\mathsf{A}}}{}^{(2)}{}^{\wedge}_{\wedge}\overline{\overline{\mathsf{B}}}{}^{(2)}$$

$$n=4, \quad (\overline{\overline{\mathsf{A}}}{}^{(2)})^{(2)} = \frac{1}{8}\overline{\overline{\mathsf{A}}}{}^{\wedge}_{\wedge}\overline{\overline{\mathsf{A}}}{}^{\wedge}_{\wedge}\overline{\overline{\mathsf{A}}}{}^{\wedge}_{\wedge}\overline{\overline{\mathsf{A}}} = 3\det\overline{\overline{\mathsf{A}}}\,\overline{\overline{\mathsf{I}}}{}^{(4)}$$

Unit dyadics

$$\overline{\overline{\mathsf{I}}}^{(2)} = (\sum \mathbf{e}_i \boldsymbol{\varepsilon}_i)^{(2)} = \sum_{i<j} \mathbf{e}_{ij} \boldsymbol{\varepsilon}_{ij}$$

$$\overline{\overline{\mathsf{I}}}^{(p)} = (\sum \mathbf{e}_i \boldsymbol{\varepsilon}_i)^{(p)} = \sum \mathbf{e}_J \boldsymbol{\varepsilon}_J, \quad J = \{i_1 i_2 \cdots i_p\}, \; i_1 < i_2 \cdots < i_p$$

$$\overline{\overline{\mathsf{I}}}^{(n)} = \mathbf{e}_N \boldsymbol{\varepsilon}_N = \frac{\mathbf{k}_N \boldsymbol{\kappa}_N}{\mathbf{k}_N | \boldsymbol{\kappa}_N}$$

$$\overline{\overline{\mathsf{I}}}^{(2)} | (\mathbf{a} \wedge \mathbf{b}) = (\overline{\overline{\mathsf{I}}} | \mathbf{a}) \wedge (\overline{\overline{\mathsf{I}}} | \mathbf{b}) = \mathbf{a} \wedge \mathbf{b}$$

$$\overline{\overline{\mathsf{I}}}^{(p)} | (\mathbf{a}_1 \wedge \mathbf{a}_2 \wedge \cdots \wedge \mathbf{a}_p) = \mathbf{a}_1 \wedge \mathbf{a}_2 \wedge \cdots \mathbf{a}_p$$

$$\mathrm{tr}(\overline{\overline{\mathsf{A}}} {}^{\wedge}_{\wedge} \overline{\overline{\mathsf{B}}}) = (\overline{\overline{\mathsf{A}}} {}^{\wedge}_{\wedge} \overline{\overline{\mathsf{B}}}) || \overline{\overline{\mathsf{I}}}^{(2)T}$$

$$n = 3, \quad \overline{\overline{\mathsf{I}}}^{(2)} = \mathbf{e}_{12} \boldsymbol{\varepsilon}_{12} + \mathbf{e}_{23} \boldsymbol{\varepsilon}_{23} + \mathbf{e}_{31} \boldsymbol{\varepsilon}_{31}, \quad \overline{\overline{\mathsf{I}}}^{(3)} = \mathbf{e}_{123} \boldsymbol{\varepsilon}_{123}$$

$$n = 4, \quad \overline{\overline{\mathsf{I}}}^{(2)} = \mathbf{e}_{12} \boldsymbol{\varepsilon}_{12} + \mathbf{e}_{23} \boldsymbol{\varepsilon}_{23} + \mathbf{e}_{31} \boldsymbol{\varepsilon}_{31} + (\mathbf{e}_1 \boldsymbol{\varepsilon}_1 + \mathbf{e}_2 \boldsymbol{\varepsilon}_2 + \mathbf{e}_3 \boldsymbol{\varepsilon}_3) {}^{\wedge}_{\wedge} \mathbf{e}_4 \boldsymbol{\varepsilon}_4$$

$$n = 4, \quad \overline{\overline{\mathsf{I}}}^{(3)} = \mathbf{e}_{123} \boldsymbol{\varepsilon}_{123} + (\mathbf{e}_{12} \boldsymbol{\varepsilon}_{12} + \mathbf{e}_{23} \boldsymbol{\varepsilon}_{23} + \mathbf{e}_{31} \boldsymbol{\varepsilon}_{31}) {}^{\wedge}_{\wedge} \mathbf{e}_4 \boldsymbol{\varepsilon}_4$$

Multivectors and unit dyadics

$$(\mathbf{a} \wedge \mathbf{b}) \lfloor \overline{\overline{\mathsf{I}}}^T = \mathbf{ba} - \mathbf{ab} = \overline{\overline{\mathsf{I}}} \rfloor (\mathbf{a} \wedge \mathbf{b})$$

$$(\mathbf{a} \wedge \mathbf{b} \wedge \mathbf{c}) \lfloor \overline{\overline{\mathsf{I}}}^T = (\mathbf{a} \wedge \mathbf{b})\mathbf{c} + (\mathbf{b} \wedge \mathbf{c})\mathbf{a} + (\mathbf{c} \wedge \mathbf{a})\mathbf{b} = \overline{\overline{\mathsf{I}}}^{(2)} \rfloor (\mathbf{a} \wedge \mathbf{b} \wedge \mathbf{c})$$

$$(\mathbf{a} \wedge \mathbf{b} \wedge \mathbf{c}) \lfloor \overline{\overline{\mathsf{I}}}^{(2)T} = \mathbf{a}(\mathbf{b} \wedge \mathbf{c}) + \mathbf{b}(\mathbf{c} \wedge \mathbf{a}) + \mathbf{c}(\mathbf{a} \wedge \mathbf{b}) = \overline{\overline{\mathsf{I}}} \rfloor (\mathbf{a} \wedge \mathbf{b} \wedge \mathbf{c})$$

$$(\mathbf{a} \wedge \mathbf{b} \wedge \mathbf{c} \wedge \mathbf{d}) \lfloor \overline{\overline{\mathsf{I}}}^T = \mathbf{d} \wedge ((\mathbf{a} \wedge \mathbf{b})\mathbf{c} + (\mathbf{b} \wedge \mathbf{c})\mathbf{a} + (\mathbf{c} \wedge \mathbf{a})\mathbf{b})$$

$$-(\mathbf{a} \wedge \mathbf{b} \wedge \mathbf{c})\mathbf{d} = -\overline{\overline{\mathsf{I}}}^{(3)T} \rfloor (\mathbf{a} \wedge \mathbf{b} \wedge \mathbf{c} \wedge \mathbf{d})$$

$$(\mathbf{a} \wedge \mathbf{b} \wedge \mathbf{c} \wedge \mathbf{d}) \lfloor \overline{\overline{\mathsf{I}}}^{(2)T} = (\mathbf{a} \wedge \mathbf{b})(\mathbf{c} \wedge \mathbf{d}) - (\mathbf{a} \wedge \mathbf{c})(\mathbf{b} \wedge \mathbf{d}) + (\mathbf{a} \wedge \mathbf{d})(\mathbf{b} \wedge \mathbf{c})$$

$$+(\mathbf{b} \wedge \mathbf{c})(\mathbf{a} \wedge \mathbf{d}) - (\mathbf{b} \wedge \mathbf{d})(\mathbf{a} \wedge \mathbf{c}) + (\mathbf{c} \wedge \mathbf{d})(\mathbf{a} \wedge \mathbf{b}) = \overline{\overline{\mathsf{I}}}^{(2)} \rfloor (\mathbf{a} \wedge \mathbf{b} \wedge \mathbf{c} \wedge \mathbf{d})$$

$$= \{(\mathbf{b} \wedge \mathbf{c})\mathbf{a} + (\mathbf{c} \wedge \mathbf{a})\mathbf{b} + (\mathbf{a} \wedge \mathbf{b})\mathbf{c}\} \wedge \mathbf{d} - \mathbf{d} \wedge \{\mathbf{a}(\mathbf{b} \wedge \mathbf{c}) + \mathbf{b}(\mathbf{c} \wedge \mathbf{a}) + \mathbf{c}(\mathbf{a} \wedge \mathbf{b})\}$$

$$(\mathbf{a} \wedge \mathbf{b} \wedge \mathbf{c} \wedge \mathbf{d}) \lfloor \overline{\overline{\mathsf{I}}}^{(3)T} = (\mathbf{a}(\mathbf{b} \wedge \mathbf{c}) + \mathbf{b}(\mathbf{c} \wedge \mathbf{a}) + \mathbf{c}(\mathbf{a} \wedge \mathbf{b})) \wedge \mathbf{d}$$

$$-\mathbf{d}(\mathbf{a} \wedge \mathbf{b} \wedge \mathbf{c}) = -\overline{\overline{\mathsf{I}}} \rfloor (\mathbf{a} \wedge \mathbf{b} \wedge \mathbf{c} \wedge \mathbf{d})$$

$$(\mathbf{A} \wedge \mathbf{B}) \lfloor \overline{\overline{\mathsf{I}}}^T = \mathbf{A} \wedge \overline{\overline{\mathsf{I}}} \rfloor \mathbf{B} + \mathbf{B} \wedge \overline{\overline{\mathsf{I}}} \rfloor \mathbf{A}, \quad \mathbf{A}, \mathbf{B} \in \mathbb{E}_2$$

$$(\mathbf{A} \wedge \mathbf{B}) \lfloor \overline{\overline{\mathsf{I}}}^{(2)T} = \mathbf{AB} + \mathbf{BA} - (\mathbf{A} \lfloor \overline{\overline{\mathsf{I}}}^T) {}^{\wedge}_{\wedge} (\mathbf{B} \lfloor \overline{\overline{\mathsf{I}}}^T)$$

$$\mathbf{a} \wedge (\mathbf{k}_N \lfloor \overline{\overline{\mathsf{I}}}^T) = \mathbf{k}_N \mathbf{a}, \qquad (\overline{\overline{\mathsf{I}}} \rfloor \mathbf{k}_N) \wedge \mathbf{a} = \mathbf{a} \mathbf{k}_N \quad \mathbf{a} \in \mathbb{E}_1$$

$$\mathbf{A} \wedge (\mathbf{k}_N \lfloor \overline{\overline{\mathsf{I}}}^{(2)T}) = \mathbf{k}_N \mathbf{A}, \quad (\overline{\overline{\mathsf{I}}}^{(2)} \rfloor \mathbf{k}_N) \wedge \mathbf{A} = \mathbf{A} \mathbf{k}_N, \quad \mathbf{A} \in \mathbb{E}_2$$

Special cases for bivectors

$$n = 3, \quad \mathbf{A} \wedge \bar{\bar{\mathsf{I}}} \rfloor \mathbf{B} + \mathbf{B} \wedge \bar{\bar{\mathsf{I}}} \rfloor \mathbf{A} = 0, \quad \mathbf{A} \lfloor \bar{\bar{\mathsf{I}}}^T \wedge \mathbf{A} = 0$$

$$n = 3, \quad (\mathbf{A} \lfloor \bar{\bar{\mathsf{I}}}^T) {}^{\wedge}_{\wedge} (\mathbf{B} \lfloor \bar{\bar{\mathsf{I}}}^T) = \mathbf{AB} + \mathbf{BA}, \quad (\mathbf{A} \lfloor \bar{\bar{\mathsf{I}}}^T)^{(2)} = \mathbf{AA}$$

$$n = 4, \quad \mathbf{A} \lfloor \bar{\bar{\mathsf{I}}}^T \wedge \mathbf{B} + \mathbf{B} \lfloor \bar{\bar{\mathsf{I}}}^T \wedge \mathbf{A} = \bar{\bar{\mathsf{I}}} \rfloor (\mathbf{A} \wedge \mathbf{B})$$

$$n = 4, \quad \mathbf{A} \lfloor \bar{\bar{\mathsf{I}}}^T \wedge \mathbf{A} = \frac{1}{2} \bar{\bar{\mathsf{I}}} \rfloor (\mathbf{A} \wedge \mathbf{A})$$

Double multiplications

$$\mathrm{tr}(\bar{\bar{\mathsf{I}}}^{(p)}) = \bar{\bar{\mathsf{I}}}^{(p)} || \bar{\bar{\mathsf{I}}}^{(p)T} = C_p^n = \frac{n!}{p!(n-p)!}$$

$$\mathrm{tr}(\bar{\bar{\mathsf{A}}} {}^{\wedge}_{\wedge} \bar{\bar{\mathsf{B}}}) = (\mathrm{tr}\, \bar{\bar{\mathsf{A}}})(\mathrm{tr}\, \bar{\bar{\mathsf{B}}}) - \mathrm{tr}(\bar{\bar{\mathsf{A}}} | \bar{\bar{\mathsf{B}}})$$

$$(\bar{\bar{\mathsf{A}}} {}^{\wedge}_{\wedge} \bar{\bar{\mathsf{B}}}) || (\bar{\bar{\mathsf{C}}} {}^{\wedge}_{\wedge} \bar{\bar{\mathsf{D}}})^T =$$

$$= (\bar{\bar{\mathsf{A}}} || \bar{\bar{\mathsf{C}}}^T)(\bar{\bar{\mathsf{B}}} || \bar{\bar{\mathsf{D}}}^T) + (\bar{\bar{\mathsf{A}}} || \bar{\bar{\mathsf{D}}}^T)(\bar{\bar{\mathsf{B}}} || \bar{\bar{\mathsf{C}}}^T) - (\bar{\bar{\mathsf{A}}} | \bar{\bar{\mathsf{D}}}) || (\bar{\bar{\mathsf{B}}} | \bar{\bar{\mathsf{C}}})^T - (\bar{\bar{\mathsf{A}}} | \bar{\bar{\mathsf{C}}}) || (\bar{\bar{\mathsf{B}}} | \bar{\bar{\mathsf{D}}})^T$$

$$(\bar{\bar{\mathsf{A}}} {}^{\wedge}_{\wedge} \bar{\bar{\mathsf{B}}}) \lfloor\lfloor \bar{\bar{\mathsf{C}}}^T = (\bar{\bar{\mathsf{A}}} || \bar{\bar{\mathsf{C}}}^T) \bar{\bar{\mathsf{B}}} + (\bar{\bar{\mathsf{B}}} || \bar{\bar{\mathsf{C}}}^T) \bar{\bar{\mathsf{A}}} - \bar{\bar{\mathsf{A}}} | \bar{\bar{\mathsf{C}}} | \bar{\bar{\mathsf{B}}} - \bar{\bar{\mathsf{B}}} | \bar{\bar{\mathsf{C}}} | \bar{\bar{\mathsf{A}}}$$

$$\bar{\bar{\mathsf{A}}} \rfloor\rfloor \bar{\bar{\mathsf{I}}}^{(2)T} = (\mathrm{tr}\, \bar{\bar{\mathsf{A}}}) \bar{\bar{\mathsf{I}}}^T - \bar{\bar{\mathsf{A}}}^T$$

$$\bar{\bar{\mathsf{A}}}^{(2)} \lfloor\lfloor \bar{\bar{\mathsf{I}}}^T = (\mathrm{tr}\, \bar{\bar{\mathsf{A}}}) \bar{\bar{\mathsf{A}}} - \bar{\bar{\mathsf{A}}}^2$$

$$\bar{\bar{\mathsf{A}}}^{(p)} {}^{\wedge}_{\wedge} \bar{\bar{\mathsf{I}}}^{(n-p)} = [\mathrm{tr}\, \bar{\bar{\mathsf{A}}}^{(p)}] \bar{\bar{\mathsf{I}}}^{(n)}$$

$$\bar{\bar{\mathsf{I}}}^{(n)} \lfloor\lfloor \bar{\bar{\mathsf{I}}}^{(n-p)T} = \bar{\bar{\mathsf{I}}}^{(p)}, \quad 0 < p < n$$

Inverse dyadics

$$\det \bar{\bar{\mathsf{A}}} = \mathrm{tr}\, \bar{\bar{\mathsf{A}}}^{(n)}$$

$$\bar{\bar{\mathsf{A}}}^{-1} = \frac{\bar{\bar{\mathsf{I}}}^{(n)} \lfloor\lfloor \bar{\bar{\mathsf{A}}}^{(n-1)T}}{\det \bar{\bar{\mathsf{A}}}}, \quad \bar{\bar{\mathsf{A}}} \in \mathbb{E}_1 \mathbb{F}_1$$

$$\bar{\bar{\mathsf{A}}}^{-1} = \frac{\boldsymbol{\kappa}_N \boldsymbol{\kappa}_N \lfloor\lfloor \bar{\bar{\mathsf{A}}}^{(n-1)T}}{\boldsymbol{\kappa}_N \boldsymbol{\kappa}_N || \bar{\bar{\mathsf{A}}}^{(n)}}, \quad \bar{\bar{\mathsf{A}}} \in \mathbb{E}_1 \mathbb{E}_1$$

$$(\bar{\bar{\mathsf{A}}}^{(2)})^{-1} = (\bar{\bar{\mathsf{A}}}^{-1})^{(2)} = \bar{\bar{\mathsf{A}}}^{(-2)} \; (\mathrm{def})$$

$$(\mathbf{e}_N \lfloor \bar{\bar{\mathsf{I}}}^{(p)T})^{-1} = (-1)^{p(n-p)} \boldsymbol{\varepsilon}_N \lfloor \bar{\bar{\mathsf{I}}}^{(n-p)}$$

$$(\bar{\bar{\mathsf{A}}} | \bar{\bar{\mathsf{B}}})^{-1} = \bar{\bar{\mathsf{B}}}^{-1} | \bar{\bar{\mathsf{A}}}^{-1}$$

$$(\bar{\bar{\mathsf{A}}} + \mathbf{a}\boldsymbol{\alpha})^{-1} = \bar{\bar{\mathsf{A}}}^{-1} - \frac{\bar{\bar{\mathsf{A}}}^{-1} | \mathbf{a}\boldsymbol{\alpha} | \bar{\bar{\mathsf{A}}}^{-1}}{1 + \boldsymbol{\alpha} | \bar{\bar{\mathsf{A}}}^{-1} | \mathbf{a}}, \quad \bar{\bar{\mathsf{A}}} \in \mathbb{E}_1 \mathbb{F}_1$$

$$(\overline{\overline{\mathsf{A}}} + \mathbf{AB})^{-1} = \overline{\overline{\mathsf{A}}}^{-1} - \frac{\overline{\overline{\mathsf{A}}}^{-1}|\mathbf{AB}|\overline{\overline{\mathsf{A}}}^{-1}}{1 + \mathbf{B}|\overline{\overline{\mathsf{A}}}^{-1}|\mathbf{A}}, \quad \overline{\overline{\mathsf{A}}} \in \mathbb{E}_2\mathbb{E}_2$$

$$n = 3, \quad \overline{\overline{\mathsf{A}}}^{-1} = \frac{1}{\mathrm{Det}\overline{\overline{\mathsf{A}}}}(\overline{\overline{\mathsf{A}}}^T \rfloor\rfloor \overline{\overline{\mathsf{I}}}^{(3)})^{(2)}, \quad \overline{\overline{\mathsf{A}}} \in \mathbb{E}_2\mathbb{F}_2,$$

$$n = 3, \quad \mathrm{Det}\,\overline{\overline{\mathsf{A}}} = \det(\overline{\overline{\mathsf{A}}}^T \rfloor\rfloor \overline{\overline{\mathsf{I}}}^{(3)}) = (\overline{\overline{\mathsf{A}}}^T \rfloor\rfloor \overline{\overline{\mathsf{I}}}^{(3)})^{(3)} || \overline{\overline{\mathsf{I}}}^{(3)T}, \quad \overline{\overline{\mathsf{A}}} \in \mathbb{E}_2\mathbb{F}_2$$

$$n = 4, \quad (\overline{\overline{\mathsf{A}}} + \mathbf{e}_4\mathbf{a} + \mathbf{b}\mathbf{e}_4 + c\mathbf{e}_4\mathbf{e}_4)^{-1} = \overline{\overline{\mathsf{A}}}^{-1} + \frac{(\overline{\overline{\mathsf{A}}}^{-1}|\mathbf{b} - \varepsilon_4)(\mathbf{a}|\overline{\overline{\mathsf{A}}}^{-1} - \varepsilon_4)}{c - \mathbf{a}|\overline{\overline{\mathsf{A}}}^{-1}|\mathbf{b}}$$

Metric and Hodge dyadics

3D Euclidean space

$$\overline{\overline{\mathsf{G}}}_{\mathbf{E}} = \mathbf{e}_1\mathbf{e}_1 + \mathbf{e}_2\mathbf{e}_2 + \mathbf{e}_3\mathbf{e}_3, \quad \overline{\overline{\Gamma}}_{\mathbf{E}} = \overline{\overline{\mathsf{G}}}_{\mathbf{E}}^{-1} = \varepsilon_1\varepsilon_1 + \varepsilon_2\varepsilon_2 + \varepsilon_3\varepsilon_3$$

$$\overline{\overline{\mathsf{G}}}_{\mathbf{E}}^{(2)} = \mathbf{e}_{12}\mathbf{e}_{12} + \mathbf{e}_{23}\mathbf{e}_{23} + \mathbf{e}_{31}\mathbf{e}_{31}, \quad \overline{\overline{\Gamma}}_{\mathbf{E}}^{(2)} = \varepsilon_{12}\varepsilon_{12} + \varepsilon_{23}\varepsilon_{23} + \varepsilon_{31}\varepsilon_{31}$$

$$\overline{\overline{\mathsf{G}}}_{\mathbf{E}}^{(3)} = \mathbf{e}_{123}\mathbf{e}_{123}, \quad \overline{\overline{\Gamma}}_{\mathbf{E}}^{(3)} = \varepsilon_{123}\varepsilon_{123}$$

$$\overline{\overline{\mathsf{H}}}_{\mathbf{E}1} = \overline{\overline{\mathsf{H}}}_{\mathbf{E}2}^{-1} = \mathbf{e}_{12}\varepsilon_3 + \mathbf{e}_{23}\varepsilon_1 + \mathbf{e}_{31}\varepsilon_2 = \mathbf{e}_{123}\lfloor\overline{\overline{\Gamma}}_{\mathbf{E}} = \overline{\overline{\mathsf{G}}}_{\mathbf{E}}^{(2)}\rfloor\varepsilon_{123}$$

$$\overline{\overline{\mathsf{H}}}_{\mathbf{E}2} = \overline{\overline{\mathsf{H}}}_{\mathbf{E}1}^{-1} = \mathbf{e}_1\varepsilon_{23} + \mathbf{e}_2\varepsilon_{31} + \mathbf{e}_3\varepsilon_{12} = \overline{\overline{\mathsf{G}}}_{\mathbf{E}}\rfloor\varepsilon_{123} = \mathbf{e}_{123}\lfloor\overline{\overline{\Gamma}}_{\mathbf{E}}^{(2)}$$

$$\mathbf{e}_i \wedge \overline{\overline{\mathsf{H}}}_{\mathbf{E}2} = \overline{\overline{\mathsf{H}}}_{\mathbf{E}1} \wedge \varepsilon_i$$

$$\mathbf{a} \wedge \overline{\overline{\mathsf{H}}}_{\mathbf{E}1}|\mathbf{b} = \mathbf{e}_{123}(\mathbf{a}|\overline{\overline{\Gamma}}_{\mathbf{E}}|\mathbf{b}), \quad \mathbf{A} \wedge \overline{\overline{\mathsf{G}}}_{\mathbf{E}}^{(2)}|\mathbf{B} = \mathbf{e}_{123}(\mathbf{A}|\overline{\overline{\Gamma}}_{\mathbf{E}}^{(2)}|\mathbf{B})$$

4D Minkowski space

$$\overline{\overline{\mathsf{G}}}_{\mathbf{M}} = \overline{\overline{\mathsf{G}}}_{\mathbf{E}} - \mathbf{e}_4\mathbf{e}_4, \quad \overline{\overline{\Gamma}}_{\mathbf{M}} = \overline{\overline{\Gamma}}_{\mathbf{E}} - \varepsilon_4\varepsilon_4$$

$$\overline{\overline{\mathsf{G}}}_{\mathbf{M}}^{(2)} = \overline{\overline{\mathsf{G}}}_{\mathbf{E}}^{(2)} - \overline{\overline{\mathsf{G}}}_{\mathbf{E}} {}^\wedge_\wedge \mathbf{e}_4\mathbf{e}_4, \quad \overline{\overline{\Gamma}}_{\mathbf{M}}^{(2)} = \overline{\overline{\Gamma}}_{\mathbf{E}}^{(2)} - \overline{\overline{\Gamma}}_{\mathbf{E}} {}^\wedge_\wedge \varepsilon_4\varepsilon_4$$

$$\overline{\overline{\mathsf{G}}}_{\mathbf{M}}^{(3)} = \overline{\overline{\mathsf{G}}}_{\mathbf{E}}^{(3)} - \overline{\overline{\mathsf{G}}}_{\mathbf{E}}^{(2)} {}^\wedge_\wedge \mathbf{e}_4\mathbf{e}_4, \quad \overline{\overline{\Gamma}}_{\mathbf{M}}^{(3)} = \overline{\overline{\Gamma}}_{\mathbf{E}}^{(3)} - \overline{\overline{\Gamma}}_{\mathbf{E}}^{(2)} {}^\wedge_\wedge \varepsilon_4\varepsilon_4$$

$$\overline{\overline{\mathsf{G}}}_{\mathbf{M}}^{(4)} = -\overline{\overline{\mathsf{G}}}_{\mathbf{E}}^{(3)} {}^\wedge_\wedge \mathbf{e}_4\mathbf{e}_4, \quad \overline{\overline{\Gamma}}_{\mathbf{M}}^{(4)} = -\overline{\overline{\Gamma}}_{\mathbf{E}}^{(3)} {}^\wedge_\wedge \varepsilon_4\varepsilon_4$$

$$\overline{\overline{\mathsf{H}}}_{\mathbf{M}1} = \mathbf{e}_4 \wedge \overline{\overline{\mathsf{H}}}_{\mathbf{E}1} + \mathbf{e}_{123}\varepsilon_4 = \mathbf{e}_{1234}\lfloor\overline{\overline{\Gamma}}_{\mathbf{M}}$$

$$\overline{\overline{\mathsf{H}}}_{\mathbf{M}2} = -\overline{\overline{\mathsf{H}}}_{\mathbf{M}2}^{-1} = -\overline{\overline{\mathsf{H}}}_{\mathbf{E}1} \wedge \varepsilon_4 - \mathbf{e}_4 \wedge \overline{\overline{\mathsf{H}}}_{\mathbf{E}2} = \mathbf{e}_{1234}\lfloor\overline{\overline{\Gamma}}_{\mathbf{M}}^{(2)}$$

$$\overline{\overline{\mathsf{H}}}_{\mathbf{M}3} = \overline{\overline{\mathsf{H}}}_{\mathbf{M}1}^{-1} = \overline{\overline{\mathsf{H}}}_{\mathbf{E}2} \wedge \varepsilon_4 + \mathbf{e}_4\varepsilon_{123} = \mathbf{e}_{1234}\lfloor\overline{\overline{\Gamma}}_{\mathbf{M}}^{(3)}$$

$$\overline{\overline{\mathsf{H}}}_{\mathbf{M}3}|\overline{\overline{\mathsf{H}}}_{\mathbf{M}1} = \overline{\overline{\mathsf{I}}}, \quad \overline{\overline{\mathsf{H}}}_{\mathbf{M}1}|\overline{\overline{\mathsf{H}}}_{\mathbf{M}3} = \overline{\overline{\mathsf{I}}}^{(3)}$$

Medium dyadics

3D Euclidean space

$$\mathbf{D} = \overline{\overline{\epsilon}}|\mathbf{E} + \overline{\overline{\xi}}|\mathbf{H}, \quad \mathbf{B} = \overline{\overline{\zeta}}|\mathbf{E} + \overline{\overline{\mu}}|\mathbf{H}$$

$$\text{Hodge dyadics} \quad \overline{\overline{\epsilon}},\ \overline{\overline{\mu}},\ \overline{\overline{\xi}}, \overline{\overline{\zeta}}\ \in \mathbb{F}_2\mathbb{E}_1$$

$$\text{Inverses} \quad \overline{\overline{\epsilon}}^{-1},\ \overline{\overline{\mu}}^{-1},\ \overline{\overline{\xi}}^{-1},\ \overline{\overline{\zeta}}^{-1}\ \in \mathbb{F}_1\mathbb{E}_2$$

$$\mathbf{e}_{123}\lfloor\mathbf{D} = \overline{\overline{\epsilon}}_g|\mathbf{E} + \overline{\overline{\xi}}_g|\mathbf{H}, \qquad \mathbf{e}_{123}\lfloor\mathbf{B} = \overline{\overline{\zeta}}_g|\mathbf{E} + \overline{\overline{\mu}}_g|\mathbf{H}$$

$$\text{Metric (``Gibbsian'') dyadics} \quad \overline{\overline{\epsilon}}_g,\ \overline{\overline{\mu}}_g,\ \overline{\overline{\xi}}_g,\ \overline{\overline{\zeta}}_g\ \in \mathbb{E}_1\mathbb{E}_1$$

$$\text{Inverses} \quad \overline{\overline{\epsilon}}_g^{-1},\ \overline{\overline{\mu}}_g^{-1},\ \overline{\overline{\xi}}_g^{-1},\ \overline{\overline{\zeta}}_g^{-1}\ \in \mathbb{F}_1\mathbb{F}_1$$

$$\text{Relations} \quad \overline{\overline{\epsilon}}_g = \mathbf{e}_{123}\lfloor\overline{\overline{\epsilon}}, \quad \overline{\overline{\epsilon}} = \boldsymbol{\varepsilon}_{123}\lfloor\overline{\overline{\epsilon}}_g, \quad \text{etc}$$

$$\text{Inverse relations} \quad \overline{\overline{\epsilon}}^{-1} = \overline{\overline{\epsilon}}_g^{-1}\rfloor\mathbf{e}_{123}, \quad \overline{\overline{\epsilon}}_g^{-1} = \overline{\overline{\epsilon}}^{-1}\rfloor\boldsymbol{\varepsilon}_{123}, \quad \text{etc}$$

4D Minkowski space ($N = \mathbf{1234}$)

$$\boldsymbol{\Psi} = \overline{\overline{\mathsf{M}}}|\boldsymbol{\Phi}, \qquad \boldsymbol{\Phi} = \overline{\overline{\mathsf{M}}}^{-1}|\boldsymbol{\Psi}$$

$$\text{Hodge dyadics} \quad \overline{\overline{\mathsf{M}}}, \overline{\overline{\mathsf{M}}}^{-1} \in \mathbb{F}_2\mathbb{E}_2,$$

$$\mathbf{e}_N\lfloor\boldsymbol{\Psi} = \overline{\overline{\mathsf{M}}}_g|\boldsymbol{\Phi}, \qquad \mathbf{e}_N\lfloor\boldsymbol{\Phi} = (\mathbf{e}_N\mathbf{e}_N\lfloor\lfloor\overline{\overline{\mathsf{M}}}_g^{-1})|\boldsymbol{\Psi}$$

$$\text{Metric (``modified'') dyadics} \quad \overline{\overline{\mathsf{M}}}_g,\ (\mathbf{e}_N\mathbf{e}_N\lfloor\lfloor\overline{\overline{\mathsf{M}}}_g^{-1}) \in \mathbb{E}_2\mathbb{E}_2,\ \overline{\overline{\mathsf{M}}}_g^{-1} \in \mathbb{F}_2\mathbb{F}_2$$

$$\text{Relations} \quad \overline{\overline{\mathsf{M}}}_g = \mathbf{e}_N\lfloor\overline{\overline{\mathsf{M}}}, \quad \overline{\overline{\mathsf{M}}} = \boldsymbol{\varepsilon}_N\lfloor\overline{\overline{\mathsf{M}}}_g$$

$$\text{Inverse relations} \quad \overline{\overline{\mathsf{M}}}^{-1} = \overline{\overline{\mathsf{M}}}_g^{-1}\rfloor\mathbf{e}_N, \quad \overline{\overline{\mathsf{M}}}_g^{-1} = \overline{\overline{\mathsf{M}}}^{-1}\rfloor\boldsymbol{\varepsilon}_N$$

Bi-anisotropic medium

$$\overline{\overline{\mathsf{M}}} = \overline{\overline{\alpha}} + \overline{\overline{\epsilon}}' \wedge \mathbf{e}_4 + \boldsymbol{\varepsilon}_4 \wedge \overline{\overline{\mu}}^{-1} + \boldsymbol{\varepsilon}_4 \wedge \overline{\overline{\beta}} \wedge \mathbf{e}_4$$

$$\overline{\overline{\epsilon}}' = \overline{\overline{\epsilon}} - \overline{\overline{\xi}}|\overline{\overline{\mu}}^{-1}|\overline{\overline{\zeta}}, \quad \overline{\overline{\alpha}} = \overline{\overline{\xi}}|\overline{\overline{\mu}}^{-1}, \quad \overline{\overline{\beta}} = -\overline{\overline{\mu}}^{-1}|\overline{\overline{\zeta}}$$

$$\overline{\overline{\epsilon}} = \overline{\overline{\epsilon}}' - \overline{\overline{\alpha}}|\overline{\overline{\mu}}|\overline{\overline{\beta}}, \quad \overline{\overline{\xi}} = \overline{\overline{\alpha}}|\overline{\overline{\mu}}, \quad \overline{\overline{\zeta}} = -\overline{\overline{\mu}}|\overline{\overline{\beta}}$$

$$\overline{\overline{\mathsf{M}}}_g = \overline{\overline{\epsilon}}_g{}^{\wedge}_{\wedge}\mathbf{e}_4\mathbf{e}_4 - (\mathbf{e}_{123}\lfloor\overline{\overline{\mathsf{I}}}^T + \mathbf{e}_4 \wedge \overline{\overline{\xi}}_g)|\overline{\overline{\mu}}_g^{-1}|(\overline{\overline{\mathsf{I}}}\rfloor\mathbf{e}_{123} - \overline{\overline{\zeta}}_g \wedge \mathbf{e}_4)$$

$$= -\mathbf{e}_{123}\mathbf{e}_{123}\lfloor\lfloor\overline{\overline{\mu}}_g^{-1} - \mathbf{e}_4\wedge\overline{\overline{\xi}}_g|\overline{\overline{\mu}}_g^{-1}\rfloor\mathbf{e}_{123} + \mathbf{e}_{123}\lfloor\overline{\overline{\mu}}_g^{-1}|\overline{\overline{\zeta}}_g\wedge\mathbf{e}_4 + (\overline{\overline{\epsilon}}_g - \overline{\overline{\xi}}_g|\overline{\overline{\mu}}_g^{-1}|\overline{\overline{\zeta}}_g){}^{\wedge}_{\wedge}\mathbf{e}_4\mathbf{e}_4$$

$$\overline{\overline{\epsilon}}' = \varepsilon_{123}\lfloor(\overline{\overline{\epsilon}}_g - \overline{\overline{\xi}}_g|\overline{\overline{\mu}}_g^{-1}|\overline{\overline{\zeta}}_g), \quad \overline{\overline{\mu}} = \varepsilon_{123}\lfloor\overline{\overline{\mu}}_g$$

$$\overline{\overline{\alpha}} = \varepsilon_{123}\lfloor(\overline{\overline{\xi}}_g|\overline{\overline{\mu}}_g^{-1})\rfloor\mathbf{e}_{123}, \quad \overline{\overline{\beta}} = -\overline{\overline{\mu}}_g^{-1}|\overline{\overline{\zeta}}_g$$

$$\overline{\overline{\mathsf{M}}} = \varepsilon_{123}\lfloor\overline{\overline{\epsilon}}_g \wedge \mathbf{e}_4 + (\varepsilon_{123}\lfloor\overline{\overline{\xi}}_g + \varepsilon_4 \wedge \overline{\overline{\mathsf{I}}}^T)|\overline{\overline{\mu}}_g^{-1}|(\overline{\overline{\mathsf{I}}}\rfloor\mathbf{e}_{123} - \overline{\overline{\zeta}}_g \wedge \mathbf{e}_4)$$

$$\overline{\overline{\mathsf{M}}}^{-1} = \overline{\overline{\alpha}}_1 + \overline{\overline{\epsilon}}_1' \wedge \mathbf{e}_4 + \varepsilon_4 \wedge \overline{\overline{\mu}}_1^{-1} + \varepsilon_4 \wedge \overline{\overline{\beta}}_1 \wedge \mathbf{e}_4,$$

$$\overline{\overline{\epsilon}}_1' = -(\overline{\overline{\mu}}^{-1} - \overline{\overline{\beta}}|\overline{\overline{\epsilon}}'^{-1}|\overline{\overline{\alpha}})^{-1},$$

$$\overline{\overline{\mu}}_1^{-1} = -(\overline{\overline{\epsilon}}' - \overline{\overline{\alpha}}|\overline{\overline{\mu}}|\overline{\overline{\beta}})^{-1},$$

$$\overline{\overline{\alpha}}_1 = \overline{\overline{\mu}}|\overline{\overline{\beta}}|\overline{\overline{\mu}}_1^{-1} = (\overline{\overline{\alpha}} - \overline{\overline{\epsilon}}'|\overline{\overline{\beta}}^{-1}|\overline{\overline{\mu}}^{-1})^{-1},$$

$$\overline{\overline{\beta}}_1 = \overline{\overline{\epsilon}}'^{-1}|\overline{\overline{\alpha}}|\overline{\overline{\epsilon}}_1' = (\overline{\overline{\beta}} - \overline{\overline{\mu}}^{-1}|\overline{\overline{\alpha}}^{-1}|\overline{\overline{\epsilon}}')^{-1}$$

$$\mathbf{e}_N\mathbf{e}_N\lfloor\lfloor\overline{\overline{\mathsf{M}}}_g^{-1} = -\overline{\overline{\mu}}_g{}^{\wedge}_{\wedge}\mathbf{e}_4\mathbf{e}_4 + (\mathbf{e}_{123}\lfloor\overline{\overline{\mathsf{I}}}^T - \mathbf{e}_4 \wedge \overline{\overline{\zeta}}_g)|\overline{\overline{\epsilon}}_g^{-1}|(\overline{\overline{\mathsf{I}}}\rfloor\mathbf{e}_{123} + \overline{\overline{\xi}}_g \wedge \mathbf{e}_4)$$

$$= \mathbf{e}_{123}\mathbf{e}_{123}\lfloor\lfloor\overline{\overline{\epsilon}}_g^{-1} - \mathbf{e}_4 \wedge \overline{\overline{\zeta}}_g|\overline{\overline{\epsilon}}_g^{-1}\rfloor\mathbf{e}_{123} + \mathbf{e}_{123}\lfloor\overline{\overline{\epsilon}}_g^{-1}|\overline{\overline{\xi}}_g \wedge \mathbf{e}_4 - (\overline{\overline{\mu}}_g - \overline{\overline{\zeta}}_g|\overline{\overline{\epsilon}}_g^{-1}|\overline{\overline{\xi}}_g){}^{\wedge}_{\wedge}\mathbf{e}_4\mathbf{e}_4$$

Appendix B
Solutions to Problems

In this Appendix we give suggestions, hints or final results to aid in solving some of the problems at the ends of different sections.

1.2.1 Writing $\boldsymbol{\alpha}_j = \sum A_{jk}\boldsymbol{\beta}_k$, from $\mathbf{a}_i|\boldsymbol{\alpha}_j = \delta_{ij}$ we obtain the matrix A_{jk} as the inverse of the matrix $B_{kj} = \mathbf{a}_j|\boldsymbol{\beta}_k$.

1.2.2 Write $\overline{\overline{\mathsf{A}}}$ in the form $\overline{\overline{\mathsf{I}}}|\overline{\overline{\mathsf{A}}}$, where the unit dyadic is of the form (1.11).

1.3.1 From $(a_ib_j - a_jb_i)\mathbf{e}_{ij} = 0$ we have $a_ib_j - a_jb_i = 0$ for all i, j. Assuming, for example, $a_1 \neq 0$, we have $b_j = a_j(b_1/a_1)$ for all j and, hence, $\mathbf{b} = \mathbf{a}(b_1/a_1)$.

1.3.2 (a)

$$(\mathbf{a}\wedge\mathbf{b})|(\boldsymbol{\alpha}\wedge\boldsymbol{\beta}) = \sum_{i<j}\sum_{k<\ell}(a_ib_j - b_ia_j)(\alpha_k\beta_\ell - \beta_k\alpha_\ell)(\mathbf{e}_{ij}|\boldsymbol{\varepsilon}_{k\ell})$$

$$= \sum_{i<j}(a_ib_j - b_ia_j)(\alpha_i\beta_j - \beta_i\alpha_j)$$

$$= \sum_{i<j}[(a_i\alpha_i)(b_j\beta_j) + (b_i\beta_i)(a_j\alpha_j)] - \sum_{i<j}[(a_i\beta_i)(\beta_j\alpha_j) + (b_i\alpha_i)(a_j\beta_j)]$$

$$= \sum_{i=1}^{n}\sum_{j=1}^{n}(a_i\alpha_i)(b_j\beta_j)(1-\delta_{ij}) - \sum_{i=1}^{n}\sum_{j=1}^{n}(a_i\beta_i)(b_j\alpha_j)(1-\delta_{ij})$$

$$= \sum_{i=1}^{n} \sum_{j=1}^{n} [(a_i\alpha_i)(b_j\beta_j) - (a_i\beta_i)(b_j\alpha_j)] = (\mathbf{a}|\boldsymbol{\alpha})(\mathbf{b}|\boldsymbol{\beta}) - (\mathbf{a}|\boldsymbol{\beta})(\mathbf{b}|\boldsymbol{\alpha}).$$

(b) Since the result must involve terms containing all vector and dual-vector quantities, it must most obviously be a linear combination of $(\boldsymbol{\alpha}|\mathbf{a})(\boldsymbol{\beta}|\mathbf{b})$ and $(\boldsymbol{\alpha}|\mathbf{b})(\boldsymbol{\beta}|\mathbf{a})$. Because of the antisymmetry of the wedge product, the result must be a multiple of the difference $(\boldsymbol{\alpha}|\mathbf{a})(\boldsymbol{\beta}|\mathbf{b}) - (\boldsymbol{\alpha}|\mathbf{b})(\boldsymbol{\beta}|\mathbf{a})$. It is natural to choose the unknown multiplying factor so that we have $\mathbf{e}_{12}|\boldsymbol{\varepsilon}_{12} = 1$, whence the identity (1.36) is obtained.

1.3.3 When $A_{31} \neq 0$, one possibilty is

$$\mathbf{A} = (\mathbf{a}_1 - \mathbf{a}_2 A_{23}/A_{31}) \wedge (\mathbf{a}_2 A_{12} - \mathbf{a}_3 A_{31}).$$

1.3.4 $\mathbf{A}$ can be expressed as $\mathbf{A} = \sum A_{ij}\mathbf{e}_i \wedge \mathbf{e}_j + \sum A_{i4}\mathbf{e}_i \times \mathbf{e}_4$ where $i, j = 1, 2, 3$. The first term is a three-dimensional bivector which from the previous problem can be expressed as a simple bivector.

1.3.5 Start from

$$(\mathbf{a} \wedge \mathbf{b} \wedge \mathbf{c})|(\boldsymbol{\alpha} \wedge \boldsymbol{\beta} \wedge \boldsymbol{\gamma}) = \boldsymbol{\alpha}|(\mathbf{a}(\mathbf{b} \wedge \mathbf{c}) + \mathbf{b}(\mathbf{c} \wedge \mathbf{a}) + \mathbf{c}(\mathbf{a} \wedge \mathbf{b}))|(\boldsymbol{\beta} \wedge \boldsymbol{\gamma}),$$

which can be expressed as

$$(\mathbf{a} \wedge \mathbf{b} \wedge \mathbf{c})\lfloor\boldsymbol{\alpha} = \boldsymbol{\alpha}|(\mathbf{a}(\mathbf{b} \wedge \mathbf{c}) + \mathbf{b}(\mathbf{c} \wedge \mathbf{a}) + \mathbf{c}(\mathbf{a} \wedge \mathbf{b}))$$

$$= (\boldsymbol{\alpha}|\mathbf{a})(\mathbf{b} \wedge \mathbf{c}) + \boldsymbol{\alpha}|(\mathbf{bc} - \mathbf{cb}) \wedge \mathbf{a} = (\boldsymbol{\alpha}|\mathbf{a})(\mathbf{b} \wedge \mathbf{c}) - (\boldsymbol{\alpha}\rfloor(\mathbf{b} \wedge \mathbf{c})) \wedge \mathbf{a}.$$

Denoting $\boldsymbol{\alpha}$ by $\boldsymbol{\beta}$ and replacing $\mathbf{b} \wedge \mathbf{c}$ by the bivector $\mathbf{C}$ leads to

$$\mathbf{a} \wedge (\boldsymbol{\beta}\rfloor\mathbf{C}) = \boldsymbol{\beta}\rfloor(\mathbf{a} \wedge \mathbf{C}) - \mathbf{C}(\mathbf{a}|\boldsymbol{\beta}).$$

1.3.6 Obviously we have $\mathbf{a}|\boldsymbol{\alpha} = \mathbf{a}|(\mathbf{a}\rfloor\boldsymbol{\Phi}) = (\mathbf{a} \wedge \mathbf{a})|\boldsymbol{\Phi} = 0$. Using (1.50), we can write for any dual vector $\boldsymbol{\beta}$

$$\mathbf{a}\rfloor(\boldsymbol{\alpha} \wedge \boldsymbol{\beta}) = \boldsymbol{\alpha}(\mathbf{a}|\boldsymbol{\beta}) - \boldsymbol{B}(\mathbf{a}|\boldsymbol{\alpha}) = \boldsymbol{\alpha}(\mathbf{a}|\boldsymbol{\beta}).$$

Taking such a $\boldsymbol{\beta}$ satisfying $\mathbf{a}|\boldsymbol{\beta} \neq 0$, we can express

$$\boldsymbol{\alpha} = \frac{\mathbf{a}\rfloor(\boldsymbol{\alpha} \wedge \boldsymbol{\beta})}{\mathbf{a}|\boldsymbol{\beta}}, \quad \Rightarrow \quad \boldsymbol{\Phi} = \frac{\boldsymbol{\alpha} \wedge \boldsymbol{\beta}}{\mathbf{a}|\boldsymbol{\beta}}.$$

Thus, $\boldsymbol{\Phi}$ is of the form $\boldsymbol{\alpha} \wedge \boldsymbol{\gamma}$ where $\boldsymbol{\gamma}$ is some dual vector.

1.4.1 Apply $\mathbf{bc}||(\boldsymbol{\beta\gamma} - \boldsymbol{\gamma\beta}) = (\mathbf{b} \wedge \mathbf{c})|(\boldsymbol{\beta} \wedge \boldsymbol{\gamma})$ to

$$\mathbf{abc}|||(\boldsymbol{\alpha\beta\gamma} + \cdots) = (\mathbf{a}|\boldsymbol{\alpha})(\mathbf{bc}||(\boldsymbol{\beta\gamma} - \boldsymbol{\gamma\beta})) + (\mathbf{a}|\boldsymbol{\beta})(\mathbf{bc}||(\boldsymbol{\gamma\alpha} - \boldsymbol{\alpha\gamma}) + \cdots$$

$$= (\mathbf{a}|\boldsymbol{\alpha})(\mathbf{b} \wedge \mathbf{c})|(\boldsymbol{\beta} \wedge \boldsymbol{\gamma}) + (\mathbf{a}|\boldsymbol{\beta})(\mathbf{b} \wedge \mathbf{c})|(\boldsymbol{\gamma} \wedge \boldsymbol{\alpha}) + (\mathbf{a}|\boldsymbol{\gamma})(\mathbf{b} \wedge \mathbf{c})|(\boldsymbol{\alpha} \wedge \boldsymbol{\beta}).$$

The last two terms can be manipulated as

$$(\mathbf{a}|\boldsymbol{\beta})(\mathbf{b}\wedge\mathbf{c})|(\boldsymbol{\gamma}\wedge\boldsymbol{\alpha}) - (\mathbf{a}|\boldsymbol{\gamma})(\mathbf{b}\wedge\mathbf{c})|(\boldsymbol{\beta}\wedge\boldsymbol{\alpha}) = (\mathbf{a}|(\boldsymbol{\beta}\boldsymbol{\gamma} - \boldsymbol{\gamma}\boldsymbol{\beta})\wedge\boldsymbol{\alpha})|(\mathbf{b}\wedge\mathbf{c})$$

$$= -((\mathbf{a}\rfloor(\boldsymbol{\beta}\wedge\boldsymbol{\gamma}))\wedge\boldsymbol{\alpha})|(\mathbf{b}\wedge\mathbf{c}) = -(\mathbf{a}\rfloor(\boldsymbol{\beta}\wedge\boldsymbol{\gamma}))|(\boldsymbol{\alpha}\rfloor(\mathbf{b}\wedge\mathbf{c})).$$

1.4.3 Every bivector can be represented in the form $\mathbf{A} = \mathbf{a}\wedge\mathbf{b} + \mathbf{c}\wedge\mathbf{d}$. $\mathbf{A}\wedge\mathbf{A} = 2\mathbf{a}\wedge\mathbf{b}\wedge\mathbf{c}\wedge\mathbf{d} = 0$. Thus, the four vectors are linearly dependent. Take, for example, $\mathbf{d} = \alpha\mathbf{a} + \beta\mathbf{b} + \gamma\mathbf{c}$, then $\mathbf{A} = (\mathbf{a} + \beta\mathbf{c})\wedge(\mathbf{b} - \alpha\mathbf{c})$.

1.4.9 Replace $\boldsymbol{\alpha}$ and $\mathbf{a}_N$ by ε_i and $\mathbf{e}_N$, respectively. Modifying the rule in the above exercise, we have

$$\varepsilon_{K(i)}|(\varepsilon_i\rfloor\mathbf{e}_N) = (\varepsilon_{K(i)}\wedge\varepsilon_i)|\mathbf{e}_N = (-1)^{n-i}\varepsilon^N|\mathbf{e}_N = (-1)^{n-i},$$

from which we conclude $\varepsilon_i\rfloor\mathbf{e}_N = (-1)^{n-i}\mathbf{e}_{K(i)}$. Now writing $\boldsymbol{\alpha} = \sum\alpha_i\varepsilon_i$, $\mathbf{a} = \sum a_j\mathbf{e}_j$, we obtain

$$(\boldsymbol{\alpha}\rfloor\mathbf{e}_N)\wedge\mathbf{a} = \sum_{i,j}\alpha_i a_j(-1)^{n-i}\mathbf{e}_{K(i)}\wedge\mathbf{e}_j = \sum_i\alpha_i a_i(-1)^{n-i}\mathbf{e}_{K(i)}\wedge\mathbf{e}_i$$

$$= \sum_i(\alpha_i a_i)\mathbf{e}_N = (\boldsymbol{\alpha}|\mathbf{a})\mathbf{e}_N.$$

Finally, because $\mathbf{a}_N = (\mathbf{a}_N|\varepsilon_N)\mathbf{e}_N$, $\mathbf{e}_N$ can be replaced by $\mathbf{a}_N$.

1.4.10 We can assume $\mathbf{a} = \mathbf{e}_1$ a vector in a base system $\{\mathbf{e}_i\}$. Expanding $\mathbf{A} = \sum A_{ij}\mathbf{e}_{ij} = \sum_{i<j}(A_{ij} - A_{ji})\mathbf{e}_{ij}$, we have $\mathbf{e}_1\wedge\mathbf{A} = \sum_{1<i<j}(A_{ij} - A_{ji})\mathbf{e}_1\wedge\mathbf{e}_i\wedge\mathbf{e}_j = 0$ which implies $A_{ij} - A_{ji} = 0$ for $1 < i < j$. Thus, the bivector $\mathbf{A}$ must be of the form $\mathbf{A} = \sum_{1<j}(A_{1j} - A_{j1})\mathbf{e}_{1j} = \mathbf{e}_1\wedge\mathbf{b}$, $b = \sum_{1<j}(A_{1j} - A_{j1})\mathbf{e}_j$.

1.4.13 The result of the second problem can be written as

$$\boldsymbol{\alpha}\rfloor(\mathbf{a}\wedge\mathbf{b}\wedge\mathbf{c}\wedge\mathbf{d}) = \mathbf{a}\wedge[\boldsymbol{\alpha}|(\mathbf{b}(\mathbf{c}\wedge\mathbf{d}) + \mathbf{c}(\mathbf{d}\wedge\mathbf{b}) + \mathbf{d}(\mathbf{b}\wedge\mathbf{c}))] - (\mathbf{b}\wedge\mathbf{c}\wedge\mathbf{d})(\boldsymbol{\alpha}|\mathbf{a}).$$

Now from (1.85) the term in square brackets can be expressed as

$$\boldsymbol{\alpha}|(\mathbf{b}(\mathbf{c}\wedge\mathbf{d}) + \mathbf{c}(\mathbf{d}\wedge\mathbf{b}) + \mathbf{d}(\mathbf{b}\wedge\mathbf{c})) = \boldsymbol{\alpha}\rfloor(\mathbf{b}\wedge\mathbf{c}\wedge\mathbf{d}).$$

Thus

$$\mathbf{a}\wedge[\boldsymbol{\alpha}\rfloor(\mathbf{b}\wedge\mathbf{c}\wedge\mathbf{d})] = \boldsymbol{\alpha}\rfloor(\mathbf{a}\wedge\mathbf{b}\wedge\mathbf{c}\wedge\mathbf{d}) + (\mathbf{b}\wedge\mathbf{c}\wedge\mathbf{d})(\boldsymbol{\alpha}|\mathbf{a}),$$

which gives the result when replacing $\mathbf{b}\wedge\mathbf{c}\wedge\mathbf{d}\to\mathbf{C}$ and $\mathbf{a}\to\mathbf{b}$. For $n = 3$ the quadrivector $\mathbf{b}\wedge\mathbf{C}$ vanishes and the identity is simplified to

$$\mathbf{b}\wedge[\boldsymbol{\alpha}\rfloor\mathbf{C}] = \mathbf{C}(\boldsymbol{\alpha}|\mathbf{b}).$$

1.4.15 From (1.82) we get

$$(\beta \wedge \gamma)\rfloor(\mathbf{a} \wedge \mathbf{b} \wedge \mathbf{c}) = \mathbf{a}(\mathbf{b} \wedge \mathbf{c})|(\beta \wedge \gamma) + ((\mathbf{bc} - \mathbf{cb}) \wedge \mathbf{a})|(\beta \wedge \gamma),$$

whose last term can be expanded as

$$((\mathbf{bc} - \mathbf{cb}) \wedge \mathbf{a})|(\beta \wedge \gamma) = -((\mathbf{bc} - \mathbf{cb})\rfloor(\beta \wedge \gamma))|\mathbf{a}$$

$$= (\mathbf{bc} - \mathbf{cb})|(\beta\gamma - \gamma\beta)|\mathbf{a} = -((\mathbf{b} \wedge \mathbf{c})\lfloor(\beta\gamma - \gamma\beta))|\mathbf{a}$$

$$= (\mathbf{b} \wedge \mathbf{c})\lfloor((\beta \wedge \gamma)\lfloor\mathbf{a}) = -(\mathbf{b} \wedge \mathbf{c})\lfloor(\mathbf{a}\rfloor(\beta \wedge \gamma)).$$

1.4.16 Start from the formula ($\mathbf{\Gamma}$ is a dual trivector)

$$(\mathbf{a} \wedge \mathbf{b} \wedge \mathbf{c} \wedge \mathbf{d})|(\alpha \wedge \mathbf{\Gamma}) = ((\mathbf{a} \wedge \mathbf{b} \wedge \mathbf{c} \wedge \mathbf{d})\lfloor\alpha)|\mathbf{\Gamma} = ...$$

to get

$$(\mathbf{a} \wedge \mathbf{b} \wedge \mathbf{c} \wedge \mathbf{d})\lfloor\alpha = (\mathbf{c} \wedge \mathbf{d}) \wedge (\mathbf{ba} - \mathbf{ab})|\alpha + (\mathbf{a} \wedge \mathbf{b}) \wedge (\mathbf{dc} - \mathbf{cd})|\alpha$$

and apply the bac cab rule

1.4.17 Expanding

$$\mathbf{\Gamma} = \sum \Gamma_{ij}\varepsilon_{ij}, \quad i < j, \quad \mathbf{e}_N\lfloor\varepsilon_{ij} = (-1)^{i-1}(-1)^{j-2}\mathbf{e}_{K(ij)}.$$

Thus, only terms $a_i\mathbf{e}_i + a_j\mathbf{e}_j$ of $\mathbf{a}$ affect the left-hand side which gives

$$\mathbf{a} \wedge (\mathbf{e}_N\lfloor\mathbf{\Gamma}) = \sum \Gamma_{ij}(-1)^{i-1}(-1)^{j-2}(a_i\mathbf{e}_i + a_j\mathbf{e}_j) \wedge \mathbf{e}_{K(ij)}$$

$$= \sum \Gamma_{ij}[-(-1)^{j-1}a_i\mathbf{e}_{K(j)} + (-1)^{i-1}a_j\mathbf{e}_{K(i)}]$$

$$= \sum \Gamma_{ij}[-(-1)^{j-1}a_i\mathbf{e}_{K(j)} + (-1)^{i-1}a_j\mathbf{e}_{K(i)}]$$

$$= \sum \Gamma_{ij}[-a_i\mathbf{e}_N\lfloor\varepsilon_j + a_j\mathbf{e}_N\lfloor\varepsilon_i].$$

On the other hand, from the bac cab rule we have

$$\mathbf{\Gamma}\lfloor\mathbf{a} = \sum \Gamma_{ij}\varepsilon_{ij}\lfloor(a_i\mathbf{e}_i + a_j\mathbf{e}_j) = \sum \Gamma_{ij}\mathbf{e}_N\lfloor(\varepsilon_j a_i - \varepsilon_i a_j).$$

1.4.18 Take $\mathbf{A} \to \mathbf{b} \wedge \mathbf{c}$,

$$(\mathbf{a} \wedge \mathbf{b} \wedge \mathbf{c})\lfloor\mathbf{\Gamma} = (\mathbf{a}(\mathbf{b} \wedge \mathbf{c}) + \mathbf{b}(\mathbf{c} \wedge \mathbf{a}) + \mathbf{c}(\mathbf{a} \wedge \mathbf{b}))|\mathbf{\Gamma}$$

$$= \mathbf{a}(\mathbf{A}|\mathbf{\Gamma}) + (\mathbf{bc} - \mathbf{cb}) \wedge \mathbf{a}|\mathbf{\Gamma} = \mathbf{a}(\mathbf{A}|\mathbf{\Gamma}) + (\mathbf{bc} - \mathbf{cb})|(\mathbf{a}\rfloor\mathbf{\Gamma})$$

$$= \mathbf{a}(\mathbf{A}|\mathbf{\Gamma}) - (\mathbf{b} \wedge \mathbf{c})\lfloor(\mathbf{a}\rfloor\mathbf{\Gamma}) = \mathbf{a}(\mathbf{A}|\mathbf{\Gamma}) + (\mathbf{a}\rfloor\mathbf{\Gamma})\rfloor(\mathbf{b} \wedge \mathbf{c}).$$

1.4.19 Set $p = q + 1$, $\mathbf{a}_1 = \mathbf{a}$ and $\mathbf{a}^q = \mathbf{a}_2 \wedge \cdots \wedge \mathbf{a}_p = \mathbf{a}_{K(1)}$,

$$\text{for } i > 1, \quad \mathbf{a}_{K(i)} = (-1)^p\mathbf{a}_{K(1i)} \wedge \mathbf{a}_1 = -(-1)^q\mathbf{a}_{K(1i)} \wedge \mathbf{a}$$

with $\mathbf{a}_{K(1i)} = \mathbf{a}_2 \wedge \cdots \wedge \mathbf{a}_{i-1} \wedge \mathbf{a}_{i+1} \wedge \cdots \wedge \mathbf{a}_p$. Using the expansion again, we have

$$\mathbf{a}_{K(1)} \lfloor \boldsymbol{\alpha} = \boldsymbol{\alpha}| \sum_{i=2}^{p} (-1)^i \mathbf{a}_i \mathbf{a}_{K(1i)},$$

$$\boldsymbol{\alpha}| \sum_{i=2}^{p} (-1)^i \mathbf{a}_i \mathbf{a}_{K(i)} = -(-1)^q \boldsymbol{\alpha}| \sum_{i=2}^{p} (-1)^i \mathbf{a}_i \mathbf{a}_{K(1i)} \wedge \mathbf{a}$$

$$= -(-1)^q (\mathbf{a}_{K(1)} \lfloor \boldsymbol{\alpha}) \wedge \mathbf{a} = -(-1)^q (\mathbf{a}^q \lfloor \boldsymbol{\alpha}) \wedge \mathbf{a} = (\boldsymbol{\alpha} \rfloor \mathbf{a}^q) \wedge \mathbf{a}.$$

Thus, we can finally write

$$(\mathbf{a} \wedge \mathbf{a}^q) \lfloor \boldsymbol{\alpha} = \boldsymbol{\alpha}| \left(\mathbf{a}_1 \mathbf{a}_{K(1)} - \sum_{i=2}^{p} (-1)^i \mathbf{a}_i \mathbf{a}_{K(i)} \right)$$

$$= (\boldsymbol{\alpha}|\mathbf{a})\mathbf{a}^q - (\boldsymbol{\alpha} \rfloor \mathbf{a}^q) \wedge \mathbf{a} = (\mathbf{a}|\boldsymbol{\alpha})\mathbf{a}^q - \mathbf{a} \wedge (\mathbf{a}^q \lfloor \boldsymbol{\alpha}).$$

1.4.20 Applying the previous identity

$$(\mathbf{a} \wedge \mathbf{b} \wedge \mathbf{a}^p) \lfloor \boldsymbol{\alpha} = (\mathbf{a}|\boldsymbol{\alpha})\mathbf{b} \wedge \mathbf{a}^p - \mathbf{a} \wedge ((\mathbf{b} \wedge \mathbf{a}^p) \lfloor \boldsymbol{\alpha}))$$

and the same to the last term, we expand

$$(\mathbf{a} \wedge \mathbf{b} \wedge \mathbf{a}^p) \lfloor \boldsymbol{\alpha} = (\mathbf{a}|\boldsymbol{\alpha})\mathbf{b} \wedge \mathbf{a}^p - \mathbf{a} \wedge ((\mathbf{b}|\boldsymbol{\alpha})\mathbf{a}^p - \mathbf{b} \wedge (\mathbf{a}^p \lfloor \boldsymbol{\alpha}))$$

$$= \boldsymbol{\alpha}|(\mathbf{ab} - \mathbf{ba}) \wedge \mathbf{a}^p + (\mathbf{a} \wedge \mathbf{b}) \wedge (\mathbf{a}^p \lfloor \boldsymbol{\alpha}))$$

$$= -(\boldsymbol{\alpha} \rfloor (\mathbf{a} \wedge \mathbf{b})) \wedge \mathbf{a}^p + (\mathbf{a} \wedge \mathbf{b}) \wedge (\mathbf{a}^p \lfloor \boldsymbol{\alpha})).$$

Replacing $\mathbf{a} \wedge \mathbf{b}$ by $\mathbf{A}$ leads to

$$(\mathbf{A} \wedge \mathbf{a}^p) \lfloor \boldsymbol{\alpha} = -(\boldsymbol{\alpha} \rfloor \mathbf{A}) \wedge \mathbf{a}^p + \mathbf{A} \wedge (\mathbf{a}^p \lfloor \boldsymbol{\alpha})$$

$$= (\mathbf{A} \lfloor \boldsymbol{\alpha}) \wedge \mathbf{a}^p + \mathbf{A} \wedge (\mathbf{a}^p \lfloor \boldsymbol{\alpha}).$$

1.4.21 From the anticipated identity we have

$$(\mathbf{a}^q \wedge \mathbf{b} \wedge \mathbf{a}^{p-1}) \lfloor \boldsymbol{\alpha} = (\mathbf{a}^q \lfloor \boldsymbol{\alpha}) \wedge \mathbf{b} \wedge \mathbf{a}^{p-1} + (-1)^q \mathbf{a}^q \wedge ((\mathbf{b} \wedge \mathbf{a}^{p-1}) \lfloor \boldsymbol{\alpha}).$$

The last term expanded through the same identity gives

$$(-1)^q \mathbf{a}^q \wedge ((\mathbf{b} \wedge \mathbf{a}^{p-1}) \lfloor \boldsymbol{\alpha}) = (-1)^q \mathbf{a}^q \wedge ((\boldsymbol{\alpha}|\mathbf{b})\mathbf{a}^{p-1} - \mathbf{b} \wedge (\mathbf{a}^{p-1} \lfloor \boldsymbol{\alpha})).$$

Also we can expand

$$(\mathbf{b} \wedge \mathbf{a}^q) \lfloor \boldsymbol{\alpha} = (\boldsymbol{\alpha}|\mathbf{b})\mathbf{a}^q - \mathbf{b} \wedge (\mathbf{a}^q \lfloor \boldsymbol{\alpha}) = (\boldsymbol{\alpha}|\mathbf{b})\mathbf{a}^q + (-1)^q (\mathbf{a}^q \lfloor \boldsymbol{\alpha}) \wedge \mathbf{b}.$$

Put together these give

$$(\mathbf{a}^{q+1} \wedge \mathbf{a}^{p-1}) \lfloor \boldsymbol{\alpha} = (\mathbf{a}^q \wedge \mathbf{b} \wedge \mathbf{a}^{p-1}) \lfloor \boldsymbol{\alpha}$$

$$= (\mathbf{a}^q\lfloor\boldsymbol{\alpha}) \wedge \mathbf{b} \wedge \mathbf{a}^{p-1} + (-1)^q \mathbf{a}^q \wedge ((\boldsymbol{\alpha}|\mathbf{b})\mathbf{a}^{p-1}) - \mathbf{b} \wedge (\mathbf{a}^{p-1}\lfloor\boldsymbol{\alpha})$$
$$= (-1)^q((-1)^q(\mathbf{a}^q\lfloor\boldsymbol{\alpha}) \wedge \mathbf{b} + (\boldsymbol{\alpha}|\mathbf{b})\mathbf{a}^q) \wedge \mathbf{a}^{p-1} - (-1)^q(\mathbf{a}^q \wedge \mathbf{b}) \wedge (\mathbf{a}^{p-1}\lfloor\boldsymbol{\alpha})$$
$$= (-1)^q((\mathbf{b} \wedge \mathbf{a}^q)\lfloor\boldsymbol{\alpha}) \wedge \mathbf{a}^{p-1} - (-1)^q(\mathbf{a}^q \wedge \mathbf{b}) \wedge (\mathbf{a}^{p-1}\lfloor\boldsymbol{\alpha})$$
$$= ((\mathbf{a}^q \wedge \mathbf{b})\lfloor\boldsymbol{\alpha}) \wedge \mathbf{a}^{p-1} + (-1)^{q+1}(\mathbf{a}^q \wedge \mathbf{b}) \wedge (\mathbf{a}^{p-1}\lfloor\boldsymbol{\alpha})$$
$$= (\mathbf{a}^{q+1}\lfloor\boldsymbol{\alpha}) \wedge \mathbf{a}^{p-1} + (-1)^{q+1}\mathbf{a}^{q+1} \wedge (\mathbf{a}^{p-1}\lfloor\boldsymbol{\alpha}).$$

1.4.23 The rule can be proved if it is valid for basis multivector as

$$\varepsilon_{ij} \wedge (\varepsilon_N\lfloor\mathbf{e}_{k\ell}) = \varepsilon_N(\varepsilon_{ij}|\mathbf{e}_{k\ell}) = \varepsilon_N\delta_{ik}\delta_{j\ell},$$

because from bilinearity it is then valid for any $\mathbf{A} = \sum A_{k\ell}\mathbf{e}_{k\ell}$ and $\boldsymbol{\Phi} = \sum \Phi_{ij}\varepsilon_{ij}$. Here we have assumed $i < j$ and $k < \ell$. The required result is obtained by expanding

$$\varepsilon_{ij} \wedge (\varepsilon_N\lfloor\mathbf{e}_{k\ell}) = (-1)^{k+\ell-1}\varepsilon_{ij} \wedge \varepsilon_{K(k\ell)}$$
$$= (-1)^{i+j-1}\varepsilon_{ij} \wedge \varepsilon_{K(ij)}\delta_{ik}\delta_{j\ell} = \varepsilon_N\delta_{ik}\delta_{j\ell}.$$

1.4.24 Because $(\mathbf{b}\wedge\mathbf{a})|(\mathbf{a}\rfloor\gamma) = \mathbf{b}|((\mathbf{a}\wedge\mathbf{a})\rfloor\gamma) = 0$ for any vector $\mathbf{b}$, $\mathbf{a}\rfloor(\mathbf{a}\rfloor\gamma) = 0$. Invoking (1.90), we can write for any dual vector $\boldsymbol{\alpha}$

$$\mathbf{a}\rfloor(\boldsymbol{\alpha} \wedge \boldsymbol{\Phi}) = \boldsymbol{\alpha} \wedge (\mathbf{a}\rfloor\boldsymbol{\Phi}) + \boldsymbol{\Phi}(\mathbf{a}|\boldsymbol{\alpha}) = \boldsymbol{\Phi}(\mathbf{a}|\boldsymbol{\alpha}),$$

whence $\mathbf{a}\rfloor\boldsymbol{\Phi} = 0$ implies

$$\boldsymbol{\Phi} = \frac{\mathbf{a}\rfloor(\boldsymbol{\alpha} \wedge \boldsymbol{\Phi})}{\mathbf{a}|\boldsymbol{\alpha}}, \quad \Rightarrow \quad \gamma = \frac{\boldsymbol{\alpha} \wedge \boldsymbol{\Phi}}{\mathbf{a}|\boldsymbol{\alpha}}.$$

1.5.1

$$\mathbf{A} + \mathbf{B} = \mathbf{a} \wedge \mathbf{c} + \mathbf{b} \wedge \mathbf{c} = (\mathbf{a} + \mathbf{b}) \wedge \mathbf{c}.$$

Summing bivectors is reduced to summing vectors. If the bivectors are in the same plane, their sum is either sum of areas or their difference depending on the orientation which can be taken as the sign of the area.

2.1.1 Use $(\overline{\overline{\mathsf{A}}} + \overline{\overline{\mathsf{B}}})^{\wedge}_{\wedge}(\overline{\overline{\mathsf{A}}} + \overline{\overline{\mathsf{B}}}) = \overline{\overline{\mathsf{A}}}^{\wedge}_{\wedge}\overline{\overline{\mathsf{A}}} + \overline{\overline{\mathsf{B}}}^{\wedge}_{\wedge}\overline{\overline{\mathsf{B}}} + 2\overline{\overline{\mathsf{A}}}^{\wedge}_{\wedge}\overline{\overline{\mathsf{B}}}$.

2.1.10 From the identity of problem 2.1.2 we have

$$2\mathrm{tr}(\overline{\overline{\mathsf{I}}}^{\wedge}_{\wedge}\overline{\overline{\mathsf{A}}}^{(2)}) = n(\mathrm{tr}\,\overline{\overline{\mathsf{A}}})^2 + 2\mathrm{tr}\,\overline{\overline{\mathsf{A}}}^2 - n\mathrm{tr}\,\overline{\overline{\mathsf{A}}}^2 - 2(\mathrm{tr}\overline{\overline{\mathsf{A}}})^2$$
$$= (n-2)((\mathrm{tr}\,\overline{\overline{\mathsf{A}}})^2 - \mathrm{tr}\,\overline{\overline{\mathsf{A}}}^2).$$

Comparing with (2.34) gives the identity. For $\overline{\overline{\mathsf{A}}} = \overline{\overline{\mathsf{I}}}$ and $\overline{\overline{\mathsf{I}}}^{\wedge}_{\wedge}\overline{\overline{\mathsf{I}}}^{(2)} = 3\overline{\overline{\mathsf{I}}}^{(3)}$ we obtain

$$3\,\mathrm{tr}\,\overline{\overline{\mathsf{I}}}^{(3)} = 3C_3^n = 3\frac{n!}{3!(n-3)!} = (n-2)\mathrm{tr}\,\overline{\overline{\mathsf{I}}}^{(2)} = (n-2)C_2^n = \frac{n!}{2!(n-3)!}.$$

2.1.11 Using the identity $\overline{\overline{\mathsf{A}}}^T\rfloor\rfloor(\overline{\overline{\mathsf{B}}}{}^\wedge_\wedge\overline{\overline{\mathsf{C}}}) = \cdots$, we can write (see previous solution)

$$\mathrm{tr}(\overline{\overline{\mathsf{I}}}^T\rfloor\rfloor\overline{\overline{\mathsf{A}}}^{(2)}) = \mathrm{tr}(\overline{\overline{\mathsf{A}}}(\overline{\overline{\mathsf{A}}}||\overline{\overline{\mathsf{I}}}^T)) - \mathrm{tr}\overline{\overline{\mathsf{A}}}^2 = 2\mathrm{tr}\overline{\overline{\mathsf{A}}}^{(2)}.$$

2.2.8 Starting from

$$\overline{\overline{\mathsf{A}}}^{(3)}|(\mathbf{a}\wedge\mathbf{b}c) = (\overline{\overline{\mathsf{A}}}|\mathbf{a})\wedge(\overline{\overline{\mathsf{A}}}|\mathbf{b})\wedge(\overline{\overline{\mathsf{A}}}|\mathbf{c}),$$

multiplying this by $(\boldsymbol{\alpha}\wedge\boldsymbol{\beta}\wedge\boldsymbol{\gamma})|$, and using the determinant rule, we obtain

$$\overline{\overline{\mathsf{A}}}^{(3)}||(\boldsymbol{\alpha}\mathbf{a}{}^\wedge_\wedge\boldsymbol{\beta}\mathbf{b}{}^\wedge_\wedge\boldsymbol{\gamma}\mathbf{c}) =$$

$$= \boldsymbol{\gamma}\mathbf{c}||((\overline{\overline{\mathsf{A}}}||\boldsymbol{\alpha}\mathbf{a})(\overline{\overline{\mathsf{A}}}||\boldsymbol{\beta}\mathbf{b})\overline{\overline{\mathsf{A}}} + \overline{\overline{\mathsf{A}}}|\mathbf{a}\boldsymbol{\alpha}|\overline{\overline{\mathsf{A}}}|\mathbf{b}\boldsymbol{\beta}|\overline{\overline{\mathsf{A}}} + \overline{\overline{\mathsf{A}}}|\mathbf{b}\boldsymbol{\beta}|\overline{\overline{\mathsf{A}}}|\mathbf{a}\boldsymbol{\alpha}|\overline{\overline{\mathsf{A}}}$$
$$-(\overline{\overline{\mathsf{A}}}||\boldsymbol{\alpha}\mathbf{a})\overline{\overline{\mathsf{A}}}|\mathbf{b}\boldsymbol{\beta}|\overline{\overline{\mathsf{A}}} - (\overline{\overline{\mathsf{A}}}||\boldsymbol{\beta}\mathbf{b})\overline{\overline{\mathsf{A}}}|\mathbf{a}\boldsymbol{\alpha}|\overline{\overline{\mathsf{A}}} - \boldsymbol{\alpha}\mathbf{a}||(\overline{\overline{\mathsf{A}}}|\mathbf{b}\boldsymbol{\beta}|\overline{\overline{\mathsf{A}}})\overline{\overline{\mathsf{A}}}).$$

Because this expression is linear in both $\mathbf{a}\boldsymbol{\alpha}$ and $\mathbf{b}\boldsymbol{\beta}$, we can replace these by the same dyadic $\overline{\overline{\mathsf{I}}}$. After dropping $\boldsymbol{\gamma}\mathbf{c}$, the remaining incomplete product becomes

$$\overline{\overline{\mathsf{A}}}^{(3)}\lfloor\lfloor(\overline{\overline{\mathsf{I}}}^T{}^\wedge_\wedge\overline{\overline{\mathsf{I}}}^T) = (\overline{\overline{\mathsf{A}}}||\overline{\overline{\mathsf{I}}}^T)(\overline{\overline{\mathsf{A}}}||\overline{\overline{\mathsf{I}}}^T)\overline{\overline{\mathsf{A}}} + \overline{\overline{\mathsf{A}}}|\overline{\overline{\mathsf{A}}}|\overline{\overline{\mathsf{A}}}$$
$$+\overline{\overline{\mathsf{A}}}|\overline{\overline{\mathsf{A}}}|\overline{\overline{\mathsf{A}}} - (\overline{\overline{\mathsf{A}}}||\overline{\overline{\mathsf{I}}}^T)\overline{\overline{\mathsf{A}}}|\overline{\overline{\mathsf{A}}} - (\overline{\overline{\mathsf{A}}}||\overline{\overline{\mathsf{I}}}^T)\overline{\overline{\mathsf{A}}}|\overline{\overline{\mathsf{A}}} - (\overline{\overline{\mathsf{A}}}|\overline{\overline{\mathsf{A}}})||\overline{\overline{\mathsf{I}}}^T\overline{\overline{\mathsf{A}}}$$

and it leads to

$$\overline{\overline{\mathsf{A}}}^{(3)}\lfloor\lfloor\overline{\overline{\mathsf{I}}}^{(2)T} = \overline{\overline{\mathsf{A}}}^3 - (\mathrm{tr}\,\overline{\overline{\mathsf{A}}})\overline{\overline{\mathsf{A}}}^2 + \frac{1}{2}[(\mathrm{tr}\,\overline{\overline{\mathsf{A}}})^2 - \mathrm{tr}(\overline{\overline{\mathsf{A}}}^2)]\overline{\overline{\mathsf{A}}}.$$

The last term can be identified as $\mathrm{tr}\,\overline{\overline{\mathsf{A}}}^{(2)}\overline{\overline{\mathsf{A}}}$. For $n = 3$ we have $\overline{\overline{\mathsf{A}}}^{(3)} = \overline{\overline{\mathsf{I}}}^{(3)}\mathrm{tr}\,\overline{\overline{\mathsf{A}}}^{(3)} = \overline{\overline{\mathsf{I}}}^{(3)}\det\,\overline{\overline{\mathsf{A}}}$, and replacing $\overline{\overline{\mathsf{A}}}$ by $\overline{\overline{\mathsf{I}}}$ in the above formula we have $\overline{\overline{\mathsf{I}}}^{(3)}\lfloor\lfloor\overline{\overline{\mathsf{I}}}^{(2)T} = \overline{\overline{\mathsf{I}}}$, whence the Cayley–Hamilton equation follows.

2.3.1

$$(\overline{\overline{\mathsf{A}}} - \lambda\overline{\overline{\mathsf{I}}})^{(n)} = ((a-\lambda)\overline{\overline{\mathsf{I}}} - \mathbf{a}\boldsymbol{\alpha})^{(n)}$$
$$= (a-\lambda)^n\overline{\overline{\mathsf{I}}}^{(n)} - (a-\lambda)^{n-1}\overline{\overline{\mathsf{I}}}^{(n-1)}{}^\wedge_\wedge\mathbf{a}\boldsymbol{\alpha}$$
$$= (a-\lambda)^{(n-1)}(a - \lambda - \mathbf{a}|\boldsymbol{\alpha})\overline{\overline{\mathsf{I}}}^{(n)} = 0.$$

From this we conclude that $\lambda_o = a$ is an $(n-1)$-fold eigenvalue and $\lambda_1 = a + \mathbf{a}|\boldsymbol{\alpha} \neq \lambda_o$ is a single eigenvalue. Left and right eigenvectors $\mathbf{x}_1, \boldsymbol{\xi}_1$ are obtained from the formula (2.141) valid for single eigenvalues

$$\boldsymbol{\xi}_1\mathbf{x}_1 = (\mathbf{a}\boldsymbol{\alpha} - \mathbf{a}|\boldsymbol{\alpha}\overline{\overline{\mathsf{I}}})^{(n-1)}\rfloor\rfloor\overline{\overline{\mathsf{I}}}^{(n)T}$$

$$= (-\mathbf{a}|\boldsymbol{\alpha})^{n-2}(\mathbf{a}\boldsymbol{\alpha}{}^\wedge_\wedge\overline{\overline{\mathsf{I}}}^{(n-2)})\rfloor\rfloor\overline{\overline{\mathsf{I}}}^{(n)T} + (-\mathbf{a}|\boldsymbol{\alpha})^{n-1}\overline{\overline{\mathsf{I}}}^{(n-1)})\rfloor\rfloor\overline{\overline{\mathsf{I}}}^{(n)T}.$$

Omitting the scalar coefficient, we can further expand

$$\boldsymbol{\xi}_1\mathbf{x}_1 = (\mathbf{a}|\boldsymbol{\alpha})\overline{\overline{\mathsf{I}}}^T - \mathbf{a}\boldsymbol{\alpha}\rfloor(\overline{\overline{\mathsf{I}}}^{(n-2)}\rfloor\rfloor\overline{\overline{\mathsf{I}}}^{(n)T} = (\mathbf{a}|\boldsymbol{\alpha})\overline{\overline{\mathsf{I}}}^T - \mathbf{a}\boldsymbol{\alpha}\rfloor\overline{\overline{\mathsf{I}}}^{(2)T}.$$

Here we have applied one of the identities. Applying the other one, the eigenvector expression finally becomes

$$\boldsymbol{\xi}_1 \mathbf{x}_1 = \boldsymbol{\alpha}\mathbf{a}.$$

The eigenvectors corresponding to the eigenvalue $\lambda_o = a$ form an $(n-1)$ dimensional subspace of vectors $\mathbf{x}_o$ and dual vectors ξ_o satisfying $\mathbf{x}_o|\boldsymbol{\alpha} = 0$ and $\xi_o|\mathbf{a} = 0$.

2.3.2

$$\overline{\overline{\mathsf{I}}}^{(n-1)} {}^\wedge_\wedge \mathbf{a}\boldsymbol{\alpha} = \sum_{i,j,k} a_j \alpha_k \mathbf{e}_{K(i)} \varepsilon_{K(i)} {}^\wedge_\wedge \mathbf{e}_j \varepsilon_k$$

$$= \sum_i a_i \alpha_i \mathbf{e}_{K(i)} \varepsilon_{K(i)} {}^\wedge_\wedge \mathbf{e}_i \varepsilon_i = (\mathbf{a}|\boldsymbol{\alpha}) \mathbf{e}_N \varepsilon_N.$$

2.3.3

$$\det(\overline{\overline{\mathsf{A}}} - \lambda \overline{\overline{\mathsf{I}}}) = \mathrm{tr}((a-\lambda)\overline{\overline{\mathsf{I}}} + \mathbf{a}\boldsymbol{\alpha})^{(n)}$$

$$= (a-\lambda)^n \mathrm{tr}\, \overline{\overline{\mathsf{I}}}(n) + (a-\lambda)^{n-1} \mathrm{tr}(\overline{\overline{\mathsf{I}}}^{(n-1)} {}^\wedge_\wedge \mathbf{a}\boldsymbol{\alpha}) = 0,$$

$$\mathrm{tr}\overline{\overline{\mathsf{I}}}^{(n)} = 1, \quad \mathrm{tr}(\overline{\overline{\mathsf{I}}}^{(n-1)} {}^\wedge_\wedge \mathbf{a}\boldsymbol{\alpha}) = (\mathbf{a}|\boldsymbol{\alpha}) \mathrm{tr}\overline{\overline{\mathsf{I}}}^{(n)} = 0,$$

$$\Rightarrow \det(\overline{\overline{\mathsf{A}}} - \lambda \overline{\overline{\mathsf{I}}}) = (a-\lambda)^n = 0.$$

All eigenvalues are the same, $\lambda = a$. To find eigenvectors $\mathbf{x}$ from

$$(\overline{\overline{\mathsf{A}}} - \lambda \overline{\overline{\mathsf{I}}})|\mathbf{x} = (\mathbf{a}\boldsymbol{\alpha})|\mathbf{x} = 0,$$

any $\mathbf{x}$ satisfying $\boldsymbol{\alpha}|\mathbf{x} = 0$ will do. They form an $(n-1)$ dimensional subspace in $\mathbb{E}_1$.

2.4.1 From $(\overline{\overline{\mathsf{I}}} + \mathbf{a}\boldsymbol{\alpha})|\overline{\overline{\mathsf{B}}} = \overline{\overline{\mathsf{B}}} + \mathbf{a}(\boldsymbol{\alpha}|\overline{\overline{\mathsf{B}}}) = \overline{\overline{\mathsf{I}}}$ we conclude that $\overline{\overline{\mathsf{B}}}$ must be of the form $\overline{\overline{\mathsf{I}}} - \mathbf{a}\boldsymbol{\beta}$ with $\boldsymbol{\beta} = \boldsymbol{\alpha}|\overline{\overline{\mathsf{B}}} = \boldsymbol{\alpha} - (\boldsymbol{\alpha}|\mathbf{a})\boldsymbol{\beta}$. From this, $\boldsymbol{\beta}$ can be solved as $\boldsymbol{\beta} = \boldsymbol{\alpha}/(1 + \boldsymbol{\alpha}|\mathbf{a})$.

2.4.2 The dyadic identity

$$\overline{\overline{\mathsf{I}}}^{(n)} \lfloor\lfloor \overline{\overline{\mathsf{I}}}^{(n-1)T} = \overline{\overline{\mathsf{I}}}$$

can be found by expanding

$$(\overline{\overline{\mathsf{I}}}^{(n)} \lfloor\lfloor \overline{\overline{\mathsf{I}}}^{(n-1)T})||\boldsymbol{\alpha}\mathbf{a} = \overline{\overline{\mathsf{I}}}^{(n)} ||(\overline{\overline{\mathsf{I}}}^{(n-1)T} {}^\wedge_\wedge \boldsymbol{\alpha}\mathbf{a})$$

$$= (\mathbf{a}|\boldsymbol{\alpha}) \overline{\overline{\mathsf{I}}}^{(n)} || \overline{\overline{\mathsf{I}}}^{(n)T} = \mathbf{a}|\boldsymbol{\alpha} = \overline{\overline{\mathsf{I}}}||\boldsymbol{\alpha}\mathbf{a},$$

which is valid for any $\mathbf{a}, \boldsymbol{\alpha}$. Similarly, we can derive the identity

$$\overline{\overline{\mathsf{I}}}^{(n)} \lfloor\lfloor \overline{\overline{\mathsf{I}}}^{(n-2)T} = \overline{\overline{\mathsf{I}}}^{(2)}.$$

2.4.3 Because tr $\bar{\bar{\mathsf{I}}}^{(p)}$ equals the number of terms of $\bar{\bar{\mathsf{I}}}^{(p)}$, i.e., the dimension of p-vector space $C_p^n = n!/(n-p)!p!$, the latter identity comes directly from the definition of $\bar{\bar{\mathsf{I}}}^{(q)} = \bar{\bar{\mathsf{I}}}{}^{\wedge}_{\wedge}\bar{\bar{\mathsf{I}}}{}^{\wedge}_{\wedge}\cdots{}^{\wedge}_{\wedge}\bar{\bar{\mathsf{I}}}/q!$. Assuming that $\bar{\bar{\mathsf{I}}}^{(n)}\lfloor\lfloor\bar{\bar{\mathsf{I}}}^{(n-p)T} = \alpha\bar{\bar{\mathsf{I}}}^{(p)}$ and operating each side by $||\bar{\bar{\mathsf{I}}}^{(p)T}$ gives

$$[\bar{\bar{\mathsf{I}}}^{(n)}\lfloor\lfloor\bar{\bar{\mathsf{I}}}^{(n-p)T}]||\bar{\bar{\mathsf{I}}}^{(p)T} = \bar{\bar{\mathsf{I}}}^{(n)}||(\bar{\bar{\mathsf{I}}}^{(n-p)T}{}^{\wedge}_{\wedge}\bar{\bar{\mathsf{I}}}^{(p)T}) = \alpha\mathrm{tr}\bar{\bar{\mathsf{I}}}^{(p)},$$

from which we obtain $\alpha = 1$.

2.4.5

$$\bar{\bar{\mathsf{I}}}^{(p)}{}^{\wedge}_{\wedge}\bar{\bar{\mathsf{I}}}^{(n-p)} = \frac{1}{q!}(\bar{\bar{\mathsf{I}}}{}^{\wedge}_{\wedge}\bar{\bar{\mathsf{I}}}{}^{\wedge}_{\wedge}\cdots{}^{\wedge}_{\wedge}\bar{\bar{\mathsf{I}}}){}^{\wedge}_{\wedge}\frac{1}{(n-p)!}(\bar{\bar{\mathsf{I}}}{}^{\wedge}_{\wedge}\cdots{}^{\wedge}_{\wedge}\bar{\bar{\mathsf{I}}}) = \frac{n!}{p!(n-p)!}\bar{\bar{\mathsf{I}}}^{(n)}$$

Because $\mathrm{tr}\bar{\bar{\mathsf{I}}}^{(p)} = \sum \mathbf{e}_J|\varepsilon^J$ where J is a p-vector, it equals the dimension of the p-vector space $C_p^n = n!/p!(n-p)!$.

2.4.6 Start from

$$\bar{\bar{\mathsf{I}}}^{(2)}||(\boldsymbol{\alpha}\wedge\boldsymbol{\beta})(\mathbf{a}\wedge\mathbf{b}) = \frac{1}{2}\sum_{i,j=1}^{n}[(\mathbf{e}_i\wedge\mathbf{e}_j)|(\boldsymbol{\alpha}\wedge\boldsymbol{\beta})][(\varepsilon^i\wedge\varepsilon^j)|(\mathbf{a}\wedge\mathbf{b})].$$

Apply

$$(\mathbf{e}_i\wedge\mathbf{e}_j)|(\boldsymbol{\alpha}\wedge\boldsymbol{\beta}) = (\mathbf{e}_i\mathbf{e}_j - \mathbf{e}_j\mathbf{e}_i)||\boldsymbol{\alpha}\boldsymbol{\beta},$$

and similarly for the other term. Combine the four terms as

$$[(\mathbf{e}_i\wedge\mathbf{e}_j)|(\boldsymbol{\alpha}\wedge\boldsymbol{\beta})][(\varepsilon^i\wedge\varepsilon^j)|(\mathbf{a}\wedge\mathbf{b})] = [\boldsymbol{\alpha}|(\mathbf{e}_i\mathbf{e}_j - \mathbf{e}_j\mathbf{e}_i)\mathbf{a}|(\varepsilon^i\varepsilon^j - \varepsilon^j\varepsilon^i)]||\boldsymbol{\beta}\mathbf{b},$$

which inserted in the sum gives

$$[\bar{\bar{\mathsf{I}}}^{(2)}\lfloor\lfloor\boldsymbol{\alpha}\mathbf{a}]||\boldsymbol{\beta}\mathbf{b} = (\boldsymbol{\alpha}|\mathbf{a})(\boldsymbol{\beta}|\mathbf{b}) - (\boldsymbol{\alpha}|\mathbf{b})(\boldsymbol{\beta}|\mathbf{b}) = [(\boldsymbol{\alpha}|\mathbf{a})\bar{\bar{\mathsf{I}}} - \mathbf{a}\boldsymbol{\alpha}]||\boldsymbol{\beta}\mathbf{b}.$$

2.4.7 Let us check:

$$(1+\boldsymbol{\alpha}|\bar{\bar{\mathsf{A}}}^{-1}|\mathbf{a})\bar{\bar{\mathsf{B}}}^{-1}|\bar{\bar{\mathsf{B}}}$$

$$= (1+\boldsymbol{\alpha}|\bar{\bar{\mathsf{A}}}^{-1}|\mathbf{a})(\bar{\bar{\mathsf{I}}} + \bar{\bar{\mathsf{A}}}^{-1}\cdot\mathbf{a}\boldsymbol{\alpha}) - \bar{\bar{\mathsf{A}}}^{-1}|\mathbf{a}\boldsymbol{\alpha} = (1+\boldsymbol{\alpha}|\bar{\bar{\mathsf{A}}}^{-1}|\mathbf{a})\bar{\bar{\mathsf{I}}}.$$

2.4.9 Denote

$$\mathbf{a}_1\wedge\mathbf{a}_2\wedge\cdots\wedge\mathbf{a}_p = \mathbf{a}^p, \qquad \mathbf{b}_1\wedge\mathbf{b}_2\wedge\cdots\wedge\mathbf{b}_p = \mathbf{b}^p$$

and assuming $\mathbf{b}_i = \bar{\bar{\mathsf{A}}}|\mathbf{a}_i$, $i = 1\cdots p$, we have

$$\bar{\bar{\mathsf{A}}}^{(p)}|\mathbf{a}^p = (\bar{\bar{\mathsf{A}}}|\mathbf{a}_1)\wedge(\bar{\bar{\mathsf{A}}}|\mathbf{a}_2)\wedge\cdots\wedge(\bar{\bar{\mathsf{A}}}|\mathbf{a}_p) = \mathbf{b}^p.$$

Now we can write either

$$\mathbf{a}^p = (\bar{\bar{\mathsf{A}}}^{(p)})^{-1}|\mathbf{b}^p$$

or

$$\mathbf{a}^p = (\overline{\overline{\mathsf{A}}}^{-1}|\mathbf{b}_1) \wedge (\overline{\overline{\mathsf{A}}}^{-1}|\mathbf{b}_2) \wedge \cdots \wedge (\overline{\overline{\mathsf{A}}}^{-1}|\mathbf{b}_p) = (\overline{\overline{\mathsf{A}}}^{-1})^{(p)}|\mathbf{b}^p.$$

Because this is valid for any p-vector $\mathbf{b}^p$, the dyadics $(\overline{\overline{\mathsf{A}}}^{(p)})^{-1}$ and $(\overline{\overline{\mathsf{A}}}^{-1})^{(p)}$ must be the same.

2.4.10 Applying previous identities, we have

$$(\overline{\overline{\mathsf{A}}}{}^{\wedge}_{\wedge}\overline{\overline{\mathsf{I}}}^{(n-2)})\rfloor\rfloor\overline{\overline{\mathsf{I}}}^{(n)T} = \overline{\overline{\mathsf{A}}}\rfloor(\overline{\overline{\mathsf{I}}}^{(n-2)}\rfloor\rfloor\overline{\overline{\mathsf{I}}}^{(n)T}) = \overline{\overline{\mathsf{A}}}\rfloor\overline{\overline{\mathsf{I}}}^{(2)T} = (\mathrm{tr}\,\overline{\overline{\mathsf{A}}})\overline{\overline{\mathsf{I}}}^T - \overline{\overline{\mathsf{A}}}^T.$$

2.4.11 Following the pattern of Section 2.4.2, the bivector dyadic is expanded as $\overline{\overline{\mathsf{A}}} = \mathbf{A}_1\mathbf{\Psi}_1 + \mathbf{A}_2\mathbf{\Psi}_2 + \mathbf{A}_3\mathbf{\Psi}_3$ and another dyadic is formed with the reciprocal basis bivectors and dual bivectors as

$$\overline{\overline{\mathsf{B}}} = \frac{[\mathbf{\Psi}_1\mathbf{\Psi}_2][\mathbf{A}_1\mathbf{A}_2] + [\mathbf{\Psi}_2\mathbf{\Psi}_3][\mathbf{A}_2\mathbf{A}_3] + [\mathbf{\Psi}_3\mathbf{\Psi}_1][\mathbf{A}_3\mathbf{A}_1]}{(\mathbf{A}_1\mathbf{A}_2\mathbf{A}_3)(\mathbf{\Psi}_1\mathbf{\Psi}_2\mathbf{\Psi}_3)}.$$

From orthogonality we have

$$\overline{\overline{\mathsf{A}}}|\overline{\overline{\mathsf{B}}} = \frac{\mathbf{A}_3[\mathbf{A}_1\mathbf{A}_2] + \mathbf{A}_1[\mathbf{A}_2\mathbf{A}_3] + \mathbf{A}_2[\mathbf{A}_3\mathbf{A}_1]}{(\mathbf{A}_1\mathbf{A}_2\mathbf{A}_3)} = \overline{\overline{\mathsf{I}}}^{(2)}.$$

This equals the unit bivector dyadic because it maps any bivector dyadic $\sum a_i\mathbf{A}_i$ onto itself, whence $\overline{\overline{\mathsf{B}}}$ equals the inverse of $\overline{\overline{\mathsf{A}}}$. To obtain the required representation for $\overline{\overline{\mathsf{B}}}$, we first expand the dyadic product

$$\begin{aligned}
[\mathbf{\Psi}_i\mathbf{\Psi}_j][\mathbf{A}_i\mathbf{A}_j] &= ((\mathbf{\Psi}_i\rfloor\mathbf{k}) \wedge (\mathbf{\Psi}_j\rfloor\mathbf{k}))((\mathbf{A}_i\rfloor\boldsymbol{\kappa}) \wedge (\mathbf{A}_j\rfloor\boldsymbol{\kappa})) \\
&= (\mathbf{\Psi}_i\mathbf{A}_i\rfloor\rfloor\mathbf{k}\boldsymbol{\kappa}){}^{\wedge}_{\wedge}(\mathbf{\Psi}_j\mathbf{A}_j\rfloor\rfloor\mathbf{k}\boldsymbol{\kappa}) \\
&= (\mathbf{\Psi}_i\mathbf{A}_i\rfloor\rfloor\overline{\overline{\mathsf{I}}}^{(3)}){}^{\wedge}_{\wedge}(\mathbf{\Psi}_j\mathbf{A}_j\rfloor\rfloor\overline{\overline{\mathsf{I}}}^{(3)}),
\end{aligned}$$

which, inserted in the above expression, gives

$$\overline{\overline{\mathsf{B}}} = \frac{(\overline{\overline{\mathsf{A}}}^T\rfloor\rfloor\overline{\overline{\mathsf{I}}}^{(3)})^{(2)}}{(\mathbf{A}_1\mathbf{A}_2\mathbf{A}_3)(\mathbf{\Psi}_1\mathbf{\Psi}_2\mathbf{\Psi}_3)}.$$

Finally, the determinant is obtained as

$$\begin{aligned}
&(\mathbf{A}_1\mathbf{A}_2\mathbf{A}_3)(\mathbf{\Psi}_1\mathbf{\Psi}_2\mathbf{\Psi}_3) \\
&= \mathbf{k}\boldsymbol{\kappa}||(\mathbf{A}_1\mathbf{\Psi}_1\rfloor\rfloor\boldsymbol{\kappa}\mathbf{k}){}^{\wedge}_{\wedge}(\mathbf{A}_2\mathbf{\Psi}_2\rfloor\rfloor\boldsymbol{\kappa}\mathbf{k}){}^{\wedge}_{\wedge}(\mathbf{A}_3\mathbf{\Psi}_3\rfloor\rfloor\boldsymbol{\kappa}\mathbf{k}) \\
&= \overline{\overline{\mathsf{I}}}^{(3)}||(\overline{\overline{\mathsf{A}}}\rfloor\rfloor\overline{\overline{\mathsf{I}}}^{(3)})^{(3)} = \det(\overline{\overline{\mathsf{A}}}^T\rfloor\rfloor\overline{\overline{\mathsf{I}}}^{(3)}) = \mathrm{Det}\,\overline{\overline{\mathsf{A}}}.
\end{aligned}$$

2.4.12 Applying the inverse formula derived in the previous problem for the inverse dyadic $\overline{\overline{\mathsf{A}}}^{-1}$, we can write

$$\overline{\overline{\mathsf{A}}} = \frac{1}{\mathrm{Det}\overline{\overline{\mathsf{A}}}^{-1}}(\overline{\overline{\mathsf{A}}}^{-1T}\rfloor\rfloor\overline{\overline{\mathsf{I}}}^{(3)})^{(2)},$$

whence we can identify the dyadic $\overline{\overline{\mathsf{Q}}}$ as

$$\overline{\overline{\mathsf{Q}}} = \pm\frac{1}{\sqrt{\mathrm{Det}\overline{\overline{\mathsf{A}}}^{-1}}}\overline{\overline{\mathsf{A}}}^{-1T}\rfloor\rfloor\overline{\overline{\mathsf{I}}}^{(3)}.$$

2.4.15 Start from the linear equation

$$\overline{\overline{\mathsf{M}}}|(\mathbf{X} + \mathbf{x}\wedge\mathbf{e}_4) = \mathbf{Y} + \mathbf{y}\wedge\mathbf{e}_4,$$

where $\mathbf{X}, \mathbf{Y} \in \mathbb{E}_2$ are Euclidean bivectors and $\mathbf{x}, \mathbf{y} \in \mathbb{E}_1$ are Euclidean vectors. Splitting the Minkowskian equation in two Euclidean equations as

$$\begin{aligned} \overline{\overline{\mathsf{A}}}|\mathbf{X} + \overline{\overline{\mathsf{C}}}|\mathbf{x} &= \mathbf{Y}, \\ \overline{\overline{\mathsf{B}}}|\mathbf{X} + \overline{\overline{\mathsf{D}}}|\mathbf{x} &= -\mathbf{y}, \end{aligned}$$

they can be easily solved by elimination. Expressing the result in the form

$$\mathbf{X} + \mathbf{x}\wedge\mathbf{e}_4 = \overline{\overline{\mathsf{M}}}^{-1}|(\mathbf{Y} + \mathbf{y}\wedge\mathbf{e}_4),$$

the inverse dyadic can be identified.

2.5.1 The answer is yes, which can be seen if we write the dyadic in the from

$$\overline{\overline{\mathsf{D}}} = \mathbf{aa} + \frac{\mathbf{b}+\mathbf{c}}{\sqrt{2}}\frac{\mathbf{b}+\mathbf{c}}{\sqrt{2}} - \frac{\mathbf{b}-\mathbf{c}}{\sqrt{2}}\frac{\mathbf{b}-\mathbf{c}}{\sqrt{2}}.$$

The dyadic corresponds to Minkowskian metric: $\sigma_1 = \sigma_2 = 1, \sigma_3 = -1$ and its vectors $\{\mathbf{e}_i\}$ make a basis:

$$\mathbf{a}\wedge\frac{\mathbf{b}+\mathbf{c}}{\sqrt{2}}\wedge\frac{\mathbf{b}-\mathbf{c}}{\sqrt{2}} = \frac{j}{2}\mathbf{a}\wedge\mathbf{b}\wedge\mathbf{c} \neq 0.$$

2.5.2 Start by finding the reciprocal dual basis $\{\varepsilon_i\}$ and then expressing $\overline{\overline{\Gamma}} = \sum \varepsilon_i\varepsilon_i$.

2.5.5 Expand $\overline{\overline{\mathsf{A}}} = \mathbf{A}_1\mathbf{B}_1 + \mathbf{A}_2\mathbf{B}_2 + \mathbf{A}_3\mathbf{B}_3$ and proceed similarly as in Problem 2.4.11.

2.5.8

$$\begin{aligned}
\overline{\overline{\mathsf{M}}}|\overline{\overline{\mathsf{M}}}^{-1} &= \overline{\overline{\mathsf{A}}}|(\overline{\overline{\mathsf{A}}} - \overline{\overline{\mathsf{C}}}|\overline{\overline{\mathsf{D}}}^{-1}|\overline{\overline{\mathsf{B}}})^{-1} - \overline{\overline{\mathsf{C}}}|(\overline{\overline{\mathsf{D}}} - \overline{\overline{\mathsf{B}}}|\overline{\overline{\mathsf{A}}}^{-1}|\overline{\overline{\mathsf{C}}})^{-1}|\overline{\overline{\mathsf{B}}}|\overline{\overline{\mathsf{A}}}^{-1} \\
&+\overline{\overline{\mathsf{A}}}|(\overline{\overline{\mathsf{A}}} - \overline{\overline{\mathsf{C}}}|\overline{\overline{\mathsf{D}}}^{-1}|\overline{\overline{\mathsf{B}}})^{-1}|\overline{\overline{\mathsf{C}}}|\overline{\overline{\mathsf{D}}}^{-1}\wedge\varepsilon_4 - \overline{\overline{\mathsf{C}}}|(\overline{\overline{\mathsf{D}}} - \overline{\overline{\mathsf{B}}}|\overline{\overline{\mathsf{A}}}^{-1}|\overline{\overline{\mathsf{C}}})^{-1}\wedge\varepsilon_4 \\
&+\mathbf{e}_4\wedge\overline{\overline{\mathsf{B}}}|(\overline{\overline{\mathsf{A}}} - \overline{\overline{\mathsf{C}}}|\overline{\overline{\mathsf{D}}}^{-1}|\overline{\overline{\mathsf{B}}})^{-1} - \mathbf{e}_4\wedge\overline{\overline{\mathsf{D}}}|(\overline{\overline{\mathsf{D}}} - \overline{\overline{\mathsf{B}}}|\overline{\overline{\mathsf{A}}}^{-1}|\overline{\overline{\mathsf{C}}})^{-1}|\overline{\overline{\mathsf{B}}}|\overline{\overline{\mathsf{A}}}^{-1} \\
&+\mathbf{e}_4\wedge\overline{\overline{\mathsf{B}}}|(\overline{\overline{\mathsf{A}}} - \overline{\overline{\mathsf{C}}}|\overline{\overline{\mathsf{D}}}^{-1}|\overline{\overline{\mathsf{B}}})^{-1}|\overline{\overline{\mathsf{C}}}|\overline{\overline{\mathsf{D}}}^{-1}\wedge\varepsilon_4 - \mathbf{e}_4\wedge\overline{\overline{\mathsf{D}}}|(\overline{\overline{\mathsf{D}}} - \overline{\overline{\mathsf{B}}}|\overline{\overline{\mathsf{A}}}^{-1}|\overline{\overline{\mathsf{C}}})^{-1}\wedge\varepsilon_4, \\
&= (\overline{\overline{\mathsf{I}}}^{(2)}_{\mathbf{E}} - \overline{\overline{\mathsf{C}}}|\overline{\overline{\mathsf{D}}}^{-1}|\overline{\overline{\mathsf{B}}}|\overline{\overline{\mathsf{A}}}^{-1})^{-1} - \overline{\overline{\mathsf{C}}}|\overline{\overline{\mathsf{D}}}^{-1}|\overline{\overline{\mathsf{B}}}|\overline{\overline{\mathsf{A}}}^{-1}|(\overline{\overline{\mathsf{I}}}^{(2)}_{\mathbf{E}} - \overline{\overline{\mathsf{C}}}|\overline{\overline{\mathsf{D}}}^{-1}|\overline{\overline{\mathsf{B}}}|\overline{\overline{\mathsf{A}}}^{-1})^{-1}
\end{aligned}$$

$$+\mathbf{e}_4\wedge[(\bar{\bar{\mathsf{I}}}_{\mathbf{E}}-\overline{\overline{\mathsf{B}}}|\overline{\overline{\mathsf{A}}}^{-1}|\overline{\overline{\mathsf{C}}}|\overline{\overline{\mathsf{D}}}^{-1})^{-1}|\overline{\overline{\mathsf{B}}}|\overline{\overline{\mathsf{A}}}^{-1}|\overline{\overline{\mathsf{C}}}|\overline{\overline{\mathsf{D}}}^{-1}-(\bar{\bar{\mathsf{I}}}_{\mathbf{E}}-\overline{\overline{\mathsf{B}}}|\overline{\overline{\mathsf{A}}}^{-1}|\overline{\overline{\mathsf{C}}}|\overline{\overline{\mathsf{D}}}^{-1})^{-1}]\wedge\varepsilon_4,$$

$$=\bar{\bar{\mathsf{I}}}^{(2)}_{\mathbf{E}}-\mathbf{e}_4\wedge\bar{\bar{\mathsf{I}}}_{\mathbf{E}}\wedge\varepsilon_4=\bar{\bar{\mathsf{I}}}^{(2)}_{\mathbf{E}}+\bar{\bar{\mathsf{I}}}_{\mathbf{E}}{}^{\wedge}_{\wedge}\mathbf{e}_4\varepsilon_4=\bar{\bar{\mathsf{I}}}^{(2)}_{\mathbf{M}}.$$

2.6.1

$$\bar{\bar{\mathsf{I}}}^{(n-p)}|\overline{\overline{\mathsf{H}}}_p=\mathbf{e}_N\lfloor\sum\varepsilon^J(\varepsilon^N\lfloor\mathbf{e}_J)|\overline{\overline{\mathsf{H}}}_p=\mathbf{e}_N\lfloor\sum\varepsilon^J(\varepsilon^N|(\mathbf{e}_J\wedge\overline{\overline{\mathsf{H}}}_p))$$

$$=\mathbf{e}_N\lfloor\sum\varepsilon^J(\varepsilon^N|\mathbf{e}_N)\mathbf{e}_J|\overline{\overline{\Gamma}}^{(p)}=\mathbf{e}_N\lfloor\bar{\bar{\mathsf{I}}}^{(p)T}|\overline{\overline{\Gamma}}^{(p)}.$$

2.6.2

$$\overline{\overline{\mathsf{R}}}(\theta_1)|\overline{\overline{\mathsf{R}}}(\theta_2)=(\bar{\bar{\mathsf{I}}}\cos\theta_1+\overline{\overline{\mathsf{H}}}_1\sin\theta_1)|(\bar{\bar{\mathsf{I}}}\cos\theta_2+\overline{\overline{\mathsf{H}}}_1\sin\theta_2)$$

$$=\bar{\bar{\mathsf{I}}}(\cos\theta_1\cos\theta_2-\sin\theta_1\sin\theta_2)+\overline{\overline{\mathsf{H}}}_1(\cos\theta_1\sin\theta_2+\cos\theta_2\sin\theta_1)$$

$$=\bar{\bar{\mathsf{I}}}\cos(\theta_1+\theta_2)+\overline{\overline{\mathsf{H}}}_1\sin(\theta_1+\theta_2).$$

2.6.5

$$\mathbf{e}_{1234}\lfloor\overline{\overline{\Gamma}}^{(3)}_{\mathbf{M}}=\mathbf{e}_{1234}\lfloor\varepsilon^{123}\varepsilon^{123}-\mathbf{e}_{1234}\lfloor(\varepsilon^{124}\varepsilon^{124}+\cdots)$$

$$=\mathbf{e}_4\varepsilon^{123}+\mathbf{e}_3\varepsilon^{124}+\cdots=\overline{\overline{\mathsf{H}}}_{\mathbf{M}3}.$$

3.3.1 Straightforward analysis gives

$$\mathbf{d}\wedge\phi=(\mathbf{d}x\wedge\mathbf{d}y)\partial_x\left(\frac{x}{x^2+y^2}\right)-(\mathbf{d}x\wedge\mathbf{d}y)\partial_y\left(\frac{-y}{x^2+y^2}\right)$$

$$=(\mathbf{d}x\wedge\mathbf{d}y)\left(\frac{y^2-x^2}{(x^2+y^2)^2}+\frac{x^2-y^2}{(x^2+y^2)^2}\right)=0.$$

However, since $x^2+y^2=\rho^2$ where ρ is the polar radius, this suggests expressing the one-form in polar coordinates with

$$\mathbf{d}y=\mathbf{d}(\rho\sin\varphi)=\sin\varphi\mathbf{d}\rho+\rho\cos\varphi\mathbf{d}\varphi,$$

$$\mathbf{d}x=\mathbf{d}(\rho\cos\varphi)=\cos\varphi\mathbf{d}\rho-\rho\sin\varphi\mathbf{d}\varphi,$$

whence we have quite simply $\phi=\mathbf{d}\varphi$. From this it is easy to see that the one-form is closed: $\mathbf{d}\wedge\phi=\mathbf{d}\wedge\mathbf{d}\varphi=0$. However, it is not exact, because integration around the origin gives

$$\phi|C=\mathbf{d}\varphi|C=\int_0^{2\pi}\varphi=2\pi.$$

3.4.4 Obviously, two of the eigenvectors are $\mathbf{e}_2$ and $\mathbf{e}_3$ (or their linear combinations) corresponding to the eigenvalue 1. The other two eigenvectors are linear combinations of $\mathbf{e}_1$ and $\mathbf{e}_4$. From

$$(c\mathbf{e}_1\varepsilon_1+s\mathbf{e}_4\varepsilon_1+s\mathbf{e}_1\varepsilon_4+c\mathbf{e}_4\varepsilon_4)|(\alpha\mathbf{e}_1+\beta\mathbf{e}_4)=\lambda(\alpha\mathbf{e}_1+\beta\mathbf{e}_4)$$

we obtain equations for the coefficients α, β, and λ. Solutions for the right and left eigenvectors and their eigenvalues are

$$\mathbf{e}_\pm = \frac{1}{\sqrt{2}}(\mathbf{e}_1 \pm \mathbf{e}_4) \quad \boldsymbol{\varepsilon}_\pm = \frac{1}{\sqrt{2}}(\boldsymbol{\varepsilon}_1 \pm \boldsymbol{\varepsilon}_4), \quad \lambda_\pm = e^{\pm\vartheta}.$$

The dyadic can now be written as

$$\overline{\overline{\mathsf{A}}} = \lambda_+ \mathbf{e}_+ \boldsymbol{\varepsilon}_+ + \lambda_- \mathbf{e}_- \boldsymbol{\varepsilon}_- + \mathbf{e}_2 \boldsymbol{\varepsilon}_2 + \mathbf{e}_3 \boldsymbol{\varepsilon}_3.$$

The inverse is obtained by inverting the eigenvalues. Because $1/\lambda_\pm = e^{\mp\vartheta} = \lambda_\mp$ this corresponds to the change $\vartheta \to -\vartheta$.

3.4.5 The trace and determinant become

$$\mathrm{tr}\,\overline{\overline{\mathsf{A}}} = 2c + 2s + 2 = 2e^{\vartheta} + 2, \quad \det \overline{\overline{\mathsf{A}}} = c^2 - s^2 = 1.$$

4.3.1 All are four-forms, which are scalar multiples of $\mathbf{d}x \wedge \mathbf{d}y \wedge \mathbf{d}z \wedge \mathbf{d}\tau$. The scalar can be found by multiplying by the four-vector $\mathbf{k}_N = \mathbf{e}_x \wedge \mathbf{e}_y \wedge \mathbf{e}_z \wedge \mathbf{e}_\tau$.

(a) $\mathbf{k}_N|(\boldsymbol{\Phi} \wedge \boldsymbol{\Phi}) = 2\mathbf{k}_N|(\mathbf{B} \wedge \mathbf{E} \wedge \mathbf{d}\tau) = 2\mathbf{e}_{xyz}|(\mathbf{B} \wedge \mathbf{E}) = 2(B_{xy}E_z + B_{yz}E_x + B_{zx}E_y)$ This corresponds to the Gibbsian vector expression $2\mathbf{B} \cdot \mathbf{E}$.

(b) Similarly we obtain $\boldsymbol{\Psi} \wedge \boldsymbol{\Psi} \to -2\mathbf{D} \cdot \mathbf{H}$.

(c) $\boldsymbol{\Phi} \wedge \boldsymbol{\Psi} \to \mathbf{D} \cdot \mathbf{E} - \mathbf{B} \cdot \mathbf{H}$.

4.3.2

$$\mathbf{d}_\mathbf{M} \wedge \gamma = (\mathbf{d}_\mathbf{E} + \mathbf{d}\tau \partial_\tau) \wedge (\boldsymbol{\varrho} - \mathbf{J} \wedge \mathbf{d}\tau)$$
$$= \mathbf{d}_\mathbf{E} \wedge \boldsymbol{\varrho} - \mathbf{d}_\mathbf{E} \wedge \mathbf{J} \wedge \mathbf{d}\tau - \mathbf{d}\tau \wedge \partial_\tau \boldsymbol{\varrho}.$$

Because $\mathbf{d}_\mathbf{E} \wedge \boldsymbol{\varrho}$ is a four-form in Eu3 space, it vanishes, whence

$$(\mathbf{d}_\mathbf{E} \wedge \mathbf{J} + \partial_\tau \boldsymbol{\varrho}) \wedge \mathbf{d}\tau = 0 \quad \Rightarrow \mathbf{d}_\mathbf{E} \wedge \mathbf{J} + \partial_\tau \boldsymbol{\varrho} = 0.$$

4.4.1 $\mathbf{d}\rho = (x\mathbf{d}x + y\mathbf{d}y)/\rho$. Thus, we can write

$$\mathbf{A} = f(\rho\cos\varphi)\mathbf{d}\rho \;\text{ or }\; \mathbf{A} = (xA(x)/\rho)\mathbf{d}x + (yA(x)/\rho)\mathbf{d}y,$$

where we must insert $\rho = \sqrt{x^2 + y^2}$. Writing

$$\mathbf{d} \wedge \mathbf{A} = A'(\rho\cos\varphi)(-\rho\sin\varphi)\mathbf{d}\varphi \wedge \mathbf{d}\rho$$

and

$$\mathbf{d} \wedge \mathbf{A} = \partial_y(xA(x)/\rho)\mathbf{d}y \wedge \mathbf{d}x + \partial_x(yA(x)/\rho)\mathbf{d}x \wedge \mathbf{d}y,$$

the latter expanded gives, after canceling two terms,

$$\mathbf{d} \wedge \mathbf{A} = A'(x)(y/\rho)\mathbf{d}x \wedge \mathbf{d}y,$$

which coincides with that above.

5.1.4 Since $\overline{\overline{\mu}}_g \in \mathbb{E}_1\,\mathbb{E}_1$ is a Euclidean dyadic, we can write

$$(\varepsilon_{123}\lfloor\overline{\overline{\mu}}_g)|(\overline{\overline{\mu}}_g^{-1}\rfloor\mathbf{e}_{123}) = \varepsilon_{123}\lfloor\overline{\overline{\mathsf{I}}}_{\mathbf{E}}\rfloor\mathbf{e}_{123} = \overline{\overline{\mathsf{I}}}_{\mathbf{E}}\rfloor\rfloor\overline{\overline{\mathsf{I}}}^{(3)}_{\mathbf{E}} = \overline{\overline{\mathsf{I}}}^{(2)T}_{\mathbf{E}},$$

which proves the relation $(\varepsilon_{123}\lfloor\overline{\overline{\mu}}_g)^{-1} = \overline{\overline{\mu}}_g^{-1}\rfloor\mathbf{e}_{123}$.

$$\overline{\overline{\epsilon}}' = \overline{\overline{\epsilon}} - \overline{\overline{\xi}}|\overline{\overline{\mu}}^{-1}|\overline{\overline{\zeta}} = \varepsilon_{123}\lfloor\overline{\overline{\epsilon}}_g - \varepsilon_{123}\lfloor\overline{\overline{\xi}}_g|(\varepsilon_{123}\lfloor\overline{\overline{\mu}}_g)^{-1}|(\varepsilon_{123}\lfloor\overline{\overline{\zeta}}_g)$$

$$= \varepsilon_{123}\lfloor(\overline{\overline{\epsilon}}_g - \overline{\overline{\xi}}_g|\overline{\overline{\mu}}_g^{-1}\rfloor\mathbf{e}_{123}|(\varepsilon_{123}\lfloor\overline{\overline{\zeta}}_g) = \varepsilon_{123}\lfloor(\overline{\overline{\epsilon}}_g - \overline{\overline{\xi}}_g|\overline{\overline{\mu}}_g^{-1}|\overline{\overline{\zeta}}_g)$$

$$\overline{\overline{\mu}} = \varepsilon_{123}|\overline{\overline{\mu}}_g, \quad \overline{\overline{\alpha}} = \overline{\overline{\xi}}|\overline{\overline{\mu}}^{-1} = \varepsilon_{123}\lfloor\overline{\overline{\xi}}_g|\overline{\overline{\mu}}_g^{-1}\rfloor\mathbf{e}_{123},$$

$$\overline{\overline{\beta}} = -\overline{\overline{\mu}}^{-1}|\overline{\overline{\zeta}} = -(\overline{\overline{\mu}}_g^{-1}\rfloor\mathbf{e}_{123})|(\varepsilon_{123}\lfloor\overline{\overline{\zeta}}_g) = -\overline{\overline{\mu}}_g^{-1}|\overline{\overline{\zeta}}_g.$$

Here we have also applied the rule

$$(\mathbf{a}\rfloor\varepsilon_{123})|(\mathbf{e}_{123}\lfloor\boldsymbol{\alpha}) = \mathbf{a}|\boldsymbol{\alpha},$$

which is valid for any three-dimensional Euclidean vector $\mathbf{a}$ and dual vector $\boldsymbol{\alpha}$.

5.1.7 Denoting

$$\overline{\overline{\mathsf{M}}}^{-1} = \overline{\overline{\alpha}}_1 + \overline{\overline{\epsilon}}'_1 \wedge \mathbf{e}_\tau + \mathbf{d}\tau \wedge \overline{\overline{\mu}}_1^{-1} + \mathbf{d}\tau \wedge \overline{\overline{\beta}}_1 \wedge \mathbf{e}_\tau,$$

there arise four dyadic equations:

$$\overline{\overline{\beta}}|\overline{\overline{\beta}}_1 - \overline{\overline{\mu}}^{-1}|\overline{\overline{\epsilon}}'_1 = \overline{\overline{\mathsf{I}}}^T_\epsilon, \quad \overline{\overline{\alpha}}|\overline{\overline{\epsilon}}'_1 - \overline{\overline{\epsilon}}'|\overline{\overline{\beta}}_1 = 0,$$

$$\overline{\overline{\mu}}^{-1}|\overline{\overline{\alpha}}_1 - \overline{\overline{\beta}}|\overline{\overline{\mu}}_1^{-1} = 0, \quad \overline{\overline{\alpha}}|\overline{\overline{\alpha}}_1 - \overline{\overline{\epsilon}}'|\overline{\overline{\mu}}_1^{-1} = \overline{\overline{\mathsf{I}}}^{(2)T}_{\mathbf{E}}.$$

From these we can solve:

$$\overline{\overline{\epsilon}}'_1 = -(\overline{\overline{\mu}}^{-1} - \overline{\overline{\beta}}|\overline{\overline{\epsilon}}'^{-1}|\overline{\overline{\alpha}})^{-1}, \quad \overline{\overline{\mu}}_1^{-1} = -(\overline{\overline{\epsilon}}' - \overline{\overline{\alpha}}|\overline{\overline{\mu}}|\overline{\overline{\beta}})^{-1},$$

$$\overline{\overline{\alpha}}_1 = \overline{\overline{\mu}}|\overline{\overline{\beta}}|\overline{\overline{\mu}}_1^{-1} = (\overline{\overline{\alpha}} - \overline{\overline{\epsilon}}'|\overline{\overline{\beta}}^{-1}|\overline{\overline{\mu}}^{-1})^{-1},$$

$$\overline{\overline{\beta}}_1 = \overline{\overline{\epsilon}}'^{-1}|\overline{\overline{\alpha}}|\overline{\overline{\epsilon}}'_1 = (\overline{\overline{\beta}} - \overline{\overline{\mu}}^{-1}|\overline{\overline{\alpha}}^{-1}|\overline{\overline{\epsilon}}')^{-1}.$$

5.1.8

$$\overline{\overline{\mathsf{M}}}^{(2)} = -(\overline{\overline{\alpha}}{}^\wedge_\wedge\overline{\overline{\beta}} + \overline{\overline{\epsilon}}'{}^\wedge_\wedge\overline{\overline{\mu}}^{-1}){}^\wedge_\wedge\mathbf{d}\tau\mathbf{e}_\tau = -\varepsilon_N\mathbf{e}_N((\overline{\overline{\alpha}}{}^\wedge_\wedge\overline{\overline{\beta}} + \overline{\overline{\epsilon}}'{}^\wedge_\wedge\overline{\overline{\mu}}^{-1})||\mathbf{e}_N\varepsilon_N)$$

5.1.10 In the Gibbsian representation, the permittivity and permeability are defined by $9 + 1 = 10$ parameters because the dyadic $\overline{\overline{\mu}}_g$ and the scalar a can be chosen freely. The magnetoelectric dyadics involve two arbitrary bivectors which correspond to $3 + 3 = 6$ parameters. The total, 16 parameters, equals that of a four-dimensional dyadic $\overline{\overline{\mathsf{Q}}} = \sum \mathbf{q}_i \mathbf{e}_i$, which can be defined in terms of four vectors $\mathbf{q}_i$ with four components each.

5.1.12 Substituted in (5.96) we have

$$\overline{\overline{\mathsf{M}}}_g = \overline{\overline{\mathsf{A}}}^{(2)} + c\overline{\overline{\mathsf{A}}}^{\wedge}_{\wedge}\mathbf{e}_\tau\mathbf{e}_\tau + (\mathbf{a}_1 \wedge \mathbf{a}_2)(\mathbf{b}_1 \wedge \mathbf{b}_2),$$

whence one can immediately see that $\overline{\overline{\xi}}_g = \overline{\overline{\zeta}}_g = 0$. Equation (5.69) is now valid and gives

$$\overline{\overline{\mathsf{A}}} = \frac{1}{c}\overline{\overline{\epsilon}}_g,$$

while (5.66) must now be replaced by

$$-\overline{\overline{\mu}}_g^{-1}\rfloor\rfloor\mathbf{e}_{123}\mathbf{e}_{123} = \overline{\overline{\mathsf{A}}}^{(2)} + (\mathbf{a}_1 \wedge \mathbf{a}_2)(\mathbf{b}_1 \wedge \mathbf{b}_2).$$

Recalling the expression for the inverse (5.70), this can be written in the form

$$\overline{\overline{\mu}}_g^{-1} = \gamma\overline{\overline{\epsilon}}_g^{-1T} + \boldsymbol{\alpha}\boldsymbol{\beta},$$

with some scalar γ and

$$\boldsymbol{\alpha}\boldsymbol{\beta} = -(\mathbf{a}_1 \wedge \mathbf{a}_2)\rfloor\varepsilon_{123}(\mathbf{b}_1 \wedge \mathbf{b}_2)\rfloor\varepsilon_{123}.$$

Expanding

$$\overline{\overline{\mu}}_g = \frac{1}{\gamma}\overline{\overline{\epsilon}}_g^T|(\overline{\overline{\mathsf{I}}}^T + \frac{1}{\gamma}\boldsymbol{\alpha}\boldsymbol{\beta}|\overline{\overline{\epsilon}}_g^T)^{-1}$$

and

$$(\overline{\overline{\mathsf{I}}}^T + \frac{1}{\gamma}\boldsymbol{\alpha}\boldsymbol{\beta}|\overline{\overline{\epsilon}}_g^T)^{-1} = \overline{\overline{\mathsf{I}}}^T - \frac{1}{\gamma}\boldsymbol{\alpha}\boldsymbol{\beta}|\overline{\overline{\epsilon}}_g^T + \frac{1}{\gamma^2}\boldsymbol{\alpha}\boldsymbol{\beta}|\overline{\overline{\epsilon}}_g^T|\boldsymbol{\alpha}\boldsymbol{\beta}|\overline{\overline{\epsilon}}_g^T \cdots$$

$$= \overline{\overline{\mathsf{I}}}^T - \frac{1}{\gamma}\boldsymbol{\alpha}\boldsymbol{\beta}|\overline{\overline{\epsilon}}_g^T(1 + \frac{1}{\gamma}\boldsymbol{\beta}|\overline{\overline{\epsilon}}_g^T|\boldsymbol{\alpha})^{-1},$$

we finally have

$$\overline{\overline{\mu}}_g = \frac{1}{\gamma}\overline{\overline{\epsilon}}_g^T - \frac{1}{\gamma^2}(\overline{\overline{\epsilon}}_g^T|\boldsymbol{\alpha})(\boldsymbol{\beta}|\overline{\overline{\epsilon}}_g^T)(1 + \frac{1}{\gamma}\boldsymbol{\alpha}\boldsymbol{\beta}|\overline{\overline{\epsilon}}_g^T)^{-1},$$

which is of the required form.

5.1.13 From $\overline{\overline{\mathsf{Q}}} = \sum \mathbf{a}_i\mathbf{b}_j$ we have $\overline{\overline{\mathsf{Q}}}^{(2)} = \sum_{i<j} \mathbf{a}_{ij}\mathbf{b}_{ij}$, whence following the pattern of Section 2.4 we can write

$$(\overline{\overline{\mathsf{Q}}}^{(2)})^{-1} = \sum_{i<j} \boldsymbol{\beta}_{ij}\boldsymbol{\alpha}_{ij}$$

where $\boldsymbol{\alpha}_{ij}$ and $\boldsymbol{\beta}_{ij}$ are dual bivectors reciprocal to $\mathbf{a}_{ij}$ and $\mathbf{b}_{ij}$, respectively. They are obtained from

$$\boldsymbol{\alpha}_{ij} = (-1)^{i+j-1}\mathbf{a}_{K(ij}\rfloor\boldsymbol{\alpha}_N,$$

and similarly for $\boldsymbol{\beta}_{ij}$. Thus, we can write

$$(\overline{\overline{\mathsf{Q}}}^{(2)})^{-1} = \sum_{i<j} \mathbf{b}_{K(ij)}\mathbf{a}_{K(ij)}\rfloor\rfloor\boldsymbol{\beta}_N\boldsymbol{\alpha}_N$$

$$= \sum_{i<j} \mathbf{b}_{ij}\mathbf{a}_{ij}\rfloor\rfloor\boldsymbol{\beta}_N\boldsymbol{\alpha}_N = \overline{\overline{\mathsf{Q}}}^{(2)T}\rfloor\rfloor\boldsymbol{\beta}_N\boldsymbol{\alpha}_N.$$

The last steps are valid for $n = 4$, because in this case the bi-indices $\{ij\}$ and $\{K(ij)\}$ form the same sequences in reverse order. Applying $\boldsymbol{\varepsilon}_N = \boldsymbol{\alpha}_N(\mathbf{a}_N|\boldsymbol{\varepsilon}_N) = \boldsymbol{\beta}_N(\mathbf{b}_N|\boldsymbol{\varepsilon}_N)$ we obtain

$$(\overline{\overline{\mathsf{Q}}}^{(2)})^{-1} = \frac{\overline{\overline{\mathsf{Q}}}^{(2)T}\rfloor\rfloor\boldsymbol{\varepsilon}_N\boldsymbol{\varepsilon}_N}{\overline{\overline{\mathsf{Q}}}^{(4)}||\boldsymbol{\varepsilon}_N\boldsymbol{\varepsilon}_N}.$$

Setting $\overline{\overline{\mathsf{M}}}_g = \overline{\overline{\mathsf{Q}}}^{(2)}$ and $\overline{\overline{\mathsf{M}}}_g^{(2)} = (\overline{\overline{\mathsf{Q}}}^{(2)})^{(2)} = 3\overline{\overline{\mathsf{Q}}}^{(4)}$, we can write

$$\overline{\overline{\mathsf{M}}}_g^{-1T}\rfloor\rfloor\mathbf{e}_N\mathbf{e}_N = 3\frac{(\overline{\overline{\mathsf{M}}}_g\rfloor\rfloor\boldsymbol{\varepsilon}_N\boldsymbol{\varepsilon}_N)\rfloor\rfloor\mathbf{e}_N\mathbf{e}_N}{\overline{\overline{\mathsf{M}}}_g^{(2)}||\boldsymbol{\varepsilon}_N\boldsymbol{\varepsilon}_N} = \frac{3\overline{\overline{\mathsf{M}}}_g}{\overline{\overline{\mathsf{M}}}_g^{(2)}||\boldsymbol{\varepsilon}_N\boldsymbol{\varepsilon}_N}.$$

This is of the required form (5.95) and valid for a modified medium dyadic $\overline{\overline{\mathsf{M}}}_g$ corresponding to any Q-medium.

5.1.14 The condition (5.95) leads to four three-dimensional conditions

$$A\overline{\overline{\epsilon}}_g - B\overline{\overline{\mu}}_g^T = 0,$$

$$A\overline{\overline{\xi}}_g|\overline{\overline{\mu}}_g^{-1} + B\overline{\overline{\xi}}_g^T|\overline{\overline{\epsilon}}_g^{-1T} = 0,$$

$$A\overline{\overline{\mu}}_g^{-1}|\overline{\overline{\zeta}}_g + B\overline{\overline{\epsilon}}_g^{-1T}|\overline{\overline{\zeta}}_g^T = 0,$$

$$A\overline{\overline{\xi}}_g|\overline{\overline{\mu}}_g^{-1}|\overline{\overline{\zeta}}_g - B\overline{\overline{\xi}}_g^T|\overline{\overline{\epsilon}}_g^{-1T}|\overline{\overline{\zeta}}_g^T = 0.$$

Denoting $a = A/B$, these reduce to

$$\overline{\overline{\mu}}_g = a\overline{\overline{\epsilon}}_g^T, \quad \overline{\overline{\xi}}_g = -\overline{\overline{\xi}}_g^T, \quad \overline{\overline{\zeta}}_g = -\overline{\overline{\zeta}}_g^T,$$

which equal the $\overline{\overline{\mathsf{Q}}}$-medium conditions (5.92).

5.4.2

$$\overline{\overline{\mathsf{f}}}_{\mathbf{E}} = \varrho_e\mathbf{E} + \mathbf{J}_e \wedge \overline{\overline{\mathsf{I}}}_{\mathbf{E}}^T\rfloor\mathbf{B} + \varrho_m\mathbf{H} + \mathbf{J}_m \wedge \overline{\overline{\mathsf{I}}}_{\mathbf{E}}^T\rfloor\mathbf{D}.$$

5.5.3 We first derive by inserting $\mathbf{a} = \sum a_i\mathbf{e}_i$

$$\mathbf{e}_{1234}\lfloor(\mathbf{e}_{123}\lfloor\mathbf{a}) = \sum a_i\mathbf{e}_{1234}\lfloor(\mathbf{e}_{123}\lfloor\mathbf{e}_i) = \sum a_i\mathbf{e}_{i4} = -\mathbf{e}_4 \wedge \mathbf{a},$$

and $(\boldsymbol{\alpha}\rfloor\mathbf{e}_{123})|(\varepsilon_{123}\lfloor\mathbf{a}) = \boldsymbol{\alpha}|\mathbf{a}$, $\mathbf{e}_{1234}\lfloor(\mathbf{v}e_4\wedge\boldsymbol{\alpha}) = -\mathbf{e}_{123}\lfloor\boldsymbol{\alpha}$. Expanding the different terms of $\mathbf{e}_N\lfloor\overline{\overline{\mathsf{M}}}$ with (5.24) substituted,

$$\mathbf{e}_{1234}\lfloor\overline{\overline{\alpha}} = -\mathbf{e}_\tau\wedge\overline{\overline{\xi}}_g|\overline{\overline{\mu}}_g^{-1}\rfloor\mathbf{e}_{123},$$

$$\mathbf{e}_{1234}\lfloor(\overline{\overline{\epsilon}}'\wedge\mathbf{e}_\tau) = -\mathbf{e}_\tau\wedge(\overline{\overline{\epsilon}}_g - \overline{\overline{\xi}}_g|\overline{\overline{\mu}}^{-1}|\overline{\overline{\zeta}}_g)\wedge\mathbf{e}_\tau,$$

$$\mathbf{e}_{1234}\lfloor(\mathbf{d}\tau\wedge\overline{\overline{\mu}}^{-1}) = -\mathbf{e}_{123}\mathbf{e}_{123}\lfloor\lfloor\overline{\overline{\mu}}_g^{-1},$$

$$\mathbf{e}_{1234}\lfloor(\mathbf{d}\tau\wedge\overline{\overline{\beta}}\wedge\mathbf{e}_\tau) = -\mathbf{e}_{123}\lfloor\overline{\overline{\mu}}_g^{-1}|\overline{\overline{\zeta}}_g\wedge\mathbf{e}_\tau,$$

the result is obtained by summing the expressions.

6.1.3 From (6.32)–(6.35) we first see that $\overline{\overline{\xi}} = \overline{\overline{\zeta}}$ must be satisfied, because otherwise $\overline{\overline{\xi}}_d - \overline{\overline{\zeta}}_d \neq 0$. Secondly, from (6.35) it follows that $\overline{\overline{\zeta}}$ must be a linear combination of $\overline{\overline{\epsilon}}$ and $\overline{\overline{\mu}}$, say $\overline{\overline{\xi}} = A\overline{\overline{\epsilon}} + B\overline{\overline{\mu}}$. Inserting in (6.35), we then obtain conditions for the transformation parameters Z, θ: $Z^2 = -A/B$, $\tan^2\theta = -4AB$.

6.1.4 From

$$(\overline{\overline{\alpha}} + \overline{\overline{\epsilon}}'\wedge\mathbf{e}_\tau + \mathbf{d}\tau\wedge\overline{\overline{\mu}}^{-1} + \mathbf{d}\tau\wedge\overline{\overline{\beta}}\wedge\mathbf{e}_\tau)|(\mathbf{B}_\pm + \mathbf{E}_\pm\wedge\mathbf{d}\tau)$$
$$= \pm(1/Z)(\mathbf{B}_\pm + \mathbf{E}_\pm\wedge\mathbf{d}\tau)$$

we can extract two equations

$$\overline{\overline{\alpha}}|\mathbf{B}_\pm + \overline{\overline{\epsilon}}'|\mathbf{E}_\pm = \pm(1/Z)\mathbf{B}_\pm,$$

$$\overline{\overline{\mu}}^{-1}|\mathbf{B}_\pm + \overline{\overline{\beta}}|\mathbf{E}_\pm = \mp(1/Z)\mathbf{E}_\pm$$

from which we obtain through substitution

$$\mathbf{B}_\pm = -\overline{\overline{\mu}}|(\overline{\overline{\beta}} \pm (1/Z)\overline{\overline{\mathsf{I}}}^T)|\mathbf{E}_\pm$$

and the equation for $\mathbf{E}_\pm$:

$$(\overline{\overline{\epsilon}}' - (\overline{\overline{\alpha}} \mp (1/Z)\overline{\overline{\mathsf{I}}}^{(2)T})|\overline{\overline{\mu}}|(\overline{\overline{\beta}} \pm (1/Z)\overline{\overline{\mathsf{I}}}^T))|\mathbf{E}_\pm = 0.$$

This is a (generalized) eigenvalue equation for the eigenvalue $1/Z$.

6.2.2 The condition for the reciprocal medium is $\overline{\overline{\mathsf{M}}}^T\rfloor\mathbf{e}_N = \mathbf{e}_N\lfloor\overline{\overline{\mathsf{M}}}$. When we apply the identities, this becomes

$$(\overline{\overline{\mathsf{M}}}^T\rfloor\mathbf{e}_{123})\wedge\mathbf{e}_\tau - \overline{\overline{\mathsf{M}}}^T|(\mathbf{e}_\tau\wedge\overline{\overline{\mathsf{I}}})\rfloor\mathbf{e}_{123} = \mathbf{e}_{123}\lfloor(\overline{\overline{\mathsf{I}}}^T\wedge\mathbf{e}_\tau)|\overline{\overline{\mathsf{M}}} - \mathbf{e}_\tau\wedge(\mathbf{e}_{123}\lfloor\overline{\overline{\mathsf{M}}})$$

Substituting the expansion for $\overline{\overline{\mathsf{M}}}$, after some steps this equation is split in four (space-space, space-time, time-space and time-time) parts as

$$\overline{\overline{\epsilon}}'^T\rfloor\mathbf{e}_{123} = \mathbf{e}_{123}\lfloor\overline{\overline{\epsilon}}', \quad \overline{\overline{\mu}}^{-1T}\rfloor\mathbf{e}_{123} = \mathbf{e}_{123}\lfloor\overline{\overline{\mu}}^{-1},$$

$$\overline{\overline{\alpha}}^T \rfloor \mathbf{e}_{123} = -\mathbf{e}_{123} \lfloor \overline{\overline{\beta}}, \quad \overline{\overline{\beta}}^T \rfloor \mathbf{e}_{123} = -\mathbf{e}_{123} \lfloor \overline{\overline{\alpha}}.$$

These conditions can also be expressed in terms of the three-dimensional unit dyadic $\overline{\overline{\mathsf{I}}}^{(3)}_{\mathbf{E}} = \mathbf{e}_{123}\varepsilon_{123}$ as

$$\overline{\overline{\epsilon}}'^T = \overline{\overline{\epsilon}}' \rfloor\rfloor \overline{\overline{\mathsf{I}}}^{(3)}_{\mathbf{E}}, \quad \overline{\overline{\mu}}^{-1T} = \overline{\overline{\mu}}^{-1} \rfloor\rfloor \overline{\overline{\mathsf{I}}}^{(3)}_{\mathbf{E}},,$$

$$\overline{\overline{\alpha}}^T = -\overline{\overline{\beta}} \rfloor\rfloor \overline{\overline{\mathsf{I}}}^{(3)}_{\mathbf{E}}, \quad \overline{\overline{\beta}}^T = -\overline{\overline{\alpha}} \rfloor\rfloor \overline{\overline{\mathsf{I}}}^{(3)}_{\mathbf{E}}.$$

7.1.4 Inserting $\boldsymbol{\alpha}_e = \mathbf{A}_e - \phi_e \mathbf{d}\tau$ and $\boldsymbol{\gamma}_e = \varrho_e - \mathbf{J}_e \wedge \mathbf{d}\tau$ in (7.37),

$$(\mathbf{d}|\overline{\overline{\mathsf{Q}}}|\mathbf{d})\overline{\overline{\mathsf{Q}}}|\boldsymbol{\alpha}_e = -\mu \mathbf{e}_{1234} \lfloor \boldsymbol{\gamma}_e.$$

Inserting $\overline{\overline{\mathsf{Q}}} = \overline{\overline{\mathsf{S}}} - \mu\epsilon \mathbf{e}_\tau \mathbf{e}_\tau$, we can expand

$$(\mathbf{d}|\overline{\overline{\mathsf{Q}}}|\mathbf{d})(\overline{\overline{\mathsf{S}}} - \mu\epsilon \mathbf{e}_\tau \mathbf{e}_\tau)|(\mathbf{A}_e - \phi_e \mathbf{d}\tau) = (\mathbf{d}|\overline{\overline{\mathsf{Q}}}|\mathbf{d})(\overline{\overline{\mathsf{S}}}|\mathbf{A}_e + \mu\epsilon \mathbf{e}_\tau \phi_e)$$

$$= -\mu \mathbf{e}_{1234} \lfloor (\varrho_e - \mathbf{J}_e \wedge \mathbf{d}\tau) = \mu \mathbf{e}_\tau (\mathbf{e}_{123}|\varrho_e) - \mu \mathbf{e}_{123} \lfloor \mathbf{J}_e.$$

Splitting the spatial and temporal parts and denoting

$$\Box^2 = \mathbf{d}|\overline{\overline{\mathsf{Q}}}|\mathbf{d} = \mathbf{d}|\overline{\overline{\mathsf{S}}}|\mathbf{d} - \mu\epsilon\partial_\tau^2 = \nabla^2 - \mu\epsilon\partial_\tau^2$$

we obtain the wave equations

$$\Box^2 \overline{\overline{\mathsf{S}}}|\mathbf{A}_e = -\mu \mathbf{e}_{123} \lfloor \mathbf{J}_e, \quad \Box^2 \phi_e = \frac{1}{\epsilon} \mathbf{e}_{123}|\varrho_e.$$

7.2.5 Starting from the bivector equation

$$\mathbf{X} = \overline{\overline{\mathsf{M}}}_g|\Xi = \overline{\overline{\mathsf{Q}}}^{(2)}|\Xi + \mathbf{A}(\mathbf{B}|\Xi),$$

we solve $\Xi = \overline{\overline{\mathsf{M}}}_g^{-1}|\mathbf{X}$ to find the inverse dyadic. Denoting

$$(\overline{\overline{\mathsf{Q}}}^{-1})^{(2)} = (\overline{\overline{\mathsf{Q}}}^{(2)})^{-1} = \overline{\overline{\mathsf{Q}}}^{(-2)}$$

for brevity and multiplying the equation with $(\mathbf{B}|\overline{\overline{\mathsf{Q}}}^{(-2)})|$ we have

$$\mathbf{B}|\overline{\overline{\mathsf{Q}}}^{(-2)}|\mathbf{X} = (1 + \mathbf{B}|\overline{\overline{\mathsf{Q}}}^{(-2)}|\mathbf{A})\mathbf{B}|\Xi,$$

from which we can solve $\mathbf{B}|\Xi$. Substituting it in the original equation we can solve Ξ. This gives the inverse dyadic as

$$\overline{\overline{\mathsf{M}}}_g^{-1} = \overline{\overline{\mathsf{Q}}}^{(-2)} - \frac{(\overline{\overline{\mathsf{Q}}}^{(-2)}|\mathbf{A})(\mathbf{B}|\overline{\overline{\mathsf{Q}}}^{(-2)})}{1 + \mathbf{B}|(\overline{\overline{\mathsf{Q}}}^{(-2)}|\mathbf{A}},$$

which is of the suggested form. From this we can define the dual bivectors as

$$\mathbf{\Gamma} = \frac{\overline{\overline{\mathsf{Q}}}^{(-2)}|\mathbf{A}}{\sqrt{1+\mathbf{B}|\overline{\overline{\mathsf{Q}}}^{(-2)}|\mathbf{A}}}, \quad \mathbf{\Pi} = \frac{-\mathbf{B}|\overline{\overline{\mathsf{Q}}}^{(-2)}}{\sqrt{1+\mathbf{B}|\overline{\overline{\mathsf{Q}}}^{(-2)}|\mathbf{A}}}.$$

7.3.3 The wave one-form $k(\boldsymbol{\nu})\boldsymbol{\nu}$ satisfies the equation

$$k(\boldsymbol{\nu})\boldsymbol{\nu}|\overline{\overline{\mathsf{S}}}|k(\boldsymbol{\nu})\boldsymbol{\nu} = k_o^2\mu\epsilon.$$

Expanding

$$k\boldsymbol{\nu} = k\mathbf{d}w = k_x\mathbf{d}x + k_y\mathbf{d}y + k_z\mathbf{d}z, \quad \overline{\overline{\mathsf{S}}} = \mathbf{e}_x\mathbf{e}_x + \mathbf{e}_y\mathbf{e}_y + 2\mathbf{e}_z\mathbf{e}_z,$$

the equation becomes

$$k_x^2 + k_y^2 + 2k_z^2 = k_o\mu\epsilon,$$

which describes a spheroid in k-space.

7.4.2 From (2.169) the inverse of the three-dimensional metric dyadic $\overline{\overline{\mu}}_g$ becomes

$$\overline{\overline{\mu}}_g^{-1} = \frac{\varepsilon_{123}\varepsilon_{123}\lfloor\lfloor\overline{\overline{\mu}}_g^{(2)T}}{\varepsilon_{123}\varepsilon_{123}||\overline{\overline{\mu}}_g^{(3)}} = \frac{1}{\mu}\varepsilon_{123}\varepsilon_{123}\lfloor\lfloor\overline{\overline{\mathsf{S}}}^{(2)}$$

whence

$$\overline{\overline{\mu}}_g^{-1}\rfloor\rfloor\mathbf{e}_{123}\mathbf{e}_{123} = \frac{1}{\mu}\mathbf{e}_{123}\mathbf{e}_{123}\lfloor\lfloor(\varepsilon_{123}\varepsilon_{123}\lfloor\lfloor\overline{\overline{\mathsf{S}}}^{(2)}) = \frac{1}{\mu}\overline{\overline{\mathsf{S}}}^{(2)}.$$

Applying

$$(\overline{\overline{\mathsf{S}}} - \mu\epsilon\mathbf{e}_\tau\mathbf{e}_\tau)^{(2)} = \overline{\overline{\mathsf{S}}}^{(2)} - \mu\epsilon\overline{\overline{\mathsf{S}}}{}^{\wedge}_{\wedge}\mathbf{e}_\tau\mathbf{e}_\tau + 0$$

leads to the required result.

7.4.3 Because in four-dimensional space any dual bivector $\mathbf{\Gamma}$ satisfying $\mathbf{\Gamma}\wedge\mathbf{\Gamma} = 0$ can be expressed as a simple dual bivector in terms of two dual vectors, we can write

$$\mathbf{\Phi}_o \wedge \mathbf{\Phi}_o = 0 \quad \Rightarrow \quad \mathbf{\Phi}_o = \boldsymbol{\eta}\wedge\boldsymbol{\gamma},$$

$$\mathbf{\Psi}_o \wedge \mathbf{\Psi}_o = 0 \quad \Rightarrow \quad \mathbf{\Psi}_o = \boldsymbol{\chi}\wedge\boldsymbol{\beta}.$$

From $\mathbf{\Phi}_o \wedge \mathbf{\Psi}_o = 0$ we then have $\boldsymbol{\eta}\wedge\boldsymbol{\gamma}\wedge\boldsymbol{\chi}\wedge\boldsymbol{\beta} = 0$, which means that the four dual vectors are linearly dependent and one of them can be expressed as a linear combination of the others. Writing $\boldsymbol{\chi} = A\boldsymbol{\gamma} + B\boldsymbol{\beta} + C\boldsymbol{\eta}$, we have

$$\mathbf{\Psi}_o = (A\boldsymbol{\gamma} + B\boldsymbol{\beta} + C\boldsymbol{\eta})\wedge\boldsymbol{\beta} = (A\boldsymbol{\gamma} + C\boldsymbol{\eta})\wedge\boldsymbol{\beta}.$$

On the other hand, we can also write

$$\mathbf{\Phi}_o = \boldsymbol{\eta}\wedge\boldsymbol{\gamma} = (A\boldsymbol{\gamma} + C\boldsymbol{\eta})\wedge\boldsymbol{\gamma}/C.$$

The two-forms are of the required form if we denote $\nu = A\gamma + C\eta$ and $\gamma = C\boldsymbol{\alpha}$.

7.4.5 Start as

$$\mathbf{e}_{123}|(D\mathbf{E}\wedge\mathbf{B}+\mathbf{H}\wedge\mathbf{D}/c') = D\mathbf{E}|(\mathbf{e}_{123}\lfloor\mathbf{B}) + \mathbf{H}|(\mathbf{e}_{123}\lfloor\mathbf{D})/c'$$

$$= D\mathbf{E}|\overline{\overline{\zeta}}_g|\mathbf{E} + D\mathbf{E}|\overline{\overline{\mu}}_g|\mathbf{H} + \mathbf{H}|\overline{\overline{\epsilon}}_g|\mathbf{E}/c' + \mathbf{H}|\overline{\overline{\xi}}_g|\mathbf{E}/c'.$$

When substituting the dyadic expressions, the terms with the antisymmetric dyadics $\overline{\overline{\mathsf{I}}}\rfloor\mathbf{X}_g$ and $\overline{\overline{\mathsf{I}}}\rfloor\mathbf{Z}_g$ vanish because

$$\mathbf{E}|(\overline{\overline{\mathsf{I}}}\rfloor\mathbf{Z}_g)|\mathbf{E} = (\mathbf{E}\rfloor\mathbf{Z}_g)|\mathbf{E} = (\mathbf{E}\wedge\mathbf{E})|\mathbf{Z}_g = 0.$$

The expression reduces to

$$\mathbf{E}|\mathbf{p}_1\mathbf{q}_1|\mathbf{E} - \mathbf{E}|\overline{\overline{\mathsf{A}}}^T|\mathbf{H} + \mathbf{E}|\mathbf{p}_1\mathbf{q}_2|\mathbf{H} + \mathbf{H}|\overline{\overline{\mathsf{A}}}|\mathbf{E} + \mathbf{H}|\mathbf{p}_2\mathbf{q}_1|\mathbf{E} + \mathbf{H}|\mathbf{p}_2\mathbf{q}_2|\mathbf{E}$$

$$= (\mathbf{p}_1|\mathbf{E} + \mathbf{p}_2|\mathbf{H})(\mathbf{q}_1|\mathbf{E} + \mathbf{q}_2|\mathbf{H}).$$

Because $\mathbf{E}\wedge\mathbf{B} = 0$ and $\mathbf{H}\wedge\mathbf{D} = 0$ for any plane wave, the last expression vanishes. Thus, in this medium any plane wave satisfies either of the two conditions

$$\mathbf{p}_1|\mathbf{E} + \mathbf{p}_2|\mathbf{H} = 0 \quad \text{or} \quad \mathbf{q}_1|\mathbf{E} + \mathbf{q}_2|\mathbf{H} = 0.$$

Index

About the Author

Ismo V. Lindell was born in Viipuri, Finland, in 1939. He received the Dr. Tech. (Ph.D.) degree at Helsinki University of Technology (HUT), Espoo, Finland, in 1971. Currently he is professor of Electromagnetic Theory at the HUT Electromagnetics Laboratory, which he founded in 1984. While on leave from 1996–2001, he held the position of academy professor of the Academy of Finland.

Dr. Lindell is a Fellow of IEEE since 1989 and has authored and coauthored 220 scientific papers and 11 books, including, *Methods for Electromagnetic Field Analysis* (Wiley-IEEE Press, 1992), *Electromagnetic Waves in Chiral and Bi-Isotropic Media* (Artech House, 1994) and *History of Electrical Engineering* (Otatieto, Finland 1994, in Finnish). Dr. Lindell received the IEEE S.A. Schelkunoff prize in 1987 and the IEE Maxwell Premium in 1997 and 1998. He enjoyed research fellowships at the University of Illinois (Champaign-Urbana, 1972–1973) and M.I.T. (Cambridge, 1986–1987). Since 2002, he has been a guest professor of Southwest University in Nanjing, China.